CW01151499

Campbell's Atlas of Oil and Gas Depletion

C.J. Campbell

Campbell's Atlas of Oil and Gas Depletion

Second Edition

Foreword by Alexander Wöstmann

Springer

C.J. Campbell
Staball Hill
Ballydehob, Ireland

ISBN 978-1-4614-3575-4   ISBN 978-1-4614-3576-1 (eBook)
DOI 10.1007/978-1-4614-3576-1
Springer New York Heidelberg Dordrecht London

Library of Congress Control Number: 2012952087

© Colin J. Campbell and Alexander Wöstmann 2013
This work is subject to copyright. All rights are reserved by the Publisher, whether the whole or part of the material is concerned, specifically the rights of translation, reprinting, reuse of illustrations, recitation, broadcasting, reproduction on microfilms or in any other physical way, and transmission or information storage and retrieval, electronic adaptation, computer software, or by similar or dissimilar methodology now known or hereafter developed. Exempted from this legal reservation are brief excerpts in connection with reviews or scholarly analysis or material supplied specifically for the purpose of being entered and executed on a computer system, for exclusive use by the purchaser of the work. Duplication of this publication or parts thereof is permitted only under the provisions of the Copyright Law of the Publisher's location, in its current version, and permission for use must always be obtained from Springer. Permissions for use may be obtained through RightsLink at the Copyright Clearance Center. Violations are liable to prosecution under the respective Copyright Law.
The use of general descriptive names, registered names, trademarks, service marks, etc. in this publication does not imply, even in the absence of a specific statement, that such names are exempt from the relevant protective laws and regulations and therefore free for general use.
While the advice and information in this book are believed to be true and accurate at the date of publication, neither the authors nor the editors nor the publisher can accept any legal responsibility for any errors or omissions that may be made. The publisher makes no warranty, express or implied, with respect to the material contained herein.

Printed on acid-free paper

Springer is part of Springer Science+Business Media (www.springer.com)

It is a pleasure to dedicate this work to my wife, Bobbins, our children and their families:

Julia and Jack	Simon and Oddny
Anne	Emma
Patrick	Clara
	Oscar

Foreword

In mid-2009, I received an invitation to go to Berlin (Germany) for one of the last speeches Colin was going to give in mainland Europe. On instinct, I decided to take on the 6-hour drive to see this special man and his lovely and lively lady, Bobbins, as I suspected that there would not be many more opportunities coming, seeing Colin is in his 80's now.

I had met Colin and Bobbins for the first time in 2003 in Berlin, at an ASPO conference, and had invited him to speak at the Global Peak Oil Gathering that my company organised in Koblenz, Germany the following year. This proved to be a life-changing experience for us all. Over the years I would meet Colin and Bobbins a few more times in other Peak Oil-related conferences.

To my amazement, Bobbins immediately spotted me, and Colin greeted me during the Q&A session after his speech. Feeling pleased but not really understanding, I went to Colin in the break to greet him properly, and he welcomed me with the words: 'Great to see you and glad you are here. I wanted to contact you to ask whether you want to have my data and build an honest model of oil depletion'.

It was a great honour to be asked to take on the heritage of this man's work and his half-a-century research and pioneering on Peak Oil, even if I had no idea then what it would mean.

Since then my life has not been the same. We have worked together for almost three years, and can proudly present the results.

The Atlas of Oil & Gas Depletion is a historic document and fascinating reading, still completely written by Colin, with great help from especially Sonja Fagan, who, with Irish enthusiasm and Germanic thoroughness, checked and re-checked all the data. We can now say that this is the best we have got, even though Colin admits that 'this is not an exact science'.

This book gives a neutral and unattached overview of the reality of the situation concerning the state of affairs of our global endowment of oil and gas. Even though the details are not precise in many cases, the general trends are clear. The implications are vast, and we hope that many people will take notice and rise above the obfuscation practiced in many official studies and reports.

The study will allow the readers to see the need for some serious thinking on how humanity can respond to this inevitable situation.

I am grateful for Colin's dedication and tenacity and all that it will bring forth.
Alexander Wöstmann

Alexander's Gas & Oil Connections,
Limbach, Germany
www.gasandoil.com

With many thanks to the beautiful Liss Ard Estate, near Colin's home in West Cork, Ireland, (www.lissardestate.com) for the conducive environment.

About the Author

Colin Campbell was born in 1931 in Berlin, Germany, but stayed only three months before his father, an architect, returned to his home country as his successful practice in Germany had crashed in the Depression. Colin then had a rather isolated childhood at Chapel Point, a rocky headland in Cornwall in the west of England, where his father was building the first houses of what he hoped to be a model village. But the Second World War brought that project to an end.

School days followed before he succeeded in getting into Oxford University in 1951 to read geology. He enjoyed university life greatly and stayed on to take a D.Phil (Ph.D), based on mapping a remote part of Connemara in Ireland and the interior of Borneo to which he went on a university expedition.

University days came to an end in 1957 and he went on to work for Texaco in Trinidad as a field geologist. There he came under the influence of Hans Kugler, a Swiss scientist who was one of the pioneers of micro-palaeontology and a great inspiration. In 1959, he was transferred to Colombia and had many colourful experiences mapping often bandit-infested remote areas on mule-back, and making a fossil collection to unravel Andean geology. He married Bobbins, a charming girl he had met in Trinidad, and they were later blessed with a son and daughter.

He continued to work in the oil industry in Colombia, Australia, Papua-New Guinea, the United States, Ecuador, the United Kingdom, Ireland and Norway, ending as an Executive Vice-President. One particularly relevant experience was participation in a world evaluation in 1969 which opened his eyes to the finite nature of the resource and the nature of depletion, which in turn made a deep impression. In latter years he found himself engaged in negotiations to secure oil rights in various countries and came to understand the role of politics and influence both internally within the company and in its external relations.

His formal career ended in 1989, but he accepted various consulting assignments. He also developed his long interest in depletion founding the Association for the Study of Peak Oil ("ASPO"), which now has associates in more than thirty countries. He has written seven books on the subject as well as many articles in scientific and other publications, which attracted increasing media interest. This led to participation in many conferences and presentations to governments. He and his wife now live in a village in the west of Ireland.

Preface

This is the second edition of the book which was originally published in 2009 from a study made jointly with Siobhan Heapes in the preceding year. The individual country assessments have been revised on the basis of subsequent information and insight, and some of the general conclusions have been modified in the light of recent events.

The study springs from a career in geology, whose practitioners are taught to observe and record. They are used to having to deal with incomplete information as they try to decipher the rocks which are only locally exposed. They are forced to use both imagination and relentless logic by which to link the data-points and deduce their meaning.

It is recognised that the page numbers remain as probably the only accurate ones, such is the state of confusion in public data, but, that said, the general patterns of production and depletion can be put forward with confidence. As much as anything, the book hopes to ask the right questions, so that the study can be progressively improved in the future.

Full recognition is due to Siobhan Heapes, my valiant co-author of the first edition, who spent a year working difficult spreadsheets and making the graphs which illustrate the book. She has been spared having to work on this edition, her place having been taken by Noreen Dalton and Sonja Fagan, who did equally splendid work, supported by Alexander Wöstmann.

The first edition of the Atlas was privately printed being now out of date and out of print. This revised edition has been produced in co-operation with Alexander's Gas and Oil Connections which for some years has issued a newsletter with information on the oil and gas industry around the world. It plans to compile a new, reliable and honest database of oil and gas discovery and production, from which to build realistic depletion profiles. *We therefore invite our readers to contact Alexander's Gas & Oil Connections if they can supply more accurate data.* Such a cooperative effort could lead to improved future studies that could provide a dynamic and evolving source of information on this subject which will have a vast impact on the future of Mankind, given the heavy dependence on oil-based energy in the modern world.

Thanks are due to those who have contributed so much to the Association for the Study of Peak Oil and Gas ("ASPO") and in other ways. The list is a long one, but an especial word of appreciation is due to Jean Laherrère, Roger Bentley, Kjell Aleklett, Ugo Bardi, Rui Rosa, Richard O'Rourke, Walter Ryan-Purcell, Chris Skrebowski and Werner Zittel. Their efforts have contributed to the growing awareness of the issue of Peak Oil and its far-reaching consequences, leading governments and official institutions to address it.

Ballydehob, Ireland   C.J. Campbell

Contents

Part I Introduction, Reporting, and Methodology

1 Introduction 3

2 The Estimation and Reporting of Reserves and Production 11
    Production 11
        War-Loss 11
        Metering 11
        Nomenclature 11
        Definition 11
        Supply Versus Production 12
        Gas and Gas Liquids 12
        Reporting 12
    Reserves 12
        Reporting in Practice 13

3 The Methodology of the Depletion Model 15
    Defining What to Measure 15
    Defining Regions 16
    Oil Analysis 16
        Step 1: Past Production 16
        Step 2: Study Discovery Trends 16
        Step 3: Estimate Future Production 16
        Step 4: Estimate Future Production in Known Fields (*Reserves*) 16
        Step 5: Forecast Future Production 17
        Step 6: Input Discovery by Year 17
        Step 7: Input Consumption and other Data 17
        Step 8: Compile Country Estimates into Regional and World Totals 17
    Gas Analysis 17
        Step 5a: Assess Gas Plateau 17
        Step 5b: Assess Gas Liquids 18

Part II Africa

4 Algeria 23
    Essential Features 23
    Geology and Prime Petroleum Systems 23
    Exploration and Discovery 24
    Production and Consumption 24
    The Oil Age in Perspective 24

5 Angola ... 27
- Essential Features ... 27
- Geology and Prime Petroleum Systems ... 27
- Exploration and Discovery ... 28
- Production and Consumption ... 28
- The Oil Age in Perspective ... 28

6 Cameroon ... 31
- Essential Features ... 31
- Geology and Prime Petroleum Systems ... 31
- Exploration and Discovery ... 31
- Production and Consumption ... 32
- The Oil Age in Perspective ... 32

7 Chad ... 33
- Essential Features ... 33
- Geology and Prime Petroleum Systems ... 33
- Exploration and Discovery ... 34
- Production and Consumption ... 34
- The Oil Age in Perspective ... 34

8 Congo ... 37
- Essential Features ... 37
- Geology and Prime Petroleum Systems ... 37
- Exploration and Discovery ... 37
- Production and Consumption ... 38
- The Oil Age in Perspective ... 38

9 Egypt ... 39
- Essential Features ... 39
- Geology and Prime Petroleum Systems ... 39
- Exploration and Discovery ... 39
- Production and Consumption ... 40
- The Oil Age in Perspective ... 40

10 Gabon ... 43
- Essential Features ... 43
- Geology and Prime Petroleum Systems ... 43
- Exploration and Discovery ... 44
- Production and Consumption ... 44
- The Oil Age in Perspective ... 44

11 Libya ... 47
- Essential Features ... 47
- Geology and Prime Petroleum Systems ... 48
- Exploration and Discovery ... 48
- Production and Consumption ... 48
- The Oil Age in Perspective ... 48

12 Nigeria ... 51
- Essential Features ... 51
- Geology and Prime Petroleum Systems ... 51
- Exploration and Discovery ... 52

Contents xv

    Production and Consumption 52
    The Oil Age in Perspective 52

13 Sudan 55
    Essential Features 55
    Geology and Prime Petroleum Systems 55
    Exploration and Discovery 55
    Production and Consumption 56
    The Oil Age in Perspective 56

14 Tunisia 57
    Essential Features 57
    Geology and Prime Petroleum Systems 57
    Exploration and Discovery 57
    Production and Consumption 58
    The Oil Age in Perspective 58

15 Uganda 61
    Essential Features 61
    Geology and Prime Petroleum Systems 61
    Exploration and Discovery 61
    Production and Consumption 62
    The Oil Age in Perspective 62

16 Africa Region 63
    The Oil Age in Perspective 64
    Confidence and Reliability Ranking 65
      Algeria 65
      Angola 65
      Cameroon 65
      Chad 65
      Congo 65
      Egypt 65
      Gabon 65
      Libya 65
      Nigeria 65
      Sudan 65
      Tunisia 65
      Uganda 65

Part III Asia-Pacific

17 Australia 75
    Essential Features 75
    Geology and Prime Petroleum Systems 75
    Exploration and Discovery 76
    Production and Consumption 76
    The Oil Age in Perspective 76

18 Brunei 79
    Essential Features 79
    Geology and Prime Petroleum Systems 79
    Exploration and Discovery 79

	Production and Consumption	80
	The Oil Age in Perspective	80
19	**India**	83
	Essential Features	83
	Geology and Prime Petroleum Systems	83
	Exploration and Discovery	84
	Production and Consumption	84
	The Oil Age in Perspective	84
20	**Indonesia**	87
	Essential Features	87
	Geology and Prime Petroleum Systems	87
	Exploration and Discovery	87
	Production and Consumption	88
	The Oil Age in Perspective	88
21	**Malaysia**	91
	Essential Features	91
	Geology and Prime Petroleum Systems	91
	Exploration and Discovery	92
	Production and Consumption	92
	The Oil Age in Perspective	92
22	**Pakistan**	95
	Essential Features	95
	Geology and Prime Petroleum Systems	95
	Exploration and Discovery	96
	Production and Consumption	96
	The Oil Age in Perspective	96
23	**Papua-New Guinea**	99
	Essential Features	99
	Geology and Prime Petroleum Systems	99
	Exploration and Discovery	100
	Production and Consumption	100
	The Oil Age in Perspective	100
24	**Thailand**	103
	Essential Features	103
	Geology and Prime Petroleum Systems	103
	Exploration and Discovery	103
	Production and Consumption	104
	The Oil Age in Perspective	104
25	**Vietnam**	105
	Essential Features	105
	Geology and Prime Petroleum Systems	105
	Exploration and Discovery	106
	Production and Consumption	106
	The Oil Age in Perspective	106

26	**Asia Pacific Region**	109
	The Oil Age in Perspective	110
	Confidence and Reliability Ranking	111
	Australia	111
	Brunei	111
	India	111
	Indonesia	111
	Malaysia	111
	Pakistan	111
	Papua-New Guinea	111
	Thailand	111
	Vietnam	111

Part IV   Eurasia

27	**Albania**	121
	Essential Features	121
	Geology and Prime Petroleum Systems	121
	Exploration and Discovery	121
	Production and Consumption	122
	The Oil Age in Perspective	122
28	**Azerbaijan**	125
	Essential Features	125
	Geology and Prime Petroleum Systems	125
	Exploration and Discovery	126
	Production and Consumption	126
	The Oil Age in Perspective	126
29	**China**	129
	Essential Features	129
	Geology and Prime Petroleum Systems	129
	Exploration and Discovery	130
	Production and Consumption	130
	The Oil Age in Perspective	130
30	**Croatia**	133
	Essential Features	133
	Geology and Prime Petroleum Systems	133
	Exploration and Discovery	133
	Production and Consumption	134
	The Oil Age in Perspective	134
31	**Hungary**	137
	Essential Features	137
	Geology and Prime Petroleum Systems	137
	Exploration and Discovery	138
	Production and Consumption	138
	The Oil Age in Perspective	138
32	**Kazakhstan**	141
	Essential Features	141
	Geology and Prime Petroleum Systems	141

	Exploration and Discovery	142
	Production and Consumption	142
	The Oil Age in Perspective	142
33	**Romania**	145
	Essential Features	145
	Geology and Prime Petroleum Systems	145
	Exploration and Discovery	146
	Production and Consumption	146
	The Oil Age in Perspective	146
34	**Russia**	149
	Essential Features	149
	Geology and Prime Petroleum Systems	149
	Exploration and Discovery	150
	Production and Consumption	150
	The Oil Age in Perspective	150
35	**Turkmenistan**	155
	Essential Features	155
	Geology and Prime Petroleum Systems	155
	Exploration and Discovery	156
	Production and Consumption	156
	The Oil Age in Perspective	156
36	**Ukraine**	159
	Essential Features	159
	Geology and Prime Petroleum Systems	159
	Exploration and Discovery	160
	Production and Consumption	160
	The Oil Age in Perspective	160
37	**Uzbekistan**	163
	Essential Features	163
	Geology and Prime Petroleum Systems	163
	Exploration and Discovery	164
	Production and Consumption	164
	The Oil Age in Perspective	164
38	**Eurasia Region**	167
	Confidence and Reliability Ranking	168
	Albania	168
	Azerbaijan	168
	China	168
	Croatia	169
	Hungary	169
	Kazakhstan	169
	Romania	169
	Russia	169
	Turkmenistan and Uzbekistan	169
	Ukraine	169

Part V Europe

39 Austria 179
   Essential Features 179
   Geology and Prime Petroleum Systems 179
   Exploration and Discovery 179
   Production and Consumption 180
   The Oil Age in Perspective 180

40 Denmark 183
   Essential Features 183
   Geology and Prime Petroleum Systems 183
   Exploration and Discovery 183
   Production and Consumption 184
   The Oil Age in Perspective 184

41 France 187
   Essential Features 187
   Geology and Prime Petroleum Systems 187
   Exploration and Discovery 188
   Production and Consumption 188
   The Oil Age in Perspective 188

42 Germany 191
   Essential Features 191
   Geology and Prime Petroleum Systems 191
   Exploration and Discovery 192
   Production and Consumption 192
   The Oil Age in Perspective 192

43 Italy 195
   Essential Features 195
   Geology and Prime Petroleum Systems 195
   Exploration and Discovery 195
   Production and Consumption 196
   The Oil Age in Perspective 196

44 Netherlands 199
   Essential Features 199
   Geology and Prime Petroleum Systems 199
   Exploration and Discovery 200
   Production and Consumption 200
   The Oil Age in Perspective 200

45 Norway 203
   Essential Features 203
   Geology and Petroleum Systems 203
   Exploration and Discovery 204
   Production and Consumption 204
   The Oil Age in Perspective 204

46 United Kingdom 209
   Essential Features 209
   Geology and Prime Petroleum Systems 209

	Exploration and Discovery	210
	Production and Consumption	210
	The Oil Age in Perspective	210
47	**Europe Region**	**213**
	Countries with Minor Production	214
	The Oil Age in Perspective	214
	Confidence and Reliability Ranking	215
	Austria	215
	Denmark	215
	France	215
	Germany	215
	Italy	215
	Netherlands	215
	Norway	215
	United Kingdom	215

Part VI Latin America

48	**Argentina**	**225**
	Essential Features	225
	Geology and Prime Petroleum Systems	225
	Petroleum Exploration and Discovery	226
	Petroleum Production and Consumption	226
	The Oil Age in Perspective	226
49	**Bolivia**	**229**
	Essential Features	229
	Geology and Prime Petroleum Systems	229
	Exploration and Discovery	230
	Production and Consumption	230
	The Oil Age in Perspective	230
50	**Brasil**	**233**
	Essential Features	233
	Geology and Prime Petroleum Systems	233
	Exploration and Discovery	234
	Production and Consumption	234
	The Oil Age in Perspective	234
51	**Chile**	**237**
	Essential Features	237
	Geology and Prime Petroleum Systems	237
	Exploration and Discovery	238
	Production and Consumption	238
	The Oil Age in Perspective	238
52	**Colombia**	**241**
	Essential Features	241
	Geology and Prime Petroleum Systems	241
	Exploration and Discovery	242
	Production and Consumption	242
	The Oil Age in Perspective	242

53	**Ecuador**	245
	Essential Features	245
	Geology and Prime Petroleum Systems	245
	Exploration and Discovery	246
	Production and Consumption	246
	The Oil Age in Perspective	246
54	**Mexico**	249
	Essential Features	249
	Geology and Petroleum Systems	249
	Exploration and Discovery	250
	Production and Consumption	250
	The Oil Age in Perspective	250
55	**Peru**	253
	Essential Features	253
	Geology and Prime Petroleum Systems	253
	Exploration and Discovery	254
	Production and Consumption	254
	The Oil Age in Perspective	254
56	**Trinidad**	257
	Essential Features	257
	Geology and Prime Petroleum Systems	257
	Exploration and Discovery	258
	Production and Consumption	258
	The Oil Age in Perspective	258
57	**Venezuela**	261
	Essential Features	261
	Geology and Prime Petroleum Systems	261
	Exploration and Discovery	262
	Production and Consumption	262
	The Oil Age in Perspective	262
58	**Latin America Region**	265
	Confidence and Reliability Ranking	266
	Argentina	266
	Bolivia	267
	Brasil	267
	Chile	267
	Colombia	267
	Ecuador	267
	Mexico	267
	Peru	267
	Trinidad	267
	Venezuela	267

Part VII Middle East

59	**Bahrain**	277
	Essential Features	277
	Geology and Prime Petroleum Systems	277

		Exploration and Discovery	278
		Production and Consumption	278
		The Oil Age in Perspective	278
60	**Iran**		281
		Essential Features	281
		Geology and Prime Petroleum Systems	281
		Exploration and Discovery	282
		Production and Consumption	282
		The Oil Age in Perspective	282
61	**Iraq**		285
		Essential Features	285
		Geology and Prime Petroleum Systems	285
		Exploration and Discovery	286
		Production and Consumption	286
		The Oil Age in Perspective	286
62	**Kuwait**		291
		Essential Features	291
		Geology and Prime Petroleum Systems	291
		Exploration and Discovery	292
		Production and Consumption	292
		Oil Age in Perspective	292
63	**Neutral Zone**		295
		Essential Features	295
		Geology and Petroleum Systems	295
		Exploration and Discovery	295
		Production and Consumption	296
		Oil Age in Perspective	296
64	**Oman**		299
		Essential Features	299
		Geology and Petroleum Systems	299
		Exploration and Discovery	299
		Production and Consumption	300
		The Oil Age in Perspective	300
65	**Qatar**		303
		Essential Features	303
		Geology and Prime Petroleum Systems	303
		Exploration and Discovery	303
		Production and Consumption	304
		The Oil Age in Perspective	304
66	**Saudi Arabia**		307
		Essential Features	307
		Geology and Prime Petroleum Systems	307
		Exploration and Discovery	308
		Production and Consumption	308
		The Oil Age in Perspective	309

67	**Syria**	313
	Essential Features	313
	Geology and Petroleum Systems	313
	Exploration and Discovery	313
	Production and Consumption	314
	The Oil Age in Perspective	314
68	**Turkey**	317
	Essential Features	317
	Geology and Petroleum Systems	317
	Exploration and Discovery	318
	Production and Consumption	318
	The Oil Age in Perspective	318
69	**United Arab Emirates**	321
	Essential Features	321
	Geology and Petroleum Systems	321
	Exploration and Discovery	322
	Production and Consumption	322
	The Oil Age in Perspective	322
70	**Yemen**	325
	Essential Features	325
	Geology and Petroleum Systems	325
	Exploration and Discovery	326
	Production and Consumption	326
	The Oil Age in Perspective	326
71	**Middle East Region**	329
	The Oil Age in Perspective	330

Part VIII North America

72	**Canada**	343
	Essential Features	343
	Geology and Prime Petroleum Systems	343
	Exploration and Discovery	344
	Production and Consumption	344
	The Oil Age in Perspective	344
73	**USA**	347
	Essential Features	347
	Geology and Prime Petroleum Systems	347
	Exploration and Discovery	348
	Production and Consumption	348
	The Oil Age in Perspective	349
74	**North America Region**	353
	Other Countries	354
	The Oil Age in Perspective	354
	Confidence and Reliability Ranking	354

Part IX Global Analysis and Perspective

75 The World 363
    Introduction 363
    Exploration and Discovery 363
    Production and Consumption 363

76 Non-Conventional Oil and Gas 369
    Coal and Oil Shale 369
    Bitumen, Extra-Heavy Oil and Heavy Oil 370
    Deepwater Oil and Gas 371
    Polar Oil and Gas 372
    Gas Liquids 373
    Non-Conventional Gas 373
        Coalbed Methane 373
        Tight Gas (or Shale Gas) 374
        Biogenic Gas 374
        Hydrates 374
    Conclusions 375

77 The Oil Age in Perspective 377
    Introduction 377
    The Arrival of Man and Settled Agriculture 377
    Not Plain Sailing 378
    Pre-Modern History 378
    The Steam Engine ushers in the Modern Age 379
    The Oil Industry Is born 379
    Hegemony 379
    Middle East Tension 381
    The Geopolitics of Oil 382
    Resource Nationalism 383
    A New Century Dawns 383
    Looking Ahead 385
    A New Reality 386

References 389

Units and Conversion Factors

Units

b = barrel
cf = cubic foot
k = thousand (10^3): kilo
M = million (10^6): mega
G = billion (10^9): giga
T = trillion (10^{12}): tera
kb/d = thousand barrels a day
Mba = million barrels a year
Gcf/a = billion cubic feet a year
Boe = barrels of oil equivalent (gas converted at calorific equivalent of 6:1)_

Conversion

1 mi = 1.61 km: 1 km = 0.62 mi
1 sq mi = 2.6 km^2; 1 km^2 = 0.38 sq mi
1 cf gas = 0.028 m^3: 1 m^3 of gas = 35.3 cf: 1 trillion cf = 0.18 billion barrels of oil equivalent (in approximate energy terms)

Part I

Introduction, Reporting, and Methodology

Introduction

Soaring oil prices, which surged to $147 a barrel in mid-2008, have drawn attention to the issue of supply and demand, suggesting that the present production capacity limits are being breached. It leads people to ask if we are running out of oil. The simple answer is: *Yes, we started doing that when we used the first gallon.* But the world is a long way from finally running out. What it does face, however, is the end of the *First Half of the Age of Oil*, which lasted 150 years, giving rise to extraordinary changes in the way people lived.

The Planet supported some 300 million people at the time of Christ, 2,000 years ago, and the number barely doubled over the next seventeen centuries. Most people lived hand-to-mouth rural lives. They were much dependent on local circumstances, facing famine if the harvest failed for climatic or other reasons.

The Bronze Age ended as people turned to iron and steel for better tools and weapons. At first, they used firewood to smelt the metals, before turning to coal as a more concentrated source of energy. When local surface deposits had been exhausted, the pits were deepened into regular mines, and minerals were extracted in a similar way, but the mines became subject to flooding on hitting the water table. Draining the mines led in turn to the development of steam pumps, which evolved into steam engines to power industry and locomotives. Sail gave way to steam as the Industrial Revolution opened. The new transport facilities opened up trade, which in turn led to the development of Empires as trading nations sought to capture and control markets. In parallel came the rapid growth of financial capital as banks lent more than they had on deposit, confident that *Tomorrow's Expansion* was collateral for *Today's Debt*, without realising that economic growth depended on energy supply. The Planet was perceived to hold near-limitless resources to be made available by Man's skill and ingenuity. Even so, in the smaller and simpler communities of the past, banking was regarded with some misgivings, usury having been once treated as a sin by the Catholic Church so far was removed from daily life.

Oil had been known since antiquity but it was not until the middle of the nineteenth century that wells were drilled for it, especially in Pennsylvania and on the shores of the Caspian. At first, it was primarily used as a fuel for lamps, replacing whale oil that was becoming scarce from over-whaling. That itself was a revolution for many people, adding an evening to the working day. The next step came in the 1860s when an enterprising German engineer, by the name of Nicholas Otto, found a way to insert the fuel directly into the cylinder of a steam engine, perfecting what became known as the *Internal Combustion Engine*, which was much more efficient. At first, it used benzene distilled from coal, before turning to petroleum refined from crude oil. The oil industry was born. The first automobile took to the roads in 1882, changing the world in previously unimaginable ways.

The demand for food was growing from an expanding population, which was increasingly moving to live in urban conditions. It was proving difficult to feed the people, and in the early years of the last century it was even necessary for Europe to import *guano*, composed of bird excrement, from South America in sailing ships, such was the pressure to improve soil fertility. In 1907, little more than a hundred years ago, came the first tractor, driven by the *Internal Combustion Engine*, to progressively replace the horse and oxen with which to plough the fields, adding greatly to agricultural capacity. Later, means of extracting nitrates from the

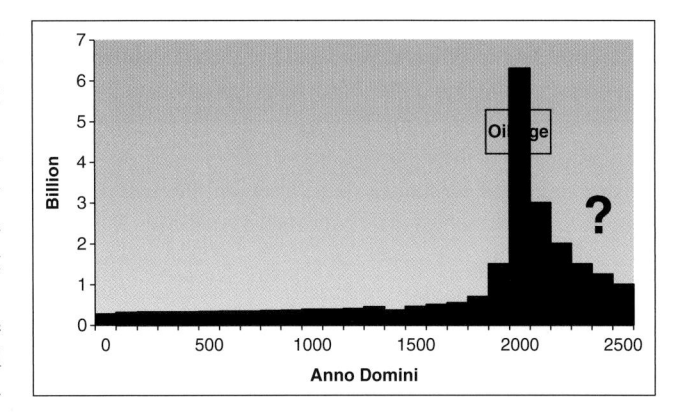

Fig. 1.1 World Population

air by electrolysis were discovered, and natural gas was used to produce nutrients, which, together with oil-based pesticides, improved crop yields greatly. Indeed, agriculture has been described as a process that turns oil into food. It is calculated that mechanisation improved farming efficiency by a factor of 40.

In mediaeval days, oil was collected from seepages and even hand-dug wells, but before the nineteenth century had closed, drilling technology, which had already been developed for salt extraction, was adapted to the oil industry. The cable-tool, consisting of no more than a bullet-shaped weight on the end of a rope which thumped its way into the earth, was followed by the more efficient rotary rig, comprising a bit on the end of a rotating shaft, allowing the search to go deeper. Great technological progress was made in all aspects of the operation.

The oil industry grew to supply this essential oil-based fuel, on which the world began to depend utterly. The early developments were concentrated in the United States, starting in Pennsylvania before major new discoveries were made in California, Oklahoma, Texas and elsewhere. But the most significant development of all came in 1908 when a well being drilled at Masjid-i-Suleiman in the foothills of the Zagros Mountains of Iran blew out, sending a plume of oil high into the sky. It had fallen upon what was to prove the world's most prolific oil province, located around the Persian Gulf. It was later found to hold almost 40% of the total endowment of the world's prime quality oil, much formed in very special conditions, 150 million years ago, in rifts that developed between the African and Eurasian continents.

While much early exploration was undertaken by the so-called wildcatters, drilling by guess and by God, it did not take long to discover the essential geological controls of source, reservoir, trap and seal. At first, petroleum geologists relied on surface observations to identify promising prospects, endowed with the rare, right combination of circumstances, but before long they developed geophysical techniques to scan the depths. Both the technology and the interpretation became ever more sophisticated, assisted in more recent years by massive computing power. Perhaps the most important development of all was a geochemical breakthrough in the 1980s which elucidated the conditions for oil generation itself, making it possible to map accurately where oil was formed and where it was not (see Fig. 1.1).

In technological terms, a major development was the semi-submersible rig, mounted on relatively stable pontoons beneath the wave-base, which opened up the continental shelves of the world to exploration, bringing in new production to replace the traditional onshore fields that were depleting. Even more elaborate floating production facilities later tapped the few deepwater areas having the necessary geological conditions to yield oil.

A coal deposit covers a wide area, and is at first mined only where the seams are thick and accessible, meaning that more becomes commercially viable if prices rise or costs fall under normal economic rules. It is effectively a matter of concentration. Oil, or to be more exact *Conventional Oil*, by contrast, is characterised by a certain polarity because it is either present in profitable abundance, or not there at all, due ultimately to the fact that it is a liquid concentrated by Nature in a few preferred locations. This particular characteristic has had an important impact on the nature of the industry itself, which has had a certain *boom or bust* character. Even a small oilfield delivers great wealth to its owners, as money effectively flows out of the ground under its own pressure. That wealth in turn brought power, allowing the more successful, or lucky, to swallow the smaller and weaker enterprises, a process that led to the formation of the great international oil companies. The first of them was the Standard Oil Empire of the Rockefellers which reached its zenith in 1911 before being dissolved by the US Government under anti-trust legislation designed to break its overweening power. Its daughters, including Esso, Mobil and Chevron to name the largest, grew up to become major international oil companies in their own right. Several of them later merged to survive in the face of contracting exploration opportunities, when they found it easier to secure their reserves by acquisition than exploration. The seven major international companies (BP, Chevron Exxon, Gulf, Mobil, Shell, Texaco), which previously controlled world oil supply, are now reduced to four by merger.

The polarity of oil is well illustrated by the discovery of the massive East Texas field in 1930. It delivered a flood of oil that depressed the price, spelling ruin to producers in other parts of the country. It forced the Government to intervene again: this time by instructing the Texas Railroad Commission to curb production to support price. Most oil in those days was moved by rail, so the Railroad Commission was in a position to exercise control. It set rules defining the number of days in a month that the wells could be produced.

While the United States dominated the world of oil during the first half of the twentieth century, important developments were taking place elsewhere, especially in the Middle East. In 1914, just before the outbreak of the First World War, the British Government took a controlling interest in the Anglo-Persian Oil Company (later BP) to obtain a secure fuel supply for the British Navy that was converting from coal to oil. The war itself led to the defeat of the Ottoman Empire, hitherto the principal power in the Middle East, which was broken up by the victorious allies into independent countries, including Saudi Arabia, Iraq and Kuwait whose oil potential had been recognised. Saudi Arabia fell to a group of American companies led by Standard of California (Chevron), being later joined by Texaco, Esso and Mobil, as

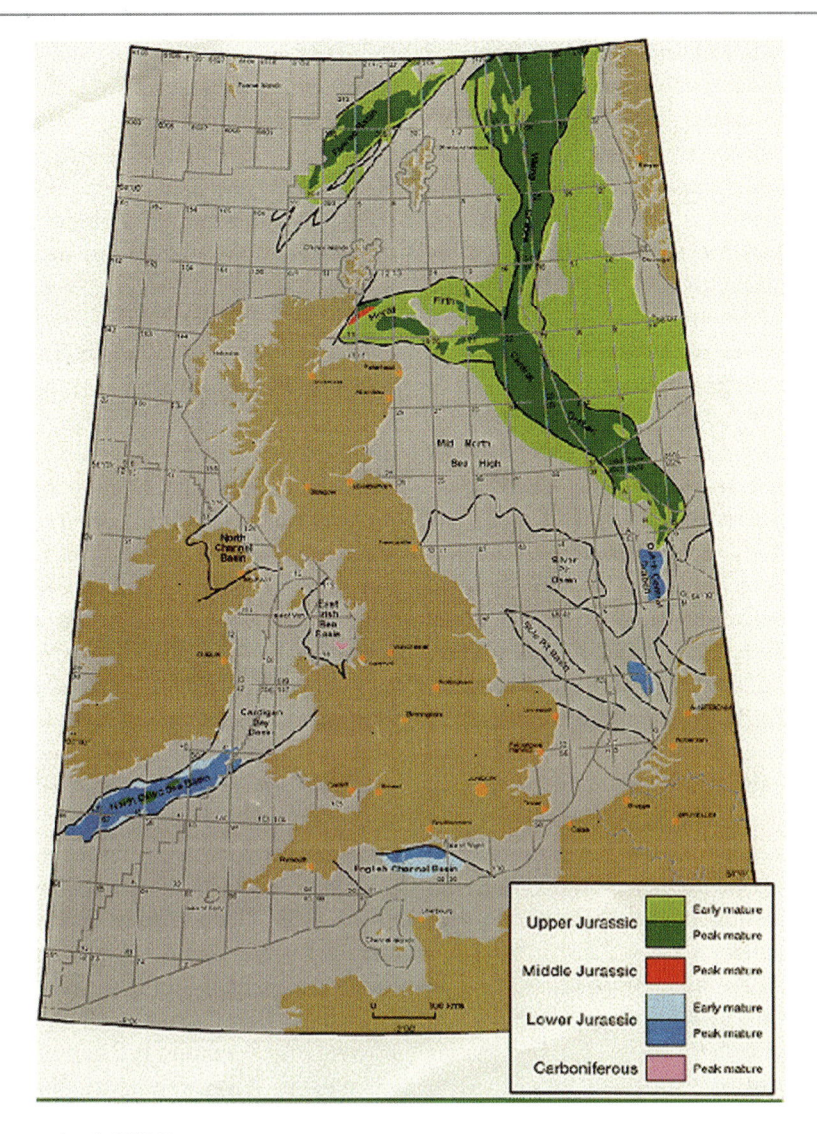

Fig. 1.2 Location of oil generation in NW Europe

owners of the Arabian American Oil Company (ARAMCO); Kuwait went to Gulf and BP, working as the Kuwait Oil Company; and the Iraq Petroleum Company was acquired by BP, Shell, Esso, Mobil, CFP of France, leaving its founder, Mr. Gulbenkian,[1] with his famous 5%. BP retained its exclusive position in Iran, which had not been part of the Ottoman Empire.

While the Middle East was to emerge as the World's most important oil province, it was by no means the only one. Mexico and Venezuela rivalled the United States in the Western Hemisphere with Shell, which had rather missed out of the Middle East carve-up, taking a strong position. The Caspian was another major oil province until eclipsed by the Bolshevik Revolution of 1917, while in the Far East, Shell was developing large finds in Borneo and Sumatra (now Indonesia).

The great wealth that flowed from oil brought with it corresponding tensions that began to manifest themselves in various countries around the world, especially as the stability imposed by the British Empire was eroded by moves to independence. Russia was the first to strike when it nationalised the holdings of foreign oil companies in 1928, followed 10 years later by Mexico, reflecting moves towards what can be termed National Socialism in the face of perceived foreign exploitation. Iran followed in 1951, when it nationalised BP's long-standing exclusive concession, which a war-weary Britain no longer had the stomach to enforce by military

---

[1] Calouste Gulbenkian, who laid the foundations for the Iraq Petroleum Company (see Yergin, 1991).

means. Perez Jimenez, the Oil Minister of Venezuela, was at first hesitant to move towards outright nationalisation, proposing instead to form an international organisation of oil producers that would curb production to support price, following the precedent of the Texas Railroad Commission. This initiative led to the formation in 1960 of the Organisation of Petroleum Exporting Countries ("OPEC") to which the following countries now belong: Algeria, Angola, Ecuador, Iran, Iraq, Kuwait, Libya, Nigeria, Qatar, Saudi Arabia, United Arab Emirates and Venezuela. But it proved to be only partially successful, and was followed by the subsequent outright nationalisations of foreign oil concessions in the major producing countries: Iraq in 1972, Kuwait in 1975, Venezuela in 1976 and Saudi Arabia in 1979. State oil companies were also formed in other countries, including Britain and Norway, often taking dominant privileged positions.

The efforts of these countries to control their oil assets did not however meet with unqualified success, for the major international oil companies moved to replace the loss of their prime supplies by stepping up exploration in new areas, especially offshore, at a time when there were still substantial new areas to bring in. This initiative was notably rewarded by the discovery of a major new province in the North Sea, where oil was found to have been generated in rifts that formed as the Atlantic opened, 150 million years ago. It is the largest new province found since the Second World War, yet, to give a sense of proportion, holds enough to meet less than 3 years of world consumption.

The international companies worked flat out in the new areas under a competitive open-market environment, which was in fact ill-suited to deal with the *boom or bust* character of their business. Prices fluctuated widely, collapsing in 1986 for 3 years during the so-called oil-glut. This put pressure on the OPEC countries, which, with their burgeoning populations, had become utterly dependent on oil revenues. The calculation of their agreed production quotas to support price became subject to difficult negotiation behind closed doors, with the prime criteria being the size of reserves and population.

They had to face the issue of calculating and reporting reserves head on. The volume of oil-bearing rock in an oilfield can be measured relatively accurately with modern methods, but the amount that can be extracted depends on many subtle characteristics, including porosity, permeability, homogeneity, water-saturation and pressure, as well as such operational elements as well density, pressure maintenance and water injection (Table 1.1).

The rules for reporting reserves had been set by the Securities and Exchange Commission (SEC) in the early days of oil in the United States. They were primarily designed to prevent fraudulent exaggeration and smiled on under-reporting as laudable commercial prudence. In short, the rules recognised *Proved Producing Reserves* for the expected future production of current wells; and *Proved Undeveloped Reserves* for the anticipated future production from infill wells before they had actually been drilled. The international companies, when still in control of Middle East production, had routinely reported reserves under this system, such being compiled in various databases, of which that maintained by the *Oil and Gas Journal* was perhaps the foremost. By all means, they may have been conservative estimates but it came as a surprise when Kuwait increased its reported reserves from 64 Gb (billion barrels) in 1984 to 90 Gb in the following year, although nothing particular had changed in the oilfields themselves. It is possible that it started reporting the total amount discovered, (*Original Reserves*) rather than the amount remaining (*Remaining Reserves*), and increased the estimated recovery factor. But, whatever the explanation, it had a strong impact on its quota, in turn affecting the other OPEC countries. Two years later, it announced a further small increase to 92 Gb, which might have reflected a genuine new find, but the move evidently exhausted the patience of its neighbours, leading them to announce massive increases to match Kuwait in various degrees. Abu Dhabi exactly matched Kuwait at 92 Gb, (up from 31 Gb); Iran went one better at 93 Gb (up from 49 Gb); while Iraq capped both with a rounded 100 Gb (up from 47 Gb). Saudi Arabia could not match Kuwait because it was already reporting more, but in 1990 announced a massive increase from 170 to 258 Gb, possibly likewise reporting *Original* rather than *Remaining Reserves*. Venezuela for its part announced a corresponding increase from 25 to 56 Gb in 1988, followed by an increase to 211 Gb in 2010, achieved by the inclusion of heavy non-conventional reserves, not hitherto counted. Whatever the explanations for these anomalous increases, it is clearly implausible that the reserves reported by these countries could have remained substantially unchanged despite subsequent production: the most extreme case being Abu Dhabi that continues to report 92 Gb, unchanged since in 1987.

The international oil companies, for their part, generally reported conservatively under the Stock Exchange rules, in effect quoting only as much as they needed to deliver a satisfactory financial result. It made commercial sense to smooth their assets to cover lean discovery years and any temporary set-back in their operations around the world. Under-reporting also significantly reduced their tax burden in some countries. The practice of under-reporting however now passes into history as the stock of giant fields, which offered the most scope for under-reporting, depletes. The erosion of the major companies' reserves was probably one of the main reasons that prompted a spate of mergers over the past decade which saw Exxon merge with Mobil; Chevron with Texaco; and BP with Amoco and Arco. Shell, for its part, did not make any major acquisition, which forced it to downgrade its reserves in 2005 in a move that caused something of a financial furore, costing the Chairman his job.

Table 1.1 Implausible reserve increases reported by OPEC countries

	Abu Dhabi	Dubai	Iran	Iraq	Kuwait	N. Zone	S.Arabia	Venezuela
1970	12	0	70	32	67	26	129	14
1980	28	1.3	58	31	65	6.1	163	18
1981	29	1.4	58	30	66	6.0	165	18
1982	31	1.4	57	30	65	5.9	164	20
1983	31	1.4	55	41	64	5.7	162	22
1984	30	1.4	51	43	64	5.6	166	25
1985	31	1.4	49	45	90	5.4	169	26
1986	30	1.4	48	44	90	5.4	169	26
1987	31	1.4	49	47	92	5.3	167	25
1988	92	4.0	93	100	92	5.2	167	56
1989	92	4.0	93	100	92	5.2	170	58
1990	92	4.0	93	100	92	5.0	258	59
1991	92	4.0	93	100	95	5.0	259	59
1992	92	4.0	93	100	94	5.0	259	63
1993	92	4.0	93	100	94	5.0	259	63
1994	92	4.3	89	100	94	5.0	259	65
1995	92	4.3	88	100	94	5.0	259	65
1996	92	4.0	93	112	94	5.0	259	65
1997	92	4.0	93	113	94	5.0	259	72
1998	92	4.0	90	113	94	5.0	259	73
1999	92	4.0	90	113	94	5.0	261	73
2000	92	4.0	90	113	94	5.0	261	77
2001	92	4.0	90	113	94	5.0	261	78
2002	92	4.0	90	113	94	5.0	259	78
2003	92	4.0	126	115	97	5.0	259	78
2004	92	4.0	125.8	115	99	5.0	259	77
2005	92	4.0	132	115	102	5.0	264	80
2006	92	4.0	136	115	102	5.0	260	80
2007	92	4.0	138	115	102	5.0	264	87
2008	92	4.0	136	115	102	5.0	264	99
2009	92	4.0	138	115	102	5.0	260	99
2010	92	4.0	137	115	102	5.0	265	211

The major companies are also selling off secondary refineries and marketing chains, evidently perceiving a looming shortage of supply, although they are understandably reluctant to admit to that in the commercial world.

These few words provide the background for addressing what becomes one of the most important issues affecting the modern world. Sufficient is now known about the origin of oil to state unequivocally that it was formed in the geological past.[2] In fact, the bulk of the world's production comes from rocks laid down in just two distinct epochs of extreme global warming, 90 and 150 million years ago, providing the conditions for the prolific algal growths that provided the organic material from which it was formed. This fact alone tells us that it is a finite resource subject to depletion. Given its central place in the economy, it is important to know not only how much is left, but to recognise that production must peak, and then decline during the *Second Half of the Age of Oil*, reflecting the constraints of Nature.

At first sight, it would seem to be an easy job to determine the status of depletion by simply consulting the databases to see how much has been produced so far, how much is left in the known fields, and then to estimate future discovery by extrapolating the past downward trend. The world has been so thoroughly explored that virtually all the accessible productive basins have been identified and explored to some degree by seismic surveys and exploratory boreholes. Most of the larger fields within them have also been identified, being too big to miss. But as we dig into the details, we find a minefield of ambiguous definitions, lax reporting practices, disinformation and confusion. As the capacity limits are

---
[2] Theories about the so-called abiotic origin of oil, as promoted by some economists, can be confidently dismissed not having contributed any material quantity to the world's stock of known oil.

breached, and shortages begin to appear, the issue becomes ever more sensitive, prompting vested interests, including those with political motives, to try to mislead, obscure and confuse. Indeed, it can be said that the skills of a detective are now called for to unearth the truth, and present a sound case sufficiently strong to stand up to withering cross-examination.

One of the reasons why the situation has become so sensitive has been the evolution of what can be described as the financial mind set. In earlier years, corporations were formed by groups of investors with a specific task in mind, such as building a canal or railway, with the payment of regular dividends being their principal reward. In daily life, the blacksmith was primarily motivated by doing a good and worthy job. The farmer ploughed his fields on a similar simple premise. Both received relatively fixed modest wages, seen as a just reward for their work. But those days are long over as the modern world finds itself motivated by financial reward in a corporate conurbation run under complex economic systems. The investment community buys and sells stocks on a massive scale, naturally having little detailed knowledge of the underlying businesses. Its skill lies in having a sharp nose for trading financial instruments and identifying the flavour of the month. An additional element is the role of world trading currencies, formerly the pound sterling and now the US dollar, which delivered a massive hidden tribute to the issuing country. For example, for many years the cost of imports of physical oil into the United States was exactly matched by the expansion of domestic credit, meaning that it effectively secured its supply for free. The credit is, at bottom, nothing more than an expression of confidence in the existing financial system, itself premised implausibly on a continuation of an abundant energy supply to support eternal economic growth. The recent financial crisis, which has left several countries virtually bankrupt, may mark a radical turning point, having been itself perhaps triggered by an oil price surge in 2008. These pressures have in turn led to riots and revolutions especially in North Africa and parts of the Middle East, which may adversely affect oil production in a vicious circle.

It was as if oil provided us with an army of unpaid and unfed slaves to do our work for us. It has been calculated that a drop of oil, weighing 1 g, yields 10,000 cal of energy, which is the equivalent of one day's hard human labour.[3] In other words, today's oil production is equivalent in energy terms to the work of 22 billion slaves. Financial control of the world has led to a certain polarisation between the wealthy West and the other countries which find themselves burdened by foreign debt as they export resources, product and profit.

Many people have come to think that it is money that makes the world go round, when in reality it is the underlying supply of cheap and largely oil-based energy, that has turned the wheels of industry, fuelled the airliners and the bombers, and generally acted as the world's blood stream.[4] Midas-like wealth flowed to those who find themselves having a controlling position in the System.

This book evaluates the current status of depletion of oil and gas, trying to assess the available evidence with which to piece together realistic estimates of the three main parameters: namely, how much has been produced so far; how much remains to produce from known fields (*Reserves*) and how much is yet to find. To simplify matters, it treats everything in terms of production to the end of century, avoiding having to worry about some insignificant tail-end. It furthermore recognises that there are many different categories of oil: some being easy, cheap and fast to extract; others, being the precise opposite. It concentrates on what can be defined as *Regular Conventional Oil*, which has provided most produced to-date and will dominate all supply far into the future. The other categories are also covered in a summary form. Gas is particularly difficult to evaluate as the data are even less reliable, and the trends less clear, but an attempt has been made. The book evaluates the details of 66 producing countries, assessing the resource base and forecasting future production in the light of the individual circumstances. The results are combined to deliver regional and world totals.

Attempts are made to place the Oil Age of each country in its political and historical perspective.[5] Certain patterns emerge from the record and deliver a message. Although it is accepted that the results of the study will be found to be a good deal less than exact, given the appallingly unreliable nature of public data, there is sufficient confidence to put it forward as a basis for planning, albeit subject to revision and improvement as new information becomes available. Indeed, in a sense it poses a challenge to the reader to investigate the situation by country and try to come forth with better information with which to build a better world assessment.

It short, it concludes that the world now faces the dawn of the *Second Half of the Age of Oil*, which will be marked by the decline of oil and all that depends upon it. The next few years, which mark the transition, will likely be a time of great international tension, with the outbreak of more resource wars as competing consumers vie with each other for access to the remaining supplies.[6] It will likely see volatile markets

---

[3] Calculated on the basis of the energy required to lift a ton of sand to a height of 2 m.

[4] See an excellent book, *The Upside of Down*, by Thomas Homer-Dixon, which reviews this subject in depth.

[5] Extensive use has been made of the national summaries provided in *The Rapid Growth of Human Populations 1750–2000* by William Stanton (Published by Multi-Science ISBN 0-906522-21-8).

[6] President Bush justified the invasion of Iraq at a Press Conference in December 2006 with the words: *our energy supply was at risk* (BBC TV). The attack on Libya in 2011 probably also has an oil subtext.

subject to vicious circles as capacity limits are breached leading to price shocks followed by recessions, falling demand and falling oil prices, which may in turn allow economic recoveries before the cycle repeats itself.

Undoubtedly there will be winners as well as losers in the new world that opens. The winners will likely be those who are better informed and plan their lives accordingly. The assumption of perpetual growth may no longer be sustainable, but sound management can make good profits even in a contracting business. There is much at stake in political terms as governments are reluctantly forced to come to terms with the new situation imposed by Nature: some may seek to prolong past postures, policies and structures, whereas others may have the sense and courage to recognise the new circumstances. There is much that can be done in the most positive of ways. The starting points are both to understand the underlying conditions and alert the people at large to what unfolds. Touching the intuitive common sense and goodwill of ordinary people may yield remarkable dividends: this is not necessarily a doomsday message, although it could easily become one if ignored.

As mentioned above, the purpose of the book is to recognise and describe the *Age of Oil*, looking not only at the physical endowment and depletion of oil and gas, but at the wider political, historical and economic developments, by which societies and countries have responded in the past and will respond in the future. The *First Half of the Oil Age* has been a time of general economic growth and technological progress although accompanied by two world wars of unparalleled intensity. We now face a transition to the *Second Half of the Age of Oil*, which will be characterised by the decline of oil and gas, and all that depends on these easy and abundant sources of energy.

It can be said that the only sure numbers in the book are the page numbers. Some estimates are clearly weaker than others, providing ammunition for the detractors who now turn to emphasise the uncertainties of date and decline rate, no longer being able to assault the reality of peak production itself. A vigorous debate surrounds the precise date of peak production, but really misses the point when what matters— and matters greatly—is the vision of the long decline that comes into sight on the other side of it.

The challenge of Peak Oil can be compared with that of crossing a mountain range. The traveller steps out of the dusty plains as he enters the foothills, where the grass is greener, before climbing the heights where the peaks are commonly covered in clouds. But eventually he passes the overall summit and begins his descent, trying to find a pathway between the cliffs and crags, later to emerge safely on the plains beyond.

It is time to ask how we might respond and ensure a better, brighter and more co-operative future for humanity. Certainly, the unparalleled oil-based economic growth of the past 150 years has carried a cost in terms of global pollution, the destruction of habitats and species, and indeed, for many, much human suffering in over-crowded conditions. This behaviour may indeed have released emissions causing the climate to change. The climate has changed many times in the geological past in response to the emissions from massive volcanic eruptions, but this may be the first, and possibly the last, time that *Homo sapiens* is responsible. It is by all means a serious subject.

The Estimation and Reporting of Reserves and Production

The reporting standards for both production and reserves are atrociously weak in many countries, so it is necessary to explain the process of analysis as adopted here in some detail. For example, the 2011 edition of the BP Statistical Review, which is widely taken as an authoritative source coming from a major oil company, reports unchanged reserves in 36 countries. Yet it is not remotely plausible that new discovery and/or an upward revision of the amount recoverable should exactly match production over the past year.

Production

Although production is generally reported more reliably than reserves, there are still major uncertainties, as well as great discrepancies between the various databases.

War-Loss

Perhaps the largest example of failed reporting has been war-loss, which was not normally reported at all although it was production in the sense that it reduced the reserves by like amount. Billions of barrels of oil went up in smoke during the invasion of Kuwait and the following Gulf War, which failed to make the production records.

Metering

In most cases, production from a field is metered at a single gathering station, although in some cases, several fields may be metered together as is the case for example for certain Maracaibo fields in Venezuela. This makes it difficult to allocate amounts to individual fields, leading to inconsistent reporting.

Nomenclature

Some oilfields straddle concession or national boundaries, having different names and reporting practices on either side. For example, the giant North Field of Qatar is known as South Pars where it enters Iranian territory. Furthermore, a cluster of closely related fields may, or may not, be given a common name, which can be particularly confusing in determining the discovery date and allocating reserves. This also affects the reporting of the so-called *New Field Wildcats*, namely exploration boreholes that if successful find a new field. In practice, boreholes drilled to test a minor extension of an existing field or a deeper reservoir are sometimes treated as *New Field Wildcats*, when they would be better designated as *New Pool Wildcats*. No doubt in some cases tax considerations or political pressures influence the reporting practices. China and the United States are particular examples of countries with a broad definition of wildcat wells.

Definition

There is confusion over what is measured. It is common practice to distinguish the so-called *conventional* from *unconventional oil*, but there is no common standard for doing so. Thus, Canada has a cut-off for *heavy oil* at 25° API,[1] whereas Venezuela prefers 22° API, with unspecified criteria being applied in most other countries. The confusion is particularly well illustrated by the fact that the *Oil & Gas Journal* reports Canadian reserves at 175 Gb whereas the equally respected *World Oil* reports 23 Gb. This book defines *regular conventional oil* as to be lighter than 17.5° API.

---

[1] API is a measure of density established by the American Petroleum Institute. Water has a density of 10° API.

Supply Versus Production

Supply is not the same as production for several reasons in addition to the issue of war-loss mentioned above. The refining process adds some 2–3% to the volume of the product. Storage, including the so-called strategic stocks, acts a buffer between production and supply, often being raised or lowered in response to price changes. Again, databases may confuse *production* with *supply*. In principle, the production numbers reported here refer to production at the wellhead.

Gas and Gas Liquids

The measurement of gas production is even more confused by the varying treatment of the non-combustible components, such as carbon dioxide or nitrogen which are present in some fields. The amounts flared, re-injected into the reservoir and used to fuel the facilities are also not consistently reported. A liquid, known as *condensate*, condenses naturally from gas at surface conditions of temperature and pressure, and may be treated as ordinary oil for most purposes. In addition, *natural gas liquids* (*NGL*), mainly pentane and butane, are produced at dedicated plants which may draw their supplies from a number of different fields, making it difficult to attribute the production to the fields concerned. Gas itself may be liquefied at very low temperature for transport from remote locations, being known as *liquefied natural gas*, with the easily confused acronym of *LNG*.

Reporting

Statistics on oil production and reserves are virtually State secrets in some countries, especially the OPEC countries, where production is subject to agreed contractual quotas. The relative amounts consumed internally and exported are also often unsure. Some databases are forced to rely on input no more accurate than the observations of ships' agents, counting the tankers as they sail from the terminals. There are many anomalies relating to consumption, which may or may not cover imports and exports. For example, the highest per capita consumption in the world is reported to be in the Netherlands Antilles, evidently not taking into account the exports from the large refinery on Aruba.

Reserves

Whatever the shortcomings of production reporting, they are small in relation to those relating to *reserves*. The estimation of reserves commences during the exploration phase. In earlier years, when exploration was less influenced by financial considerations, it was normal for companies to test geological assessments and gather information by drilling exploration boreholes, known as *wildcats*, without being overly burdened by hypothetical economic evaluations. They were generally testing large prospects, and knew that a successful outcome would be highly profitable, whatever its precise size turned out to be. Furthermore, the drilling might yield valuable information leading to other prospects or a negative assessment allowing the company to concentrate on better alternatives. The Soviets perhaps worked under the most advantageous circumstances as the exploration process was managed apart from production. They even had the luxury of being able to drill boreholes simply to gather information. It explains incidentally how they led the world in the field of geochemistry which they recognised would provide key insight into which areas were prospective and which were not.

Accordingly, in earlier years, the assessment of the potential reserves of un-drilled prospects was not a particularly vital or sophisticated business. In measuring the size of a find, the industry adopted a fairly pragmatic and sound practice, describing reserves in discoveries as *proved, probable* and *possible*, with the meanings the words carry. The oilmen of those distant days were often content with rule-of-thumb measurements, such as to assume, for example, that a normal reservoir would yield, say, 200 barrels per acre-foot: the area of the structural trap being measured in acres and the thickness of the reservoir in feet.

With the passage of time, control of the international oil companies passed to financiers and image-makers, with the exploration departments being almost relegated to the role of internal contractors. They were no longer asked to express judgment, but to produce hypothetical economic evaluations based on the most ingenious geological case that they could muster. Affiliates of the same company around the world found themselves competing with each other for exploration funds from distant boardrooms: sound technical judgment as such gave way to salesmanship.

Under this regime, the explorers would still make their best scientific estimate of the size of a prospect, by measuring the rock volume of the trap and using regional knowledge to forecast the likely reservoir thickness and properties. Thus, if a particular structure measured, for example, 3 × 5 km with an average relief of 50 m, they would compute its volume as 750 M m^3. To this, they would apply, say, a net-to-gross reservoir value of 60; porosity of 20%; oil saturation of 70%; and a recovery of 30%: together yielding potential reserves of 18.9 M m^3, or 119 million barrels. But this might not pass the competitive economic test, especially offshore or in a remote area, after the cost of wells, platforms and operating costs had been taken into account. If so, it would be a simple matter to massage the estimates to deliver an acceptable outcome. Increasing the net-to-gross to 70% and the porosity to 25%, would yield enough to jump the economic hurdle. But whatever the machinations, the outcome remained confidential to the company concerned, if not to the affiliate or department responsible.

In the event that the proposal was accepted by the company's management, and a successful borehole drilled, an entirely new reporting procedure went into place as engineers took over to plan the actual development, which, offshore and in remote locations, could involve massive investments, calling for extreme caution. It was normal for the development to proceed on a step-by-step basis with the reserves attributable to each committed phase being reported to meet tax and other mandatory requirements. The prime objective was to invest the minimum as safely as possible to secure an early payout, with the reserves supporting that objective, being reported at the point when the field development was sanctioned by the relevant national authority. Such estimates, which may be far below what the field actually contains, were commonly reported for financial and regulatory purposes, and entered the public database. They were naturally subject to upward revisions over time.

In the simplest case, the old pragmatic terms of *proved, probable* and *possible* were retained: the *proved* category being further divided into *proved producing* for the anticipated future production of current wells; and *proved undeveloped* for the anticipated yield of yet-to-be-drilled infill wells, located between the existing one, where the risks were minimal. To provide a single best estimate for the combined value, it was common to accept *proved reserves* in full; two-thirds of the *probable;* and one-third of the *possible.* Thus, if a particular field had 100 Mb of *proved*; 30 Mb of *probable* and 15 Mb of *possible*, the total would be estimated at 125 Mb.

Later, attempts were made to apply subjective probability assessments in a system that has its followers. Thus, a range of reserve values is plotted against their *subjective* probability rankings on a log-normal scale. The area under the curve gives the *mean* value; the highest point is termed the *mode;* and the *50% probability* or *median* value is identified. The *mean* value is theoretically the most reliable estimate, but the *median* value (also termed P_{50}) is commonly quoted, being broadly equated with *Proved + Probable*. It gets even more complex with the application of the so-called Monte Carlo simulation that plots every single combination of parameter. It all sounds very scientific but the ranking of probability is itself subjective, and it did not take long for the explorers, in selling their prospects to management, to realise that raising the essentially unquantifiable low probability case, raised the *mean* value. A beautiful example is provided by the US Geological Survey's assessment of un-drilled East Greenland in 2000: saying that there is a 95% probability of finding more than zero, namely at least one barrel, and a 5% probability of finding more than 112 billion barrels, which together delivered a *mean* value of 47 billion barrels that duly entered the world database. The estimates are quoted to three decimal places, implying greater than justifiable accuracy.

Modern petroleum engineering has advanced to the point of being able to make accurate estimates using sophisticated computer-based models in which a field is covered by a grid of cells, each extrapolating the several parameters based on information derived from the wells. The model is adjusted to match the actual production against the theoretical expectation. It is sometimes even possible to detect the impact on reservoir pressure of the tides in the sea above an offshore field, such is the accuracy of the measurements.

Reporting in Practice

The models described above are designed to help engineers plan an optimal development strategy, normally aimed to secure the maximum return on investment at the minimum risk. In practice, especially offshore where costly platforms have to be built, this means holding production at an optimal economic rate related to the costs of the installed facilities for as long as possible, giving an abrupt final decline. The size of the facilities has to be carefully designed to optimise results with an early payout by delivering a long plateau rather than a short peak of production. Reserves themselves are only of interest to the extent they represent return on investment. The economics are also influenced by the rules of *discounted cash flow* whereby money today is worth more than tomorrow, making it difficult for companies to conserve resources. In the past, most oil companies operated their entire investment strategy on stable and conservative oil price assumptions, envisaging for example the *gentle ramp* whereby prices would rise at a point or two above inflation rate. Any notions of finite limits, or of profiteering from shortage, were far from their considerations.

As already discussed, the management, for its part, had good reason to report the minimum reserves needed to deliver a satisfactory financial result. The practice commonly had the added advantage of reducing tax, especially in fiscal regimes, such as that of the United States, which recognise a *depletion allowance*. The OPEC countries were subject to their own pressures to report whatever was politically expedient, as described in Chap. 1. National statistics are commonly simply compilations of company reports, although some countries, such as Norway, do publish independent national evaluations.

So, in forecasting future production, it becomes necessary to use experience and judgment in estimating and evaluating the various data-sets. Published and confidential sources, commonly giving a wide range of estimates, have to be compared and analysed in the hope of reaching valid conclusions, at least in orders of magnitude. It is not an easy task. The one thing that can be said with absolute assurance is that the detailed forecasts given herein will prove to be wrong. That is not in doubt, but poses the more useful question of asking: *By how much?* This is why we propose to our readers to report back, if more valid estimations are possible or if data can be improved.

The Methodology of the Depletion Model

The foregoing discussion has described the extremely unreliable nature of public data. There are, in addition, various confidential industry databases of varying cost, access, quality and credibility. With these constraints, it becomes impossible to forecast future production with any degree of scientific exactitude. This in turn has facilitated those with motives to ignore or distort the issue of depletion, leaving the man-in-the-street relatively uninformed of what awaits him. The Economist prefers not to recognise depletion as imposed by Nature because it tends to undermine the notion of market supremacy, the bedrock of his subject.[1] It offends his near-doctrinaire *faith* that supply must meet demand in an open market, and that one resource replaces another as the need arises with new technology breaking barriers. The Politician prefers not to face the issue because it would call for unpopular fundamental shifts of policy. In practice, his interest is not served by posing a *problem* unless he has a palatable *solution*. The Investor for his part does not want to know because it undermines the credibility of the assumptions upon which his portfolio was built. But at the end of the day, we all desperately need to know because our very future is at stake. Fortunately, there is now a new awakening despite the many vested interests bent on obfuscation. The recent financial and economic crisis affecting the world underlines the importance of coming to terms with reality.

This book attempts therefore to provide, not a definitive forecast of future oil and gas production, but a sound working hypotheses for 65 producing countries, providing most of the world's supply, which are compiled into regional and world totals. It can be taken as a foundation for planning, provided that it is understood to be subject to revision as new information comes in.

As discussed in Chap. 2, *production* data are generally more reliable than *reserve* data. It is expedient therefore to build the model solely on *production to 2100*, avoiding the terms *reserves* and *ultimate recovery* so far as is possible. Setting such a timeframe avoids having to worry about some irrelevant tail-end. The construction of each country's forecast involves the steps outlined below (see also the following table).

Defining What to Measure

As already discussed, a major cause of confusion has been a failure to define clearly what to measure. The terms *conventional* and *unconventional* are in wide usage but without standard definition. Accordingly, it is well to recognise what is here termed *regular conventional oil* (>17.5° API), defined to exclude oil from coal and *shales*[2]; Bitumen; extra heavy oil (<10° API), heavy oil (10–17.5° API); deepwater oil and gas (>500 m); polar oil and gas; and natural gas liquids from gas plants. The *shales* may be subdivided into two subcategories: immature source-rocks which have to be artificially retorted to give up their oil; and rocks lacking adequate permeability to constitute a normal reservoir, from which oil and gas may nevertheless be extracted by artificial fracturing. Such production is growing rapidly in the United States, including for example that from the Barnet Shale of Texas. There are however quite some environmental hazards.

Regular conventional oil and gas have provided most oil to-date, and will dominate supply far into the future, being relatively easy, cheap and, above all, fast to produce. The greater part of this book will be dedicated to forecasting the production of this category, although the others are also covered in summary form.

---

[1] See Clarke (2007), an economist's impassioned rejection of the impact of natural depletion.

[2] The term *shale* is in common usage, but strictly speaking the rock should be termed *kukersite*.

Defining Regions

It has been found expedient for this purpose to classify the significant producing countries into seven regions as follows, listed in order of their total endowment. The lesser endowed countries within each region, some of which may be major consumers, are lumped together as *minor*, being assessed in the World evaluation in Chap. 11:

Middle East [A]: The five major producers bordering the Persian Gulf (Iran, Iraq, Kuwait, Saudi Arabia, including the Neutral Zone, and the United Arab Emirates) and the lesser producers (Bahrain, Dubai, Oman, Qatar, Syria, Turkey and the Yemen).

Eurasia [B]: The former Communist bloc of the Soviet Union, Eastern Europe and China.

North America [C]: USA and Canada.

Latin America [D]: South America, Caribbean and Mexico.

Africa [E].

Europe [F]: Austria, France, Denmark, Germany, Italy, Netherlands, Norway and United Kingdom.

Asia-Pacific [G]: Pakistan to Vietnam and Australasia (excluding China).

Minor [H]: Countries with an endowment of less than 500 million barrels.

Oil Analysis

The process of analysis goes through the following steps for each country. Table 3.1 illustrates the format of analysis (the full spreadsheet covers the period 1930–2030), which is supplemented by a range of graphs, including *creaming curves*, *derivative logistic* and *parabolic fractals* of size distribution.

Step 1: Past Production

The first step is to plot past reported production in Column B in kb/d (thousands of barrels a day[3]), from 1930 to the current year. Back issues of the *Oil & Gas Journal* are a primary source of such information, although post-1980 data are taken from the Energy Information Administration (a branch of the US Department of Energy). Any pre-1930 production is summed at the top of Column D [Cell D14]. Production in kb/d is converted into billion barrels a year (Gb/a) in Column C. The following parameters are computed, with the results of (b), (c) and (d) being dependent on step 3.

(a) Cumulative production by year in Column D
(b) The amount left to produce by year in Column E.
(c) The percentage depleted in Column F.
(d) The depletion rate (annual/future production) in Column H.

Step 2: Study Discovery Trends

The second step is to plot a graph showing annual/cumulative vs. cumulative production (*derivative logistic*), and extrapolate it to zero, yielding thereby an indication of the total which will have been produced when production ends, termed the *total*. It works well in some countries but not in others where production was for some reason artificially constrained. This estimate is compared against other plots of cumulative exploration effort (principally the drilling of exploration boreholes, known as *wildcats*) against cumulative discovery (*creaming curve*). In addition, it is useful to plot field size against rank on log–log scales and extrapolate to zero with a parabolic projection (*parabolic fractal*).

Step 3: Estimate Future Production

The third step is to take a deep breath and determine from the above considerations a reasonable estimate of the *total* to be produced to the end of the century, inserting the value in Cell C9. In most cases, production will have ended by then, so this is not far removed from what is termed *Estimated Ultimate Recovery—EUR*. It is an iterative process to evaluate the maturity of exploration; the reported reserves; and the resulting depletion rates, to eventually define a reasonable estimate. With the total in place, future production is calculated [Cell C5] by subtracting past production.

Step 4: Estimate Future Production in Known Fields (*Reserves*)

The fourth step is to take a second deep breath, and estimate the percentage of Future Production coming from known fields [Cell D6] giving, in other words, the *Reserves* in [Cell C6]. That in turn delivers the amount yet-to-find [Cell C7] in new fields and the total Discovered-to-Date [Cell C8]. It is another iterative process to check the estimate against the published data [Cells K3-7] and their average [Cell L4], as well as any other available information. It takes into account any so-called *static* reserves in Cell K8 that being the production for any period of implausible unchanged reserves, as reported by the *Oil & Gas Journal* or the *BP Statistical Review*, as well as any identified *non-conventional*, termed *other* [Cell K9].

It is important to take note of the so-called *fallow fields*, namely those in remote locations or subject to political

---

[3] See Table for units of measurement and conversion factors.

constraints, whose production may be delayed or constrained by pipeline capacity. The late development of fields in the remote Llanos Basin of Colombia is an example.

Step 5: Forecast Future Production

The fifth step is to forecast future production, considering two groups of country:
1. *Post-midpoint countries*
   If the country concerned has passed the *midpoint of depletion* (as reported in Cell G3), namely having produced more than half the total, future production is modelled to decline at the current depletion rate.
2. *Pre-midpoint countries*
   If the country has not yet reached midpoint, it is necessary to model production to midpoint in the light of local circumstances. Since most countries in this category, outside the Middle East Gulf, are close to midpoint, the assumptions are not critical to the overall assessment;

Step 6: Input Discovery by Year

The sixth step is to input discovery data for giant fields (Column I) and all fields in (Column J). This information may be drawn from public data, as for example in Norway, or from confidential sources as available. In the interest of internal consistency, the total discovery to-date is then adjusted by the so-called *growth factor* [Cell L9] to match the assessed Total Discovery [Cell C8], with annual discovery in Column K being adjusted accordingly. Exploration drilling of the so-called *New Field Wildcats* is input in Column M, with the Cumulative being given in Column N. Future exploration is also assessed, in some cases to exhaustion.

Step 7: Input Consumption and other Data

The seventh step is to input the remaining statistics, including consumption in Column G, and study the derivatives in the appropriate cells.

Step 8: Compile Country Estimates into Regional and World Totals

The eighth step is to compile the country totals into regional and world totals.

The details of the mechanics of the model are somewhat complex and take some experience to master. It is an iterative process of trial and error to check all the different parameters, to see relationships, trends and anomalies, in order to try to come up with the best solution. The results are to be seen as a working hypothesis, subject to revision and refinement as better information becomes available. That said, the estimates are believed to come close enough to the real position to form a sound basis for corporate and government planning.

The essential features of each country are described, covering the geography, geology and petroleum systems, together with the record of exploration, discovery, production and consumption. That is followed by an outline of its history, politics and general circumstances, in order to place the oil age in perspective and evaluate the country's responses to the decline of this essential energy supply. The treatment varies from country to country based on varying degrees of knowledge and insight. It is clearly not possible to provide definitive assessments, and the idea is rather to provide a framework for further discussion and evaluation.

Gas Analysis

The gas analysis follows a similar path as that of oil, but it is much more difficult as the data are less reliable and the trends less clear. All the steps are basically the same as for oil, save when forecasting future production under Step 5 which follows a different premise as described below.

Step 5a: Assess Gas Plateau

In a most general sense, it is assumed that gas production would normally follow a plateau of relatively constant production over the central part of a country's production life when it is between 30 and 70% depleted. Such a profile reflects the general oilfield practice of producing oil before associated gas and also the constraints of the pipeline capacity, which are normally set for a long plateau rather than a peak. Actual production to-date may, or may not, approximate with the plateau assumption, but it is normally assumed that future production will be held at current levels until the 70% depletion point has been reached, when it declines at the then depletion rate. Naturally, each country has to be assessed in the light of local circumstances as there may be departures due, for example, to actual or planned export pipeline capacity.

It is evident that public databases are most unreliable, with the amounts flared, re-injected, processed to liquids, or used as operating fuel, as well as the content of non-combustible gases, being far from clearly distinguished. In principle, the amount indicated is gross production from the wellhead.

Table 3.1 Illustrative abridged Spreadsheet of the analysis by country: the full sheet covering 1930–2030

COUNTRY		Austria			REGION		Europe	Date	5/10/11	2010
PRODUCTION	Gb	%	MIDPOINT		PEAK		RESERVES		WILDCATS	
Current	0.01		Gb	0.48	Date	1955	O & GJ	0.05	Av	Past	860
PAST	0.83	88%	Date	1970	Depleted	19%	WO	0.09	0.07	Future	45
FUTURE	0.12	12%	Lag	−15	Disc	1949	BP		SD	Av Disc	1.08
Known	80%	0.09	10%		GIANTS	Lag Disc	6	BGR	0.08	0.02	DEP RATE
To be found	0.02	2%	Total	0.55	Wildcat	1975	EIA	0.05	0.073	MP	3.9%
Discovered	0.93	98%	%	59%	Static Yr	3	Static	0.018	Growth	Current	4.9%
TOTAL to 2100	0.95		Last	1949	5yr Tmd	0%	Other		102%	Final	
Consume kb/d	277.94	Per Cap	12.1	Pop.	8.4	Area	0.08	Pop. Density	100	Trade	−261

DATE	PRODUCTION				CONS-UME	DEP. RATE	DISCOVERY			WILDCATS	
	Kb/d	GB/D	Cum	Y-t-p	Dep			Giant	All	Grth	Cum		Cum
Pre 1930				0.95					0.00			5
1930	0	0.000	0.00	0.95	0%			0.00	0.00	0.00	0	5
1931	0	0.000	0.00	0.95	0%		0.00%	0.00	0.00	0.00	0	5
1932	0	0.000	0.00	0.95	0%		0.00%	0.02	0.02	0.02	1	6
1933	0	0.000	0.00	0.95	0%		0.00%	0.00	0.00	0.02	0	6
1934	0	0.000	0.00	0.95	0%		0.00%	0.00	0.00	0.02	1	7
1935	0	0.000	0.00	0.95	0%		0.00%	0.00	0.00	0.02	5	12
1936	0	0.000	0.00	0.95	0%		0.01%	0.00	0.00	0.02	3	15
1937	1	0.000	0.00	0.95	0%		0.02%	0.01	0.01	0.03	1	16
1938	1	0.000	0.00	0.95	0%		0.04%	0.09	0.09	0.12	8	24
1939	3	0.001	0.00	0.95	0%		0.11%	0.02	0.02	0.14	11	35
1940	8	0.003	0.00	0.95	0%		0.31%	0.00	0.00	0.14	6	41

Obviously the profile from associated and non-associated gas will be different, but this has not been possible to address here, as the country data do not distinguish the two classes.

Step 5b: Assess Gas Liquids

A further complexity with regard to gas is the derived liquids, which fall into three classes:

Condensate: the liquids that condense naturally from gas at surface conditions of temperature and pressure, which are treated together with oil and not distinguished as such in this study.

Natural gas liquids (*NGL*): liquids, such as pentane and butane, which are extracted from gas in specialised plants, such data being taken from the EIA database.

Liquefied natural gas (*LNG*): gas which is liquefied at low temperature for export from remote locations. Data are far from complete, but information on particular projects is included where available.

The foregoing emphasises that this is far from an exact science, given particularly the appallingly unreliable public data. It does however have the advantage of being the so-called bottoms-up approach looking at the individual fields and countries as building blocks for regional and world assessments.

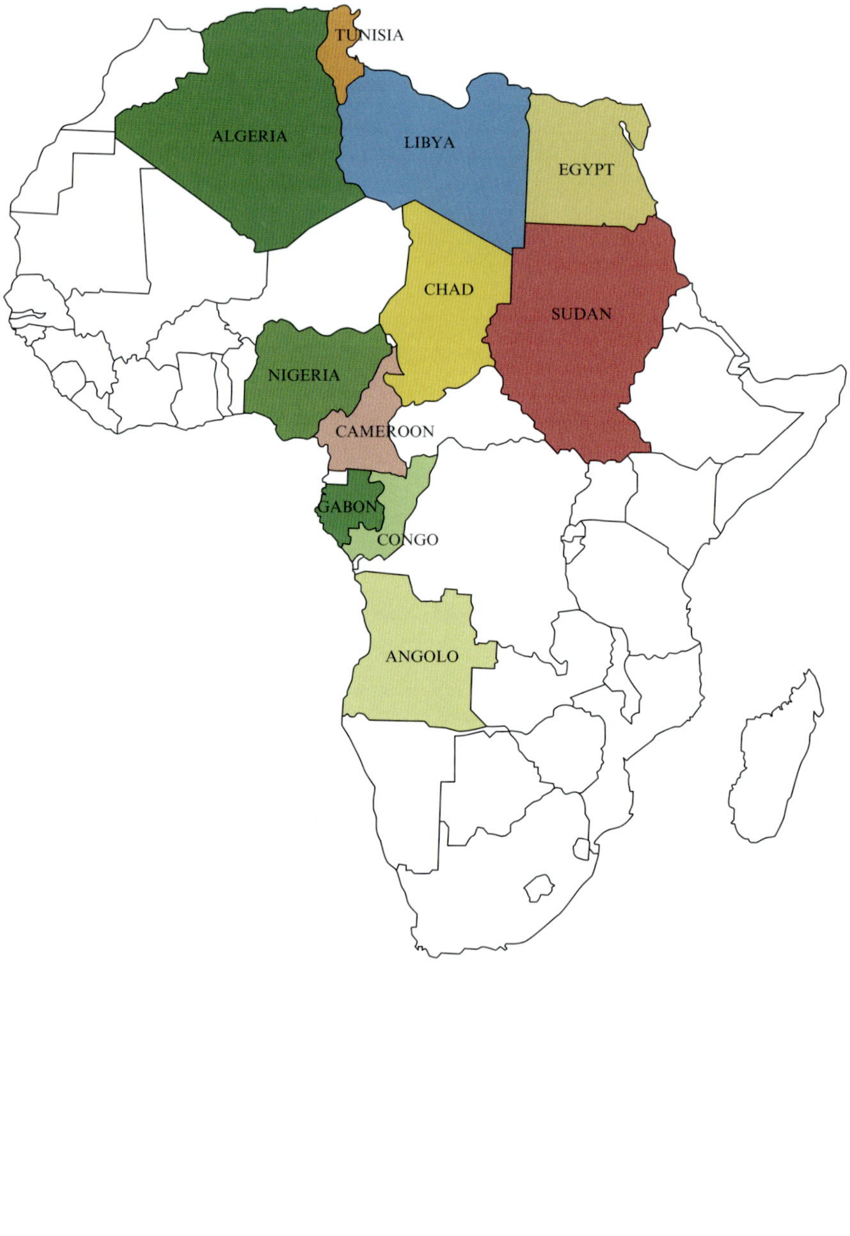
TUNISIA
ALGERIA
LIBYA
EGYPT
CHAD
SUDAN
NIGERIA
CAMEROON
GABON
CONGO
ANGOLO

Part II
Africa

Algeria

Table 4.1 Algeria regional totals (data through 2010)

Production to 2100					Peak Dates			Area	
Amount		Rate				Oil	Gas	'000 km²	
	Gb	Tcf	Date	Mb/a	Gcf/a	Discovery	1956	1957	Onshore	Offshore
PAST	21	142	2000	458	5757	Production	2007	2011	2390	80
FUTURE	13	158	2005	656	6597	Exploration	1961		Population	
Known	12	142	2010	631	6788	Consumption	Mb/a	Gcf/a	1900	5
Yet-to-Find	1.3	16	2020	396	6517	2010	114	1018	2010	36
DISCOVERED	33	284	2030	249	3139		b/a	kcf/a	Growth	7.2
TOTAL	34	300	Trade	+517	+5770	Per capita	3.2	28	Density	15

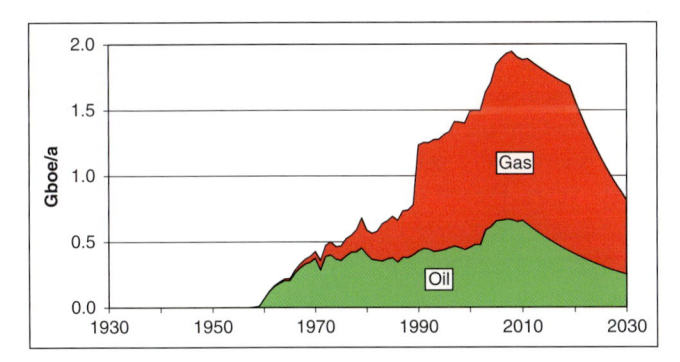

Fig. 4.1 Algeria oil and gas production 1930 to 2030

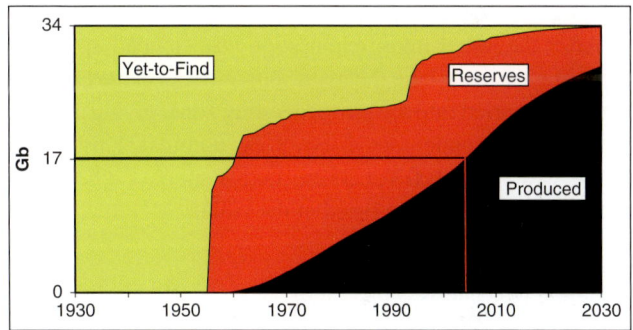

Fig. 4.2 Algeria status of oil depletion

Essential Features

Algeria covers an area of about 2.4 million km², supporting a population of about 36 million people. They are concentrated along the northern seaboard, where the capital, Algiers, is located. The Atlas Mountains in the north of the country border the Mediterranean and comprise two main ranges which are separated by a high plateau. The highest peaks rise to 2,500 m. To the south, lies the Sahara Desert, which is cut by a northerly trending divide that separates it into two arid depressions covered in sand dunes: the eastern desert lies at an altitude of about 600 m, while the western one locally drops below sea level. The Hoggar Mountains in the far south rise to as much as 3,000 m, exposing geological sequences that give a clue as to what lies beneath the deserts.

Geology and Prime Petroleum Systems

In geological terms, the Atlas Mountains form the boundary between the African and Eurasian tectonic plates, being subject to severe recurrent earthquakes. The country did not at first appear to hold much promise for oil. The mountains to the north were too disturbed, while the margins of the African Shield beneath the Saharan sands appeared less than attractive. But the perseverance of the French explorers was eventually rewarded with the discovery of a series of northerly trending Palaeozoic basins beneath the desert sands, which were evaluated by regional refraction seismic surveys and deep boreholes. The Silurian is the prime source-rock which has fed both underlying reservoirs flanking structural uplifts as well as those in the overlying sequence. Triassic salt forms an

C.J. Campbell, *Campbell's Atlas of Oil and Gas Depletion*,
DOI 10.1007/978-1-4614-3576-1_4, © Colin J. Campbell and Alexander Wöstmann 2013

important seal to the underlying accumulations. In many areas, the source-rocks have been depressed into the gas-generating window.

Additional Devonian source-rocks were found later in other areas, providing a second cycle of oil and gas discoveries, which are generally of small to moderate size.

Exploration and Discovery

The first recorded exploration borehole was drilled in 1910, but exploration remained at a very low level until after the Second World War when drilling increased rapidly to reach a peak in 1961 and more than 50 boreholes were sunk. It has since dwindled in an erratic fashion, partly for political reasons. Altogether, about 1,500 exploration boreholes have been drilled.

The effort was rewarded by a string of major discoveries in 1956 and ensuing years, dominated by the Hassi Messaoud Field, the largest field in Africa, with some 10 Gb of oil and 284 Tcf of gas, where flanking Silurian source-rocks have charged Cambrian sandstones with oil.

After the initial finds, exploration was stepped up in the adjoining basins, where the main source-rocks in the Silurian have been buried into the gas-generating zone, giving a series of major gas-condensate discoveries, dominated by Hassi R'Mel with about 100 Tcf. There are a number of other giant oilfields including those listed in the table.

Table 4.2 Major fields

	Discovery	Oil Gb	Gas Tcf
Hassi Messaoud	1956	10.3	8
Zarzaitine	1957	1.2	1.4
T-F Tabakkort	1961	0.84	8
Rhourde El Baguel	1962	1.2	2
Ourhoud	1994	2	2
Hassi Berkine S	1994	0.8	0.9

The political situation, associated with independence, led to a curtailment of exploration, the establishment of a State Oil Company (Sonatrach), and a decision to join OPEC. But later during the 1990's, the country decided to re-admit foreign companies, partly with a view to securing gas exports to both Europe and the United States. This policy was rewarded by a series of new discoveries, including several in the remote Ghadames Basin, which straddles the border with Libya.

A total of some 33 Gb of oil and 285 Tcf of gas have been found. The established areas are now at a mature state of depletion, although some possibilities for new discovery remain especially in the remote interior.

Production and Consumption

Oil production picked up rapidly in the 1950's following the major finds, to pass 1 Mb/d by 1970, since when it has increased only gradually to around 1.7 Mb/d in 2010, being constrained by pipeline capacity from the remote fields and also OPEC quota. Gas production too has risen in parallel through the 1960's and 1970's to pass 1 Tcf a year in 1979 and reach almost 6.8 Tcf in 2010. The country also produces substantial amounts of NGL and LNG.

Oil consumption stands at 114 Mb a year, leaving a surplus of 517 Mb a year for export. The substantial gas supply is largely exported by pipelines to Europe and used both to produce Natural Gas Liquids and Liquefied Natural Gas.

The Oil Age in Perspective

Algeria, which was originally occupied by Berber tribes, was invaded by Arabs during the first millennium, bringing the people into the fold of Islam. Later, the country was incorporated into the Ottoman Empire, partly as a defence against Spanish incursions. The coast became home to pirates threatening Mediterranean trade, which led France to invade in 1830, successfully subjugating the country over the next twenty years. French settlers arrived in increasing numbers to eventually dominate life in the country, which was later incorporated into the French State. Algerian soldiers fought for France in the First World War many of the survivors remained in metropolitan France, supporting their families at home with remittances. Even so, a strong thread of nationalist feeling survived to erupt during the Second World War. By 1954, the National Liberation Front initiated a phase of what would now be called terrorism, seeking to re-establish an independent Islamic State. France sent in an army to suppress them, while attempting to find a political compromise. In 1959, General de Gaulle began to bow to the inevitable, finally granting self-determination in 1962. As many as 10,000 French troops and 250,000 Muslims lost their lives in the struggle.

Many French settlers returned to the homeland after independence, and their departure crippled the local administration and much of the business life. The economy collapsed, with revenues from newly found oilfields being the only bright light. Facing these problems, the new government adopted a policy of central planning, being a form of Islamic Communism, similar to that followed in Iraq prior to the invasion. Border disputes with Morocco over the ownership of iron-ore deposits led to a short war in 1963. Remittances from the large number of Algerian workers in France continued to be an important element in the economy.

It appears that while the independent government has been moderate and pragmatic, the country retains its revolutionary traditions and outlook. In 1992, the Army intervened to suspend an election that would have returned the Islamic Salvation Front, and some 150,000 civilians lost their lives in the subsequent repression. Mr Bouteflika has been running the country since with the support of the Army, but his position has now been endorsed in new elections. It remains to be seen how the country will react to the growing movement towards pan-Arab solidarity provoked by the invasion of Iraq. The military will likely retain a strong hand in government, even with the return of a form of democratic process. Rising food prices and economic recession prompted a series of demonstrations starting in late 2010 but, although serious enough, they were less severe than in other North African countries, as the Government tried to react to the demands.

Algeria can in some respects look forward to a prosperous future, becoming one of Europe's premier sources of oil and gas, reaping the benefits of rising prices. If that were not enough, the country clearly has great potential for solar energy. Perhaps the immigrants in France, who are now denied their headscarves under a new ordinance, will be tempted to head for home in increasing numbers. On the other hand, climate change may lead to increasing desertification restricting the population to the northern seaboard. Algeria too has suffered from urbanisation and an insupportably high birth rate.

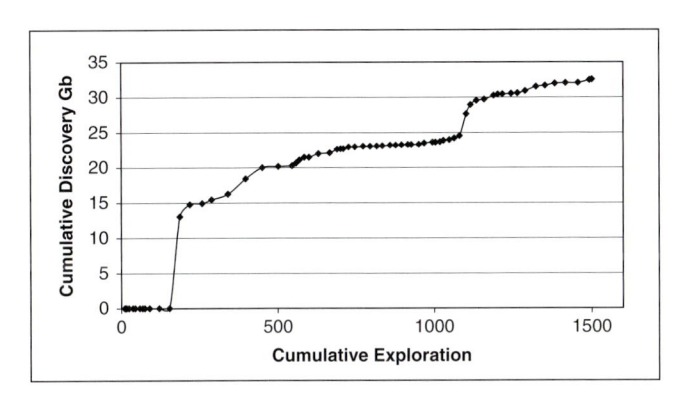

Fig. 4.3 Algeria discovery trend

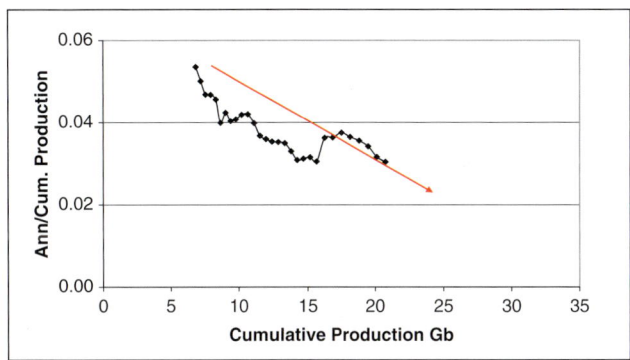

Fig. 4.4 Algeria derivative logistic

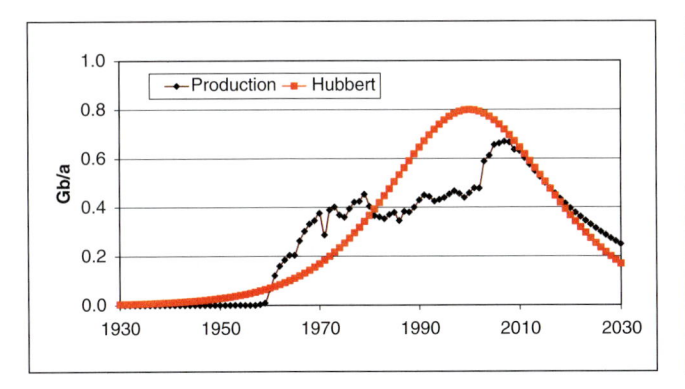

Fig. 4.5 Algeria oil production: actual and theoretical

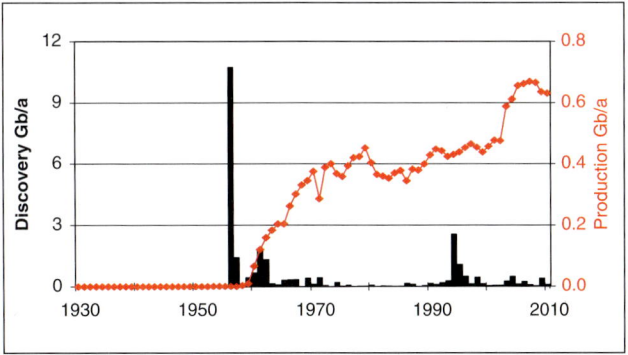

Fig. 4.6 Algeria discovery and production

Angola

Table 5.1 Angola regional totals (data through 2010)

Production to 2100						Peak Dates			Area	
Amount		Rate						Oil	Gas	'000 km²	
	Gb	Tcf	Date	Mb/a	Gcf/a	Discovery	1978	1971	Onshore	Offshore
PAST	6.3	6.7	2000	272	251	Production	2000	2015	1,250	20
FUTURE	3.7	11	2005	215	300	Exploration	1968		Population	
Known	3.3	10	2010	217	364	*Consumption*	Mb/a	Gcf/a	1900	2.3
Yet-to-Find	0.4	1.1	2020	123	400	2010	27	26	2010	19.6
DISCOVERED	9.6	17	2030	70	278		b/a	kcf/a	Growth	8.5
TOTAL	10	18	*Trade*	+190	+338	Per capita	1.4	1.5	Density	15.7
Excludes deepwater										

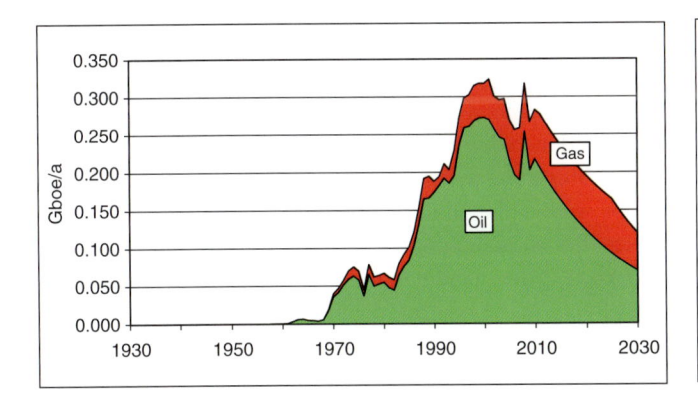

Fig. 5.1 Angola oil and gas production 1930 to 2030

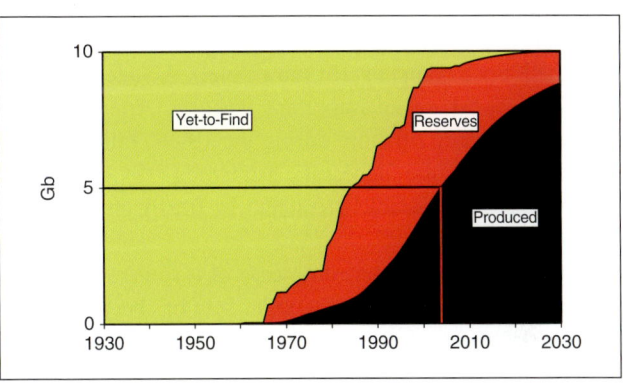

Fig. 5.2 Angola status of oil depletion

Essential Features

Angola is a former Portuguese territory covering an area of some 1.3 million km² and supporting a population of about 20 million. A coastal strip is flanked by escarpments, rising to extensive plateaux between 1,500 and 2,500 m above sea level, which in turn give way to featureless plains falling eastwards to an elevation of about 500 m. The northern part of the country is drained by the great Congo River, whereas the southern part lies in the headwaters of the Zambesi, which flows eastwards across Africa. The cold Benguela Current has led to near desert conditions along the southern coast, but most of the country is covered by tropical forest. It has substantial mineral resources of diamonds, iron, manganese, copper and phosphate in addition to petroleum.

Geology and Prime Petroleum Systems

Plate tectonic movements led to the opening of the South Atlantic during the early Cretaceous about 140 million years ago, when freshwater lakes formed in rifts. A sequence of such rifts developed progressively westwards as the continents moved apart, each having its own particular set of geological circumstances. Rich oil source-rocks were laid down both in the lower part of the sequence in the early rifts and in the upper part of the sequence in the later rifts. The sea broke into the lakes during an epoch of high sea-levels associated with mid-Cretaceous global warming. The evaporation of the seawater under these conditions led to the deposition of salt, which formed both a seal to underlying oil, and later, a glide-plane for structural slides for

C.J. Campbell, *Campbell's Atlas of Oil and Gas Depletion*,
DOI 10.1007/978-1-4614-3576-1_5, © Colin J. Campbell and Alexander Wöstmann 2013

huge rafts of limestone that slipped down the subsiding ocean floor. Oil reservoirs occur within the rifts in both sands and the rafted carbonate rocks, as well as in the Tertiary sediments that later covered the rifts. The latter mainly comprise turbidite deposits, which could be compared with a form of submarine avalanche, triggered when coastal bars ruptured under storm, hurricane or earthquake to slump into the ocean depths. Their quality as oil reservoirs was locally enhanced where the sediment was taken back into suspension by long-shore currents, which winnowed out the fine-grained material. Many of the traps for oil have a stratigraphic component being partly controlled by contemporaneous seafloor relief that served to trap the reworked turbidite sands.

Exploration and Discovery

Exploration commenced in the early 1950s when the Belgian company, PetroFina, secured rights to the onshore littoral, where it succeeded in finding a number of modest fields. Attention then turned offshore where the first wells were drilled in 1966 by a number of companies, including particularly the French company, Elf (now Total). A State Company was formed, taking an important position. Attention later turned to the deep offshore during the 1990s, following the development of the necessary technology by Petrobras, which faced comparable prospects in Brasil on the other side of the Atlantic.

Angola is thus an interesting case of a country, in which the depletion of *Regular Conventional Oil* has been followed by second cycle of deepwater oil, delivering substantial reserves that had not been anticipated in earlier years. But the deepwater area itself is now becoming mature, yielding no more than modest discoveries over the past few years, as described in Chap. 12.

A total of almost 700 exploration boreholes, excluding those in deepwater, have now been drilled, delivering about 9.6 Gb of oil and 17 Tcf of gas. Peak exploration drilling was in 1968, when almost 50 boreholes were drilled, but has now dwindled to no more than about five per year.

Production and Consumption

Oil production commenced in 1956, since when about 6.3 Gb have been produced. The production of *Regular Conventional Oil*, now standing 595 kb/d, is set to decline at about 5.5% a year. But that decline will be more than offset by growing deepwater production until it in turn reaches a peak or plateau around 2014. Oil consumption stands at a modest 27 Mb/a, meaning that the country can remain a substantial exporter for many years to come.

Gas production commenced in 1969 and has risen to 364 Gcf/a, of which 26 Gcf/a are reported to be consumed internally with the balance being presumably exported, flared or re-injected.

The Oil Age in Perspective

The country was home to warring Bantu tribes when the Portuguese arrived in 1483 to establish trading posts, largely for slaves who were exported to Brasil in their thousands. The trade continued illegally after the abolishment of slavery in 1836. Portuguese settlement had progressed over the centuries, bringing Christianity and order to the tribal interior. The formal boundaries of the country were set in 1926, and the capital, Luanda, was at its prime a delightful town with an almost European atmosphere.

Authoritarian colonial rule was administered by Portugal until 1961 when revolt broke out in the north, accompanied by massacres and reprisals. The conflict gradually spread over the rest of the country until 1975, when Portugal decided to withdraw from its African territories following a radical change of government at home. A Communist-led movement, the MPLA, succeeded in taking power in the main cities by force of arms, establishing close ties with Cuba and the Soviet Union. Cuban troops were called in to help maintain order, and did indeed protect the onshore oilfields, operated by PetroFina. As many as 300,000 Portuguese nationals, some of whose families had been living in Angola for generations, had to return to their home country. However, a rival movement (UNITA) calling for total independence, challenged the Government, leading to a protracted form of civil war that continued into the 1990s. By 1995, as many as 8,000 UN peacekeepers were in the country that was facing appalling conditions, made worse by the widespread use of land-mines, provided by unscrupulous Western arms dealers. Average life expectancy fell to 41 years, and more than a million people lost their lives. No doubt the revenues from the burgeoning offshore oil industry have enflamed the political conflict, as different factions seek to get their hands on it. The few surviving old people must look back on the Colonial period as a golden age.

Angola's oil revenues have risen to dizzy heights following the recent price surges, which in turn is likely to lead to further intractable political problems, associated with many well-lined pockets and foreign bank accounts. It will also be a key source of world oil, which possibly explains the US interest in establishing a military base on the islands of Sao Tome and Principe in adjoining waters to the north. It is noteworthy in this connection that the leaders of Britain and the United States have stated:

5 Angola

We have identified a number of key oil and gas producers in the West Africa area on which our two governments and major oil and gas companies could cooperate to improve investment conditions, good governance, social and political stability, and thus underpin long-term security of supply (*http://politics.guardian. co.uk/foreignaffairs/story/0,11538,1084958,00.html*).

It sounds as if they contemplate using the cloak of democracy at the point of the bayonet by which to impose market economics aimed at taking the region's oil. Angola however reacted by joining OPEC in 2007, possibly hoping to find allies. Deepwater production will likely come to an end by around 2040, with the tail end of the *Regular Conventional Oil* production dragging on for about another 10 years beyond that. Angola will then have to revert to agricultural self-sufficiency. It might be able to achieve conditions reminiscent of those under colonial rule in the early part of the last century, but disintegration and tribal warfare is unfortunately perhaps a more likely outcome. There is ample land and natural fertility which could support the current population.

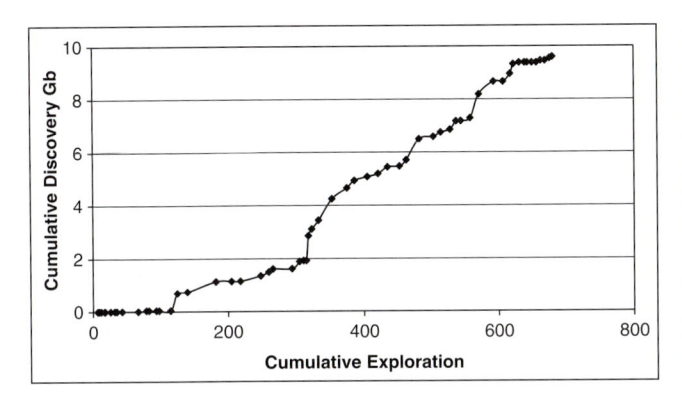

Fig. 5.3 Angola discovery trend

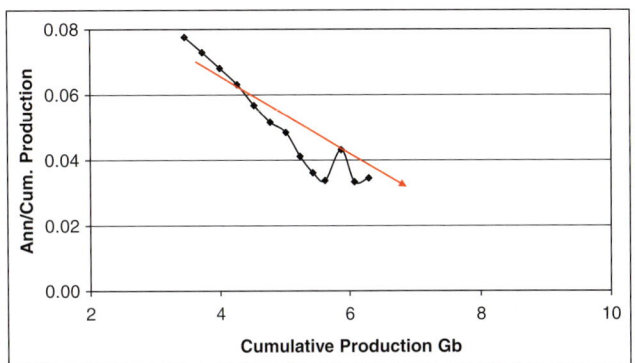

Fig. 5.4 Angola derivative logistic

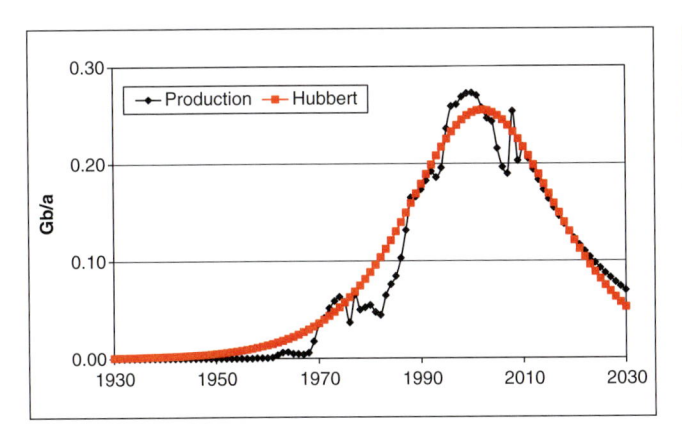

Fig. 5.5 Angola production: actual and theoretical

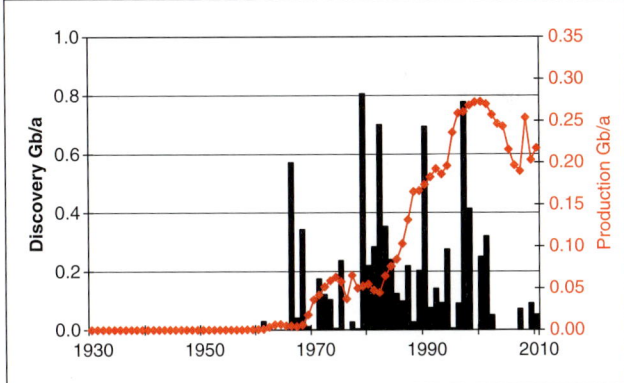

Fig. 5.6 Angola discovery and production

Cameroon

Table 6.1 Cameroon regional totals (data through 2010)

Production to 2100						Peak Dates			Area	
Amount		Rate					Oil	Gas	'000 km²	
	Gb	Tcf	Date	Mb/a	Gcf/a	Discovery	1977	1979	Onshore	Offshore
PAST	1.3	1.9	2000	31	69	Production	1985	2029	480	20
FUTURE	0.5	4.6	2005	30	65	Exploration	1977		Population	
Known	0.4	4.1	2010	24	69	*Consumption*	Mb/a	Gcf/a	1900	1.5
Yet-to-Find	0.1	0.5	2020	14	70	2010	11	0	2010	20.1
DISCOVERED	1.6	6.0	2030	9	70		b/a	kcf/a	Growth	13.4
TOTAL	1.75	6.5	*Trade*	+13	+68	Per capita	0.5	0	Density	42

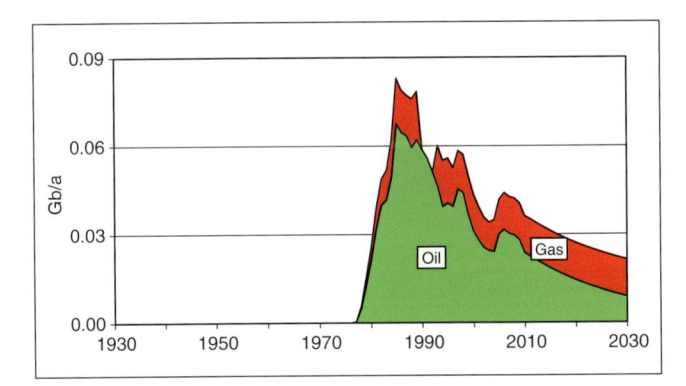

Fig. 6.1 Cameroon oil and gas production 1930 to 2030

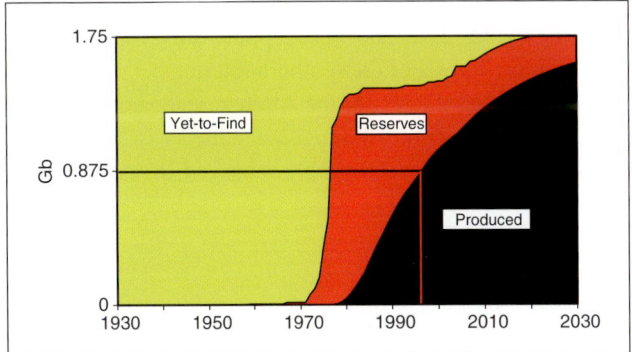

Fig. 6.2 Cameroon status of oil depletion

Essential Features

The territory, known as the Cameroons, runs inland from the Gulf of Guinea to the south of Nigeria. A coastal plain gives way to a densely forested plateau at an altitude of about 600 m. In the north, the country is cut by a mountainous volcanic belt, capped by Mount Cameroon rising to 4,000 m, which in turn gives way to high savannah approaching Lake Chad.

Geology and Prime Petroleum Systems

The geology of the Cameroons is dominated by the easterly trending rift system, caused by a major sinistral transform fault-zone, which offset the Sahara Shield from the rest of the Continent, being also represented across the Atlantic by the Amazon Basin in Brasil. This line of weakness has been the site of volcanic activity, both onshore and offshore, where it is represented by the island of Fernando Po (also known as Bioko). It marks the boundary between the Niger Delta to the north and the South Atlantic marginal basins to the south, described respectively in connection with Nigeria and Angola. The relatively high geothermal gradients associated with the rift zone have served to place shallow, relatively lean early Tertiary source-rocks in the oil generating window, but the amounts generated have been modest.

Exploration and Discovery

Exploration commenced onshore in 1953 before extending offshore 12 years later to reach a peak in 1977 when as many as 23 exploration boreholes were sunk, but has since dwindled to no more than one or two a year. It resulted in a few, mainly modest, finds of oil and gas, which peaked in 1977 and 1979, respectively. A total of 1.6 Gb of oil have been found, accompanied by almost 6 Tcf of gas. The country also has some deepwater potential, yet to be fully evaluated.

Production and Consumption

Oil production commenced in 1977, passing 100 kb/d in 1982, before reaching a peak at 185 kb/d in 1985 and declining thereafter to around 65 kb/d in 2010. Gas production followed a similar profile with a peak of 92 Gcf in 1988. Oil consumption stands at about 11 Mb a year, making the country a modest exporter. Its gas is also flared.

The Oil Age in Perspective

The Cameroons is something of a melting pot, being home to more than 200 different racial and tribal groups, coming from the south, east, and north. Some surviving pygmies remain in the forests of the southern interior.

Like much of the rest of the Atlantic seaboard of Africa, the Cameroons fell under Portuguese influence in the fifteenth century, when the coastal people managed to secure slaves from the tribal interior for export, primarily to Brasil. In 1809, a Muslim leader from the north invaded the territory seeking to expand Islam at the expense of the coastal people who had come under the influence of Christian missionaries.

In 1884, it became a German colony bringing a degree of stability and economic progress, which ended following the First World War when the Germans were expelled and the country was split into French and British administrations. The French took the greater interest, endeavoring to integrate the territory into its African empire, whereas the British treated it as a sort of extension of Nigeria. These divisions led to further conflicts. In 1960, the French sector won independence from France, being later joined by the former British sector to form the Federal Republic of Cameroon. The Government was at first dominated by Muslim leaders from northern tribes, who succeeded in crushing a rebellion by the Christian south. But later, the scales were tipped in the other direction as the Christians gained the upper hand.

Looking ahead, it would be reasonable to expect the territory to disintegrate into its several ethnic, religious, and linguistic communities, facing continuing conflict and tension. Oil revenues, which tend to distort sustainable life-patterns in countries facing these challenges, may rise in the near term due to higher world prices but production is set to decline in step with depletion. Perhaps by the end of the century a new sustainable environment for the survivors will have emerged.

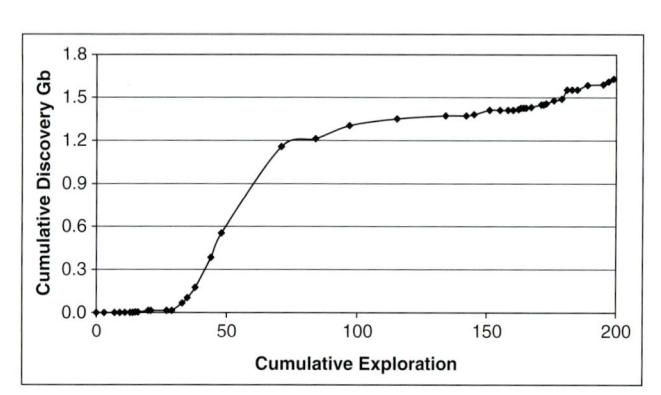

Fig. 6.3 Cameroon discovery trend

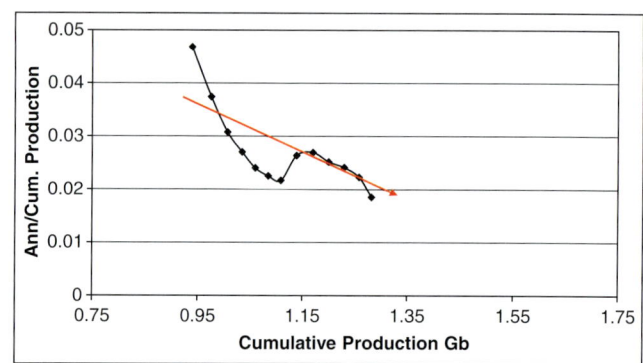

Fig. 6.4 Cameroon derivative logistic

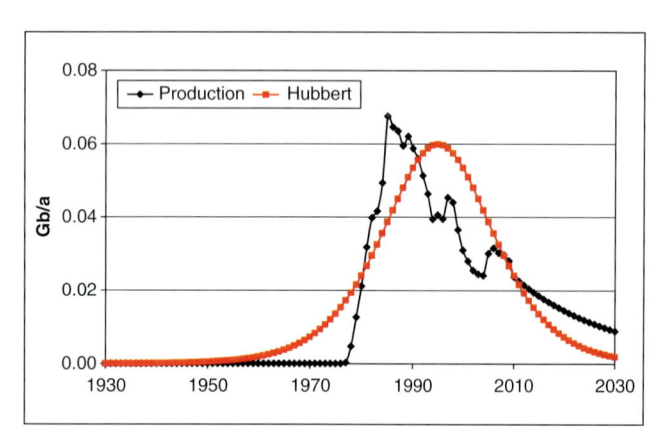

Fig. 6.5 Cameroon production: actual and theoretical

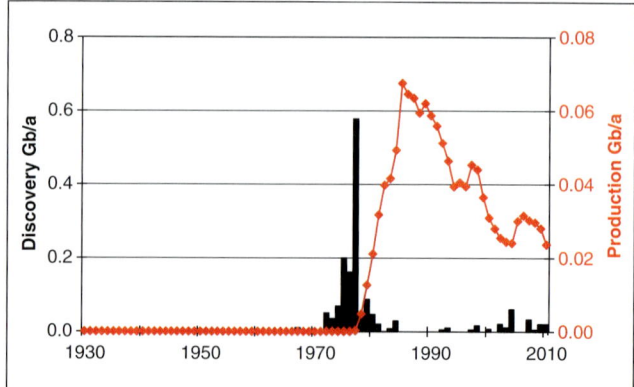

Fig. 6.6 Cameroon discovery and production

Chad

Table 7.1 Chad regional totals (data through 2010)

	Production to 2100					Peak Dates			Area	
	Amount		Rate				Oil	Gas	'000 km²	
	Gb	Tcf	Date	Mb/a	Gcf/a	Discovery	1977	1975	Onshore	Offshore
PAST	0.4	0	2000	0	0	Production	2015	2030	1290	0
FUTURE	1.6	0.5	2005	64	0	Exploration	2002		Population	
Known	1.0	0.3	2010	46	0	*Consumption*	Mb/a	Gcf/a	1900	1
Yet-to-Find	0.7	0.2	2020	50	10	2010	0.7	0	2010	11.5
DISCOVERED	1.4	0.3	2030	34	10		b/a	kcf/a	Growth	11.5
TOTAL	2.0	0.5	*Trade*	+45	0	Per capita	0.1	0	Density	8.9

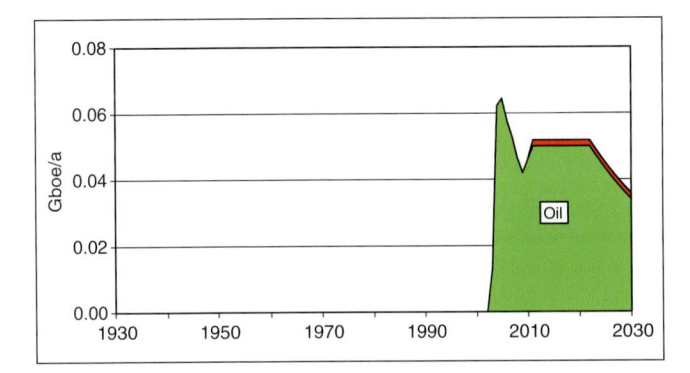

Fig. 7.1 Chad oil and gas production 1930 to 2030

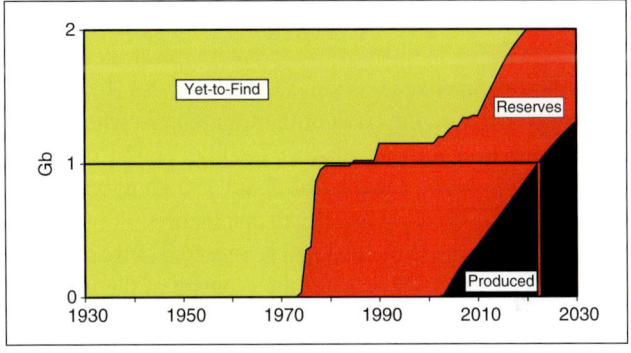

Fig. 7.2 Chad status of oil depletion

Essential Features

Chad is a landlocked country, covering some 1.2 million km² in the heart of Africa. In physiographic terms, it forms a large basin between the Tibesti Massif to the north, rising to over 3,000 m, and high plateaux to the south and east. Lake Chad, in the southwest, is a prominent feature. In the past, it was much larger than at present, being over 150 m deep, and draining over impressive falls into the Benue River, which runs westward to the Atlantic. Today, its level fluctuates greatly with climatic changes, and it almost dried up in droughts during the 1970s and 1980s.

The population of 11.5 million comprises many different tribal groups of mixed African and Arabic origins, speaking a range of local languages.

Geology and Prime Petroleum Systems

Chad is situated on a system of rifts and transcurrent faults, which are associated with the major plate boundary that cut across the great Southern Continent (Pangea), before the opening of the South Atlantic. This boundary formed the lines of weakness followed by the Niger and Amazon Rivers. Evidently, algae proliferated in these opening rifts during the Cretaceous, providing the organic material that was locally preserved and buried sufficiently to become hydrocarbon source rocks. The prime reservoirs are also of Cretaceous age, but the conditions of entrapment are complex due to late fault movements and structural inversions.

Exploration and Discovery

Chad did not represent obvious oil territory in geographic, geological or political terms, but nevertheless Conoco mounted an exploration campaign in the 1970s, possibly at the instigation of the US Government, wishing to counter Libyan pressure. Conoco later withdrew, and its place was taken by Chevron, Exxon, Total and other companies, being partly funded by the World Bank. Almost 60 exploration boreholes have been drilled so far, finding a total of about 1.35 Gb of oil, some of which is partly degraded, having a gravity of 20° API. Most lies in three fields, Bolombo, Kombe and Miandoum, in the southwest corner of the country. Further exploration is likely to be stimulated by the recent opening of an export pipeline, and is here expected to yield another 650 Mb of new discovery, with the drilling of some 20 new exploration boreholes. Little is known about the gas potential, tentatively estimated at 0.5 Tcf. Most production is presumably at present re-injected.

Production and Consumption

A 1,000 km pipeline, costing $3.7 billion, has been constructed to Kribi on the coast of the Cameroons, where oil is transhipped to a floating loading facility, 11 km offshore. The pipeline, with a capacity of about 150 kb/d, was completed in 2005, allowing exports to commence.

Production started in 2003 and is expected to reach a plateau of about 140 kb/d in 2011, the capacity of the pipeline, and continue at this level to around 2022 before declining at about 5% a year.

The Oil Age in Perspective

In earlier times, the country lay at the southern end of a trans-Saharan trade route leading to Tripoli on the Mediterranean, which was partly used for the slave trade. The territory fell within the French sphere of influence during the early years of the twentieth century, being incorporated into the Federation of French Equatorial Africa in 1910, before becoming an overseas territory of the French Republic in 1947.

A degree of independence was achieved in 1957 under the leadership of a West Indian immigrant, named Gabriel Lisette, and full independence followed in 1960. Years of conflict followed between rival partly tribal movements, reflecting tensions between the Muslim north—backed by Libya, which has physically intervened from time to time—and the more Christian south. A civil war in 1990 led to a new government under Lt. General Idriss Déby, who remains in power to this day as the Head of State. He enjoyed the support of France, which maintains a military presence in the country.

Most of the population, who are amongst the World's poorest, rely on subsistence farming and cattle-raising. The country now faces the added burden of being host to some 280,000 refugees from the Darfur region of neighbouring Sudan following recent tensions there.

Chad will enjoy a brief epoch of relative prosperity on the strength of the new oil revenues, estimated at $80 million a year, but this may serve to fuel the ambitions of the varied political factions in the country, leading to further tensions, as has been the experience in other African countries. In 2006, the President threatened to use the oil revenue to fund the military to maintain order, but this led the World Bank to briefly suspend the loans and freeze foreign banks accounts before a new agreement was reached. In April 2006, a coup to depose the President was overcome with the help of French troops stationed in the country, but another revolt broke out in early 2008.

If internal tensions were not bad enough, it is evident that the neighbouring Cameroons, if not factions within it, can exert a certain stranglehold on the country by controlling the export pipeline. The US troops protecting it may find themselves fully occupied.

As the Century passes, life in Chad will likely revert to its historical subsistence agriculture with a declining population facing disease, malnutrition, AIDS and tribal warfare, especially if the climate should change for the worse as seems to be happening. Lake Chad itself may finally dry up completely.

7 Chad

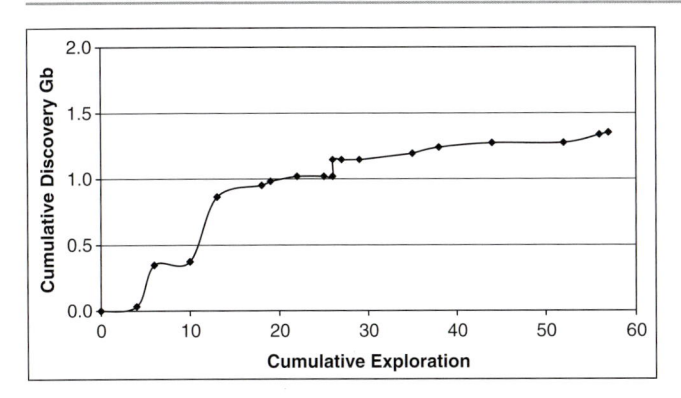

Fig. 7.3 Chad discovery trend

Fig. 7.4 Chad derivative logistic

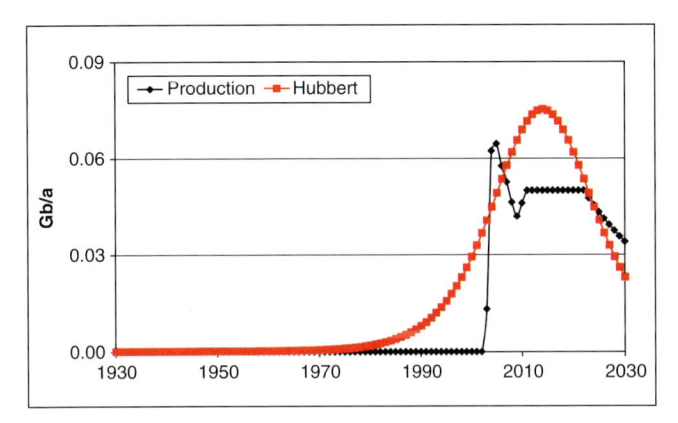

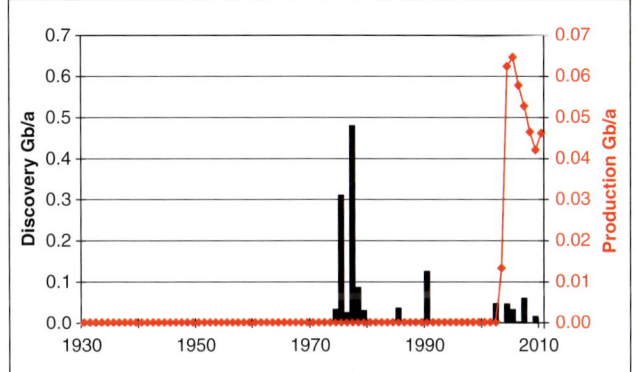

Fig. 7.5 Chad production: actual and theoretical

Fig. 7.6 Chad discovery and production

Congo

Table 8.1 Congo regional totals (data through 2010)

Production to 2100						Peak Dates			Area	
Amount		Rate					Oil	Gas	'000 km²	
	Gb	Tcf	Date	Mb/a	Gcf/a	Discovery	1984	1984	Onshore	Offshore
PAST	2.3	3.2	2000	102	127	Production	2010	2012	340	2
FUTURE	1.7	2.8	2005	83	251	Exploration	1992		Population	
Known	1.6	2.4	2010	108	296	Consumption	Mb/a	Gcf/a	1900	0.6
Yet-to-Find	0.2	0.4	2020	59	106	2010	4	33	2010	4.1
DISCOVERED	3.8	5.6	2030	32	24		b/a	kcf/a	Growth	6.8
TOTAL	4.0	6.0	Trade	+104	+265	Per capita	1	8	Density	12

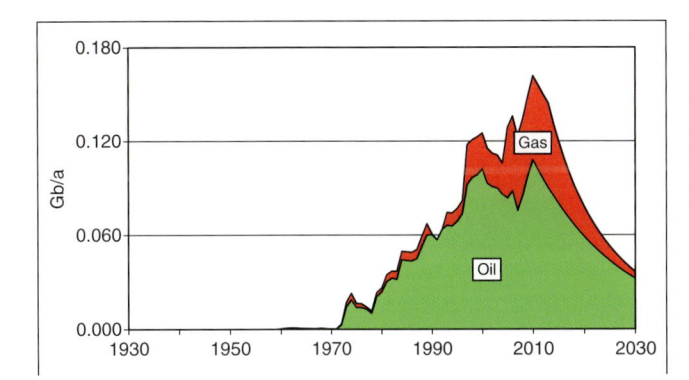

Fig. 8.1 Congo oil and gas production 1930 to 2030

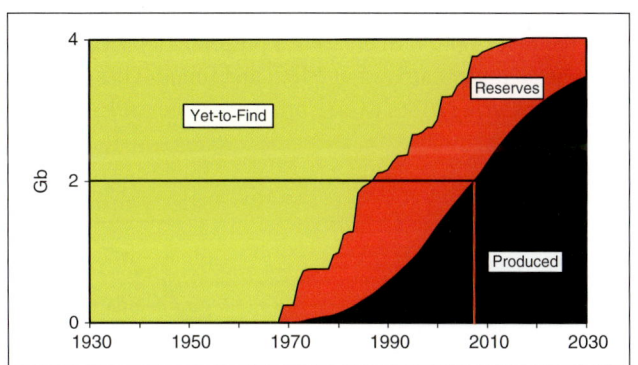

Fig. 8.2 Congo status of oil depletion

Essential Features

The Republic of the Congo, which is not to be confused with the Democratic Republic of the Congo (Zaire) comprises a long corridor, following the valley of the Congo River. A coastal plain gives way to a low range of mountains before opening into interior plains. It has a warm tropical climate supporting rain forests, but its lateritic soils give relatively low fertility. About one-third of the population of 4.1 million live in the capital, Brazzaville.

Geology and Prime Petroleum Systems

The coastal regions and territorial waters of the Republic of the Congo cover part of the prospective basin system that forms the eastern margin of the South Atlantic between Nigeria and Namibia, extending into deep water. It is divided by various arches and zones are transformed into a series of sub-basins, one of which is the Congo Basin. Rifts developed during the opening of the Atlantic during the Cretaceous. The prime source rock occurs in Lower Cretaceous (Barremian) shales, preserved beneath an important salt deposit. Oil was however locally able to migrate upwards to charge Middle Cretaceous carbonates, which have slumped down the continental shelf lubricated by the bed of salt.

Exploration and Discovery

Exploration drilling commenced onshore at a modest level in 1957, before extending offshore during the 1960s and finally into the deepwater domain during the 1990s. The peak of drilling came in 1992 when 19 boreholes were sunk, but has since declined markedly. The effort was rewarded by a numbers of modest discoveries, mainly offshore, with discovery peaking in 1984. Part of the deepwater territory is shared with Angola under a unitization agreement.

Production and Consumption

Production commenced in 1969 and probably reached its peak at 296 kb/d in 2010. It is expected to decline at almost 6% a year, ignoring the deepwater. Consumption is at a low level, meaning that the country can export about 104 Mb a year.

Gas production has also risen progressively to almost 300 Gcf a year, much of it must be re-injected or flared, but the situation is unclear. Decline is expected to begin around 2013 when some 70% will have been depleted.

The Oil Age in Perspective

Pygmies inhabited the rain forests of the northern Congo until the first Bantu farmers took a position 2,000 years ago. Portuguese traders arrived in 1482, and found it easy to persuade the warring tribes to take prisoners to be sold as slaves. The famous British explorer, Henry Stanley, arrived in 1877 penetrating the higher reaches of the Congo River. He was followed by a French adventurer, by the name of Pierre Savorgnon de Brazza, who arrived in 1880 to sign an agreement with a tribal leader. This led to French influence that eventually brought the Congo into the colony of French Equatorial Africa, with Brazzaville as its capital. It was a fairly draconian regime of native exploitation in the quest for rubber and ivory.

In 1946, the Congo became an overseas territory of France before gaining full independence as a republic in 1960. Shortly afterwards, it adopted a Communist form of government. A tense political situation followed, leading to the assassination of the President in 1977, as well as subsequent civil wars and conflicts, some reflecting deeper tribal loyalties. Control of petroleum revenues appears to have been one of the causes of conflict. The country has been run since 1997 by Mr. Sassou Nguesso, with a military background and certain socialist leanings.

It seems likely that the Congo will revert to an essentially tribal structure in the future, and the decline of oil production may be, in a sense, welcomed insofar as the control of revenues ceases to distort the political situation.

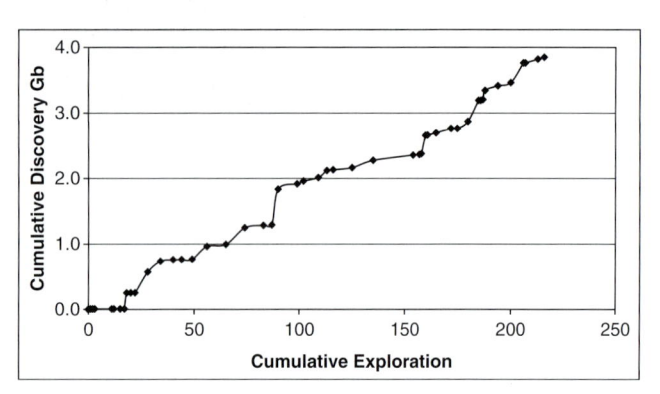

Fig. 8.3 Congo discovery trend

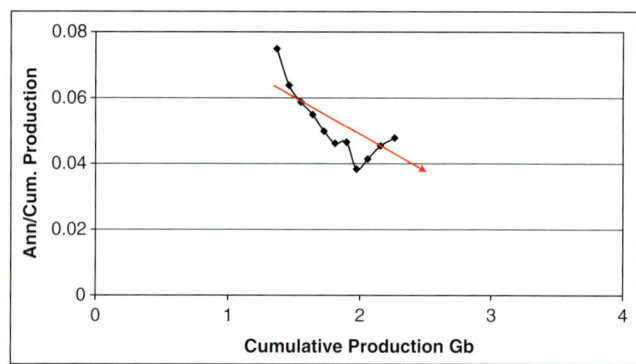

Fig. 8.4 Congo derivative logistic

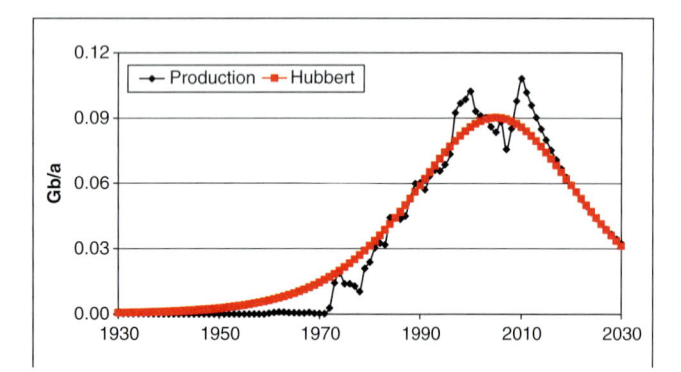

Fig. 8.5 Congo production: actual and theoretical

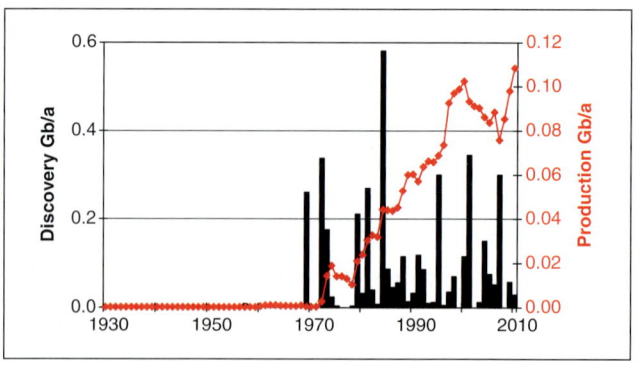

Fig. 8.6 Congo discovery and production

Egypt

Table 9.1 Egypt regional totals (data through 2010)

Production to 2100					Peak Dates			Area		
Amount		Rate				Oil	Gas	'000 km²	
	Gb	Tcf	Date	Mb/a	Gcf/a	Discovery	1965	1996	Onshore	Offshore
PAST	11	25	2000	280	860	Production	1996	2017	1001	150
FUTURE	4.2	70	2005	240	1660	Exploration	2008		Population	
Known	3.8	59	2010	191	2369	*Consumption*	Mb/a	Gcf/a	1900	10
Yet-to-Find	0.4	10	2020	123	2700	2010	270	1630	2010	82.6
DISCOVERED	15	85	2030	79	1876		b/a	kcf/a	Growth	8.3
TOTAL	15	95	*Trade*	−79	+739	Per capita	3.3	19.7	Density	82

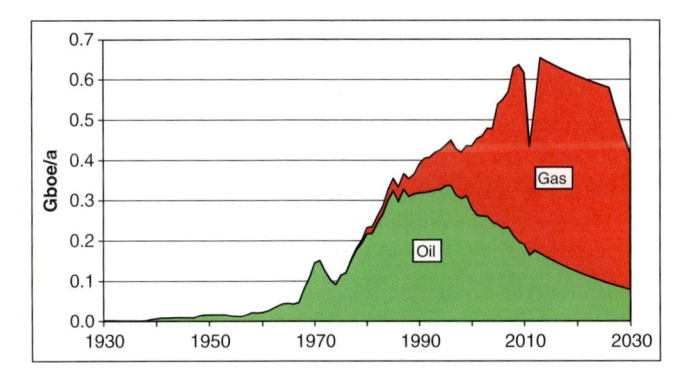

Fig. 9.1 Egypt oil and gas production 1930 to 2030

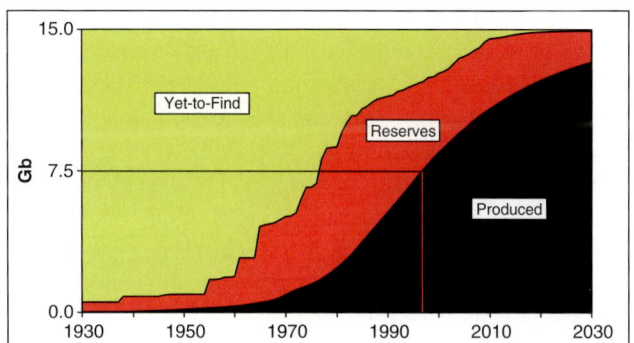

Fig. 9.2 Egypt status of oil depletion

Essential Features

Egypt covers an area of about one million km². Apart from the Nile Valley forming a long fertile strip, most of the country is barren desert. The Red Sea and Gulf of Suez separate it from Arabia, while the Mediterranean washes its northern shore. It has common frontiers with Libya to the west and the Sudan to the south. Its population of almost 83 million makes it the third most populous country in Africa after Nigeria and Ethiopia.

Geology and Prime Petroleum Systems

In geological terms, there are three main productive basins, of which by far the largest is the offshore Gulf of Suez, where oil is trapped beneath Miocene salt. Although now a very mature basin, modern technology has improved the mapping of the sub-salt plays, which may possibly lead to a few more modest finds. Another basin lies in the El Alemein area of western Egypt where Jurassic source-rocks have charged Cretaceous reservoirs in easterly trending rift zones. The third basin is the Nile Delta, which is a gas province. The Mediterranean shelf is narrow and steep. It might hold some deepwater potential, but the chances are slim.

Exploration and Discovery

Exploration commenced in the 1920s when a number of small discoveries were made, but it was not until the opening of the Gulf of Suez in the 1960s that the country's potential was realised. Amoco took a dominant position, working closely with the State oil company, to bring in the El Morgan Field in 1965 with over a billion barrels of ultimate recovery.

It was followed by the July, Belayim and Ramadan fields, which just attained 500 Mb giant status. Almost 2000 exploration boreholes have been drilled. Exploration drilling reached an extreme peak in 2008 when about 100 boreholes were drilled. But overall exploration is at a mature stage, with the larger fields well into decline. Exploration drilling is expected to draw to a close around 2030. The Nile Delta holds substantial gas reserves of about 50 Tcf and offers some further potential.

Production and Consumption

Oil production commenced in 1914 but did not rise significantly until after the Second World War with the discoveries in the Gulf of Suez. It passed 500 kb/d in 1979 to reach a peak of 922 kb/d in 1996, since when it has declined to 523 kb/d in 2010. It will likely continue to decline in the future at about 4.35% a year. Consumption has risen steeply in recently years to reach 270 Mb/a, meaning that the country has become a net importer.

Gas production commenced in 1935 and reached an early plateau at around 1 Gcf a year from 1941 to 1953 before falling steeply. A second surge of production came with the opening of the offshore in the Nile Delta to climb steeply over the last few years to reach 2 Tcf a year in 2010. Consumption has risen in parallel, with supply being dedicated to the domestic market. Along with related gas liquids, it will be an increasingly important source of energy for the population centres of Cairo and Alexandria. Both, oil and gas production is expected to fall over the next year or two as a consequence of the fall of the last government and political uncertainty.

The Oil Age in Perspective

It seems likely that *Homo sapiens,* as he first stepped out of Africa, followed the fertile Nile Valley, which has been a focus of population ever since, being the site of early civilisations marked by the famous Pyramids and tombs that still attract the attentions of archaeologists.

Alexander the Great conquered Egypt in 332 BC, founding Alexandria, the second largest city, which became a centre of early learning. Egypt became a Roman colony in 32 BC, following the Battle of Actium, when its celebrated Queen, Cleopatra, endeared herself to the Roman conqueror, Mark Anthony. In 642, it was taken by Arabs, who introduced Islam and their language, founding Cairo, the present capital, in 973. Saladin overthrew a previous Shi'ite dynasty in 1171, restoring the Sunni faith. The country was later invaded by the Ottoman Turks in 1517, who operated a fairly delegated administration, entrusted to the surviving Mameluke leaders despite their military defeat. Napoleon led a short-lived French invasion in 1798, before Muhammed Ali, an Ottoman ruler of Albanian origin, established a new strong government that embarked on foreign conquests, including what is now Saudi Arabia and Syria.

British interest in Egypt stemmed from the Suez Canal, which was built by the French in 1869, providing a strategic short cut to India. An additional interest was the production of cotton by which to supply the textile mills of Lancashire. Rivalry with France for control of Egypt ended in 1904 when France withdrew in exchange for British recognition of French claims to Morocco.

Britain occupied Egypt when Turkey entered the First World War on the German side, declaring it to be a Protectorate. There were various moves to independence in the inter-war years with the establishment of puppet regimes but the country remained under British control. In the Second World War, it saw the first decisive defeat of German forces at the Battle of El-Alamein in 1942. British influence dwindled after the war leaving the country more or less independent under King Farouk.

In 1948, Egypt, together with Syria, Jordan and Iraq, launched an attack in Palestine to oppose the creation of the State of Israel, but they were roundly defeated. That setback indirectly led a group of Army officers, under Col. Nasser, to take control of government in 1952, declaring a republic one year later. Having failed to secure international finance, Nasser decided to nationalise the Suez Canal in 1956 to fund the construction of the Aswan Dam on the Nile, which was desperately needed to control irrigation and provide hydroelectricity. Israel took this opportunity to strike in what might have been a contrived attack, giving Britain and France a pretext to send in a military force to protect the canal. But they were ignominiously made to withdraw under the United States and Russian pressure.

Col. Nasser emerged as an imaginative and strong leader of the Arab world, combining with Syria in 1958 to form the United Arab Republic, which was however dissolved 3 years later by Syria. The continuing perceived threat by Israel led to various interventions and alliances over the ensuing years, culminating in the Six-Day War in 1967 when Israel launched a preemptive strike. As many as 10,000 Egyptians were killed in the hostilities, and Nasser resigned, only to die 3 years later.

He was succeeded by President Sadat, who, with Syria, launched the so-called Yom-Kippur War in 1973 as a surprise attack on Israel. Although not a total military victory, it did bring Israel to the conference table. The Oil Shock of 1974 was a related event, as several Arab countries tried to counter US support for Israel by briefly restricting oil exports to that country. It led to a certain more even-handed

US position under President Carter, who now began to court friendship with Egypt, making substantial "aid" payments. However, the failure to resolve the plight of the Palestinians led to simmering unrest, which resulted in the assassination of Sadat in 1981 by a group of activists opposed to conciliation.

Hosni Mobarak replaced Sadat as President and pursued a moderate policy, concentrating on economic development, on which progress has been countered by the effects of an exploding population. There remained an undercurrent of deep frustration by those seeking to restore Arab confidence. It exploded in January 2011 in the face of an economic recession caused by growing oil imports and higher oil prices, and led to the removal of President Mobarak. The country remains in a relatively unstable condition without a new strong government.

The country has become a net importer of oil as domestic production continues to fall. But gas production can be maintained, and the country is blessed with a high level of solar radiation, which could help provide for its energy needs over the next few decades. Looking ahead it would be reasonable to expect growing Egyptian pressure on Libya, its oil-rich neighbour, which may take the form of close cooperation or, if that fails, outright hostility. It might provide a foundation for new Arab power-base following the frustrated dreams of Col. Nasser.

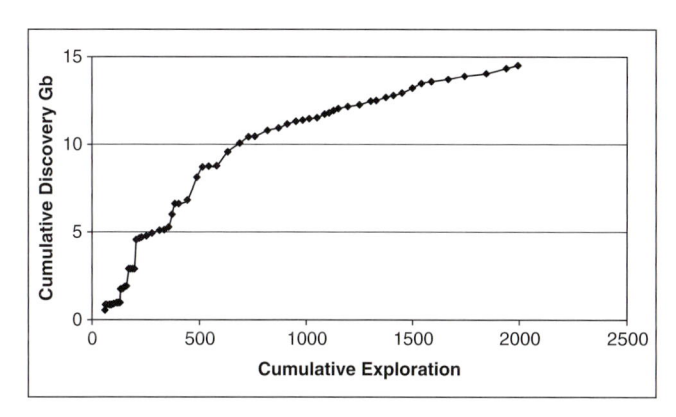

Fig. 9.3 Egypt discovery trend

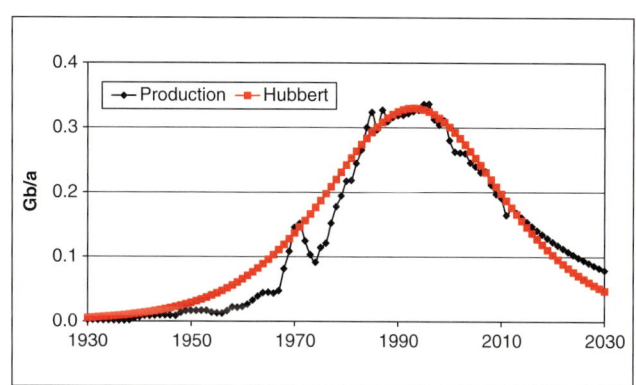

Fig. 9.4 Egypt derivative logistic

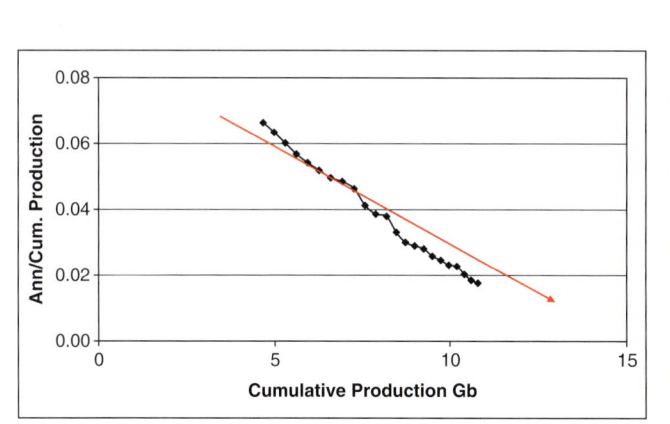

Fig. 9.5 Egypt production: actual and theoretical

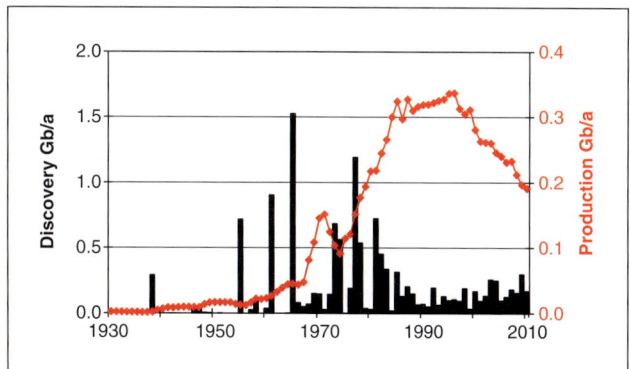

Fig. 9.6 Egypt discovery and production

Gabon

Table 10.1 Gabon regional totals (data through 2010)

	Production to 2100					Peak Dates			Area	
	Amount		Rate				Oil	Gas	'000 km²	
	Gb	Tcf	Date	Mb/a	Gcf/a	Discovery	1985	1965	Onshore	Offshore
PAST	3.6	2.9	2000	115	84	Production	1997	1995	270	80
FUTURE	1.4	0.6	2005	97	74	Exploration	1989		*Population*	
Known	1.3	0.5	2010	83	73	*Consumption*	Mb/a	Gcf/a	1900	0.4
Yet-to-Find	0.1	0.1	2020	47	24	2010	6.6	3	2010	1.5
DISCOVERED	4.9	3.4	2030	26	9		b/a	kcf/a	Growth	3.8
TOTAL	5.0	3.5	*Trade*	+77	+70	Per capita	4.4	2	Density	5.6

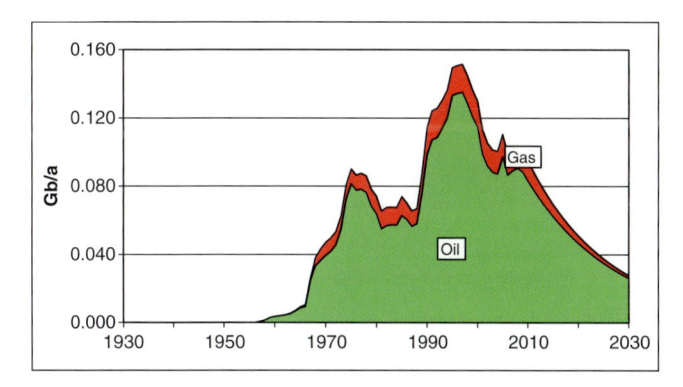

Fig. 10.1 Gabon oil and gas production 1930 to 2030

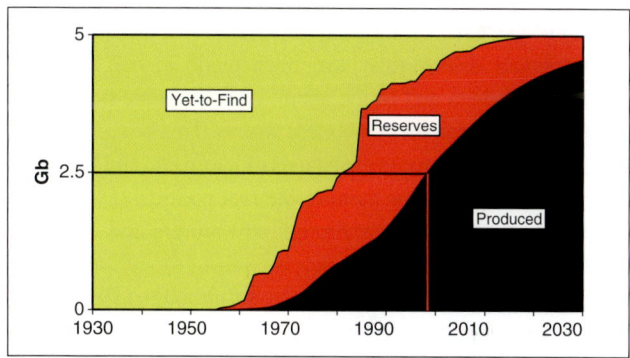

Fig. 10.2 Gabon status of oil depletion

Essential Features

Gabon covers an area of some 270,000 km², straddling the Equator on the west coast of Africa. Partly dissected plateaux in the interior, rising to some 600–1,000 m above sea level, give way to a fairly narrow coastal strip, washed by the northward flowing Benguela Current. The country supports a population of 1.5 million, belonging to about ten different tribal groups who originally spoke Bantu languages before French became the *lingua franca*. Many live in Libreville, the capital, as well as Port Gentil. Gabon is bordered by the Congo to the south and east, while the Cameroons and the enclave of Equatorial Guinea lie to the north. Offshore are the islands of Sao Tome and Principe, in which the United States has taken a strategic interest.

The land is mainly under a cover of tropical rain forests, which have been profitably exploited by the timber industry since the 1970s. The construction of railways in the 1980s also opened up mineral deposits, including uranium and manganese, of which it is one of the world's largest producers. A major iron ore deposit awaits development.

Geology and Prime Petroleum Systems

In geological terms, Gabon lies on a rift zone that developed as the South Atlantic opened during the early Cretaceous. The first deposits to be laid down were in lakes, and included hydrocarbon source-rocks. The rifts were temporarily invaded by the sea, which was subject to evaporation leading to the deposition of salt. It not only sealed the deeper sequence, but also gave rise to subsequent salt-induced structures, offering traps of oil. A new cycle of deposition followed, also with the early deposition of hydrocarbon source-rocks during the mid-Cretaceous, and lasted into the Tertiary period, when the increased gradient of the continental slope gave rise to turbidity currents. Both pre- and post-salt plays have now been thoroughly evaluated.

Exploration and Discovery

Petroleum exploration commenced onshore after the Second World War and was soon rewarded by the discovery of a number of small to modest fields before the giant Rabi-Konga Field was found in 1985 with about 800 Mb. Later, exploration moved offshore and was again rewarded by a number of moderately sized fields in the area south of Port Gentil. A total of about 730 exploration boreholes have been drilled, to deliver a total of about 4.9 Gb, of which some 3.6 Gb have been produced. Accordingly, the country is at a mature stage of exploration with little scope for significant new discovery. About 3.4 Tcf of gas have been found in parallel with the oil.

Deepwater discoveries have been made in neighbouring Equatorial Guinea and to the south, opening some hopes that the play may extend into Gabonese waters. This brings particular strategic importance to the islands of Sao Tome and Principe, which probably exposes them to the risk political disturbance and sedition, sponsored by foreign adventurers and interests. But deepwater discovery depends on a most exceptional combination of geological circumstances, so, although Gabon has a relatively good address in the light of neighbouring finds, it is far from sure that its deepwater will deliver.

Production and Consumption

Oil production commenced in 1957 and reached a peak of 370 kb/d in 1997 which was close to the midpoint of depletion. It is now declining at a depletion rate of 5.6% a year. The country has modest gas resources, mainly used for local electricity generation, with reported reserves standing at about 0.3 Tcf.

The Oil Age in Perspective

The Portuguese explored the Gabon Estuary in 1472, being followed by French, Dutch and British traders, many active in the slave trade during the eighteenth and early nineteenth centuries. The French successfully negotiated rights with the local chieftains around 1840 in an effort to curb the slave trade, whereupon the territory was administered by French naval officers. Later, it became part of the French Congo before being given independent colonial status in 1910 as part of French Equatorial Africa. It was occupied by Free French forces in the Second World War, becoming an independent overseas French territory in 1946.

Having passed through the status of an autonomous republic in 1958, it became fully independent in 1960. The first President, Leon M'ba, was succeeded by Omar Bongo in 1967 under whose dictatorial reign the country enjoyed a 20-year epoch of relative stability and prosperity, largely funded by oil revenues. It was a full member of OPEC from 1975 to 1995. A decline in oil revenues, consequent upon a fall in oil price in the mid-1980s, led to some political unrest and tensions, which required French military intervention to protect French nationals and property. The country has maintained its close ties with France. President Bongo was returned to power in 2005, but died in 2009. His son was the declared winner of subsequent elections, but this led to violent protests and riots, followed by a coalition government in 2010.

It looks as if Gabon will not escape the tensions affecting most other African countries, and as oil revenues decline in the years ahead, it is a matter of concern to find that Gabon's indigenous food production meets less than 20% of the country's needs.

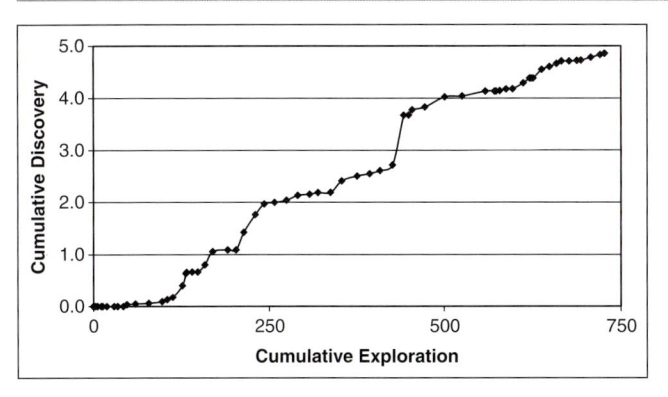

Fig. 10.3 Gabon discovery trend

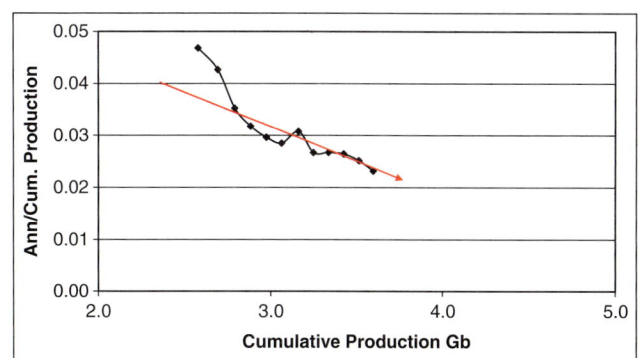

Fig. 10.4 Gabon derivative logistic

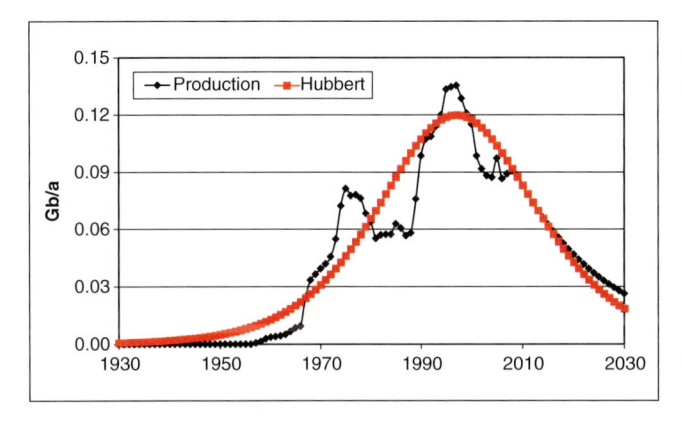

Fig. 10.5 Gabon production: actual and theoretical

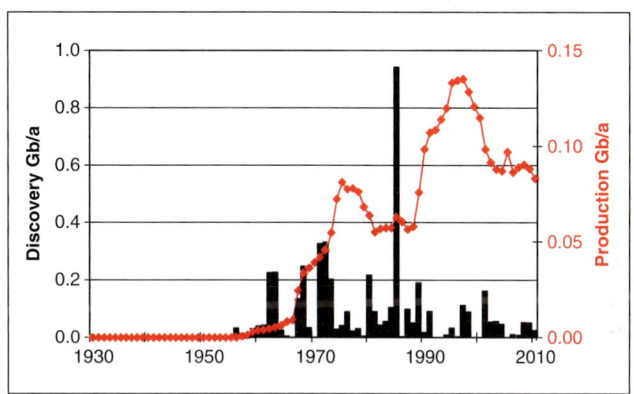

Fig. 10.6 Gabon discovery and production

Libya

Table 11.1 Libya regional totals (data through 2010)

	Production to 2100					Peak Dates			Area	
	Amount		Rate				Oil	Gas	'000 km²	
	Gb	Tcf	Date	Mb/a	Gcf/a	Discovery	1961	1977	Onshore	Offshore
PAST	27	21	2000	515	358	Production	1970	2030	1770	150
FUTURE	28	74	2005	596	696	Exploration	1963		Population	
Known	25	59	2010	602	1069	*Consumption*	Mb/a	Gcf/a	1900	0.5
Yet-to-Find	2.8	15	2020	455	1276	2010	105	242	2010	6.4
DISCOVERED	52	80	2030	376	1600		b/a	kcf/a	Growth	12.8
TOTAL	55	95	*Trade*	+497	+827	Per capita	16.5	38	Density	3.6

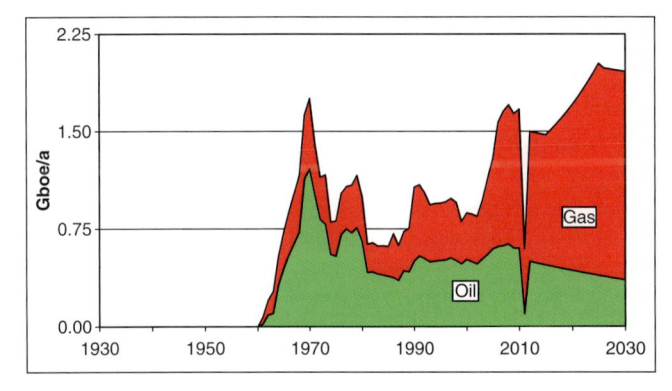

Fig. 11.1 Libya oil and gas production 1930 to 2030

Fig. 11.2 Libya status of oil depletion

Essential Features

Libya covers an area of 1.77 million km², supporting a population of about 6 million, who live mainly along the Mediterranean seaboard. It has common frontiers with Egypt to the east, Algeria and Tunisia to the west, and Chad, Sudan and Niger to the south. A mountain range, known as the Akhdar, rises to 900 m in the northeast, but is flanked by true deserts and rocky arid plateaux over most of the rest of the country.

Oil exploration has led to the identification of an extensive fossil aquifer beneath the desert sands, which dates from before the last Ice Age.

Field	Date	Gb
Amal	1959	4.5
Beda	1959	1.0
Nasser	1959	2.0
Defa	1960	2.0
Gialo	1961	3.5
Sarir	1961	6.0
Waha	1961	1.0
A-Nafoora	1965	2.0
Intisar	1967	2.3
Bu Attifel	1968	1.5

Geology and Prime Petroleum Systems

Most of Libya's oil comes from the Sirte Basin, flanking a Gulf of the same name, but there have also been a number of isolated finds in the interior of the country, including those in the remote Ghadames Basin, which straddles the frontier with Algeria, as much as 600 km inland. The Sirte Basin covers an area of some 300,000 km^2, comprising a series of north-westerly trending rifts. It contains rich Upper Cretaceous and Paleocene source-rocks, which have charged reefal reservoirs, located both on the contemporaneous structural highs and in the overlying Eocene. Lower Cretaceous sandstones form additional reservoirs along the interior margin of the basin, as locally do fractured Cambro-Ordovician quartzites. High heat-flow led to early generation.

Exploration and Discovery

Exploration drilling commenced in 1957 with a minor find, and expanded rapidly in the ensuing years with as many as 144 exploration boreholes being drilled in 1963, an all-time peak. The effort was rewarded by a string of giant fields, listed in the table.

Substantial amounts of associated gas, totalling some 80 Tcf, were found in parallel.

Exploration drilling has however declined markedly since the 1980s, partly for political reasons. It will presumably cease during the present political crisis but may pick up in the future if stability returns, which is perhaps somewhat unlikely to happen soon.

Production and Consumption

Libya is a fairly mature province, although depletion has been slowed by political factors. There is a certain offshore potential awaiting evaluation, and the remote interior basins remain relatively unknown. Future discovery is here estimated at almost 2.75 Gb.

Gas has been produced from the deeper parts of the Sirte Basin, at the rate of 1,000 Gcf/a in 2010, prior to the political crisis, from reported reserves of almost 59 Tcf. There is evidently scope for exports to populous Egypt, if and when new pipelines are constructed and political circumstances permit.

The Oil Age in Perspective

The coastal strip of Libya, previously known as Cyrenaica, was settled by Phoenicians and Greeks, and became an important source of grain for the Roman Empire. It later fell under the control of Egyptian dynasties, which in turn led to its nominal incorporation in the Ottoman Empire, but for much of the last millennium it was a sparsely populated and inhospitable backwater of no great interest to anyone.

The country's modern history opened in 1911, when it was invaded by Italy. Initial resistance was soon subdued, and the country was settled by Italian peasants, being fully incorporated into the Italian State in 1939. In the Second World War, a German army under General Rommel advanced through Libya in 1942, only to be defeated at the Battle of El-Alamein. The Italian population was evacuated during the retreat, leaving the country to its Arab indigenous people, mainly belonging to the Senussi tribe. It became effectively a British Protectorate during and after the Second World War, before being granted full independence in 1951 under King Idris, a well-disposed ageing Senussi leader. The principal export of the country at the time was scrap-iron left over from the battlefields. Visitors remarked on the high incidence of one-legged inhabitants: the victims of the many minefields left over from the war.

The fortunes of Libya changed quite literally with the discovery of oil in 1957, which ushered in a period of economic expansion and even prosperity. A Petroleum Law had been passed in 1955 to pave the way for the entry of British and American oil companies, including Exxon, which made the first major strike at Zelten in the Sirte Basin, 150 km in from the Mediterranean coast. It found light crude, which compared favourably with Middle East supplies, especially as Libya was closer to the European markets. As is so often the case, discovery followed discovery as a new prolific trend was opened up, so that by the end of the 1960s, the country's production briefly passed 3 Mb/d, exceeding even that of Saudi Arabia. Most of the major companies concentrated on the heart of the Sirte Basin, while BP headed into the interior to bring in the remote Serir Field, a giant with almost 5 Gb of oil, which was found in 1961. It contains oil with a high wax content that may block pipelines if they are shut down by political tensions, as has recently been the case. The opening of Libya gave a particular opportunity to the smaller oil companies, such as Occidental and Oasis (a consortium of Marathon, Amerada and Conoco), which had been largely excluded from the Middle East.

But the new production drowned Europe in a flood of cheap oil, which depressed world prices, being one of the factors prompting the creation of OPEC, which Libya joined in 1962. Libyan oil took on even more strategic value following the Six-Day War in 1967, although that further inflamed Arab nationalist passions reacting to Israel's occupation of Palestine. The scene was set for a coup d'etat to replace the ageing, pro-Western, King Idris, and on 1 September 1 1969, a group of officers, led by Colonel Muammar al-Qaddafi, declared a Republic.

The new regime at first sought unions with neighbouring countries, including Egypt, the Sudan, Tunisia and later Morocco, resurrecting the notion of a greater Arab nation, but the efforts ended in failure even triggering a brief war with Egypt in 1977. Col. Qaddafi changed the Constitution to create what was termed the Popular Islamic Socialist State to represent the mass of people, although he remained firmly in control. In fact, this formula did not differ greatly from the Communist regimes in several countries, or the Ba'athist Party of Iraq.

In 1971, Libya nationalised the holdings of BP as a gesture in its support of Islamic and pan-Arab power, establishing a national oil company, which now controls some 60% of the country's production.

The US support for Israel led to deteriorating relations during the 1980s, which culminated in the US aerial attack in 1986. It aimed to assassinate Qaddafi, but killed his daughter instead. This may have prompted a retaliatory act of planting a bomb on an American airliner in 1988, which exploded over the Scottish town of Lockerbie with tragic loss of life, although a similar outcry did not accompany the death of Quaddafi's daughter. (Various alternative explanations for the bombing have been advanced, some seeing it as a contrived incident to strengthen anti-Arab feeling in relation to the Israeli conflict.) Libya too is said to have encouraged revolutionary movements in many countries, including the Irish Republican Army. It was as a consequence made subject to the US trade sanctions which have been in force since 1996. They exclude the US companies from operating, but several European companies continued to work there satisfactorily. In 2003, Libya negotiated an end of sanctions with the payment of three billion dollars to the relatives of the Lockerbie victims.

Born in 1942, Col. Qaddafi became a mellowing dictator showing some signs of rapprochement with the Western powers, which may however have suffered a setback with the invasion of Iraq that has understandably inflamed Arab passions everywhere. A revolution in Tunisia in early 2011 sparked unrest throughout North Africa and parts of the Middle East. The people of eastern Libya, who largely belong to the Senussi tribe, to which the previous leader, King Idris, belonged, rose in rebellion, to be brutally suppressed by Qaddafi forces. Later, the United States, Britain and France led NATO military forces under a UN resolution to support the rebels and oust Col. Qaddafi from power. It may be supposed that the country's oil endowment prompted this intervention, as other humanitarian crises as in Syria or Rwanda did not trigger a response. Col. Qaddafi was finally killed but at the time of writing has yet to be replaced by an effective government. Western contractors and oil companies are now lining up to try to make money from Libya's oil, repairing the damage caused by the attack.

The flush oil production from the early giant fields is coming to an end as is the easy wealth that flowed from it. It is too soon to evaluate the consequences of the current political tensions and foreign intervention, but looking ahead it may be time for the country to think of replanting and tending the olive groves introduced by the Italian colonists during the early years of the last century because there is not much else to do in the barren deserts. The country also probably faces increasing pressure from Egypt whose vastly greater population will desperately need access to its remaining oil and gas. That threat apart, the modest size of its population augers well for the future of the country.

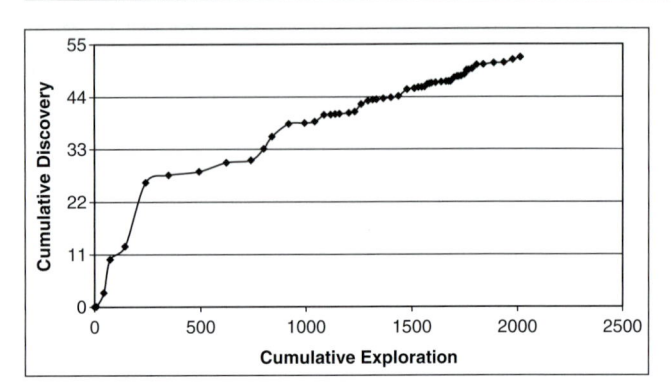

Fig. 11.3 Libya discovery trend

Fig. 11.4 Libya derivative logistic

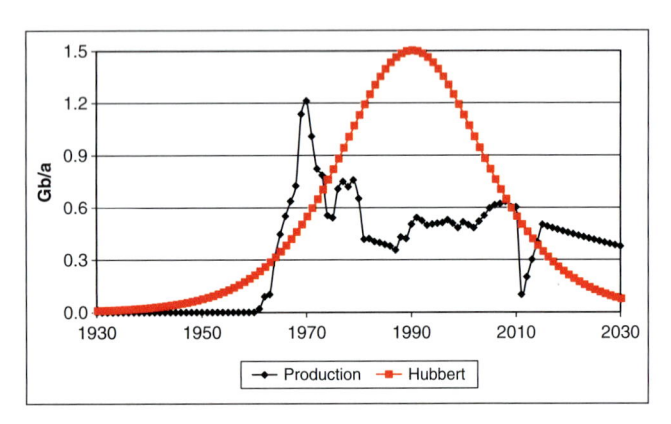

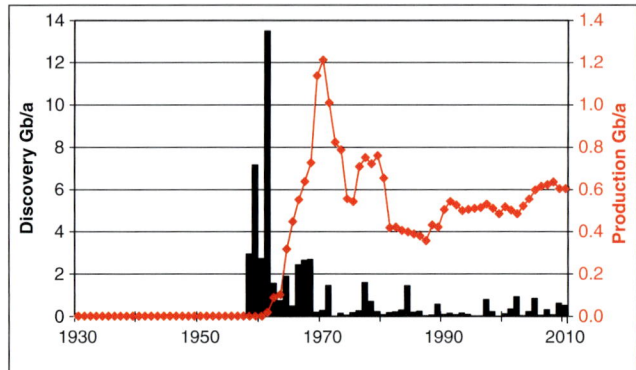

Fig. 11.5 Libya production: actual and theoretical

Fig. 11.6 Libya discovery and production

Nigeria

Table 12.1 Nigeria regional totals (data through 2010)

Production to 2100					Peak Dates			Area	
Amount		Rate				Oil	Gas	'000 km²	
	Gb	Tcf	Date	Mb/a	Gcf/a	Discovery	1968	1967	Onshore	Offshore
PAST	28	50	2000	790	1231	Production	2005	2029	930	800
FUTURE	27	170	2005	946	1918	Exploration	1967		Population	
Known	24	153	2010	455	2392	*Consumption*	Mb/a	Gcf/a	1900	16
Yet-to-Find	2.7	17	2020	440	3500	2010	102	176	2010	162
DISCOVERED	52	203	2030	369	3500		b/a	kcf/a	Growth	10.1
TOTAL	55	220	*Trade*	+353	+2216	Per capita	0.6	1.1	Density	175

Note: Excludes non-conventional deepwater areas

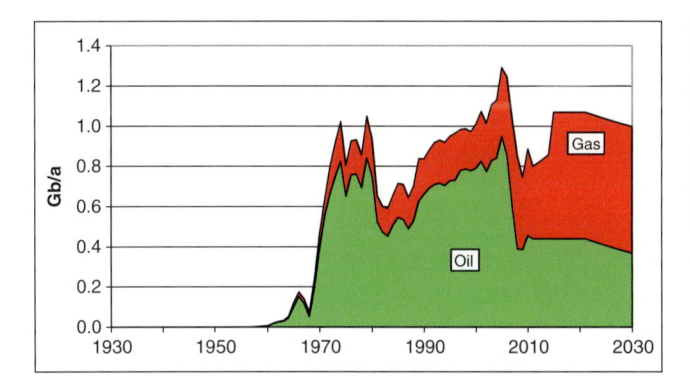

Fig. 12.1 Nigeria oil and gas production 1930 to 2030

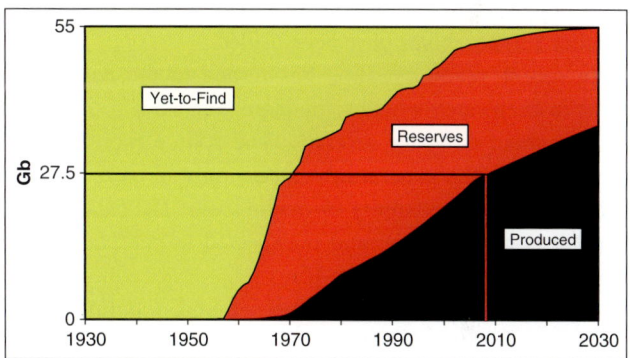

Fig. 12.2 Nigeria status of oil depletion

Essential Features

Nigeria covers 930,000 km², extending from the deserts of the north, which border the Sahara, to tropical rain forests of the south. Most of the country is relatively flat-lying with rivers flowing through shallow valleys in extensive plains, although more mountainous terrain builds in the south on the border with the Cameroons. The country is drained by the Niger and Benue rivers, which converge in an extensive delta of freshwater swamps and mangroves. With 162 million inhabitants, it is the most populous country in Africa, also having one of the highest population densities. Lagos, the capital, is home to some eight million people. Nigeria joined OPEC in 1971, and has played a prominent part in the organisation.

Geology and Prime Petroleum Systems

In geological terms, Nigeria lies on a suture cutting the African Shield. It was responsible for the *bulge of Africa* and for a line of weakness, caused by faulting, which is followed by the Niger River that has built up a large delta in the south of the country. A Mesozoic trough, known as the Benue Trough underlies the delta running inland.

The prospective part of the delta covers an area of about 120,000 km², distributed equally onshore and offshore, and is made up of about 10,000 m of Eocene to Recent clastic sediments. Paralic shales near the base of the sequence comprise the principal source-rock. Low thermal conductivity means that the oil window is deeper than normal, with generation starting at a depth of about 3,000 m. Migration

C.J. Campbell, *Campbell's Atlas of Oil and Gas Depletion*,
DOI 10.1007/978-1-4614-3576-1_12, © Colin J. Campbell and Alexander Wöstmann 2013

commenced during the Miocene, mainly following extensional faults. Reservoirs are provided by multiple deltaic sands with excellent characteristics. Growth faults and roll-over anticlines form the principal traps.

Exploration and Discovery

Shell and BP commenced pioneering exploration for oil in the Niger Delta in 1937 but operations were suspended during the War, so that the first discovery was not made until 1957. Progress was slow as exploration relied on seismic surveys through the difficult swamps of the delta, but gradually the potential was recognised.

It is a prolific basin with about 250 producing fields, including 13 giant fields. The first was found in 1958, and the last onshore was in 1990. It is a dispersed distribution with an almost perfect fit with the theoretical parabolic fractal.

Exploration is now at a very mature stage. Almost 1,375 exploration boreholes have been drilled but the pace of exploration is dwindling as fewer and fewer prospects remain to be tested. Exploration will likely end around 2035, after another 130 boreholes have been drilled.

In recent years, attention has turned to the delta-front lying in deepwater, where some large finds have been made, adding about 6 Gb of reserves. The source rocks in this domain reach the generating window only in areas of high heat flow in the proximity of major faults. Deepwater production is now starting, and is expected to peak at around 1.4 Mb/d in 2017 (see Chap. 12)

Some 16 giant fields have been found, totalling about 13 Gb, or one-fifth of the total endowment, but it is generally a dispersed geological environment with a large number of small to moderately sized fields as is typical of a deltaic environment.

Nigeria remains an important element in Shell's global portfolio, but BP was less fortunate having had its rights sequestered in 1979 for allegedly supplying South Africa when that country was embargoed. Its production became a foundation for a newly created State company. BP did later return in partnership with Statoil. Most of the other international companies are represented in the country.

About 200 Tcf of gas has been discovered, but much has been flared in the absence of a market.

Production and Consumption

Oil production commenced in 1958, rising to pass a million barrels a day by 1970 before reaching an early peak in 1979 at 2.3 Mb/d, of which about 40% was offshore. It then declined, partly in response to OPEC constraints, to a low of 1.2 Mb/d in 1983, before recovering in recent years to an overall peak in 2005. Approximately half the endowment has now been produced but production is expected to plateau to about 2020 in view of the very low depletion rate, presumably reflecting the constraints of operating a large number of small fields in a difficult environment. Consumption is running at 102 Mb/a, making the country a substantial exporter.

Associated gas production rose in parallel with oil, much being flared. There are now plans to build a gas export pipeline to connect with the Algerian system to eventually supply Europe.

The Oil Age in Perspective

Nigeria was previously tribal territory dominated by the Muslim Hausas in the north and the Ibo in the south. Portuguese and British slave-trading stations were established in the seventeenth and eighteenth Centuries, before Britain brought the territory under relatively benign colonial rule in 1906, which led to an epoch of progress and stability.

Independence was granted, or it could almost be said, imposed, in 1960, as part of Britain's withdrawal from Empire, whereupon ancient tribal rivalries erupted, leading to a military government from 1966 to 1979. It was dominated by Islamic Hausa and Fulani people, who were responsible for a massacre of up to 30,000 Ibo Christians in the southeastern part of the country. That prompted the development of a separatist movement under Lt. Col. Ojukwu, who declared an independent Republic of Biafra in May 1967, plunging the country into a civil war in which millions died, partly from starvation. This conflict was fuelled to some extent by the rising oil revenues from the Niger Delta region to which the indigenous people felt entitled to a preferential claim.

A brief period of civilian government followed from 1979 to 1983 when a new military government took control, following a coup. There have been several subsequent attempts at democratic government, but they were short-lived. General Abacha came to power in 1993 imposing a particularly brutal and corrupt regime but met his end in somewhat mysterious circumstances 5 years later. He was succeeded by General Obasanjo following a form of election.

Nigeria is well known for the high level of government corruption. Even the internet abounds with offers of oil supply contracts and requests to foreigners to facilitate the transfer of funds to overseas accounts. Ethnic and religious tensions within the country continue, partly exacerbated by the distribution of oil revenues between the different communities. The situation appears to be deteriorating with increased strife, including the occasional abduction of foreign oil workers.

There have been some hints that the United States seeks a special relationship with Nigeria for access to its oil, possibly encouraging it to withdraw from OPEC.

Nigeria will continue to be an important source of world oil for the next two decades, especially as its deepwater production comes in. It will also earn high revenues as oil prices rise higher. Further strife and violence will likely erupt when it is found that falling oil revenue can no longer support the large population. The country is already a substantial net food importer. In 50 years or so, the daily life for the survivors may be little different from the sustainable patterns the country knew in the nineteenth century.

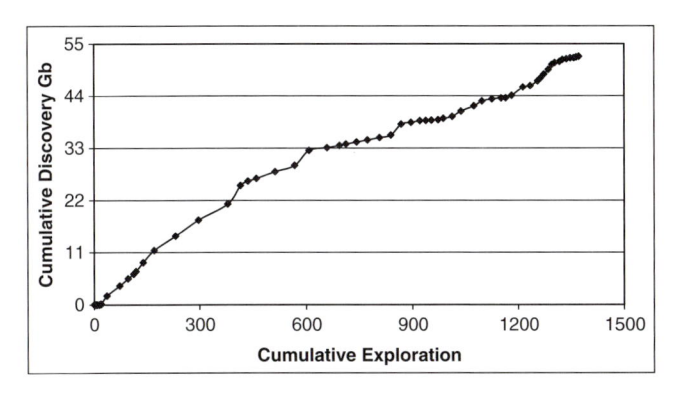

Fig. 12.3 Nigeria discovery trend

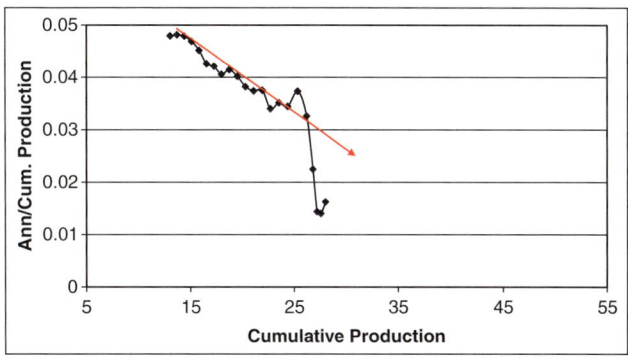

Fig. 12.4 Nigeria derivative logistic

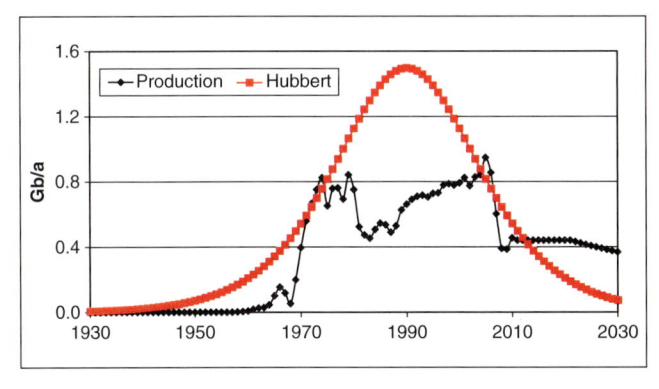

Fig. 12.5 Nigeria production: actual and theoretical

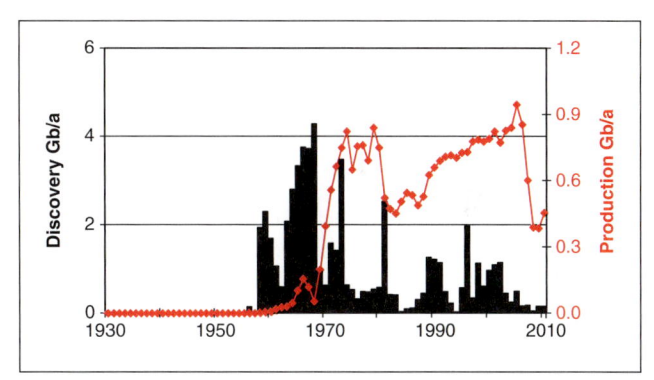

Fig. 12.6 Nigeria discovery and production

Sudan

Table 13.1 Sudan regional totals (data through 2010)

Production to 2100				Peak Dates			Area	
Amount		Rate				Oil	Gas	'000 km²	
	Gb	Tcf	Date	Mb/a	Gcf/a	Discovery	2003	2003	Onshore	Offshore
PAST	1.5	0	2000	68	0	Production	2013	2020	2,520	130
FUTURE	4.5	2	2005	128	0	Exploration	2002 || Population	
Known	3.7	1	2010	187	30	Consumption	Mb/a	Gcf/a	1900	3
Yet-to-Find	1	1	2020	175	49	2010	36	0	2010	44.6
DISCOVERED	5	1	2030	91	57		b/a	kcf/a	Growth	14.9
TOTAL	6	2	Trade	+151	+30	Per capita	0.8	0	Density	18

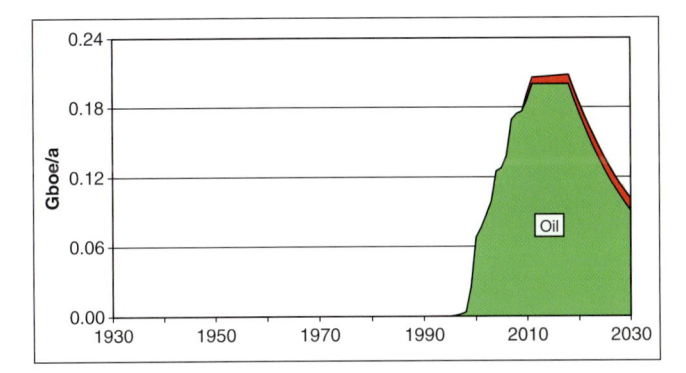

Fig. 13.1 Sudan oil and gas production 1930 to 2030

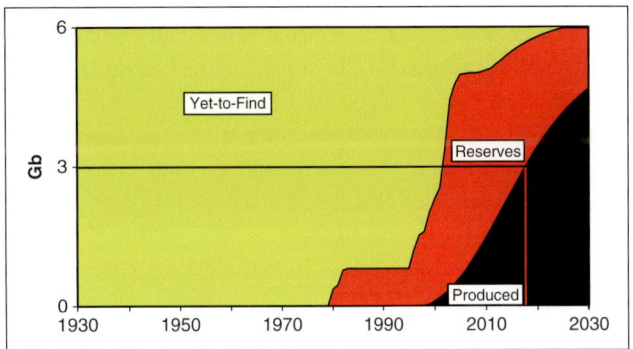

Fig. 13.2 Sudan status of oil depletion

Essential Features

The Sudan comprises a vast area of plains and plateaux, covering 2.6 million km² in the eastern part of the continent, which are drained by the upper reaches of the Nile River. It is flanked to the east by the Red Sea, and to the southwest by a high plateau of Darfur, which is capped by volcanic peaks rising to 3,000 m. Most of the county faces a hot, arid climate and barren soils, although higher rainfall is found in the mountainous south.

Geology and Prime Petroleum Systems

The Central African rift system, already described in connection with Nigeria and Chad, extends eastward into the southern Sudan. The prime petroleum systems are in Mesozoic and Tertiary rifts.

Exploration and Discovery

Exploration commenced in 1961, led by the Italian company Agip, which found no more than minor gas-condensate deposits near the Red Sea. Chevron then took up rights to the interior in 1975, making the first significant discovery in 1980 in the Upper Nile region. Exploration then ceased following an attack on Chevron's operations in 1990, causing the company to withdraw. Efforts resumed in the late 1990s, when a number of independent companies, some having ties with the Government, and Chinese companies came in. The overall peak of exploration drilling came in 2002 when 26 boreholes were sunk, but has since fallen to one or two a year, due in part to the political strife. These two exploration campaigns were rewarded by a number of modest discoveries, totalling some 5 Gb of oil.

Production and Consumption

A pipeline to the Red Sea coast was completed in 1999 allowing the production of about 500 kb/d, which accounts for almost three-quarters of the country's export earnings. Barring further political tension, it should be able to maintain this level until around 2020, close to the midpoint of depletion, whereupon production is likely to decline at about 6 % a year. With consumption of around 36 Mb/a, the country can remain a modest exporter at about 150 Mb/a until then. Gas production has yet to commence, and the forecast of future production remains most uncertain.

The Oil Age in Perspective

The territories of what is now Sudan have been settled since the earliest times, and for much of their history have faced alternating conflict and assimilation with the people of the Lower Nile, now Egypt. Somewhat surprisingly, parts of the country were converted to Christianity in the first millennium, but later came under increasing Islamic influences. In 1820, Muhammad Ali, the Ottoman ruler of Egypt, backed by Britain, invaded the territory, developing a trade in slaves and ivory. Five years later, this led to a successful rising against the British administration, in which General Gordon died. The British reacted by a fresh invasion under Lord Kitchenor, which defeated the Islamic forces at the famous Battle of Omdurman in 1899. Thereafter, the Sudan became a British colony closely linked with the administration of Egypt.

Moves to independence resumed in the 1920s, and were finally successful in 1956, when independence was declared, but the eternal conflict between the Arab/Muslim north and the Black/Christian south remained not far below the surface, leading to civil war in the 1980s. The problems were compounded when one of the movements adopted Communism and began to receive military support from the Soviet Union. A partial resolution of the dispute was achieved in 2005, which led to a settlement whereby Southern Sudan seceded in 2011. Meanwhile, proxy wars and other disputes festered in the tribal lands of Darfur in the remote western interior, which were exacerbated by climate changes, imposing famine on the region. At root, there is a long standing tension between wandering nomadic herdsmen and settled farmers, but it has escalated in recent years, forcing many of the inhabitants to flee into neighbouring Chad.

Looking ahead, it seems likely that conflict over the control of Sudan's new oil wealth will continue to influence its political landscape, probably leading to more conflict. Much of the land is arid and incapable of supporting the current population.

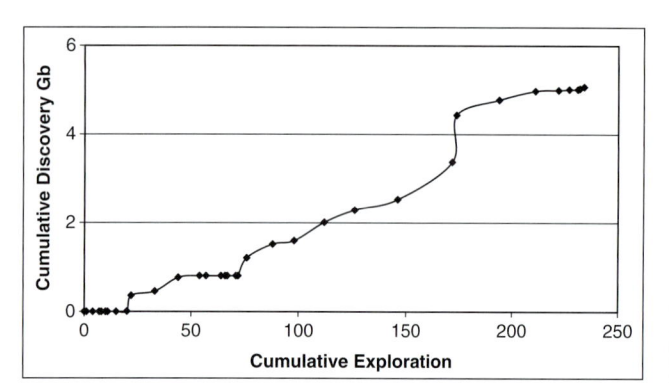

Fig. 13.3 Sudan discovery trend

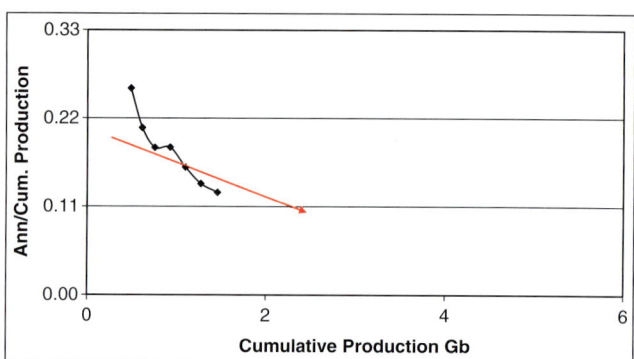

Fig. 13.4 Sudan derivative logistic

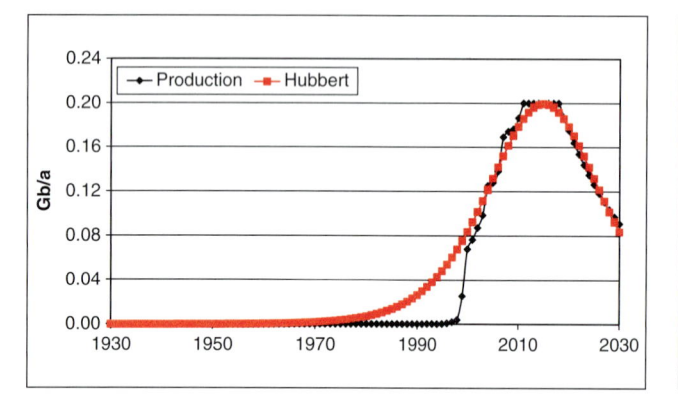

Fig. 13.5 Sudan production: actual and theoretical

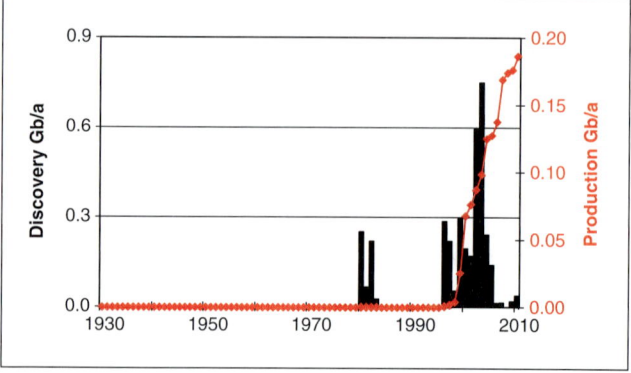

Fig. 13.6 Sudan discovery and production

Tunisia

Table 14.1 Tunisia regional totals (data through 2010)

Production to 2100						Peak Dates			Area	
Amount			Rate				Oil	Gas	'000 km²	
	Gb	Tcf	Date	Mb/a	Gcf/a	Discovery	1964	1974	Onshore	Offshore
PAST	1.5	2.2	2000	29	83	Production	1983	2025	164	130
FUTURE	0.8	2.8	2005	27	111	Exploration	1981		Population	
Known	0.7	2.6	2010	29	109	Consumption	Mb/a	Gcf/a	1900	1.8
Yet-to-Find	0.1	0.3	2020	19	126	2010	31	116	2010	10.7
DISCOVERED	2.1	4.7	2030	14	53		b/a	kcf/a	Growth	5.9
TOTAL	2.3	5	Trade	−1	−7	Per capita	3	11	Density	65

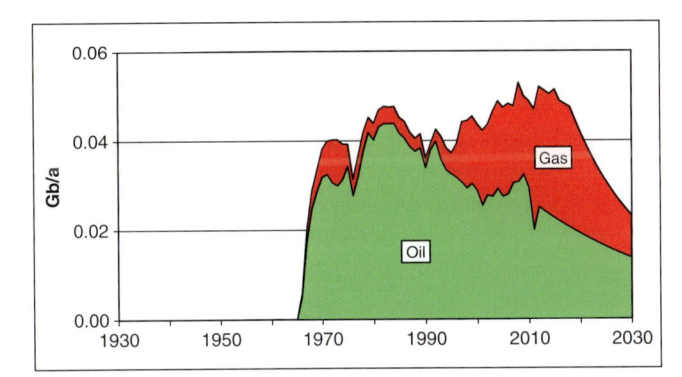

Fig. 14.1 Tunisia oil and gas production 1930 to 2030

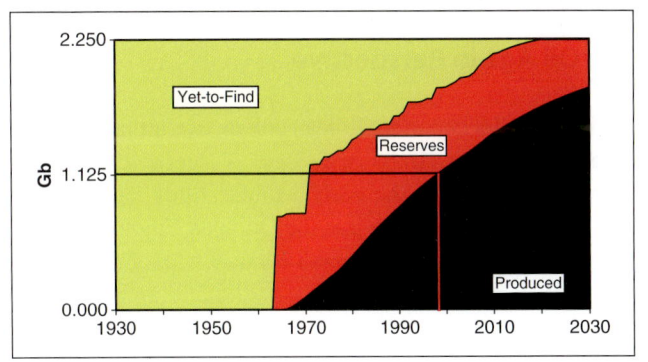

Fig. 14.2 Tunisia status of oil depletion

Essential Features

Tunisia, which covers an area of 164,000 km², lies at the eastern end of the Atlas Mountains of North Africa, being sandwiched between Algeria to the west and Libya to the east. The mountains are of moderate relief, with various ranges rising to about 1,500 m. They are flanked by an arid plateau, passing into the Sahara Desert to the south. The Mediterranean coast in the north is highly dissected but passes eastwards to a fertile coastal strip, bordering the Gulf of Tunis.

Geology and Prime Petroleum Systems

The Atlas Mountains overlie a plate-boundary bordering the continent of Africa. Ancient rocks of the shield rocks of the interior are overlain northward by younger sediments, which are strongly deformed in the Atlas Mountains. The principal area of interest from a petroleum standpoint is an important Mesozoic basin to the south, straddling the boundary with Algeria, which has yielded the giant El Borma Field. But other finds have been made in rifts in the coastal region to the east and adjoining offshore waters. The prime source-rock development appears to be in the mid-Cretaceous, although other leaner sources have been identified in the Lower-Middle Jurassic and the Lower Eocene. Structural deformation is relatively complex giving fault traps and related features.

Exploration and Discovery

The first exploration borehole was drilled in 1917, but the main thrust came after the Second World War, with 15 boreholes being sunk in 1954. Exploration lapsed during the 1960s following independence, but picked up subsequently to reach a peak in 1981 when 33 boreholes were drilled. The first major discovery was in 1964 when the giant El Borma field with 650 Mb of oil and 1 Tcf of gas was found. The subsequent discoveries have been modest in size. In total, just over 2 Gb of oil and almost 5 Tcf of gas have been found, distributed about equally in onshore and offshore areas.

Production and Consumption

Oil production effectively commenced in 1966 and rose steadily to a peak of 120 kb/d in 1983 before declining gradually. Consumption stands at 31 Mb/a making the country a modest importer, but on a rising trend.

Gas production rose in parallel reaching a peak in 2010 of 109 Gcf/a. This was due to a significant number of discoveries over the past decade. Production is now expected to plateau at around 150 Gcf/a for the next decade, slowly declining thereafter.

The Oil Age in Perspective

Tunisia was the important Roman colony of Carthage, before being invaded by Arabs in 647 AD. A degree of national identity emerged in the early Middle Ages, and largely survived its incorporation into the Ottoman Empire in 1574. Tensions arose in the nineteenth century when France took neighbouring Algeria, and the Ottomans sought to impose direct rule. Tunisia became a centre of piracy, affecting trade in the Mediterranean, which in part prompted France to invade in 1881, establishing a Protectorate that lasted until 1956. The more fertile parts of the country in the northeast were subject to some French settlement, and there was a degree of economic development. Nevertheless, new moves to independence began to emerge during the 1920s.

The situation changed with the occupation of France during the Second World War, when the leaders of the independence movement, including Habib Bourguiba, formed a certain alliance with Italy. British and American forces secured a major victory in 1943, re-imposing rule by the French Government in exile. Independence was finally granted in 1955, and Bourguiba became President, two years later, imposing authoritarian rule. He was succeeded in a bloodless coup in 1987 by Zine el Abidini ben Ali who ran the country in an authoritarian manner until 2011 when he was forced to flee in the face of serious demonstrations precipitated by growing unemployment and soaring food prices. The country had enjoyed a degree of prosperity relative to other countries in Africa, and faced the consequences of the world recession. The rising in Tunisia triggered other upheavals through Egypt and Libya, spreading to the Middle East.

Tunisia seems relatively well placed to face the Second Half of the Age of Oil with a modest population and plenty of scope to develop solar energy. It is a Muslim country not facing any particular ethnic or religious conflicts.

14 Tunisia

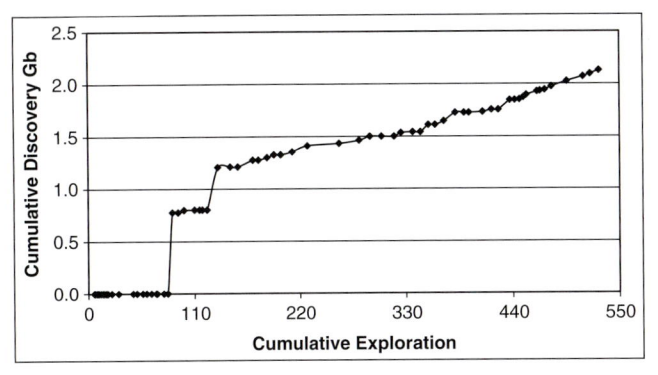

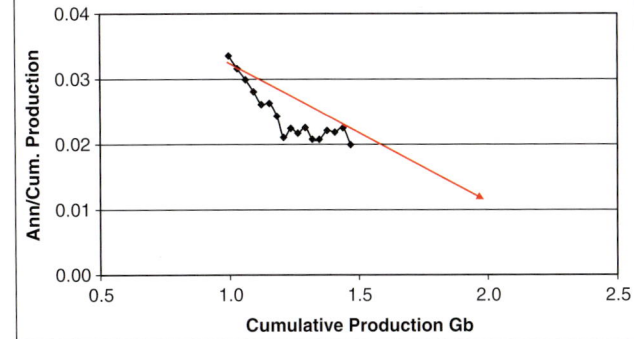

Fig. 14.3 Tunisia discovery trend

Fig. 14.4 Tunisia production: actual and theoretical

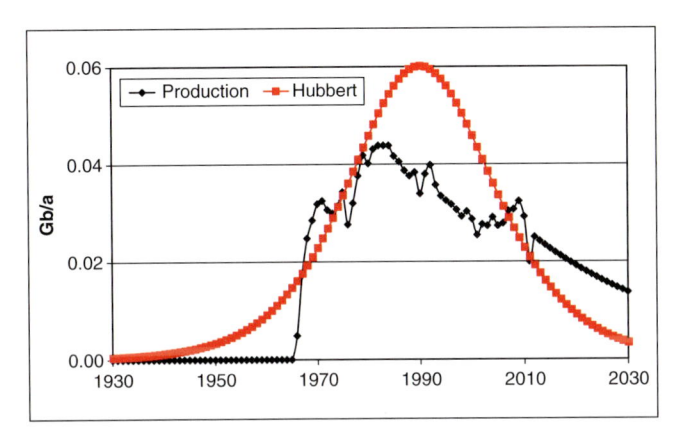

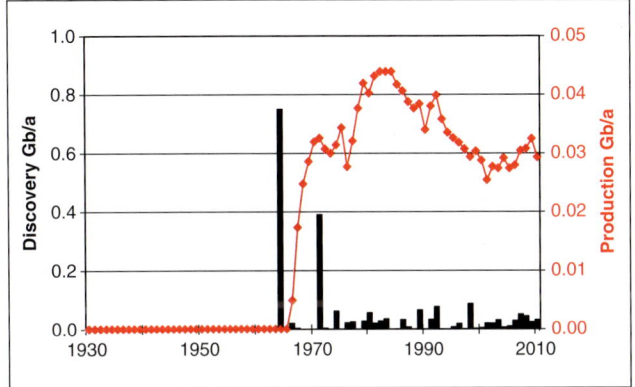

Fig. 14.5 Tunisia production: actual and theoretical

Fig. 14.6 Tunisia derivative logistic

Uganda

Table 15.1 Uganda regional totals (data through 2010)

Production to 2100					Peak Dates			Area		
Amount		Rate				Oil	Gas	'000 km²	
	Gb	Tcf	Date	Mb/a	Gcf/a	Discovery	2009	2015	Onshore	Offshore
PAST	0.0	0.0	2000	0	0	Production	2030	2023	240	0
FUTURE	2.0	0.5	2005	0	0	Exploration	2010		*Population*	
Known	1.6	0.2	2010	0	0	*Consumption*	Mb/a	Gcf/a	1900	2.0
Yet-to-Find	0.4	0.3	2020	50	24	2010	5.1	0	2010	34.5
DISCOVERED	1.6	0.2	2030	50	16		b/a	kcf/a	Growth	17.3
TOTAL	2.0	0.5	*Trade*	−5	0	Per capita	0.1	0	Density	143

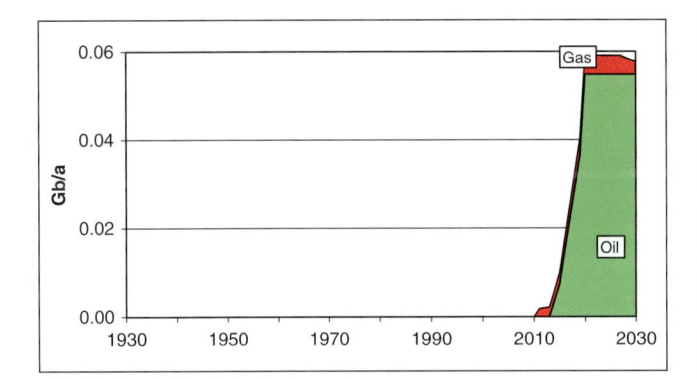

Fig. 15.1 Uganda oil and gas production 1930 to 2030

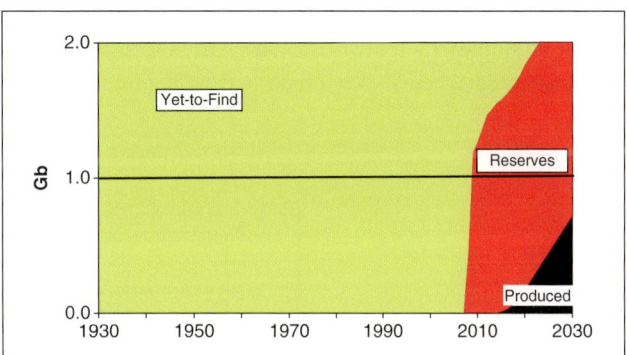

Fig. 15.2 Uganda status of oil depletion

Essential Features

Uganda, which covers an area of 240,000 km², is a land-locked country in the central part of Africa, lying between Tanzania and Kenya to the east, Zaire to the west and the Sudan to the north. Much of the country lies at an altitude of about 1,100 m, declining northward. Lake Victoria, the largest in Africa, straddles its southeastern border, and drains northwestwards into Lake Albert and the headwaters of the Nile. There are extensive marshlands around some of the lakes. The altitude moderates the equatorial climate. The main rainfall is in the south, leaving barren near-desert conditions in the north.

Geology and Prime Petroleum Systems

Most of Uganda is made up of crystalline rocks of the African shield, but rifting as the continent broke up gave rise to Mesozoic basins containing hydrocarbon source-rocks in the western part of the country around Lake Albert. Major associated uplifts also gave the impressive Rwenzori mountain ranges in the south, rising to 5,100 m. Little is known in detail about the recently found petroleum system around Lake Albert but it may be assumed that Cretaceous source rocks have charged overlying sandstone reservoirs in fault-controlled structures.

Exploration and Discovery

In regional terms, Uganda was not perceived to have much in the way of oil prospects. But a relatively small exploration company, by the name of Tullow, took an interest a few years ago in the prospects beneath Lake Albert, which were evaluated by marine seismic surveys showing some promise. The first exploration well was drilled in 2003 with encouraging results, which led to the first discovery in 2006. Almost 30 exploration wells have now been drilled making various subsequent discoveries reported to amount to about 1.6 Gb. A preliminary estimate of the total to be produced stands at around 2 Gb, but it is too soon to place much reliance upon it.

Production and Consumption

Production is expected to commence around 2014 and may rise to a plateau of about 150 kb/d determined by the size of the off-take pipeline. Consumption stands at a modest 5.1 Mb/a, with one of the lowest per capita values of any country in Africa at no more than 0.1 barrels a head.

The Oil Age in Perspective

The early inhabitants of Uganda were hunter-gatherers, but evolved into Bantu speaking people who developed settled agriculture around the lakes, and also found out how to make tools and weapons from iron. Arab traders moved in during the early nineteenth century, before a degree of control of the region was taken by the British East Africa Company. That was cemented when it was made a British protectorate in 1894, with formal boundaries being defined in 1914, but it remained a substantially tribal territory under various chieftains until full independence was gained in 1962.

Milton Obote became the first Prime Minister before assuming presidential control in 1967. He remained in power until 1971 when Idi Amin seized power, assuming dictatorial powers, leading to the expulsion of non-African traders who played a dominant part in the country's economy. His rule ended in 1979 when Tanzanian forces, backed by Ugandan exiles, invaded the country, reinstating Obote. But he was again deposed in 1985, to be replaced by Yoweni Museveni. Much political strife and tension remains, leading to substantial loss of life.

Looking ahead, Uganda seems moderately well placed to face the future as its lakes and fertile country can support a modest population, with the new oil finds providing a welcome source of revenue over the next decade or so. But it will likely revert to localism with the re-emergence of various tribal groups, which will no doubt continue to face conflict between each other.

Oil developments are at too early a stage to produce the normal set of illustrative graphs.

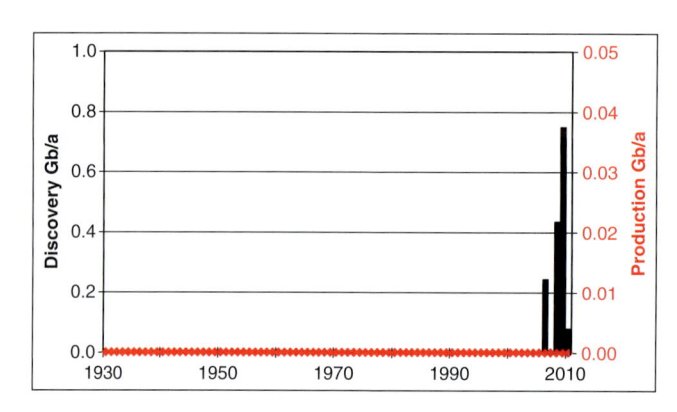

Fig. 15.3 Uganda discovery and production

Africa Region

Table 16.1 Africa regional totals (data through 2010)

Production to 2100						Peak Dates			Area	
Amount			Rate				Oil	Gas	'000 km²	
	Gb	Tcf	Date	Mb/a	Gcf/a	Discovery	1961	1957	Onshore	Offshore
PAST	104	257	2000	2660	8821	Production	2005	2019	12640	155
FUTURE	88	495	2005	3084	11672	Exploration	1981		Population	
Known	78	434	2010	2574	13560	*Consumption*	Mb/a	Gcf/a	1900	44
Yet-to-Find	10	62	2020	1956	14803	2010	1204	3556	2010	434
DISCOVERED	182	690	2030	1402	10683		b/a	kcf/a	Growth	9.8
TOTAL	192	752	*Trade*	+1563	+10836	Per capita	1.2	3.5	Density	34.3

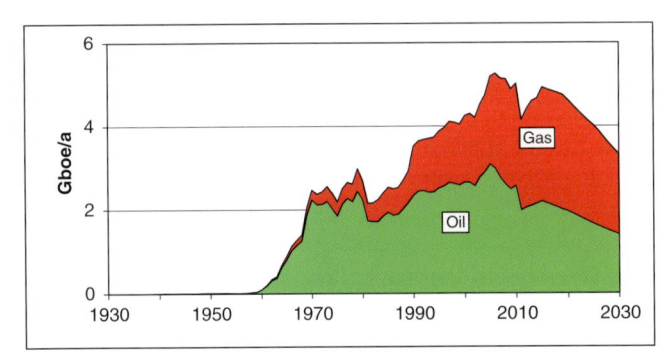

*Note: Data refer to main producing countries only, **except for consumption and trade data** which include other countries.*

Fig. 16.1 Africa oil and gas production 1930 to 2030

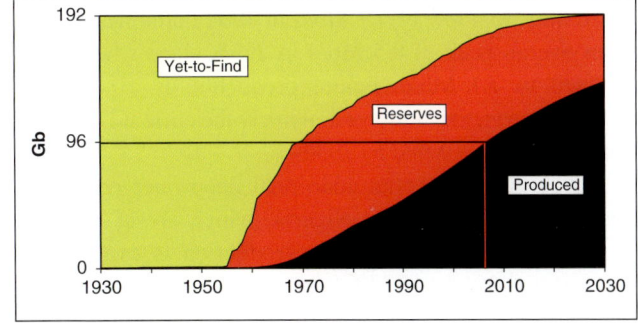

Fig. 16.2 Africa status of oil depletion

The boundaries of the Region are straightforward being defined by the physical limits of the African Continent, although there are a number of associated islands, of which the largest is Madagascar. It covers an area of 32 million km², with the major oil-producing regions — as reported here — covering 12.64 million km². The continent is built primarily of the ancient rocks of the African Shield, which are cut by major trans current faults and rifts, locally affected by volcanic activity, and flanked by some petroleum-bearing sedimentary basins.

The tropical rain forests of the equatorial regions give way to widespread deserts, principally the Sahara in the north and the Kalahari in the south. Mt. Kilimanjaro, which is a volcanic peak associated with the rift valleys of East Africa, rises to almost 6,000 m, while Lake Assal in Djibouti, bordering the Red Sea, is actually below sea-level. The Continent is drained by several major rivers, principally the Nile, the Congo and the Zambezi, flowing respectively northwards, westwards and eastwards. Lake Chad acts as a catchment for the semi-arid central interior, itself emptying into the Niger River.

In addition to the twelve main oil producers of Africa, whose oil and gas position is summarised in the table above, are a large number of other countries as listed below. Those indicated (*) have minor amounts of oil and/or gas and are evaluated together in the world assessment in Chapter 11 as

C.J. Campbell, *Campbell's Atlas of Oil and Gas Depletion*,
DOI 10.1007/978-1-4614-3576-1_16, © Colin J. Campbell and Alexander Wöstmann 2013

the amounts are too small to model meaningfully. Some are, however, significant consumers of oil and gas.

The Continent supports a population of just under a billion people, living in some sixty countries, many of which are however somewhat artificial constructions, inherited from colonial administrations, being subject to internal tribal and other conflicts. The population density varies widely, but averages at about 28/km², having a life expectancy of 53 years and a fertility rate of 5 children per woman.

The total amount of *Regular Conventional Oil* to be produced to the end of this Century is estimated to be a rounded 192 Gb, of which 104 Gb have been produced through 2010, leaving around 90 Gb for the future, including an estimated 10 Gb from new discovery. Three principal producers, Libya, Nigeria and Algeria, which are members of OPEC, hold about 80% of the estimated future production.

Considering the main producing countries, small-scale production commenced in the 1930's but grew rapidly in the 1960's as major fields in Libya came on stream. It then fluctuated as a decline in Libya, due to political factors, was only partly offset by increases in Nigeria and Algeria. Production was also supplemented from a number of other smaller producers. There was an overall early peak in 1979 of 6.7 Mb/d. Production then fell sharply, due to OPEC quota constraints, to a low of 4.7 Mb/d in 1982 before recovering to an overall peak of 8.5 Mb/d in 2005, since when it has declined to 7.1 Mb/d. Future production is expected to decline at just under 3% a year, namely the current Depletion Rate, to about 3.8 Mb/d by 2030.

Africa has substantial additional deepwater resources, especially in Angola and Nigeria, which are at an early stage of depletion. Deepwater production is expected to peak around 2017 at 2 Mb/d and then decline steeply (see Chapter 12).

The total production of all liquids is accordingly expected to reach a peak of about 9 Mb/d in 2015 before declining to about half that amount in 2030. In a certain sense, the addition of the deepwater will make the tensions of the transition to the Second Half of the Oil Age even more difficult by giving this brief yet unsustainable surge. The current difficult political situation in some of the principal producing countries, especially Libya, gives an anomalous fall in production for the region.

It is very difficult to analyse the gas situation, but a preliminary estimate suggests that almost 700 Tcf have been found, of which 257 Tcf have been produced, with some 62 Tcf yet to be found. Current production stands at 13.5 Tcf/a, and is expected to plateau at around this level through to 2030. The related production of Natural Gas Liquids is expected to remain at about 500 Mb/a. A major new gas province has been found off Mozambique and adjoining countries but it is too soon to evaluate the details.

Table 16.2 Other countries

*Benin	G.Bissea	Rwanda
Botswana	*Ivory Coast	*S.Africa
Burk.Faso	Kenya	Senegal
Burundi	Lesotho	Sierra Leone
C.Africa R.	Liberia	Somalia
*D.R.Congo	*Madagascar	Swaziland
Djibouti	Malawi	Tanzania
*Eq.Guinea	Mali	Togo
Eritrea	*Mauritania	W.Sahara
Ethiopia	*Morocco	Zambia
Gambia	*Mozambique	Zimbabwe
*Ghana	Namibia	
Guinea	Niger	
(*) = with minor production		

Oil consumption for the entire continent has been growing consistently from 530 kb/d in 1965 to about 1200 Mb/a at present. Per capita consumption stands at about 1.2 barrels per year, which is extremely low, when compared for example with the United States at 21 b/a but similar to that of countries like India (0.95) or Indonesia (1.95). Africa is accordingly in a position to remain an exporter for the next few decades, political circumstances permitting. Much gas is flared, and the statistics on consumption seem to be particularly unreliable, but production is reported to be running at about 2.8 Tcf/a making the continent a substantial exporter, both directly to Europe and in the form of LNG and NGL.

The Oil Age in Perspective

Africa supports a rich diversity of animal life and is considered to be the birthplace of *Homo sapiens*. The Nile Valley enjoyed an early advanced culture, typified by the well-known Pyramids, but most of the continent remained substantially tribal. The European powers, principally, Britain, France and Portugal, established trading stations along the western coast, while the Ottoman Empire extended dominion over the northern part of the continent. The predominant foreign trade for many centuries was slavery, with Britain supplying its plantations in North America and the Caribbean, the Portuguese supplying Brasil, and Arab traders working the east coast, especially Zanzibar.

By the late 19th Century, the trading arrangements gave way to formal empires, run principally by Britain and France, which established territorial administrations, defining boundaries that did not necessarily match tribal or ethnic distributions. Christian missionaries worked the central and southern parts of the continent, while Islam spread from the north, providing the banners for many subsequent tensions. South Africa was substantially settled by Dutch and later British immigrants, as the indigenous population at the time was small. Rhodesia (now Zimbabwe), Kenya and Algeria also received substantial European settlement during the 19th and 20th Centuries, but many Europeans subsequently left following independence.

The colonial epoch ended with the Second World War, leading towards independence. For the most part, it can be said that independence brought authoritarian governments, being accompanied in many cases by continuing tensions, conflicts, and even, in the case of Rwanda, wholesale tribal genocide.

The social structure of the Continent is one of extremes from the downtrodden masses, many facing famine, disease, and turbulence, to the war lords and powerful elite whose income in large measure derives from the exploitation of natural resources, including gold, diamonds, minerals and petroleum. This prompts a somewhat ironic conclusion: namely, that the decline in oil supply over the next decades may come as something of a blessing, insofar as it will reduce the power of the powerful yet leave the suffering masses, who barely consume oil, not materially worse off. The foreign debt, which did so much to impoverish the African is being forgiven, presumably because the creditors prefer forgiveness to default. The global financial system is in any case now in turmoil. Presumably, cash crops for export will decline along with the import of military equipment for the war lords, leaving the ordinary African to recover the sustainable way of life he knew before the Oil Age. A major fall in population seems inevitable over the years ahead until a more sustainable life-style for the survivors can be adopted.

Confidence and Reliability Ranking

The table lists the standard deviation of the range of published data on reserves. A relative confidence ranking in the validity of the underlying assessment is also given: the lower the Surprise Factor the less likely is the need for revision. There may be additional deepwater discoveries in other countries, including Guinea.

Algeria Algeria has been well explored, although there may be scope for some positive surprises in the remote interior. Gas supply is partly constrained by pipeline and LNG facilities, but could well rise in the future. It is an OPEC country, so the reporting of its reserves may be subject to political factors.

Angola Angola, apart from the deepwater, has been thoroughly explored, leaving little scope for surprises. The deepwater domain is at an earlier stage of depletion, but most of the larger fields have probably been found.

Cameroon The Cameroons is a small country, which has been fully explored.

Table 16.3 Africa: Range of Reported Reserves and Scope for Surprise

	Standard Deviation Public Reserve Data		Surprise Factor	
	Oil	Gas	Oil	Gas
Algeria	0.0	0.05	3	6
Angola	1.79	2.26	1	2
Cameroon	0	2.04	1	2
Chad	0	–	4	6
Congo	0.17	0.68	1	2
Egypt	0.31	8.04	2	3
Gabon	0.93	0.79	1	2
Libya	1.49	0.76	7	8
Nigeria	0	1.15	4	6
Sudan	0.93	0.19	9	10
Tunisia	0.10	2.23	1	2
Uganda	–	–	7	7

Scale 1–10

Chad Chad probably offers more scope for exploration with the opening of the export pipeline, although its political situation remains difficult.

Congo The Congo, apart from its deepwater areas, has been thoroughly explored.

Egypt Egypt has been thoroughly explored, but might offer some unexpected deepwater potential, primarily for gas.

Gabon Gabon may have some unrealised deepwater potential, but is otherwise at a mature state of depletion.

Libya Libya is a large country, which has been subject to political constraints, being also a member of OPEC. The size of its reserves is subject to much uncertainty, and the scope for new discovery may be larger than recognised here, especially as the offshore is now being opened, but on balance it is probable that the more prospective and prolific areas have already been exploited.

Nigeria Nigeria is also a member of OPEC, so that doubts attach to the figure given for existing reported reserves. But it has been thoroughly explored, such that virtually all the larger fields have almost certainly been found. It offers new deepwater potential, which may offer surprises.

Sudan The Sudan is at a relatively early stage of exploration, being also subject to serious political tensions, so the assessment is a good deal less than sure.

Tunisia Tunisia has limited potential and is at a mature stage of depletion offering little scope for surprise.

Uganda Uganda's oil potential has only recently been determined and so there is scope for surprises although they are unlikely to be significant as only a small part of the country is prospective.

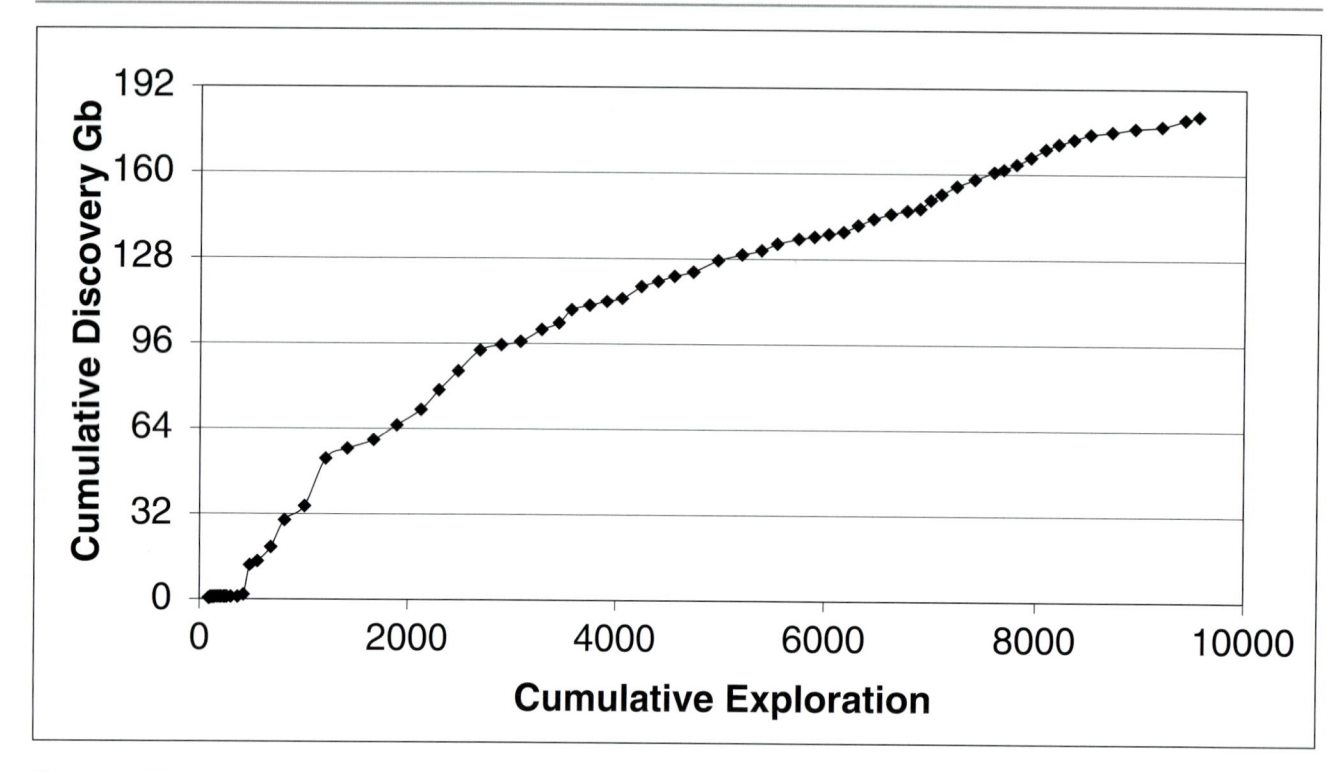

Fig. 16.3 Africa discovery trend

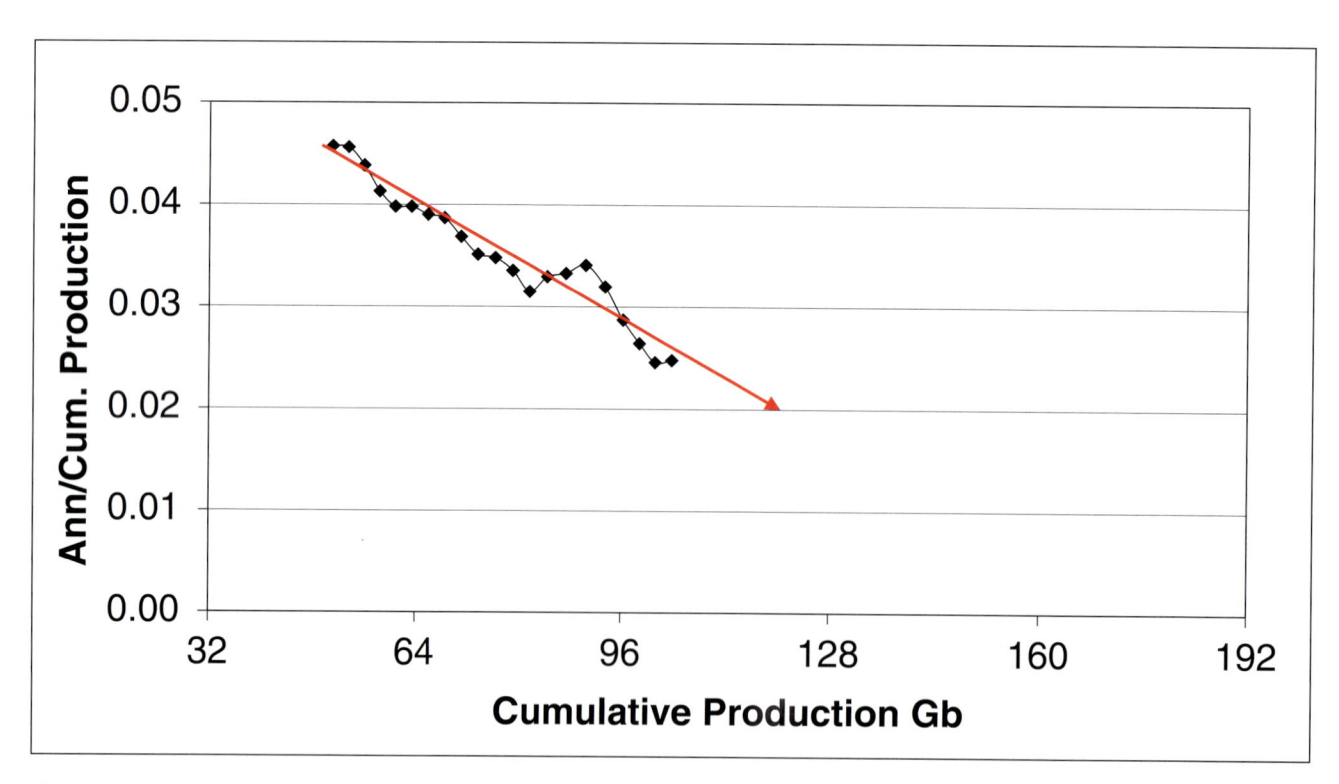

Fig. 16.4 Africa derivative logistic

16 Africa Region

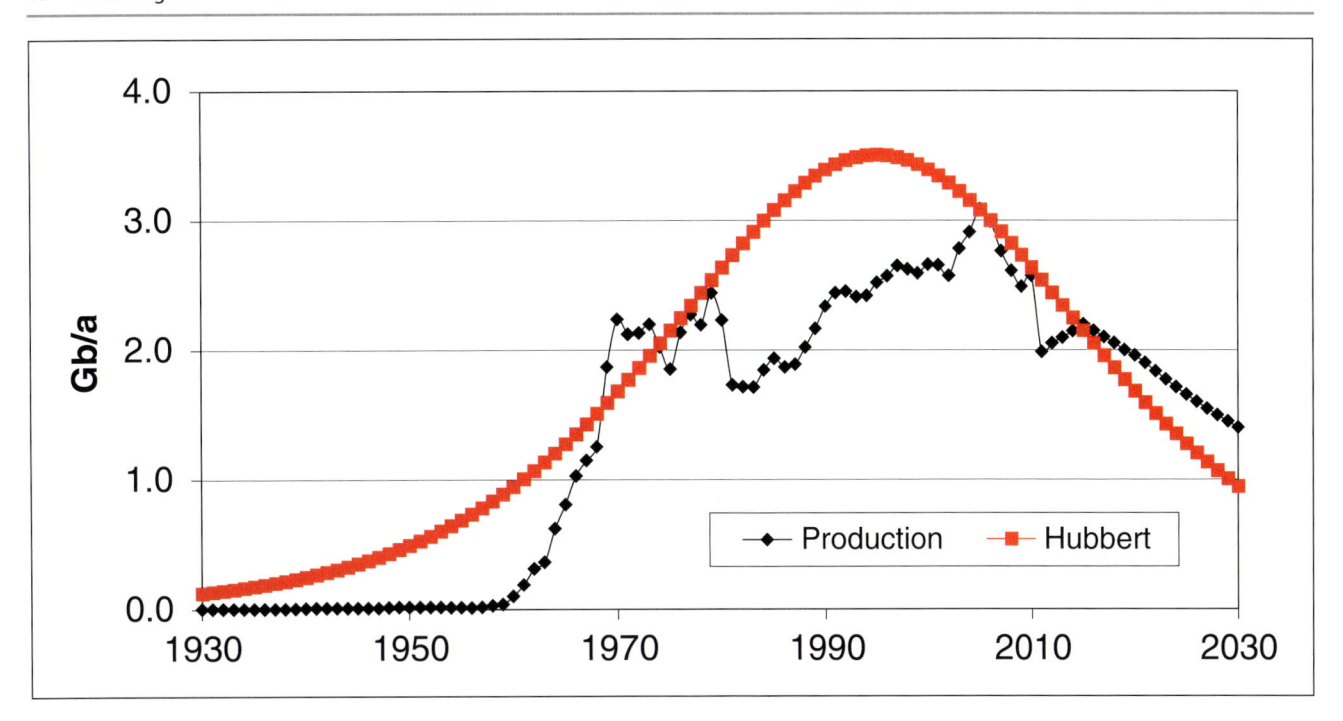

Fig. 16.5 Africa production: actual and theoretical

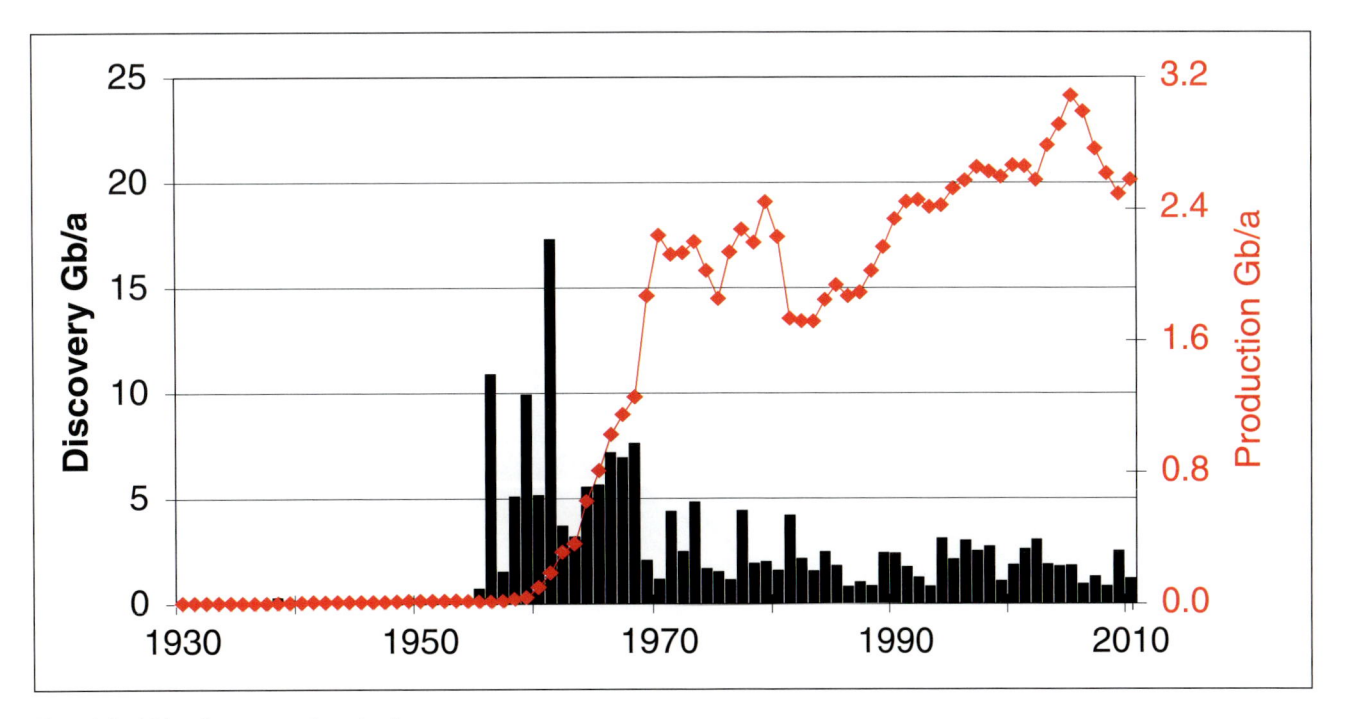

Fig. 16.6 Africa discovery and production

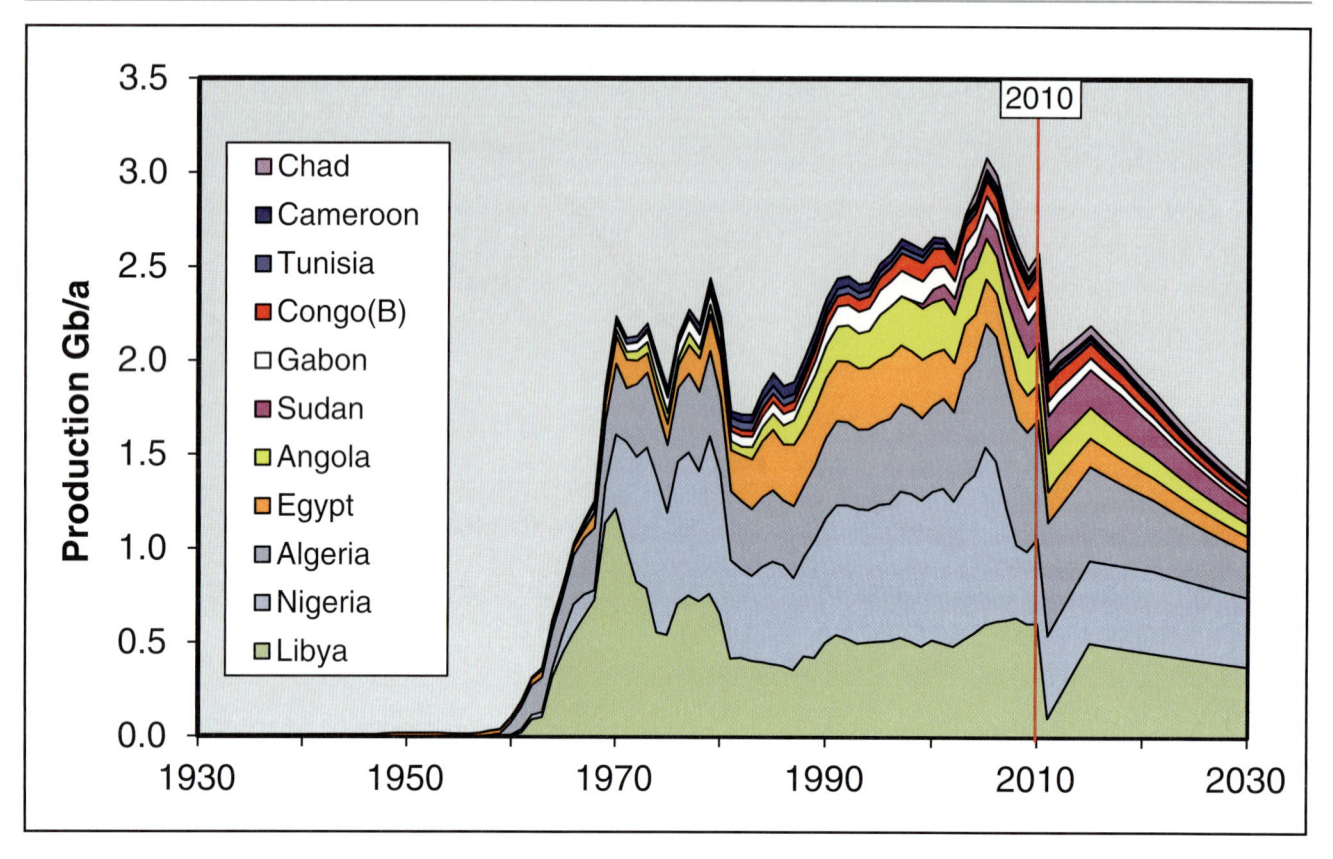

Fig. 16.7 Africa oil production

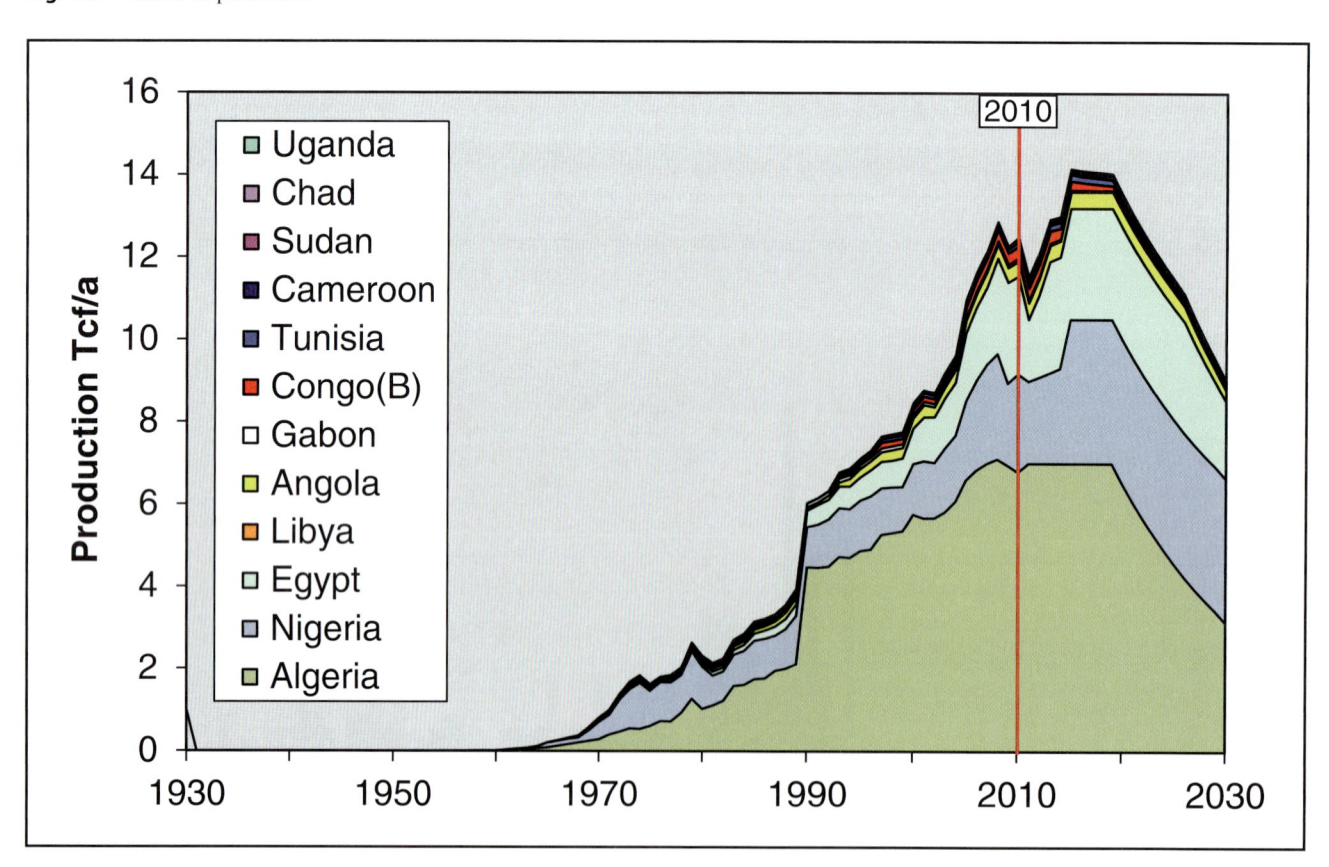

Fig. 16.8 Africa gas production

Table 16.4 Africa: Oil Resource Base

		\multicolumn{9}{c	}{PRODUCTION TO 2100}							
		\multicolumn{15}{c	}{Regular Conventional Oil}
		\multicolumn{7}{c	}{KNOWN FIELDS}			\multicolumn{7}{c	}{Revised 05/10/2011}
	Region	Present		Past			Reported Reserves		%	Future	Total Found	NEW FIELDS	ALL FUTURE	Total	DEPLETION			PEAKS		
Country		Kb/d 2010	Gb/a 2010	Gb	5 yr Trend	Disc 2010	Average	Deductions Static	Other	Disc.	Gb	Gb	Gb	Gb	Gb	Rate	%	Mid-Point	Expl	Disc	Prod
Libya	A	1650	0.60	27.46	0%	0.51	44.90	0.00	0.00	95%	24.78	52.25	2.75	27.54	55.0	2.14%	50%	2010	1963	1961	1970
Nigeria	A	1247	0.46	28.04	–9%	0.15	37.00	0.00	–6.43	95%	24.27	52.30	2.70	26.96	55.0	1.66%	51%	2009	1967	1968	2005
Algeria	A	1729	0.63	20.76	–1%	0.09	12.19	1.93	0.00	96%	11.92	32.68	1.32	13.24	34.0	4.55%	61%	2004	1970	1956	2007
Egypt	A	523	0.19	10.80	–3%	0.17	4.23	0.83	0.00	97%	3.78	14.58	0.42	4.20	15.0	4.35%	72%	1995	2008	1965	1996
Angola	A	595	0.22	6.29	2%	0.05	11.10	0.00	–11.24	96%	3.34	9.63	0.37	3.71	10.0	5.54%	63%	2004	1968	1978	2000
Gabon	A	228	0.08	3.60	–1%	0.02	2.68	0.35	0.00	97%	1.26	4.86	0.14	1.40	5.0	5.44%	72%	1998	1989	1985	1997
Sudan	A	511	0.19	1.46	7%	0.04	6.02	0.71	0.00	85%	3.63	5.09	0.91	4.54	6.0	3.95%	24%	2013	2002	2003	2013
Congo	A	296	0.11	2.26	5%	0.03	1.79	0.37	–0.50	96%	1.56	3.83	0.17	1.74	4.0	5.86%	57%	2007	1992	1984	2009
Tunisia	A	80	0.03	1.47	1%	0.03	0.49	0.06	0.00	95%	0.66	2.13	0.12	0.78	2.3	3.61%	65%	1998	1981	1964	1983
Chad	A	126	0.05	0.38	–4%	0.00	2.00	0.19	0.00	68%	0.97	1.35	0.65	1.62	2.0	2.77%	19%	2021	2002	1977	2015
Cameroon	A	65	0.02	1.28	–5%	0.02	0.20	0.08	0.80	93%	0.35	1.63	0.12	0.47	1.8	4.82%	73%	1996	1977	1977	1985
Uganda	A	0	0.00	0.00	0%	0.08	1.00	0.00	0.00	80%	1.60	1.60	0.40	2.00	2.0	0.00%	0%	2030	2010	2009	2030
AFRICA	A	7051	2.57	103.8	–3%	1.19	121.90	4.52	–17.37	95%	78	182	10	88	192	2.84%	54%	2006	1981	1961	2005

Table 16.5 Africa: Gas Resource Base

		\multicolumn{15}{c	}{PRODUCTION TO 2100}	
AFRICA		\multicolumn{15}{c	}{Conventional Gas}	2010
		\multicolumn{7}{c	}{KNOWN FIELDS}						Revised		15/11/11
	Region	Present			Past	Discovery		Reserves		FUTURE	TOTAL	FUTURE	TOTAL	TOTAL	DEPLETION			PEAKS		
Country		Tcf/a 2010	FIP 2008	Gboe/a 2010	5yr Tcf Trend	Tcf 2010	% Disc.	Reported Average	Deduct Non-Con	KNOWN Tcf	FOUND Tcf	FINDS Tcf	FUTURE Tcf	Tcf	Current Rate	%	Mid-Point	Expl	Disc	Prod
Algeria	E	6.79	3.46	1.22	142.45	0.04	1.06	95%	159.03	0.00	141.79	284	15.75	157.55	300.00	4.13%	47%	2011	1961	1957	2011
Angola	E	0.36	0.33	0.07	6.71	0.02	0.13	94%	8.94	–8.00	10.16	17	1.13	11.29	18.00	3.12%	37%	2016	1968	1971	2015
Cameroon	E	0.07	0.06	0.01	1.95	0.00	0.08	93%	5.95	0.00	4.10	6	0.46	4.55	6.50	1.50%	30%	2025	1977	1979	2029
Chad	E	0.00	0.00	0.00	0.00	N/A	0.00	65%	0.00	0.00	0.32	0	0.17	0.50	0.50	0.00%	0%	2030+	2002	1975	2030
Congo (B)	E	0.30	0.26	0.05	3.22	0.02	0.07	93%	3.75	0.00	2.36	6	0.42	2.78	6.00	9.69%	54%	2011	1992	1984	2012
Egypt	E	2.37	0.10	0.43	25.36	0.07	0.84	89%	72.49	0.00	59.19	85	10.45	69.64	95.00	3.29%	27%	2019	1985	1996	2019
Gabon	E	0.07	0.07	0.01	2.87	0.00	0.02	96%	1.41	0.00	0.50	3	0.13	0.63	3.50	10.50%	82%	1994	1989	1965	1995
Libya	E	2.39	0.48	0.19	21.38	0.02	0.95	85%	67.66	0.94	0.62	1	0.15	73.62	0.00	1.43%	24%	2027	1963	1965	2030
Nigeria	E	2.39	1.18	0.43	50.44	0.02	0.98	92%	185.73	–13.73	152.61	203	16.96	169.56	220.00	1.39%	23%	2029	1966	1967	2029
Sudan	E	0.03	0.00	0.01	0.00	0.00	0.01	50%	3.11	0.00	1.00	1	1.00	2.00	2.00	1.50%	0%	2030	2002	2003	2030
Tunisia	E	0.11	0.01	0.02	2.16	–0.01	0.08	94%	3.03	0.00	2.56	5	5.00	2.84	5.00	3.68%	43%	2012	1981	1974	2015
Uganda	E	0.00	0.00	0.00	0.00		0.07	40%	0.13	0.00	0.20	0	0.30	0.50	0.50	0.00%	0%	2030	2009	2015	2023
AFRICA	E	13.56	5.73	2.44	256.57	0.02	4.31	92%	511.22	–21.73	433.69	690.26	61.74	495.43	752.00	2.66%	34%	2018	1963	1957	2018

Table 16.6 Africa: Oil Production Summary

	Regular Conventional Oil Production				
Mb/d	2000	2005	2010	2020	2030
Libya	1.41	1.63	1.65	1.25	1.03
Nigeria	2.17	2.59	1.25	1.21	1.01
Algeria	1.25	1.80	1.73	1.09	0.68
Egypt	0.77	0.66	0.52	0.34	0.22
Angola	0.75	0.59	0.60	0.34	0.19
Gabon	0.32	0.27	0.23	0.13	0.07
Sudan	0.19	0.35	0.51	0.48	0.25
Congo	0.28	0.23	0.30	0.16	0.09
Tunisia	0.08	0.08	0.08	0.05	0.04
Chad	0.00	0.18	0.13	0.14	0.09
Cameroon	0.08	0.08	0.07	0.04	0.02
Uganda	0.00	0.00	0.00	0.15	0.15
AFRICA	7.29	8.45	7.05	5.36	3.84

Table 16.7 Africa: Gas Production Summary

	Gas Production				
Tcf/a	2000	2005	2010	2020	2030
Algeria	5.76	6.60	6.79	6.52	3.14
Angola	0.25	0.30	0.36	0.40	0.28
Cameroon	0.07	0.07	0.07	0.07	0.07
Chad	0.00	0.00	0.00	0.01	0.01
Congo (B)	0.13	0.25	0.30	0.11	0.02
Egypt	0.86	1.66	2.37	2.70	1.88
Gabon	0.08	0.07	0.07	0.02	0.01
Libya	0.36	0.70	1.07	1.28	1.60
Nigeria	1.23	1.92	2.39	3.50	3.50
Sudan	0.00	0.00	0.03	0.05	0.06
Tunisia	0.08	0.11	0.11	0.13	0.05
Uganda	0.00	0.00	0.00	0.02	0.02
AFRICA	8.82	11.67	13.56	14.80	10.68

Table 16.8 Africa : Oil and Gas Production and Consumption, Population and Density

AFRICA	OIL				GAS				POPULATION & AREA					
	Production	Consumption		Trade	Production	Consumption		Trade	Population		Area			Density
Country		p/capita		(+)			p/capita	(+)		Growth	Mkm2			
Major	Mb/a	Mb/a	b/a	Mb/a		Gcf/a			M	Factor	Onshore	Off	Total	
Algeria	631	113.88	3.2	517	6788	1018	28	5770	36	7.2	2.380	0.080	2.46	15
Angola	217	27.01	1.4	190	364	26	1	338	20	8.5	1.270	15.661	7.88	16
Cameroon	24	10.95	0.5	13	69	1	0	68	20	13.4	0.475	0.020	0.50	42
Chad	46	0.73	0.1	45	0	0	0	0	12	11.5	1.284	0.000	1.28	9
Congo	108	4.02	1.0	104	298	33	8	265	4	6.8	0.342	0.002	0.34	12
Egypt	191	270.10	3.3	−79	2369	1630	20	739	83	8.3	1.001	0.145	1.15	82
Gabon	83	6.57	4.4	77	73	3	2	70	2	3.8	0.268	0.080	0.35	6
Libya	602	105.49	16.5	497	1069	242	38	827	6	12.8	1.760	0.150	1.91	4
Nigeria	455	101.84	0.6	353	2392	176	1	2216	162	10.1	0.924	0.800	1.72	175
Sudan	187	35.77	0.8	151	30	0	0	30	45	14.9	2.506	0.129	2.63	18
Tunisia	29	30.66	2.9	−1	109	116	11	−7	11	5.9	0.164	0.125	0.29	65
Uganda	0	5.11	0.1	−5	0	0	0	0	35	17.3				
Sub-total	2574	712	1.6	1861	13561	3245	7	10316	434	9.8	12.4	17.2	20.51	40
Other								0						
Benin		9.1	1.01	−9	0	0	0.0	0	9.0		0.11			0.01
Botswana		5.5	3.04	−5	0	0	0.0	0	1.8		0.60			0.33
Burkino F.		3.3	0.22	−3	0	0	0.0	0	14.8		0.27			0.02
Burundi		1.1	0.13	−1	0	0	0.0	0	8.5		0.03			0.00
C.African R.		0.7	0.17	−1	0	0	0.0	0	4.3		0.62			0.14
D.R.Congo	7.3	4.0	0.06	3	298	0	0.0	298	62.6		2.35			0.04
Djibuti		4.4	5.48	−4	0	0	0.0	0	0.8		0.02			0.03
Eq. Guinea	133.59	0.4	0.74	133	299	56	112.0	243	0.5		0.50			1.00
Eritrea		2.2	0.45	−2	0	0	0.0	0	4.9		0.12			0.02
Ethiopia		17.2	0.22	−17	0	0	0.0	0	77.1		1.13			0.01
Gambia		0.7	0.49	−1	0	0	0.0	0	1.5		1.46			0.97
Ghana	2.19	21.9	0.95	−20	0	4.2	0.2	−4	23.0		0.24			0.01
Guinea		3.3	0.33	−3	0	0	0.0	0	10.1		0.25			0.02
Guinea Bis.		1.1	0.65	−1	0	0	0.0	0	1.7		0.04			0.02
Ivory Coast	32.39	9.1	0.45	23	57	56.5	2.8	0	20.2		0.32			0.02
Kenya		28.5	0.77	−28	0	0	0.0	0	36.9		0.58			0.02
Lesotho		0.7	0.41	−1	0	0	0.0	0	1.8		0.03			0.02
Liberia		1.6	0.43	−2	0	0	0.0	0	3.8		0.11			0.03
Madagascar		8.0	0.44	−8	0	0	0.0	0	18.3		0.59			0.03
Malawi		2.9	0.22	−3	0	0	0.0	0	13.1		0.12			0.01
Mali		2.2	0.18	−2	0	0	0.0	0	12.3		1.24			0.10
Mauritania	11.17	7.3	2.35	4	0	0	0.0	0	3.1		1.03			0.33
Morocco	0.18	76.3	2.41	−76	2.8	20.1	0.6	−17	31.7		0.45			0.01
Mozambique		6.2	0.30	−6	110.18	2.8	0.1	107	20.4		0.80			0.04
Namibia		8.8	4.17	−9	0	0	0.0	0	2.1		0.83			0.39
Niger		2.2	0.15	−2	0	0	0.0	0	14.2		1.27			0.09
Rwanda		2.2	0.24	−2	0	0	0.0	0	9.3		0.03			0.00
S. Africa	7.36	201.9	4.21	−194	34.25	142	3.0	−107	47.9		1.22			0.03
Senegal		15.0	1.21	−15	1.76	1.76	0.1	0	12.4		0.20			0.02
Sierra Leone		3.3	0.62	−3	0	0	0.0	0	5.3		0.07			0.01
Somalia		1.8	0.20	−2	0	0	0.0	0	9.1		0.64			0.07
Swaziland		1.5	1.30	−1	0	0	0.0	0	1.1		0.02			0.02
Tanzania		13.9	0.36	−14	27.89	27.5	0.7	0	38.7		0.95			0.02
Togo		8.4	1.27	−8	0	0	0.0	0	6.6		0.06			0.01
Uganda		5.1	0.18	−5	0	0	0.0	0	28.5		0.24			0.01
W.Sahara		0.7	1.46	−1	0	0	0.0	0	0.5		0.27			0.53
Zambia		5.8	0.51	−6	0	0	0.0	0	11.5		0.75			0.07
Zimbabwe		4.0	0.30	−4	0	0	0.0	0	13.3		0.39			0.03
Sub-Total	194.18	492	0.84	−298	830	311	0.53	520	582.7		19.90			0.12
AFRICA ALL	2768	1204	1.2	1563	14391	3556	3.50	10836	1017		32.27			20.2

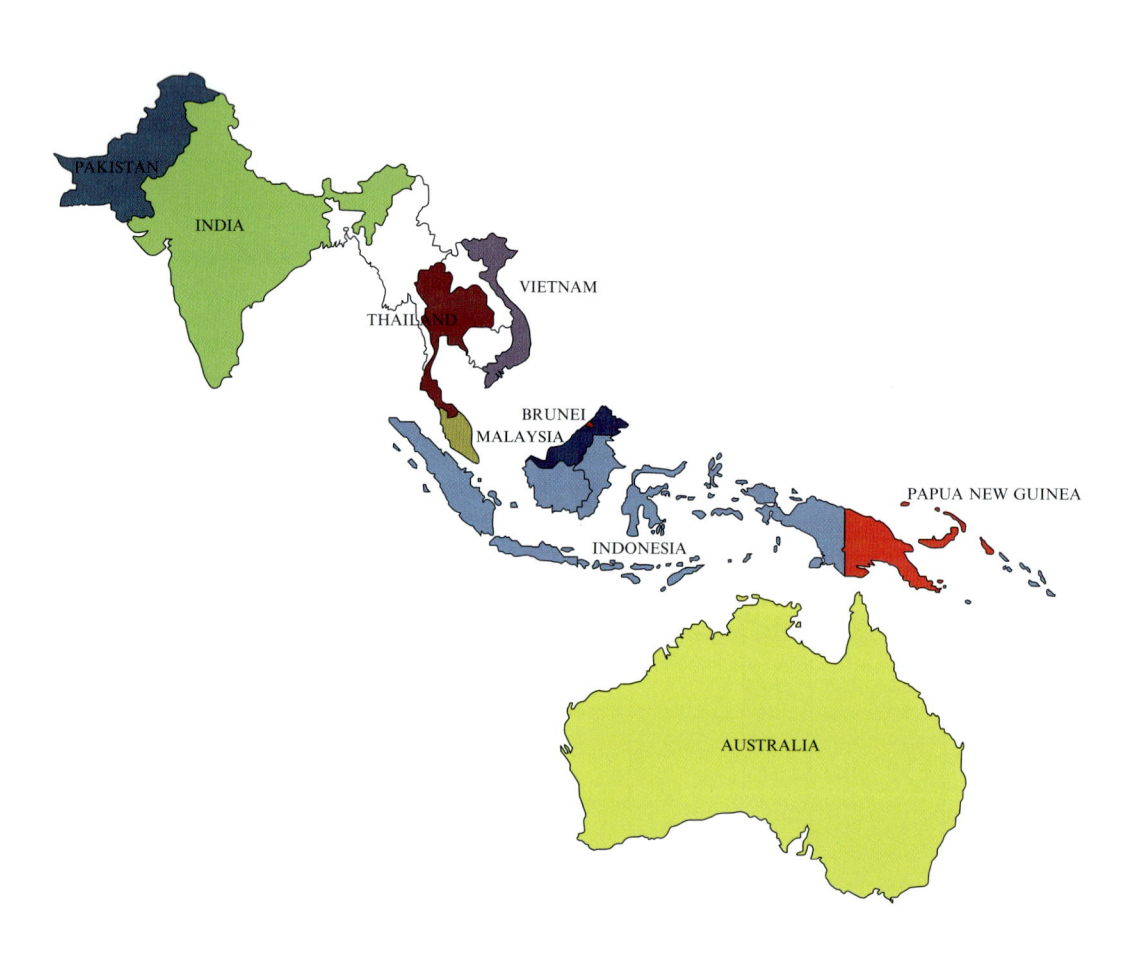

PAKISTAN
INDIA
VIETNAM
THAILAND
BRUNEI
MALAYSIA
INDONESIA
PAPUA NEW GUINEA
AUSTRALIA

Part III

Asia-Pacific

Australia

Table 17.1 Australia regional totals (data through 2010)

	Production to 2100					Peak Dates			Area	
Amount			Rate				Oil	Gas	'000 km²	
	Gb	Tcf	Date	Mb/a	Gcf/a	Discovery	1967	1971	Onshore	Offshore
PAST	7.2	32	2000	260	1,170	Production	2000	2025	7,771	2,560
FUTURE	4.8	188	2005	160	1,446	Exploration	1985		*Population*	
Known	4.1	150	2010	160	1,730	*Consumption*	Mb/a	Gcf/a	1900	3
Yet-to-Find	0.7	38	2020	110	4,487	2010	351	1,141	2010	22.4
DISCOVERED	11	182	2030	80	5,000		b/a	kcf/a	Growth	7.3
TOTAL	12	220	*Trade*	−192	+589	Per capita	16	51	Density	3

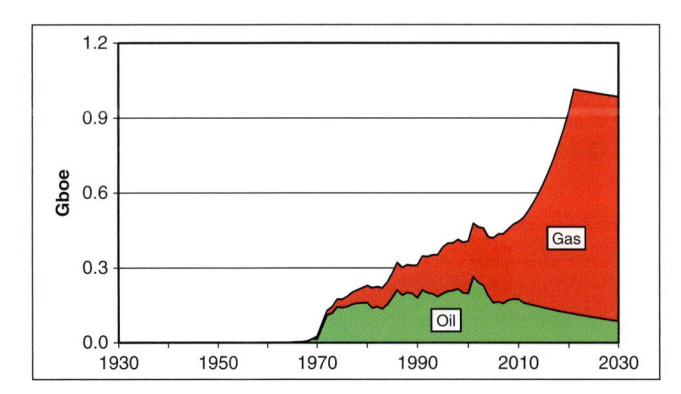

Fig. 17.1 Australia oil and gas production 1930 to 2030

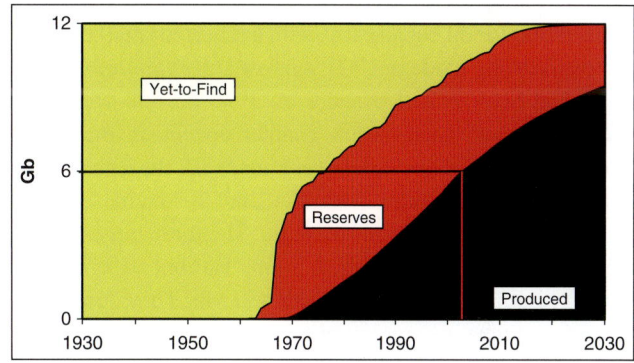

Fig. 17.2 Australia status of oil depletion

Essential Features

Australia is a continent in the southern hemisphere, covering an area of almost 7.8 million km². The interior forms a huge barren depression of arid lands and deserts, separated from the eastern seaboard by what is known as the Great Dividing Range, rising to 2,200 m, which itself is a mature landscape of gorges, plateaus and subsidiary mountain spurs. Most of the population of 22 million live in the fertile temperate south-eastern corner of the continent, where the largest town, Sydney, is located. The Great Barrier Reef, the world's largest coral reef, borders the northeast coast, while the large island of Tasmania lies off the south coast.

The country has a substantial mining industry, especially of gold and coal, but is largely agricultural, being well known as a major source of wool and wheat, although threatened by the risks of periodic droughts.

Geology and Prime Petroleum Systems

Australia formed part of the Permian super-continent, known as Pangea, before it began drifting southeastwards during the Triassic, about 180 million years ago. It reached its present position some 50 million years ago, when Antarctica split off to continue its geotectonic voyage to the South Pole. The early separation has given Australia a

C.J. Campbell, *Campbell's Atlas of Oil and Gas Depletion*,
DOI 10.1007/978-1-4614-3576-1_17, © Colin J. Campbell and Alexander Wöstmann 2013

unique flora and fauna. It is noteworthy that fossils of some of the first forms of life on the planet have been found in Australia.

Much of the continent is made up of ancient shield rocks, but they are flanked by two Tertiary petroleum systems: the Bass Strait Basin off the south coast and the extensive NW Shelf. There are, in addition, interior Palaeozoic basins of minor potential. The NW Shelf forms the passive margin of the continent facing the contact with the Eurasian Plate, bordering the Indonesian island arc. It is made up of a thick sequence of Mesozoic and Tertiary sediments. Several rather lean source-rock intervals have been identified in the Mesozoic sequence, but in many areas lie below the oil-generating window, explaining the preponderance of gas-condensate finds.

Australia and Indonesia have also agreed to share the so-called Timor Gap, within the Indonesian arc, where some discoveries have been made.

Exploration and Discovery

Exploration for oil began early, with a reported small onshore discovery being made in 1900. Almost 160 exploration boreholes had been drilled onshore by 1930 despite very little encouragement. Then, a new chapter opened in the 1960s when important discoveries were made in the Bass Strait Tertiary Basin between Australia and Tasmania, and on Barrow Island off Western Australia. The three largest fields were Kingfish (1967) with 1.2 Gb, Halibut (1967) with 850 Mb and Mackerel (1969) with 450 Mb. They stimulated renewed interest in exploration generally, resulting in a number of finds both in onshore Palaeozoic basins and in other marginal basins. The last campaign came with the opening of the huge NW Shelf.

Exploration is now at a mature stage, and to judge from the discovery trend and field-size distribution is unlikely to deliver more than about 700 Mb in new finds. A total of almost 5,000 exploration boreholes have been drilled so far, which is a relatively high number. The peak of exploration was in 1985 when 186 boreholes were drilled. The number has since declined to about 50 a year, and is expected to continue to do so as the list of viable prospects dwindles, coming to an estimated end around 2035. Australia also has some deepwater potential, but exploration is at an early stage.

Substantial deposits of oil shale have been identified in Queensland, but have so far proved uneconomic to produce despite Government subsidy.

Production and Consumption

There was some earlier production that escaped the records, but significant production commenced in the 1960s and reached a peak of 722 kb/d in 2000, some 33 years after peak discovery. It has since declined to 436 kb/d, being set to continue to decline at about 3% a year.

About 180 Tcf of gas has been discovered, of which 32 Tcf have been produced. Production stands at about 1.7 Tcf/a. Assuming that production rises at 10% a year, it could reach a plateau of about 5 Tcf/a lasting from 2020 to 2040 before a final fall. Such a depletion profile would give a total of 220 Tcf, assuming future discovery of almost 37.5 Tcf. The gas also yields a substantial amount of gas-liquids, contributing about half the total liquid production by 2010. It is evident that Australia will increasingly rely on its substantial gas resources which are at an early stage of depletion.

Oil consumption stands at 351 Mb/a, meaning that Australia already has to import about one quarter of its needs, a percentage set to increase in the future as production declines. All of its gas production is consumed internally, apart from that used for LNG production.

The Oil Age in Perspective

Aboriginal people have lived in Australia for over 40,000 years. They developed a system of land use and management that used all parts of the continent sustainably. They had a complicated ceremonial style of life before contact and conflict with European settlers decimated their population and cultures within only a couple of centuries.

Dutch navigators started to put Australia on European maps in 1606. The British explorer, James Cook, followed in 1770, effectively bringing the territory into the British Empire. When Britain was no longer able to ship convicts to America after its independence, it began to establish penal colonies in Australia. The first convoy of 11 ships with 759 convicts arrived in Botany Bay in 1788. Life was harsh but in 1810 came the introduction of the merino sheep yielding its special wool, and sheep-farming prospered. Freed convicts and settlers began to build a new society during the nineteenth century, encouraged in part by the discovery of substantial gold deposits in 1851. The discrete early colonies were brought together as a Commonwealth in 1901.

Immigration continued during the twentieth century, predominantly from Britain, such that Australia became a loyal member of the British Commonwealth, making major

17 Australia

military contributions to the allied cause in two world wars. Its population of 22 million is concentrated in the principal cities, leaving an empty harsh interior. Politically, the country seems to be moving towards a Republican status, as its population gradually loses ties with their original homelands.

Unlike the land-use patterns of the Aboriginal people, many of Australia's European-derived farming systems are inherently unsustainable, as they do not work in harmony with the local ecosystems. Farmers are now experiencing serious problems from soil erosion, declining water quality, loss of biodiversity and rising salinity. Even so, Australia, and particularly the temperate island of Tasmania, are well placed to face the Second Half of the Age of Oil, having a relatively low population density and huge coal deposits, as well as ample solar energy. The country also has substantial uranium deposits with which to support a nuclear industry. It does at the same time face serious risks from climate change causing droughts that would adversely affect its agricultural capacity, especially in the semi-arid regions. Probably the main threat it faces is from unsustainable immigration pressures from Indonesia and the East in general.

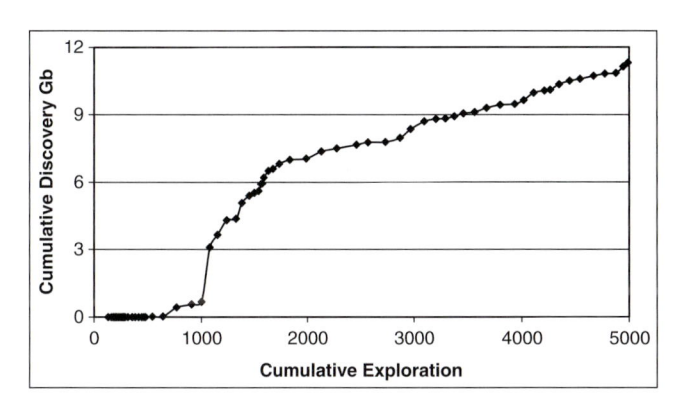

Fig. 17.3 Australia discovery trend

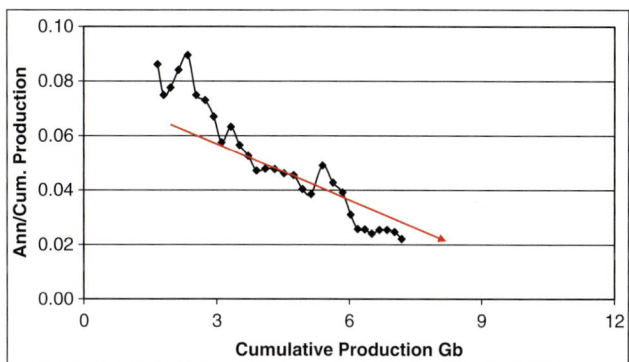

Fig. 17.4 Australia derivative logistic

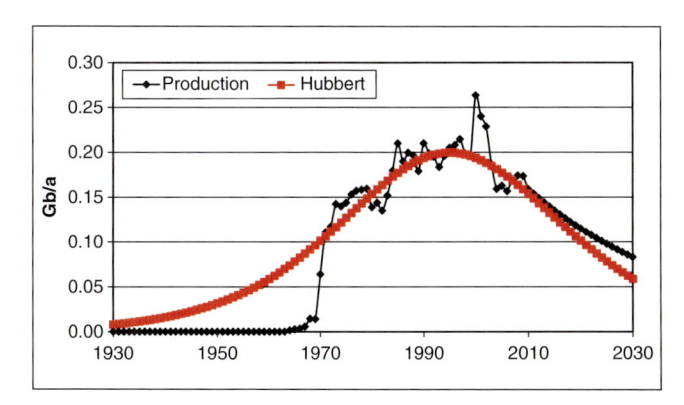

Fig. 17.5 Australia production: actual and theoretical

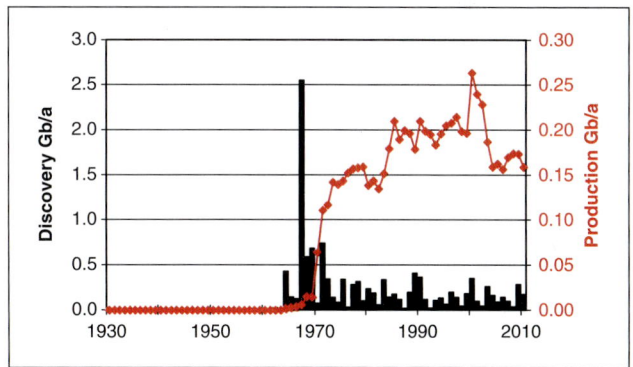

Fig. 17.6 Australia discovery and production

Brunei

Table 18.1 Brunei regional totals (data through 2010)

	Production					Peak Dates			Area	
Amount			Rate				Oil	Gas	'000 km²	
	Gb	Tcf	Date	Mb/a	Gcf/a	Discovery	1929	1963	Onshore	Offshore
PAST	3.5	16	2000	70	416	Production	1979	2006	10	0.15
FUTURE	1.0	9	2005	72	474	Exploration	1975		*Population*	
Known	0.9	8.6	2010	50	440	*Consumption*	Mb/a	Gcf/a	1900	0.01
Yet-to-Find	0.1	0.5	2020	31	302	2010	6	105	2010	0.4
DISCOVERED	4.4	25	2030	19	171		b/a	kcf/a	Growth	40
TOTAL	4.5	25	*Trade*	+43	+335	Per capita	16	263	Density	69

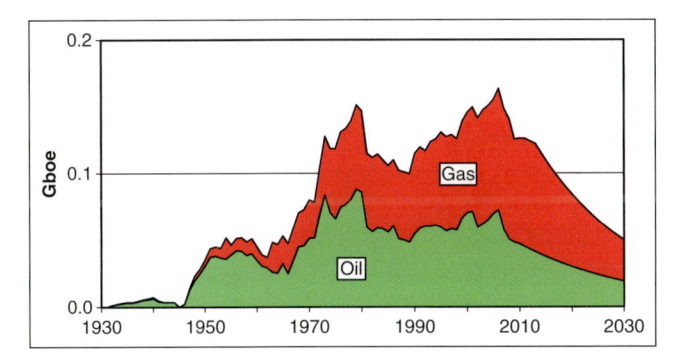

Fig. 18.1 Brunei oil and gas production 1930 to 2030

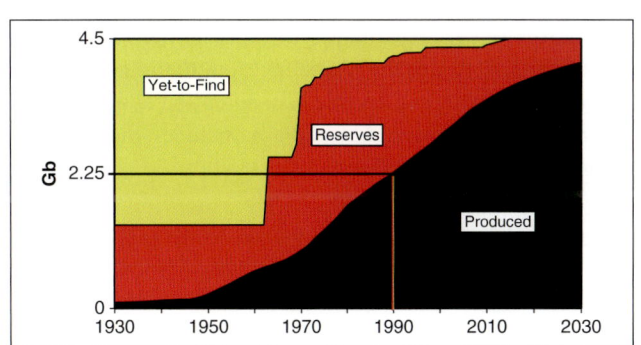

Fig. 18.2 Brunei status of oil depletion

Essential Features

Brunei is a small independent State on the north coast of Borneo. It comprises two tracts of tropical rain forest, meeting in Brunei Bay. The eastern tract is mountainous, whereas the remainder is low lying. Together they cover an area of 5,700 km² and support a population of 400,000. The country also has sovereignty to extensive prospective tracts of the adjoining South China Sea. The mighty Baram River, which drains the interior of Borneo, marks the country's western boundary, while Brunei Bay in the east forms a natural harbour.

Geology and Prime Petroleum Systems

Brunei lies on the northwestern flank of the Borneo geosyncline, which is made up of a series of progressively younger Tertiary basins flanking a Cretaceous core of submarine volcanic and siliceous rocks. The sequences become progressively less deformed towards the coast, where gently folded Mio-Pliocene strata are to be found, offering traps for oil generated lower in the Tertiary sequence.

Exploration and Discovery

Shell took up rights in the territory and adjoining parts of Sarawak in the early years of the last Century, having been encouraged by the occurrence of oil seepages. Exploration was rewarded by the discovery of the Miri Field in Sarawak in 1910, to be followed in 1928 by the giant Seria Field in Brunei, with about 1.3 Gb of oil. The offshore was opened during the 1960s and 1970s, yielding two early giant fields: Ampa SW in 1963 and Champion in 1969, each holding about a billion barrels. They were in turn followed by a series of smaller fields in the 100–200 Mb range (Iron Duke, Fairley, Magpie), as well as by some gas finds. A total of 138 exploration boreholes have been drilled, finding almost 4.4 Gb of oil, of which 3.5 Gb have been produced. In addition, some 25 Tcf of gas has been found, of which about 16 Tcf have been produced. Exploration is now at a very mature stage, such that future oil discovery, apart from whatever the deepwater may yield, is unlikely to exceed about 100 Mb.

Production and Consumption

Production commenced in the 1920s yielding a series of peaks as the different fields were opened up. The first came in 1956 at 115 kb/d; the second in 1979 at 241 kb/d and the third was passed in 2006 at 198 kb/d. The midpoint of depletion was passed in 1990, suggesting that production is now set to decline at the current depletion rate of 5% a year. Consumption stands at no more than 6 Mb/a, albeit at a high per-capita level of 16 b/a, meaning that the country can continue to support exports for many years to come.

Associated gas has been in production since the early 1930s. It rose to a peak of 505 Gcf in 2006 and is now expected to plateau at slightly below that level until 2015. About 70% of the resource will have been depleted by then, heralding a steep decline. As is so often the case, reported gas statistics seem implausible. Consumption at around 105 Gcf/a seems excessive, but perhaps includes the amounts liquefied at the Bintula Plant.

The Oil Age in Perspective

Brunei's early history is obscure, but by the Middle Ages, it had come under the influence of the kingdom of Java. The Portuguese explorer, Magellan, anchored in Brunei Bay in 1521 when the Sultan of Brunei had dominion over most of Borneo as well as several other islands of what is now Indonesia.

His empire subsequently declined due to internal conflicts, and in 1841, his successor had to appeal to Sir James Brooke, a British adventurer who had arrived in Borneo waters in a warship, to help put down a native rising. He not only put down the rising, but managed to take possession of most of the territory which became Sarawak. He and his descendants ran it as a personal estate, being known as the White Rajahs of Sarawak. The poor Sultan was left with no more than a small enclave, known as Brunei, but by an ironic twist of fate it turned out to be the richest part of territory, having prolific oil deposits.

Brunei became a British Protectorate in 1888 with the Sultan becoming little more than a figurehead. It was occupied, along with the rest of Borneo, by Japanese troops in the Second World War, being liberated in 1945 by British forces. The ensuing years saw a gradual transfer of power from the British Crown to the Sultan and an elected Legislative Council, as oil wealth began to flow into the territory. It became a fully independent Islamic Sultanate in 1984, being recognised by Britain, Malaysia and Indonesia.

The massive oil and gas revenues flowing into the country have made the Sultan one of the world's richest men. His subjects do not fare badly either. The State clearly has a golden future if it can defend its independence, which may become increasingly difficult as Malaysia faces increasing economic difficulties during the Second Half of the Age of Oil. It is never easy to be a rich man in a crowd of beggars.

18 Brunei

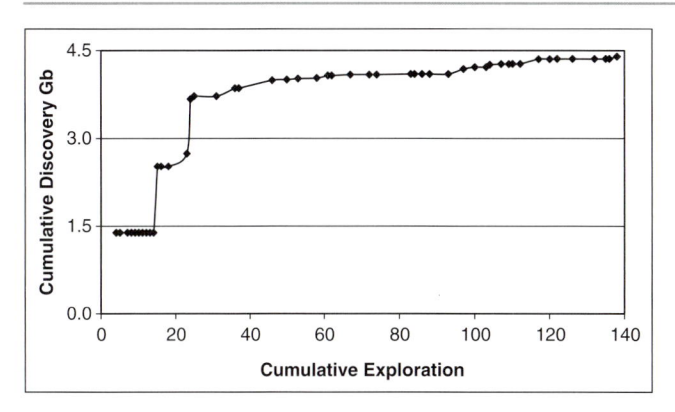

Fig. 18.3 Brunei discovery trend

Fig. 18.4 Brunei derivative logistic

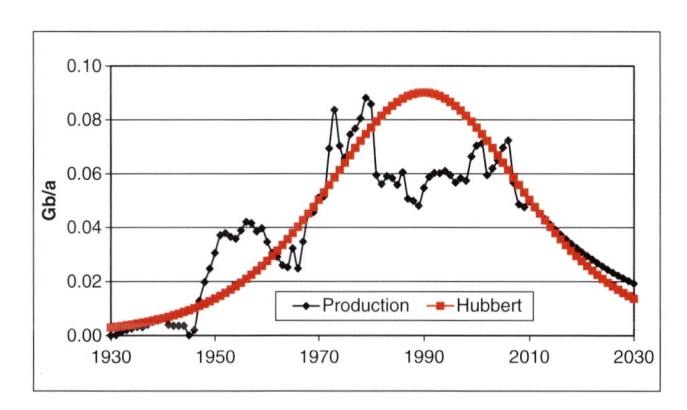

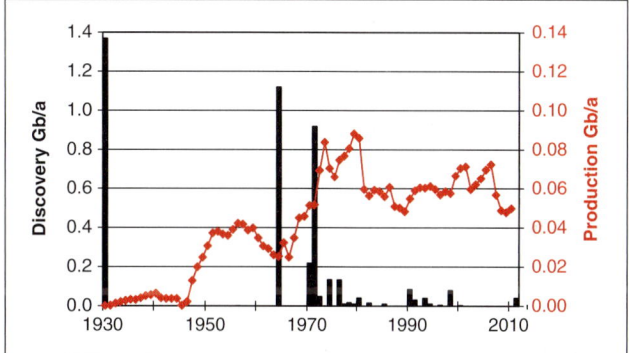

Fig. 18.5 Brunei production: actual and theoretical

Fig. 18.6 Brunei discovery and production

India

Table 19.1 India regional totals (data through 2010)

Production						Peak Dates			Area	
Amount			Rate				Oil	Gas	'000 km²	
	Gb	Tcf	Date	Mb/a	Gcf/a	Discovery	1974	2004	Onshore	Offshore
PAST	7.9	24	2000	236	910	Production	1995	2023	3300	410
FUTURE	5.1	56	2005	243	1,190	Exploration	1991		*Population*	
Known	4.9	45	2010	275	1,883	*Consumption*	Mb/a	Gcf/a	1900	230
Yet-to-Find	0.3	11	2020	136	2,000	2010	1,161	2,277	2010	1,189
DISCOVERED	13	69	2030	97	1,450		b/a	kcf/a	Growth	5.2
TOTAL	13	80	Trade	−887	−394	Per capita	1	2	Density	360

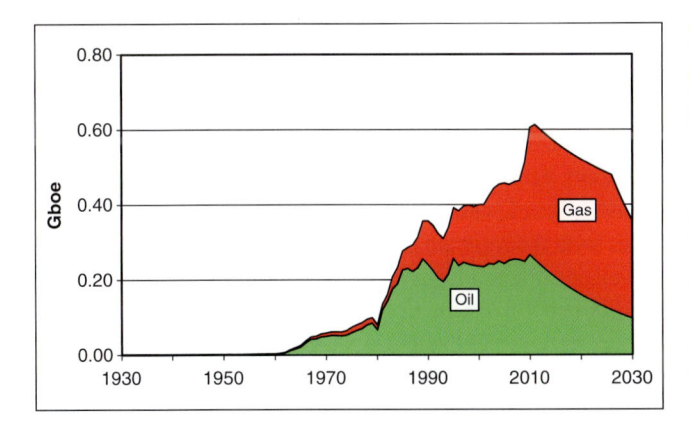

Fig. 19.1 India oil and gas production 1930 to 2030

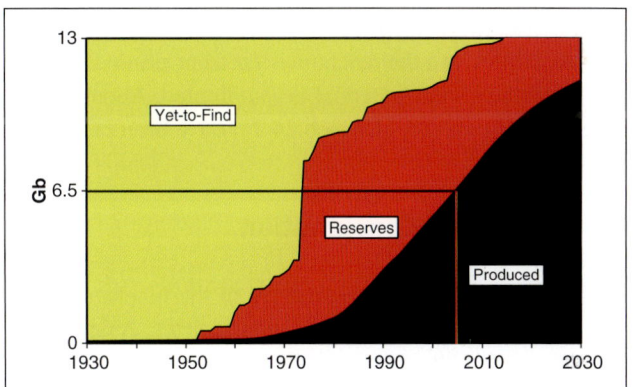

Fig. 19.2 India status of oil depletion

Essential Features

India was the cornerstone of the British Empire, but the territory was divided, following independence in 1947, with the secession of what is now Pakistan, Bangladesh, Bhutan and Nepal, leaving Assam as a somewhat isolated enclave in the northeast, which itself has several active separatist movements.

The Republic of India, as formally constituted, covers an area of some 3.3 million km², making it the seventh largest country in the World. Geographically, it is divided into a mountainous north, flanking the Himalayan Range; the North Indian Plain, drained by the Indus and Ganges Rivers; the mountainous Brahmaputra Valley of Assam in the northeast, and the Deccan Plateau in the south, which itself is flanked by the Ghat mountain ranges, locally rising to around 3,000 m. Its climate is characterised by three seasons: hot and wet from June to September; cool and dry from October to February and hot and dry from February to June. But they are subject to marked annual variations, spelling famine if the rains are late or weak, or flooding in the opposite case. Much of the country is forested. It supports a population of 1.1 billion people, making it the second most populous country in the world.

Geology and Prime Petroleum Systems

India forms a segment of the ancient southern continent of Gondwanaland that moved northwards to collide with the Eurasian Plate, some 50 million years ago. Most of the country is underlain by ancient rocks of the Indian Shield,

C.J. Campbell, *Campbell's Atlas of Oil and Gas Depletion*,
DOI 10.1007/978-1-4614-3576-1_19, © Colin J. Campbell and Alexander Wöstmann 2013

capped by extensive Permian lava flows of the so-called Deccan Traps. But it is flanked by several Cretaceous-Tertiary rift basins where early mid-Cretaceous source-rocks have charged overlying sandstones.

Exploration and Discovery

The first discovery was made in Assam in 1889, but significant finds did not come until the opening of the offshore in the 1970s, especially on the Bombay High, off the west coast, where some 2.5 Gb of oil was found in 1974. The industry is dominated by the State Company, ONGC, although some small foreign private firms are also active. Almost 2,000 exploration boreholes have been drilled, finding almost 13 Gb of oil, of which 8 Gb have been produced. Exploration drilling peaked in 1991 when 96 exploration boreholes were drilled, but is now down to less than half that number. A fairly high level of activity is likely to continue, as the country is in desperate need of oil, but is unlikely to be rewarded by more than perhaps another 250 million barrels, mainly in small fields. Some interest is now being devoted to deepwater possibilities, but the outcome is far from assured.

The country's gas potential is also limited. About 70 Tcf have been discovered, of which 24 Tcf have been produced.

Production and Consumption

Onshore oil production commenced in the nineteenth century in Assam, but its fields were soon depleted. A new phase of production opened after the Second World War, passing 100 kb/d in 1967. The start of the offshore production in 1976 delivered an overall peak of 703 kb/d in 1995, since when it has declined slightly. At the current Depletion Rate of 5%, production is set to fall to about 500 kb/d by 2018 and 446 kb/d by 2020. Consumption stands at 1,161 Mb/a, giving the country a large and growing need for imports, which will be increasingly difficult to obtain. This readily explains why State-backed Indian companies are taking up rights overseas in, for example, the Sudan, Libya, Iran and Venezuela.

Gas production stands at about 1.9 Tcf/a, a level which can probably be maintained until around 2020. It is also a minor importer of gas from Pakistan. The country has substantial coal deposits, although some have a high arsenic content which has caused serious environmental damage in the past.

The Oil Age in Perspective

India (which included Pakistan and Bangladesh prior to 1948) has had a very long history, with the earliest records of the Indus Civilisation going back more than 4,000 years. That was followed by the so-called Aryans, so admired by the Nazis, who spread out from Central Asia to populate India as well as Europe and intervening territories. Later Greek, Roman, Arab and Turkish influences came to India, with the subsequent growth of sundry kingdoms, whose fortunes waxed and waned with the passage of history. The people enjoyed an advanced culture embracing many religions, principally Buddhism, which itself evolved and split into diverse sects. Arab invasions and raids brought the Muslim faith particularly to northern and western India from the twelfth century onwards. The great Mughal Empire, lasting for 200 years from 1526, effectively unified the subcontinent, bringing an age of relative affluence and stability, as well as the growth of trade with Europe, but it finally disintegrated from conflicts between the nobility.

The Portuguese navigator Vasco da Gama landed in 1498, paving the way for the establishment of Goa as a Portuguese territory. The Dutch and French also had a presence, but it was the British who finally made India the jewel of their Empire. British influence started with the East India Company that secured a trade monopoly in 1600, and later demanded military and political support, becoming an early kleptocracy as its functionaries amassed great wealth. Tea plantations were established in the early nineteenth century, especially in the hill country of Assam, and tea became a major source of export earnings as Europeans developed a taste for it. British control was achieved gradually by a series of alliances with the separate principalities making up the country, as well as through military engagements (one notable General was named Sir Colin Campbell). The pinnacle of British power came in the latter part of the nineteenth century, and seems to have enjoyed wide support from the people at large. Indian regiments under British officers were raised, playing heroic parts in the both World Wars. But stirrings of independence developed in the early twentieth century, receiving some sympathy in the mother country. The movement was led by Mahatma Gandhi (1869–1948), who preached tolerance and non-violence. The eclipse of the British Empire in the Second World War and the ensuing socialist regime paved the way for Indian independence, which was granted in 1947. It saw the partition of the country into mainly Hindu and Muslim territories, the latter becoming Pakistan, but it cost the lives of more than a million people in various factional massacres.

The new government, led by Mr Nehru, faced a continuation of communal conflict resulting from partition and economic dislocation. That was shortly followed by the outbreak of an undeclared war with Pakistan over the status of Kashmir, with its predominantly Muslim population, which found itself on the wrong side of the dividing line. The country has also faced some boundary disputes with China.

India has evidently proved to be a difficult place to govern. Nehru's daughter, Indira Gandhi, came to power after the death of her father. She proved to have an iron will and an autocratic style, taking a non-aligned position between the opposing powers in the Cold War, but was shot dead in 1984 by two Sikh guards following a dispute with the Sikh minority. She was succeeded by her son Rajiv, who was in turn assassinated by a Tamil suicide bomber in 1991. His Italian-born widow, Sonja, might have come to power recently with adequate political support, but perhaps wisely stepped aside in 2004 for Manmohan Singh, a gentle economist, educated at both Oxford and Cambridge, who is the first Sikh to hold the office. India has been a nuclear power since 1991.

The country has recently enjoyed something of an economic boom, based in part on services run through the Internet. Western manufacturers have also set up to benefit from cheap labour. It is however likely to be a short-lived chapter of relative prosperity, as imported energy becomes at first expensive and then in short supply. An economic downturn will likely impinge on an already fragile political structure, rendered even more difficult by the country's huge population of more than a billion.

How India will fare during the Second Half of the Oil Age is hard to predict, but disintegration is a likely outcome, as people revert to their old communal and religious identities, a process which will probably be accompanied by much bloodshed and suffering. Clearly, the present population far exceeds the carrying capacity of the land, but Indians can draw upon an ancient history and, benign spirituality that may help in finding a sustainable path.

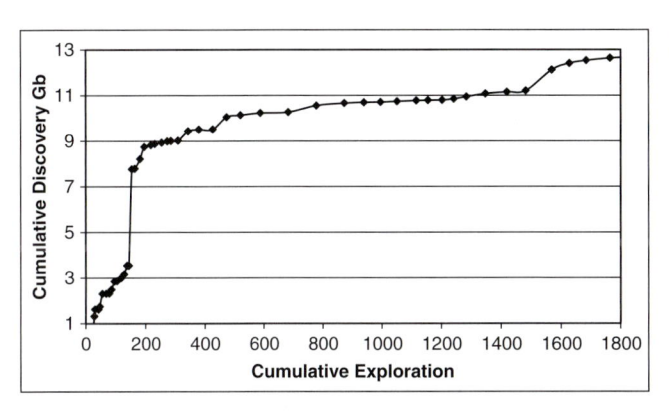

Fig. 19.3 India discovery trend

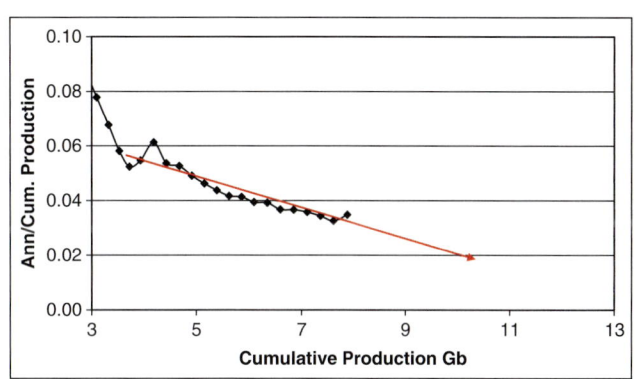

Fig. 19.4 India derivative logistic

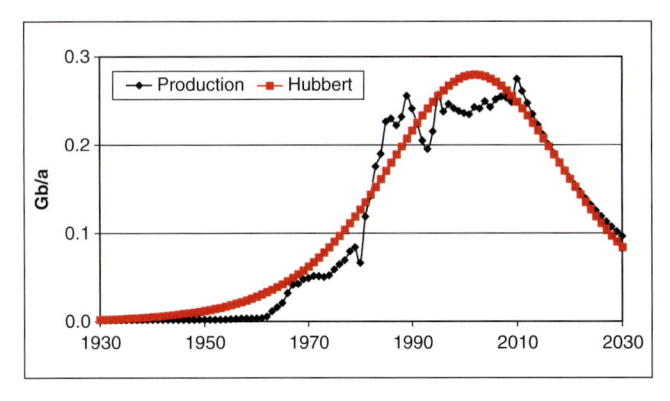

Fig. 19.5 India production: actual and theoretical

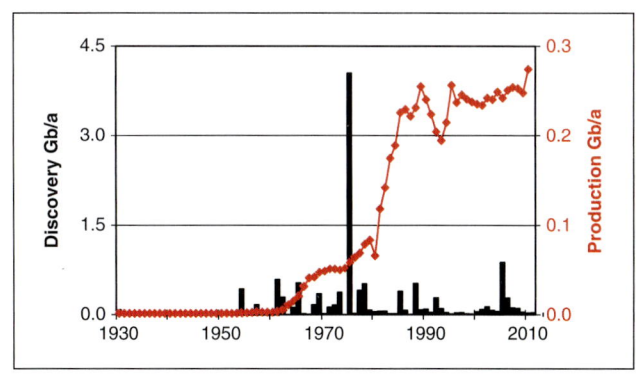

Fig. 19.6 India discovery and production

Indonesia

Table 20.1 Indonesia regional totals (data through 2010)

Production						Peak Dates			Area	
Amount			Rate				Oil	Gas	'000 km²	
	Gb	Tcf	Date	Mb/a	Gcf/a	Discovery	1944	1973	Onshore	Offshore
PAST	24	78	2000	521	2,901	Production	1979	1998	1910	1375
FUTURE	8.1	142	2005	389	2,984	Exploration	1983		Population	
Known	7.3	121	2010	344	3,406	Consumption	Mb/a	Gcf/a	1900	40
Yet-to-Find	0.8	21	2020	227	3,500	2010	472	1,460	2010	236
DISCOVERED	31	199	2030	150	3,500		b/a	kcf/a	Growth	5.8
TOTAL	32	220	Trade	−127	+1946	Per capita	2.0	6.0	Density	123

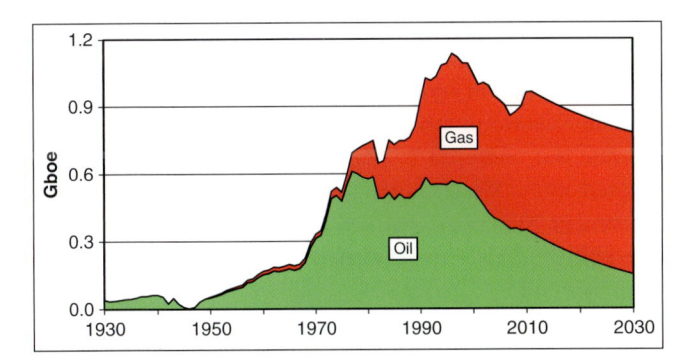

Fig. 20.1 Indonesia oil and gas production 1930 to 2030

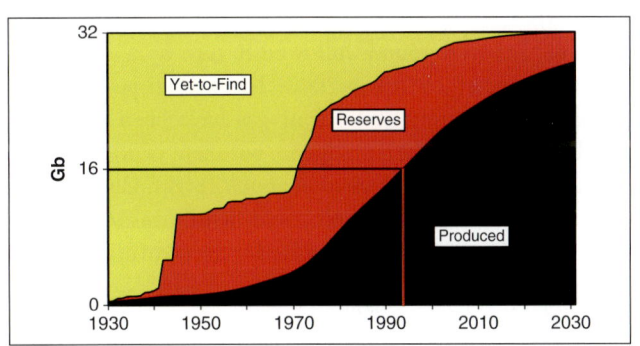

Fig. 20.2 Indonesia status of oil depletion

Essential Features

Indonesia is an archipelago, stretching for about 3,000 km from Asia to Australasia, and including the large islands of Java and Sumatra, as well as much of Borneo. Tropical rain forest is the natural vegetation, but this is now being depleted by logging and conversion to croplands, largely for biofuels. The country has a diverse ethnic population of some 236 million, which has increased sixfold over the past Century. It is predominantly Muslim, but about 3% are of Chinese origin, who have traded and settled in the area for centuries.

Geology and Prime Petroleum Systems

Much of the country is strongly deformed and volcanic, so that its petroleum prospects are confined to a few well-known Tertiary sedimentary basins in Sumatra, S.E. Borneo and locally in Irian Jaya, as well as on an extensive continental shelf. The Tertiary sequences attain a considerable thickness and include both source and reservoir rocks. The territory has several impressive active volcanoes, as for example Mount Semeru and Mount Bromo in eastern Java, and is subject to earthquakes and tsunamis, some having had disastrous consequences.

Exploration and Discovery

Exploration is at a mature stage having commenced in the nineteenth century. Some 4,500 exploration boreholes have been drilled. Peak exploration was in 1983 when 151 boreholes were drilled, but the number has now fallen to less than 40 over the past few years, as fewer prospects remain to be tested.

The first reported discovery was in 1885 following Shell's exploratory efforts in southern Borneo. There were subsequent cycles as different areas were opened up. The onshore peak came in the 1940s with the giant Duri and Minas fields in Sumatra, which were not however developed until after the Second World War. Duri contains heavy oil (20° API), being produced with a low net energy yield by steam injection, putting it on the borderline of *Non-conventional*.

The opening of the offshore in the 1970s gave another chapter with an initial peak in 1970 when 1.2 Gb of oil was found. Generally, the country is now at a mature stage of exploration with perhaps not more than about 1.2 Gb left to find in ever smaller fields in the established producing areas.

It has some *Non-conventional* deepwater oil potential, as already confirmed by Unocal's work off Borneo, but generally the source-rock conditions for such are adverse. Whereas the prolific deepwater tracts of West Africa and the Gulf of Mexico are underlain by rifts containing rich source-rocks, the possibilities in Indonesia are confined to the delta-fronts themselves that are likely to be lean and gas prone.

The gas discovery trends broadly matched those of oil. The peak onshore came in 1971 when 13 Tcf were found. The main peak of offshore discovery came two years later when 48 Tcf were found, followed in turn by a subsidiary peak offshore of 11 Tcf in 2000. The overall peak was in 1998. Approximately 200 Tcf of gas have been found, of which 78 Tcf have been produced, but there is potential for new discovery, here estimated at about 21 Tcf. Gas production stands at 3.4 Tcf/a compared with consumption at 1.5 Tcf/a, the balance being exported in the form of Liquefied Natural Gas.

Production and Consumption

Oil production commenced in 1893 and grew steadily to 170 kb/d in 1940 when it fell as a result of the Second World War. It increased again rapidly after the war to an initial peak of almost 1.7 Mb/d in 1977. It then declined to a saddle due to OPEC quota constraints before recovering to 1.6 Mb/d in 1991, when terminal decline set in at about 4% a year. At this decline rate, it fell to about 943 kb/d by 2010 and is expected to fall to 623 kb/d by 2020. Consumption is running at 472 Mb/a making the country a net importer on a rising trend.

The Oil Age in Perspective

Indonesia has had a long history having been settled by peoples from Malaya and Oceania, and it was also influenced by Arab traders in the Middle Ages. Portuguese traders arrived in 1512 wanting to monopolise the supply of nutmeg, cloves and pepper. They were followed by the Dutch in 1602. At first, they ran it as a commercial enterprise through the Dutch East Indies Company, before bringing it into formal Dutch colonial rule in 1798. It was occupied in the Second World War by Japan, whose motive for going to war was partly to secure access to its oil.

A move to independence followed under the leadership of General Sukarno, being finally granted in 1949 under less than amicable terms. The western end of New Guinea, with its very different ethnic people, was added to the new republic in 1963, later being renamed Irian Jaya. The former Portuguese territory of East Timor, with its predominantly Catholic population, was annexed in 1976, but has recently successfully seceded.

President Sukarno, who had Communist leanings, ruled in an authoritarian style until 1965 when he was ousted by General Suharto in a bloody conflict costing 500,000 lives. His rule was endorsed by popular elections in 1968, having adopted more Western-oriented policies, seeking overseas investment. Since his departure, the country has lurched from one political crisis to another under somewhat uncertain administrations.

Indonesia has had a long oil history, being the birthplace of Royal Dutch/Shell, with its early fields in Borneo. It joined OPEC in 1962, and effectively nationalised the oil industry in 1965 with the creation of a state company, Pertamina, which, in turn, entered into production-sharing contracts with foreign companies, bringing about a successful and active cooperation. It later resigned from OPEC as natural depletion meant it had no reason to limit production.

It would not be surprising if the many islands of Indonesia progressively secede from the central administration during the Second Half of the Age of Oil as the people, some of differing ethnic backgrounds, find benefits in a new regionalism. But the moves may lead to various local conflicts.

20 Indonesia

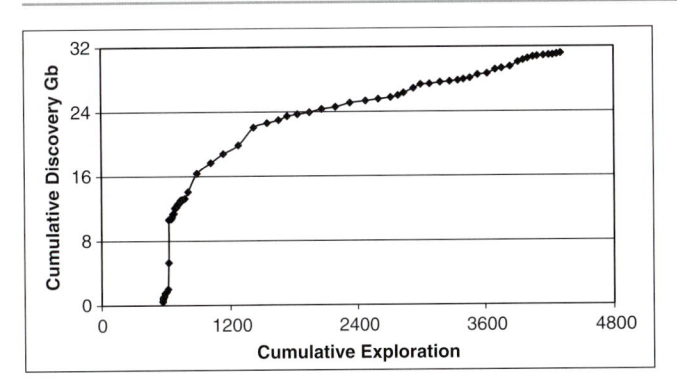

Fig. 20.3 Indonesia discovery trend

Fig. 20.4 Indonesia derivative logistic

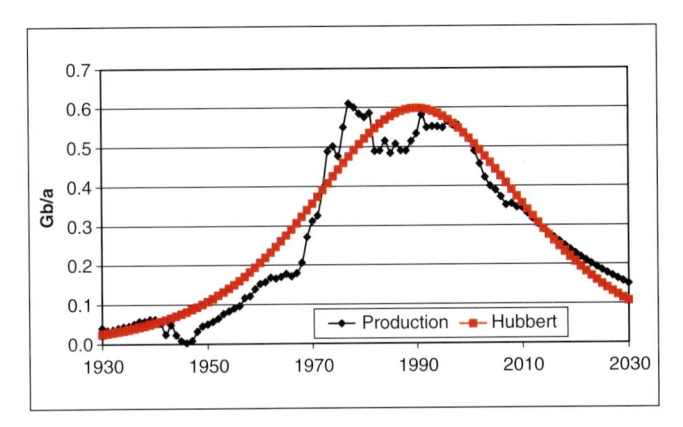

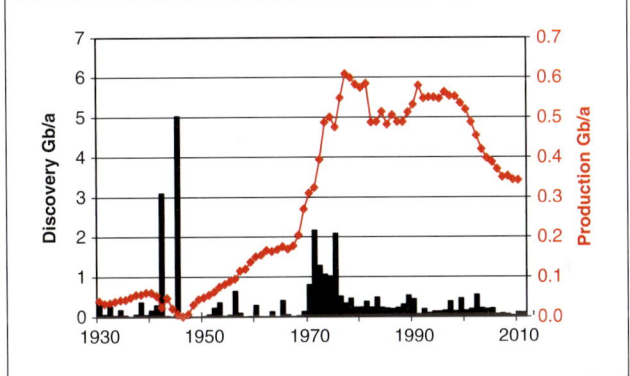

Fig. 20.5 Indonesia production: actual and theoretical

Fig. 20.6 Indonesia discovery and production

Malaysia

Table 21.1 Malaysia regional totals (data through 2010)

Production						Peak Dates			Area	
Amount			Rate				Oil	Gas	'000 km²	
	Gb	Tcf	Date	Mb/a	Gcf/a	Discovery	1971	1970	Onshore	Offshore
PAST	7.1	43	2000	252	1734	Production	2004	2029	330	220
FUTURE	4.9	83	2005	230	2680	Exploration	1970		Population	
Known	4.6	74	2010	202	2718	Consumption	Mb/a	Gcf/a	1900	2
Yet-to-Find	0.2	8	2020	135	2800	2010	205	1145	2010	29
DISCOVERED	12	117	2030	90	2102		b/a	kcf/a	Growth	14.5
TOTAL	12	125	Trade	−3	+1573	Per capita	7	40	Density	87

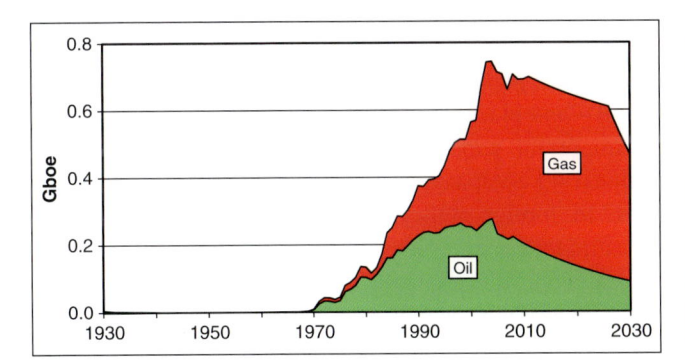

Fig. 21.1 Malaysia oil and gas production 1930 to 2030

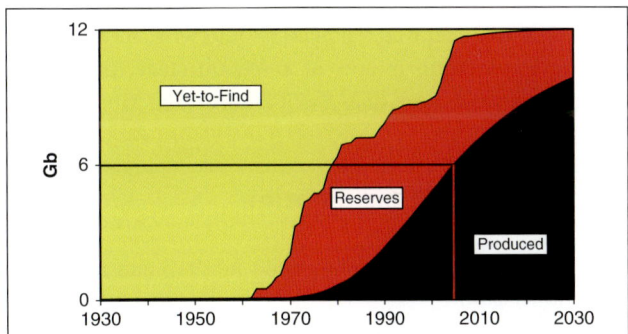

Fig. 21.2 Malaysia status of oil depletion

Essential Features

Malaysia covers an area of some 330,000 km², and controls large sectors of the Gulf of Thailand and the South China Sea, which separate the Malay Peninsula from Borneo. The Malay Peninsula itself has a mountainous core rising to 2,000 m, consisting largely of karst country, whereas the Borneo territories of Sarawak and Sabah comprise heavily forested ranges, capped by Mt Kinabalu rising to over 4,000 m, being drained by the mighty Rajang and Baram rivers. The Malays themselves are of mixed origins, coming from various places throughout Southeast Asia. They also have some Arab blood from early traders, perhaps explaining why most belong to the Muslim faith. Chinese immigrants comprise about one-third of the 29 million inhabitants, some having been there for generations. Various indigenous tribes just survive, but most have now lost their cultural identity through integration. The timber industry now decimates the forests of Borneo, destroying the habitat of the Punan indigenous people, who have been forced to exchange their blow guns and *parangs* (machetes), with which they used to procure wild boar and wild sago in the forests, for squalid lives in run-down apartments.

Geology and Prime Petroleum Systems

The onshore Malay Peninsula is formed mainly of Palaeozoic and older non-prospective rocks, while the Malay territories of onshore Borneo are made up of strongly folded Tertiary strata which are locally prospective by the coast, yielding the Miri Field in what was formerly Sarawak.

The offshore of the Gulf of Thailand and South China Sea cover prospective rift systems, where oil and gas were generated in Lower Tertiary sequences to charge overlying sandstone reservoirs.

C.J. Campbell, *Campbell's Atlas of Oil and Gas Depletion*,
DOI 10.1007/978-1-4614-3576-1_21, © Colin J. Campbell and Alexander Wöstmann 2013

Exploration and Discovery

Seepages of oil in Brunei and the adjoining parts of Sarawak attracted the interest of the Shell Company in the early years of the twentieth century, to be rewarded first by the discovery of the modest Miri Fields in Sarawak, and later by the giant Seria Field of Brunei, found in 1928 with about 1.2 Gb.

About 26 exploration boreholes had been drilled prior to 1930, mainly in Sarawak and Brunei. Exploration drilling then lapsed until after the Second World War.

Attention turned offshore during the 1960s and 1970s both in the Gulf of Thailand and off Borneo. The level of exploration has fluctuated with peaks in 1970 and 1991 when over 40 boreholes were sunk. It resulted in the discovery of almost 12 Gb of oil and 117 Tcf of gas, lying mainly in modest to small oil and gas fields, although the Seligi and Samarang fields, found in respectively 1971 and 1973, attain giant status. The overall peak of discovery was in 1971 when 1.3 Gb were found.

A giant deepwater discovery, named Kikeh, off north Borneo (Sabah) was reported in 2003, lying in 1,400 m of water, but the deepwater potential otherwise seems rather limited. Exploration is now at a mature stage, and is not expected to yield more than about another 300 Mb.

Production and Consumption

Onshore oil production commenced in 1912 and reached a peak of 15 kb/d in 1929. Offshore production followed in 1968 and reached a peak in 2004 at 755 kb/d, which represented the overall peak, since when production has declined slightly. Future production is expected to decline at almost 4% a year, ignoring any deepwater production.

Oil consumption stands at about 205 Mb/a making the country a small importer.

Gas production has risen steadily since 1968 to 2.7 Tcf/a, and is expected to hold this level to around 2026 before terminal decline sets in. Consumption stands at 1.1 Tcf/a making the country a net exporter, mainly in the form of LNG.

The Oil Age in Perspective

Various kingdoms developed in Malaya during the first millennium, and were later subject to Indian and Arab influences, leading to the eventual adoption of Islam. The Portuguese established a colony in Malacca on the west coast of the peninsula in the 1500s, which became a prominent trading port. Britain followed in 1786, leasing the island of Penang. By 1826 it had managed to unite various territories, including Singapore, into a crown colony, known as the Straits Settlements, which was administered by the East India Company.

The opening of the Suez Canal in 1869 brought increasing western influences and trade. In Borneo, an Englishman by the name of James Brooke, the so-called White Rajah, took control in 1841 of a territory, known as Sarawak, operating it as a family estate until the Second World War, while North Borneo was administered by the British North Borneo Company. Eventually, the various sultanates and territories were absorbed into the British Empire. They were occupied by Japan during the Second World War, before becoming the Federation of Malay States in 1948, attaining full independence in 1957.

However, the trading city of Singapore seceded in 1965, and the small but oil-rich Sultanate of Brunei on the coast of Borneo has also managed to retain its independence. The economy was previously built on rubber plantations—the rubber plant having been originally brought from Brasil—and placer tin mining. The oil industry now plays an important part, having seen the establishment of a national oil company, Petronas, which is expanding overseas. In foreign policy, the country has supported the Palestinian cause.

A degree of communal strife has developed in recent years between the Malays and those of Chinese and Indian origins, who were seen to have a disproportionate stake in industry and business. In the 1980s a new, somewhat authoritarian, government under Dr Mahathir brought new prosperity, although it faced a regional financial crisis in the late 1990s. He retired in 2003 after 22 years in office, to be followed by his deputy, Abdulla Badawi, who won a landslide victory in the 2004 elections. In the following year came an important decision to break the link with the US dollar.

The political life of the territories seems to have developed in a fairly orderly and stable fashion, although deep-seated ethnic divisions remain, and may well surface in the future as economic stress bites deeper. It seems likely that the Federation will disintegrate as the different territories search for survival strategies in the face of declining world trade. The natural tropical forests may recover.

21 Malaysia

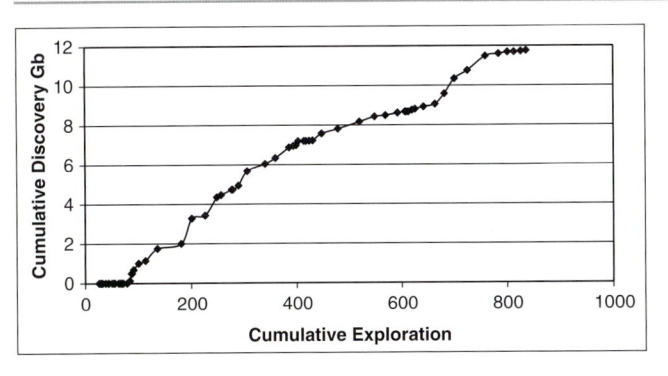

Fig. 21.3 Malaysia discovery trend

Fig. 21.4 Malaysia derivative logistic

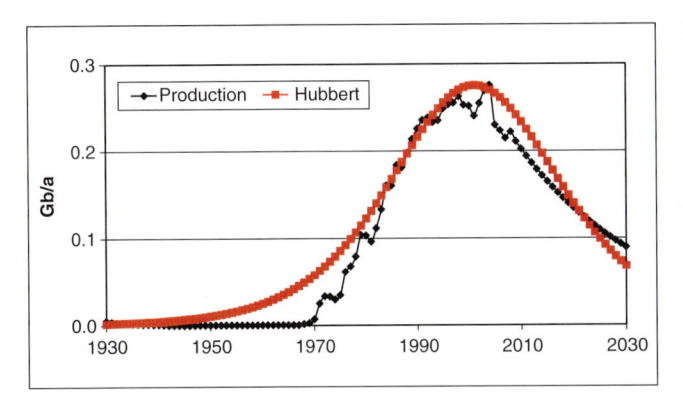

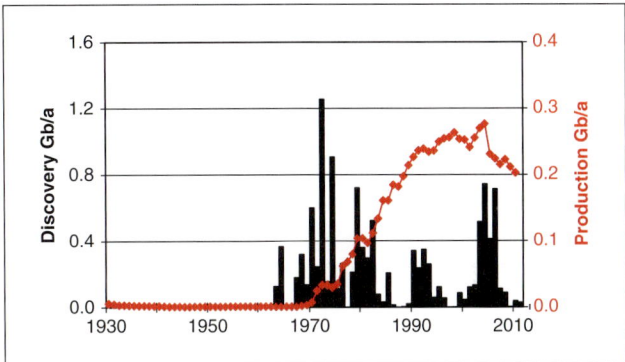

Fig. 21.5 Malaysia production: actual and theoretical

Fig. 21.6 Malaysia discovery and production

Pakistan

Table 22.1 Pakistan regional totals (data through 2010)

Production						Peak Dates			Area	
Amount			Rate				Oil	Gas	'000 km²	
	Gb	Tcf	Date	Mb/a	Gcf/a	Discovery	1984	1952	Onshore	Offshore
PAST	0.65	26	2000	20	856	Production	2005	2013	800	7
FUTURE	0.35	39	2005	24	1,215	Exploration	2003		Population	
Known	0.31	31	2010	22	1,513	*Consumption*	Mb/a	Gcf/a	1900	20
Yet-to-Find	0.03	7.8	2020	12	1,500	2010	150	1,400	2010	185
DISCOVERED	0.97	57	2030	6	888		b/a	kcf/a	Growth	9.3
TOTAL	1.00	65	*Trade*	−128	+113	Per capita	0.8	7.6	Density	231

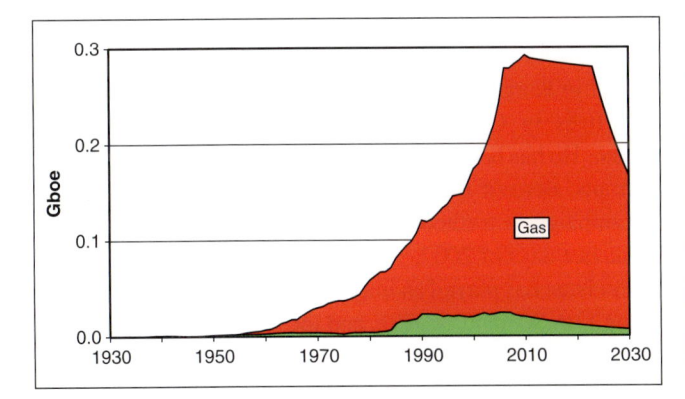

Fig. 22.1 Pakistan oil and gas production 1930 to 2030

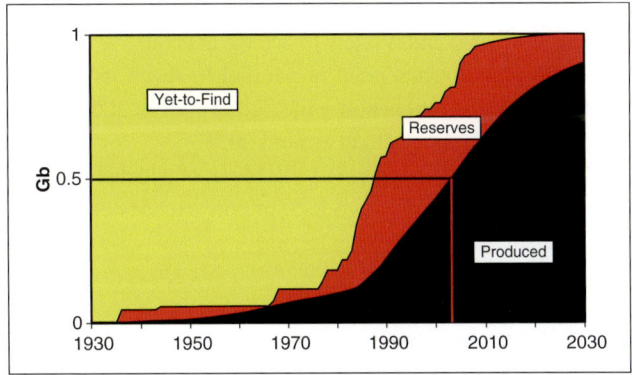

Fig. 22.2 Pakistan status of oil depletion

Essential Features

Pakistan, which was part of British India prior to independence in 1947, covers an area of 804,000 km² in the northwest of the Indian subcontinent. The fertile Indus valley, which runs through the country separates the arid hills of Baluchistan in the northwest from deserts to the southeast, while to the northeast develop the impressive mountain ranges of the Himalayan chain, rising to over 7,000 m. The country supports a population of 169 million, increasing at 2.4% a year. Some 80% of the population belong to the *Sunni* sect of Islam. The country is also home to a large number of refugees from neighbouring Afghanistan.

Geology and Prime Petroleum Systems

The Indus Valley overlies a Mesozoic and Tertiary basin flanking the Himalayan Range, which marks the collision between the Indian and Eurasian tectonic plates. The primary source-rock is in the Lower Cretaceous but additional minor Permian, Triassic and Paleocene sources have been identified. They have been locally buried deeply into the gas generating zone, being responsible for the Sui gas field. Oil and gas have been subject to mainly vertical migration filling adjoining and overlying reservoirs.

Exploration and Discovery

Exploration commenced in the nineteenth century with the first recorded exploration borehole being sunk in 1866, and continued at a modest level until after the Second World War when there was new interest, with 14 boreholes being sunk in 1958. The last decade has seen another surge of activity with drilling at times exceeding 20 boreholes a year.

Minor amounts of oil were found in early years, and the overall peak came in 1984 when 87 Mb were found. Rather more gas was discovered, with the peak of 12.6 Tcf coming in 1952. In total, almost 1 Gb of oil and 57 Tcf of gas have been found.

Production and Consumption

Oil production started around 1922, and has risen slowly to 66 kb/d in 2005, which represents the peak, with 54% of the assessed total having been produced. Oil production is currently running at 61 kb/d and is likely to decline at about 5% a year.

Gas production has risen in parallel to reach its present level of 1.5 Tcf/a, and is now expected to plateau at that level until terminal decline sets in around 2025.

Oil consumption has risen to 150 Mb/a, of which 128 Mb/a have to be imported. Gas production now exceeds consumption by a small amount.

The Oil Age in Perspective

The territory of modern Pakistan was settled in earliest times by migrants from Asia and the Middle East region, and was later incorporated into early Greek and Persian Empires, before being subject to several Arab invasions during the Middle Ages. A British trading company, known as the East Indian Company, gradually gained control of the Indian subcontinent, including what is now Pakistan, which was incorporated into the British Empire in 1858. An epoch of stability and prosperity followed although underlying conflicts between the Hindu, Muslim and Sikh communities simmered not far beneath the surface as moves towards independence built.

The British Empire was eclipsed by the Second World War, and its administration in India faced the difficulty of granting an independence that recognised the Hindu and Muslim communities. This was eventually achieved through partition into what became Pakistan and India. The dividing line was, perforce, arbitrary, and led to the wholesale migrations of communities who found themselves on the wrong side of the line. The division was accompanied by much bloodshed. Pakistan was in fact made up of two discrete widely separated territories, one of which later became Bangladesh. The conflict continues in the region of Kashmir, which is administered by India but has a largely Muslim population with ties to Pakistan.

Tensions between Pakistan and India marked the early years of independence, erupting from time to time into full-scale wars, and included the secession of what became Bangladesh. An epoch of military rule was followed by elections in 1972 which returned President Bhutto of the Pakistan People's Party. It had a broadly socialist platform, which included the nationalisation of industries. He was deposed in 1977 by General Huq and subsequently executed. The new government moved towards the introduction of Islamic law and administration, but ended with the death of General Huq in a plane crash, when Benazir Bhutto, the daughter of the earlier President was elected to power, moving the country into a more Western-influenced regime, even supporting the United States in the First Gulf War. She was in turn deposed in 1999 by a military coup which brought General Pervez Musharraf to power. He was educated in St. Patrick's High School in Karachi, a Christian establishment, and has a son who lives in California investing the new found family wealth. He has been the object of several failed assassination attempts. Benazir Bhutto returned from a self-imposed exile to campaign in forthcoming elections, but was assassinated in the last days of 2007.

Pakistan has played an important role following the events of the 11th September 2001, and the subsequent US invasion of neighbouring Afghanistan. As a result it has been the recipient of greatly increased financial aid from the United States, but evidently has a somewhat ambivalent position in relation to the conflict, as many of its citizens have sympathy with the Afghans. Another issue is its nuclear weapon capacity which it built up to match developments in India.

It is evident that the Government of Pakistan has had many difficulties since independence, facing internal conflicts of various kinds, some related to deep-seated religious and tribal dispositions. It has limited energy resources, and a population far in excess of the carrying capacity of the land. It would seem to follow that it faces some radical changes during the Second Half of the Age of Oil. It also balances on the edge of the current conflict between the United States and the peoples of Afghanistan to the north and Iran to the west. This conflict could well develop further, especially if there is another change of government back to one more representative of the ideals of the country's Islamic population. In short, the Second Half of the Age of Oil threatens to be a difficult time for Pakistan.

22 Pakistan

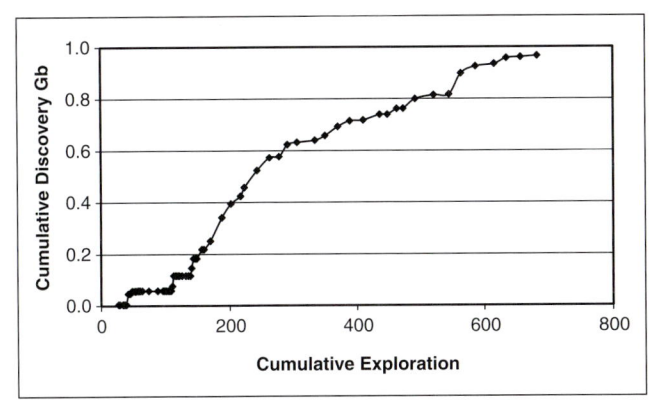

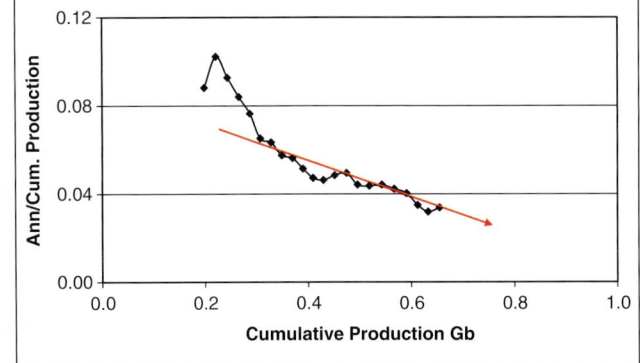

Fig. 22.3 Pakistan discovery trend

Fig. 22.4 Pakistan derivative logistic

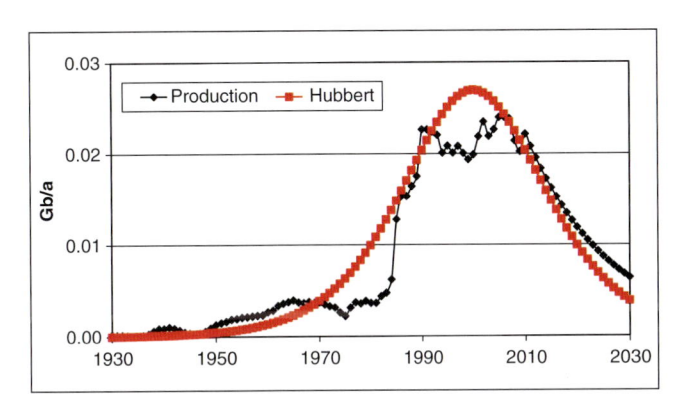

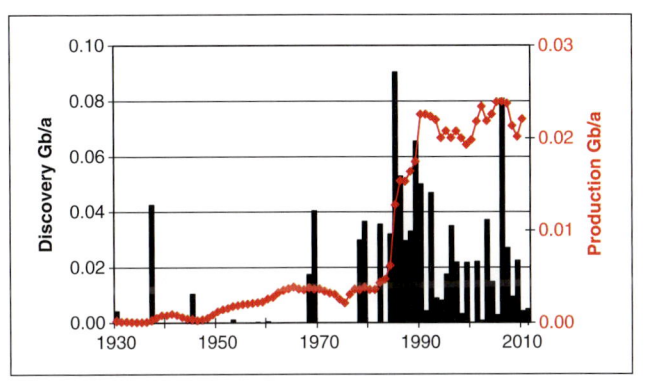

Fig. 22.5 Pakistan production: actual and theoretical

Fig. 22.6 Pakistan discovery and production

Papua-New Guinea

Table 23.1 Papua-New Guinea regional totals (data through 2010)

Production						Peak Dates			Area	
Amount			Rate				Oil	Gas	'000 km²	
	Gb	Tcf	Date	Mb/a	Gcf/a	Discovery	1986	1987	Onshore	Offshore
PAST	0.46	0.1	2000	26	4	Production	1993	2030+	460	1,200
FUTURE	0.54	20	2005	15	4	Exploration	1990		Population	
Known	0.51	18	2010	11	4	Consumption	Mb/a	Gcf/a	1900	0.8
Yet-to-Find	0.03	1.99	2020	9	300	2010	12	4	2010	6.8
DISCOVERED	0.97	18	2030	8	300		b/a	kcf/a	Growth	9.9
TOTAL	1.00	20	Trade	−1	0	Per capita	1.8	0.6	Density	15

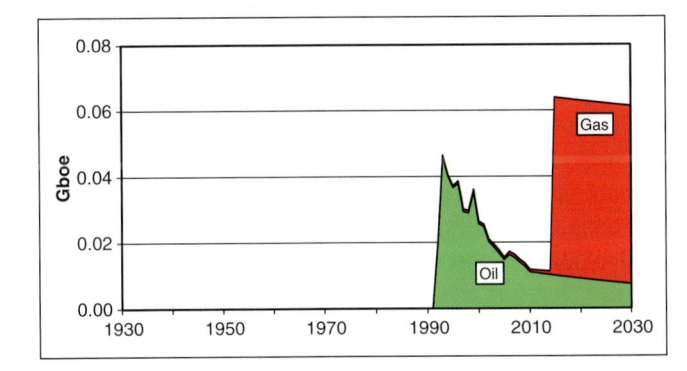

Fig. 23.1 Papua-New Guinea oil and gas production 1930 to 2030

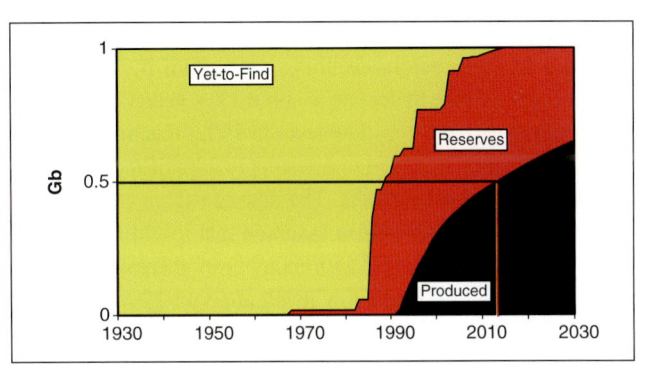

Fig. 23.2 Papua-New Guinea status of oil depletion

Essential Features

Papua-New Guinea comprises the eastern end of a large island and an adjoining archipelago lying to the north of Australia. The northern part of the island comprises New Guinea, while the southern part is known as Papua. The western end of the island is the Indonesian territory of Irian Jaya.

New Guinea and the archipelago were formerly German colonies. A mountainous interior, locally capped by active volcanoes rising to 4,500 m, is flanked by lowlands covered by tropical rain forest. An almost impenetrable terrain, known as pinnacle limestone, covers much of lowland Papua, being made up of deep clefts and gullies separating residual limestone pinnacles.

The population amounts to almost seven million, being largely made up of extremely primitive Melanesian tribal communities, leading lives little changed from the Stone Age.

Geology and Prime Petroleum Systems

A major transcurrent fault runs down the spine of the island, being locally associated with volcanic activity and mineralisation. Thick sequences of Mesozoic and Tertiary sediments fill the adjoining basins, which are flanked by important thrust-belts. Cretaceous hydrocarbon source-rocks locally crop out in the interior highlands where they give rise to seepages, but are progressively more deeply buried in the Papuan basin and its offshore extensions into the Gulf of Papua, passing into the gas-generating zone. The New Guinea Basin on the northern side of the island is much less prospective, due, in part, to the high volcanic content of the sediments. The archipelago contains some rich copper deposits.

Exploration and Discovery

Oil exploration commenced in 1923 when BP and a local company, by the name of Oil Search, took an interest. One or two boreholes were drilled a year, rising to a peak in 1990, when fifteen were sunk, but exploration has dwindled since. Offshore exploration commenced in 1967 but has remained at a very low level.

A total of almost 1 Gb of oil and 18 Tcf of gas have been discovered, with the peaks of discovery coming in respectively in 1986 and 1987.

Production and Consumption

Oil production commenced in 1991, and rose to a peak of 126 kb/d in 1993 before declining to 30 kb/d by 2010. It is expected to continue to decline at about 2% a year. Gas production commenced at a very low level in 1992, reaching 4 Gcf/a by 2010. It is now expected to rise steeply with the construction of a new pipeline to Australia. The date of such a construction has not been set, but it is here assumed that it will be in place by 2015 delivering 300 Gcf/a for many years thereafter.

Oil consumption is at a low level, at about 12 Mb/a, making the country a modest importer.

The Oil Age in Perspective

The territory was settled at a very early date by immigrants from southeast Asia, who have lived an almost unchanged lifestyle in the highlands ever since. Various European explorers visited the territory in the Middle Ages but did not settle until the Germans took the northern part of the island in 1884 in their failed quest to match the empires of France and Britain, naming it Kaiser-Wilhelms-Land and the Bismark Archipelago.

Meanwhile, the southern part of the island was taken by the Australian State of Queensland in 1883, in an action repudiated by Britain which moved to absorb the territory into the British Empire, although passing responsibility to the Dominion of Australia in 1902. Australian troops conquered the German colonies in the First World War, bringing them into the Australian sphere of influence.

Japanese troops invaded during the Second World War, but were eventually repulsed, leading to the unification of the territories after the war. Independence was granted in 1975, although the Queen of England remains as the constitutional Head of State. The country, which is substantially tribal in nature, is ill-suited to conventional democratic government, although tribal leaders from the different regions have formed a legislature. Tribal wars have characterised recent history, and such conflict is likely to continue. The native communities remain primitive and colourful with feathered head-dresses and grass skirts, but there has also been a degree of European settlement, mainly from Australia, with the development of coffee and other estates, especially in the highlands.

Bougainville, the easternmost island of the archipelago, has rich deposits of gold and copper, which encouraged a secessionist movement in the 1970s. The conflict, which claimed some 20,000 lives, was resolved in 2000, when the territory became an autonomous province.

Being a very primitive territory, Papua-New Guinea is well placed to face the Second Half of the Age of Oil with little change. The various tribal communities will no doubt retain their identity and successfully fight for survival on the lands they control.

23 Papua-New Guinea

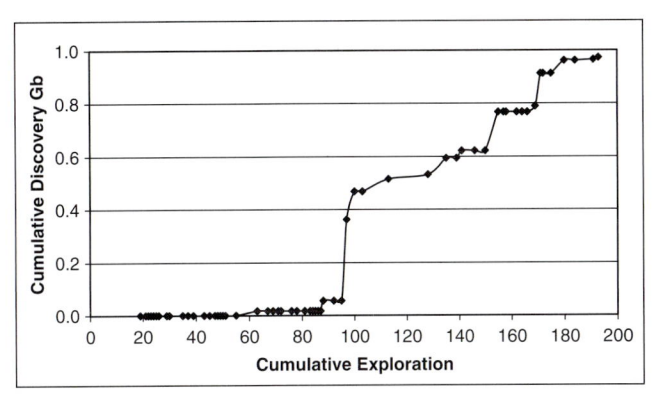

Fig. 23.3 Papua-New Guinea discovery trend

Fig. 23.4 Papua-New Guinea derivative logistic

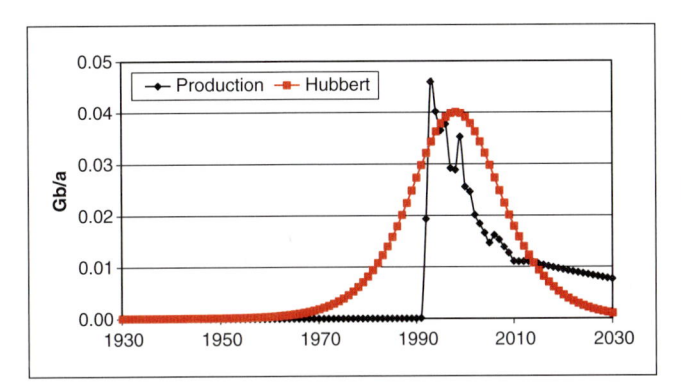

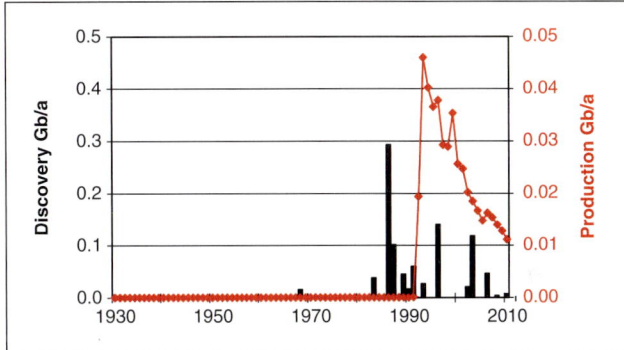

Fig. 23.5 Papua-New Guinea production: actual and theoretical

Fig. 23.6 Papua-New Guinea discovery and production

Thailand

Table 24.1 Thailand regional totals (data through 2010)

Production						Peak Dates			Area	
Amount			Rate				Oil	Gas	'000 km²	
	Gb	Tcf	Date	Mb/a	Gcf/a	Discovery	1981	1973	Onshore	Offshore
PAST	1.0	16	2000	40	713	Production	2010	2015	520	270
FUTURE	1.2	14	2005	67	925	Exploration	1983		*Population*	
Known	1.0	14	2010	88	1,364	*Consumption*	Mb/a	Gcf/a	1900	7
Yet-to-Find	0.3	0.7	2020	44	589	2010	361	1592	2010	68.1
DISCOVERED	2.0	29	2030	22	180		b/a	kcf/a	Growth	9.7
TOTAL	2.3	30	*Trade*	−272	−228	Per capita	5.3	23	Density	132

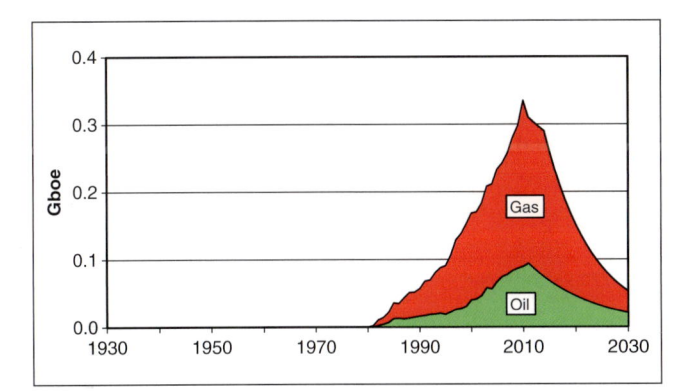

Fig. 24.1 Thailand oil and gas production 1930 to 2030

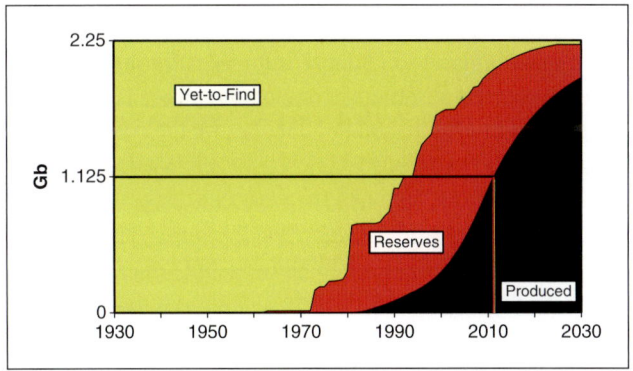

Fig. 24.2 Thailand status of oil depletion

Essential Features

Thailand, which covers an area of 520,000 km², is made up of three physiographic elements: mountains rising to 1,600 m in the north give way in turn to the Khorat Plateau and a fertile river basin that drains into the Gulf of Thailand. A narrow corridor leads south connecting additional territory on the Malay Peninsula to the west of the Gulf of Thailand. The country is bordered by Myanmar (formally Burma) to the west and by Laos and Cambodia to the east. It is primarily an agricultural country, with a population of 68 million, although industry has been developed in recent years. Its economy is also supported by tin mining and the production of gem stones, supplemented by offshore oil and gas revenue.

Geology and Prime Petroleum Systems

Most of onshore Thailand is built of granites and strongly deformed sediments which are substantially non-prospective. The Gulf of Thailand, by contrast, overlies a Mesozoic and Tertiary basin, which extends over the adjoining delta and has yielded a number of modest oil and gas discoveries.

Exploration and Discovery

The first exploration borehole was sunk in 1953, but exploration then lapsed until the 1970s. The onshore peak was in 1976 when 15 boreholes were sunk, followed by the offshore peak seven years later when 24 were drilled. It resulted in a number of modest oil and gas discoveries: oil discovery peaked in 1981 when 325 Mb was found and gas discovery peaked in 1973 when 8.25 Tcf was found.

A total of 2 Mb of oil and 29 Tcf of gas have been found.

Production and Consumption

Oil production commenced in 1958 and has since risen to its present level of 242 kb/d. It passes the midpoint of depletion in 2011, when terminal decline at about 7% a year sets in. Gas production has steadily risen in parallel to 1,364 Gcf/a in 2010, and is expected to remain at approximately this level until 2014 before terminal decline sets in.

Oil consumption stands at 361 Mb/a, making the country a net importer of almost 80% of its needs. Gas consumption also exceeds current production.

The Oil Age in Perspective

The country, previously known as Siam, has lain at the junction of Indian and Chinese influences through its early history, receiving settlers from both places. A Buddhist Kingdom was established in 1238, before the country came into its own in 1782 with a new administration based in Bangkok, the capital. It had a strong government that managed to play off the rival claims of the British and French empires, retaining independence, save for the loss of what became northern Malaya in 1909.

The country was a hereditary monarchy although the absolute powers of the king were surrendered in 1932, after which the country faced alternating military dictatorships and generally weak democratic administrations.

It was occupied by Japanese troops during the Second World War, after which it came under American influence, which led to generally more stable, democratic governments and economic development.

A military coup in 2006 brought in a new authoritarian government, which nevertheless aims to return to democracy in due course, having cleaned up widespread corruption and growing disunity amongst the diverse communities making up the country. The previous Prime Minister, Mr Thaksin, fled to England, where he offered to buy a football club for £81 million. The country has since faced much political turmoil and in 2011 the problems were compounded by serious and exceptional floods.

The country, which is already a net oil importer, faces a difficult future during the Second Half of the Age of Oil, having an agricultural capacity barely able to support its population.

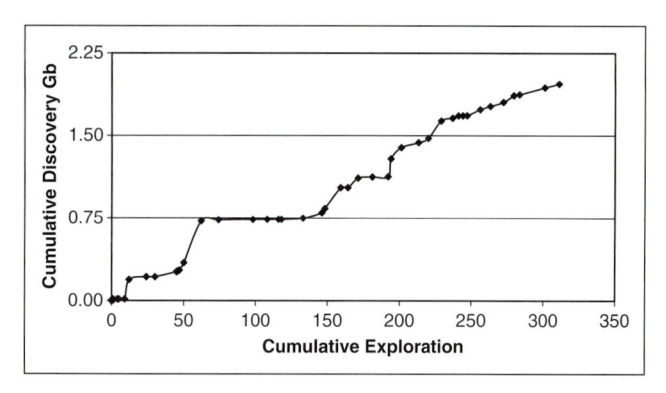

Fig. 24.3 Thailand discovery trend

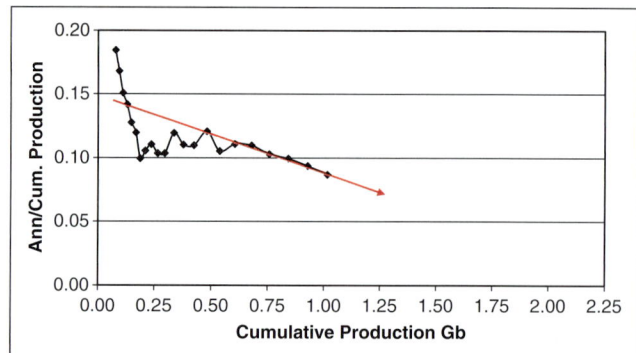

Fig. 24.4 Thailand derivative logistic

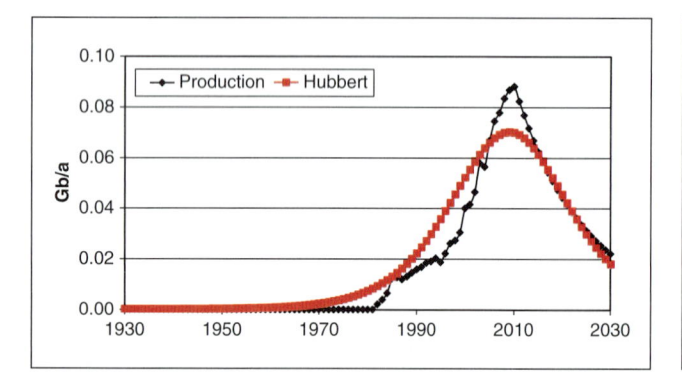

Fig. 24.5 Thailand production: actual and theoretical

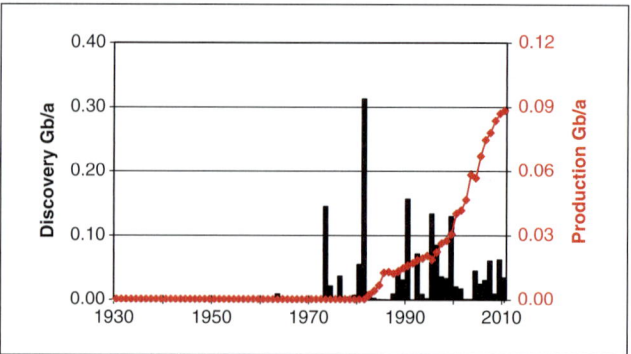

Fig. 24.6 Thailand discovery and production

Vietnam

Table 25.1 Vietnam regional totals (data through 2010)

	Production					Peak Dates			Area	
	Amount		Rate				Oil	Gas	'000 km²	
	Gb	Tcf	Date	Mb/a	Gcf/a	Discovery	1975	1995	Onshore	Offshore
PAST	2.1	2.3	2000	130	55	Production	2003	2030	330	360
FUTURE	3.4	23	2005	126	170	Exploration	1996		Population	
Known	3.1	22	2010	115	330	Consumption	Mb/a	Gcf/a	1900	10
Yet-to-Find	0.3	1.1	2020	113	538	2010	117	290	2010	89
DISCOVERED	5.2	24	2030	67	600		b/a	kcf/a	Growth	8.9
TOTAL	5.5	25	Trade	−1	+40	Per capita	1	3	Density	267

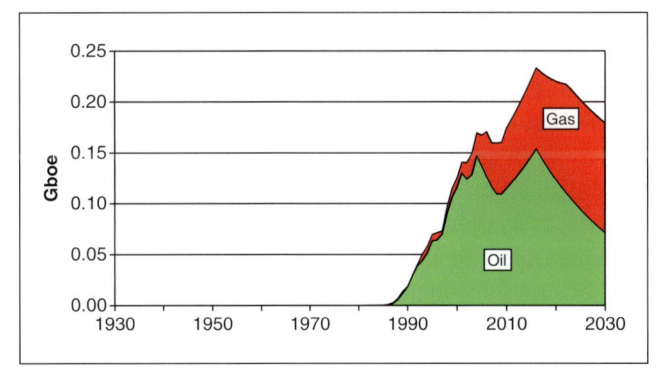

Fig. 25.1 Vietnam oil and gas production 1930 to 2030

Fig. 25.2 Vietnam status of oil depletion

Essential Features

Vietnam forms the eastern margin of Indo-China, covering some 330,000 km². Its continental shelf extends into both the South China Sea and the southern part of the Gulf of Thailand. The land is dominated by two rivers systems: the Red River in the north; and the Mekong in the south, which are separated by a coastal strip, some 50–150 km wide. The Annamese Mountains run parallel with the coast, rising to 3,100 m, and are flanked by high plateaux. The people, now numbering 89 million, are of mixed Indo-Chinese origins with a long history.

Geology and Prime Petroleum Systems

The country is broken up by a series of easterly-trending lineaments, which probably reflect ancient transcurrent fault lines. The Red River follows one such feature giving a structural depression, filled with deltaic sediments, having mainly gas prospects. Another lineament, off the Mekong delta in the south, extends eastwards to the disputed Spratley Islands. It has given rise to an uplifted area, known as the Con Son Swell, which is flanked by structural depressions. The northern one comprises a 4,000 m deep trough filled with Mio-Pliocene sediments, which have yielded several important oilfields. The oil, which is sourced in the basal Tertiary sequence, has migrated into both overlying reservoirs and, locally, into the fractured and weathered volcanic rocks below. The latter occurrence has been taken by the adherents to an abiotic origin of oil as evidence for their flawed theory, so beloved by the flat-earth economist.

Exploration and Discovery

Oil exploration commenced in the early 1970s when the small Canadian oil company, Sunningdale, secured rights from the Saigon Government whose oil policy at the time was administered by a Jesuit priest: he being perceived to know about all matters natural. Elf and Mobil later joined the search, but the rights were annulled with the formation of a national company, PetroVietnam, backed by Russian technical aid. BP and other foreign companies later responded to invitations to explore the shelf following a change of policy.

Some 241 exploration boreholes have been drilled, finding a total of 5.2 Gb of oil and some 24 Tcf of gas. Exploration drilling peaked in 1996 when 22 boreholes were sunk, but the annual rate has now dwindled to less than half that number. The principal discoveries were Bach Ho (1975) with 500 Mb; and Dai Hung (1988) and Rang Dong (1994) with about half that amount each. It does not appear to be a very prospective area, and future exploration is unlikely to yield more than about 350 Mb, mainly in small fields. Much misplaced media interest is directed at the disputed Spratley Islands, which are commonly depicted as being rich in oil when they are more likely to be no more than coral reefs resting on a crystalline uplift associated with the Con Son Swell.

Production and Consumption

Oil production commenced in 1986 and rose to a peak of 403 kb/d in 2003, before declining to 318 kb/d in 2010. It is expected to rise to another peak in 2015 before terminal decline sets in.

The prospects for gas are somewhat greater, especially in the Red River Basin and adjoining waters in the north. Gas production commenced in 1982 and is expected to rise to a plateau at about 600 Gcf/a, starting in 2023. Oil consumption stands at 117 Mb/a, meaning that the country has become a modest importer.

The Oil Age in Perspective

The territory formed part of the Chinese Empire during the first millennium before gaining its independence. Conflict erupted periodically in later years between the inhabitants of the two river systems, leading to the development of separate kingdoms: one based on Hanoi in the north, and the other on Saigon in the south. The French appeared on the scene in 1788 assisting the southern kingdom to resist attack. They returned in 1857 mounting a full-blown military invasion which led to the eventual incorporation of the territory into the French Empire. It became a rather extreme society of privileged landlords, middle men and moneylenders controlling a landless peasantry. The French language and Catholicism were introduced.

Movements towards independence took root during the early years of the twentieth century, being led by Ho Chi Minh, a sailor who settled in Paris in 1917 before joining the international Communist movement. He returned home to form the Indo-Chinese Communist Party in 1930. Various peasant revolts, in which landlords and officials were killed, were successfully put down, yet the movement grew in strength until the Second World War. At this point, the new Vichy Government of France under German occupation authorised the entry of 30,000 Japanese troops. When Japan surrendered in 1945, Ho Chi Minh emerged as a forceful Resistance leader, proclaiming the Democratic Republic of Vietnam. The French however were resolved to recover their colony, and did so in 1946 with British help. At first, they tried to reach an accommodation with the Communists, but that failed, leading to a guerrilla war ending in the defeat of the French at Dien Bien Phu in 1954, despite military aid from the United States. The country was then divided under the Geneva Accords into the Communist Democratic Republic in the north and an anti-communist regime in the south, led by Ngo Dinh Diem. The latter, a Catholic, was supported by the US financial aid, and later declined to participate in unified elections across both territories as provided by the Geneva Accords. Renewed guerrilla attacks by what became known as the Viet Cong followed. Diem was assassinated in a coup that led to greater US military involvement, which erupted into full-scale war in 1965. As many as 47,000 American soldiers lost their lives in a particularly vicious campaign, marked by the butchering of women and children, the shelling of villages and the use of napalm and a gruesome defoliant (Agent Orange). However, it was without notable success in breaking the resolve of the Viet Cong. Finally, President Nixon was forced to sue for peace in 1973, and two years later, Communist forces entered Saigon in triumph. In 1976, the Socialist Republic of Vietnam was finally proclaimed with a capital in Hanoi. But life has been difficult under the desperate conditions of this war-torn country. Doctrinaire Communism has since been gradually replaced by a more pragmatic system of government, albeit one still characterised by a high level of State control.

The country will probably continue in its present path through the Second Half of the Age of Oil, eventually finding a sustainable level of population.

25 Vietnam

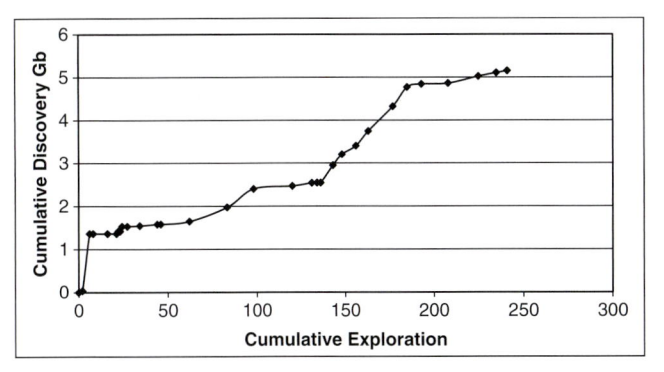

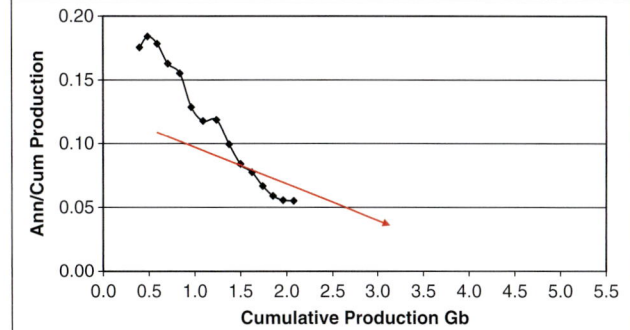

Fig. 25.3 Vietnam discovery trend

Fig. 25.4 Vietnam derivative logistic

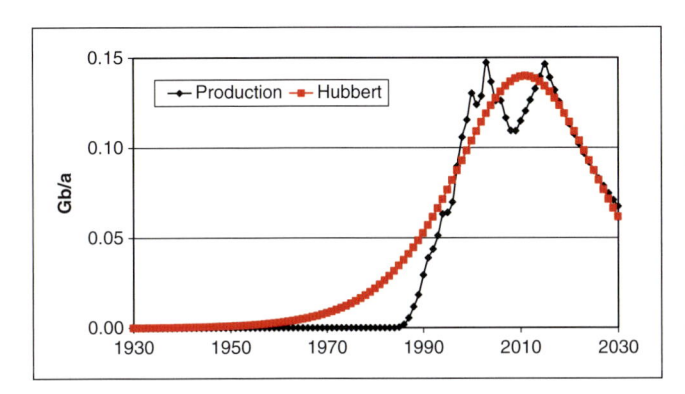

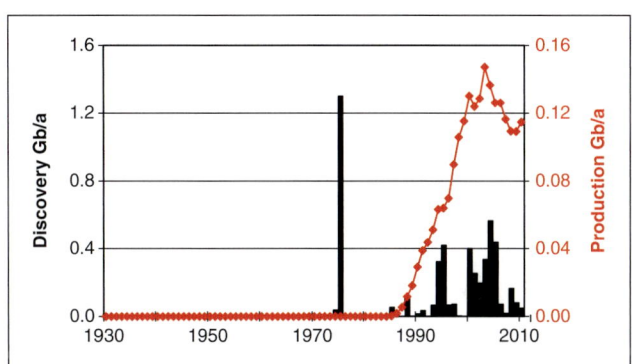

Fig. 25.5 Vietnam production: actual and theoretical

Fig. 25.6 Vietnam discovery and production

Asia Pacific Region

Table 26.1 Asia Pacific regional totals (data through 2010)

Production					Peak Dates			Area	
Amount			Rate			Oil	Gas	'000 km²	
	Gb	Tcf	Date	Mb/a	Gcf/a	Discovery	1974	1973	Onshore	Offshore
PAST	54	237	2000	1559	8759	Production	2000	2021	15100	6400
FUTURE	29	573	2005	1326	11089	Exploration	1990		Population	
Known	27	482	2010	1267	13388	*Consumption*	Mb/a	Gcf/a	1900	313
Yet-to-Find	2.8	90	2020	849	16015	2010	6336	20466	2010	1825
DISCOVERED	80	720	2030	542	14191		b/a	kcf/a	Growth	5.8
TOTAL	83	810	*Trade*	–5038	–2100	Per capita	2.6	8.5	Density	121

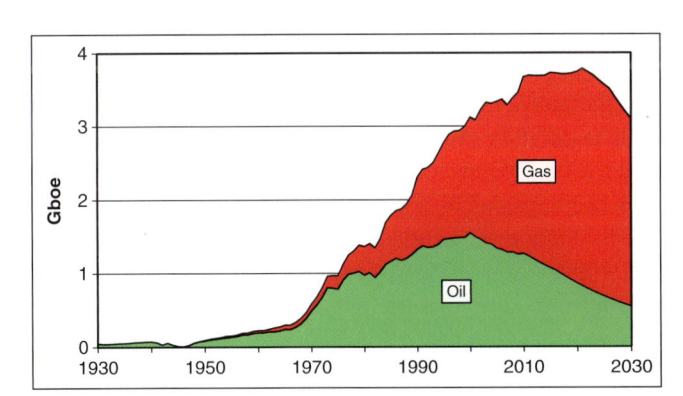

*Note: Data refer to main producing countries only, **except for consumption and trade data** which include minor countries.*

Fig. 26.1 Asia Pacific oil and gas production 1930 to 2030

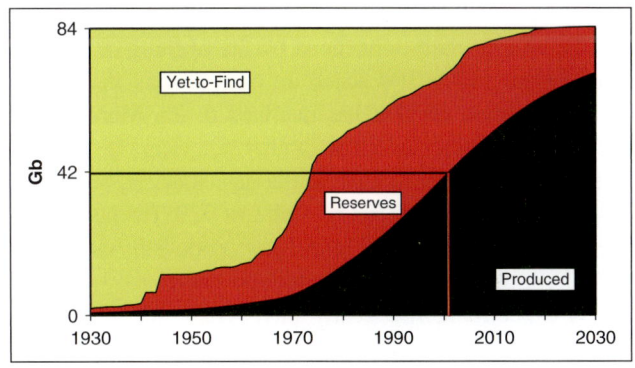

Fig. 26.2 Asia Pacific status of oil depletion

The Region, as herein defined, comprises a number of countries on the southern margin of Asia from Pakistan in the west to Vietnam in the east, as well as Australasia and the Pacific Islands. It excludes China, North Korea and the former Soviet Union, on account of their Communist background. It covers a diverse terrain of 19 million km², made up of two ancient shield areas, comprising the margins of the Siberian Shield to the north and that of Australia to the south, which are separated by mobile island arcs, largely following tectonic plate boundaries. Various, mainly Tertiary, sedimentary basins flank the shields and mobile belts, and some of them, especially offshore, are prospective for oil and gas.

The tropical rain forests cover much of the area apart from Australasia, which enjoys a more temperate climate, with extensive deserts in the interior.

In addition to the eight main oil producers, whose oil and gas position is summarised in the table above, are a large number of other countries, together with some additional small Pacific Islands, as listed below. Those indicated (*) have minor amounts of oil and/or gas which will be evaluated together with other minor producers in Chapter 11 as the amounts are too small to model meaningfully. Some are, however, significant consumers of oil and gas.

C.J. Campbell, *Campbell's Atlas of Oil and Gas Depletion*,
DOI 10.1007/978-1-4614-3576-1_26, © Colin J. Campbell and Alexander Wöstmann 2013

Table 26.2 Asia-Pacific: Minor countries in the Region (*with minor oil/gas production)

Afghanistan	*N.Zealand
*Bangladesh	*Philippines
Bhutan	Singapore
Cambodia	S. Korea
Laos	Sri Lanka
Mongolia	Taiwan
Nepal	Pacific Isles

The Region supports a population of 1.8 billion people. Their density varies widely, but averages at about 120/km^2, having a life expectancy of about 66 years and a fertility rate of 2.5 children per woman.

The total amount of *Regular Conventional Oil* to be produced by the end of this Century is estimated to be a rounded 80 Gb, of which 54 Gb have been produced through 2010, leaving 29 Gb to be produced in the future, including an estimated 3 Gb from new discovery. Indonesia is the most significant oil and gas producer, followed by India, Malaysia and Australia.

Oil production had commenced by the birth of the last century and grew to an early Peak in 1940, before falling steeply during the Second World War. It resumed after the war, being especially influenced by the opening of the offshore in the 1970s. It reached an overall Peak in 2000 at 4.3 Mb/d, since when it has declined to 3.5 Mb/d and is expected to continue to fall at about 4 % a year.

It is very difficult to analyse the gas situation, but a preliminary estimate suggests that almost 720 Tcf have been found, of which 237 Tcf have been produced, with some 90 Tcf yet-to-find. Current production stands at 13.4 Tcf/a, and is expected to rise gently in the years ahead to peak around 2020 at almost 16 Tcf/a. The related LNG production from gas plants is expected to increase in the future from about 526 kb/d at present.

Oil consumption for the region has been growing consistently to 6,336 Mb/a today, giving a per capita consumption of about 2.6 barrels per year. The region is already a net importer of 5.038 Mb/a and the trend is upward. Gas consumption is running at 20.5 Tcf/a, with the balance presumably being converted to liquids, and exported.

The Oil Age in Perspective

The region has had a long history being the melting pot between several early civilisations, dominated by India and China, and it was also home to many different races of indigenous island people. It saw the arrival of Portuguese, British, Dutch, French and Arab traders in the late Middle Ages. They paved the way for the later empires dominated by Britain, the Netherlands and France, who cemented their positions in the nineteenth century. India, which then included Pakistan and Bangladesh, became the jewel in the crown of the British Empire, while the Netherlands held dominion through Indonesia. The French Empire was the lesser of the three, but established an important position in Indo-China, which it was reluctant to lose. Australia and New Zealand with their temperate climate was subject to immigration from Europe during the nineteenth and twentieth century, becoming members of the British Commonwealth. Many Chinese traders have also settled in the region over long periods of history.

The Second World War opened a new chapter in the region's history. Japan successfully occupied much of Indo-China and Indonesia, partially in an effort to secure a much needed supply of oil, but was eventually defeated. The postwar epoch saw the demise of the earlier empires as the United States became the world's premier economic power. But the burgeoning moves to independence were accompanied by much conflict as ancient tribal and religious tensions reappeared, and frontiers were redrawn.

In the ensuing Cold War, the United States sought to strengthen its position, and counter growing moves towards Communist government, which culminated in the Korean and Vietnam wars where, in the latter case, it faced eventual defeat in 1973 after years of struggle.

The Region has enjoyed substantial economic prosperity over the past few decades, and has also seen a sixfold increase in its population over the last Century, bringing the total to about 2.4 billion, of whom about half live in India. This has been accompanied by a great expansion of urban living.

The Region is already a substantial net importer of oil, and will face increasing difficulties in meeting its demand in the future. Natural Gas will help ameliorate the transition to the Second Half of the Oil Age, but in general the region can expect growing tensions as different territories and tribal groups seek more independence in their quest for survival in the postindustrial age.

Table 26.3 Asia-Pacific: Range of Reported Reserves and Scope for Surprise

	Standard Deviation Public Reserve Data		Surprise Factor	
	Oil	**Gas**	**Oil**	**Gas**
Australia	0.45	28.3	3	10
Brunei	0.01	0.83	1	1
India	1.78	0.8	5	6
Indonesia	0.23	6.05	4	4
Malaysia	0.85	2.32	5	6
Pakistan	0.01	0.97	3	3
Papua-NG	0.06	4.12	2	3
Thailand	0.02	0.72	5	6
Vietnam	2	7.36	6	7
				Scale 1–10

Confidence and Reliability Ranking

The table lists the standard deviation of the range of published data on reserves. A relative confidence ranking in the validity of the underlying assessment is also given: the lower the Surprise Factor, the less the chance for revision.

Australia

Australia is not well endowed and has been thoroughly explored. However the gas data seem unreliable and the potential may well be exaggerated here.

Brunei

Brunei is a small, well-explored territory, offering negligible surprises.

India

India is not a very prospective area, but there remains some scope for the discovery of new marginal offshore basins and some deepwater potential.

Indonesia

The prime areas of Indonesia have been very thoroughly explored but there remains a small chance of some freak occurrences in one or more of the lesser known islands, and perhaps Irian Jaya.

Malaysia

Malaysia's prospects are entirely offshore, which by now are well known, but some surprises and a certain new deepwater potential cannot be totally excluded.

Pakistan

Pakistan has limited potential and has been thoroughly explored but some surprises at depth in the thrust-belts of Baluchistan, especially for gas, cannot be discounted.

Papua-New Guinea

Papua-New Guinea has limited potential, but there is scope for surprises in the pinnacle-limestone of the Fly River area, where seismic surveys are difficult and give poor results. The future of gas development is dependent on an export pipeline to Australia or liquefaction, as the internal demand is small.

Thailand

Thailand's prospects are offshore and have now been thoroughly investigated, but the shelf is vast, meaning that some surprise discoveries, especially for gas, could yet be made.

Vietnam

Exploration in Vietnam has been somewhat delayed due to political factors. The southern shelf is now well known but there may be more scope, especially for gas in the Red River basin of the north.

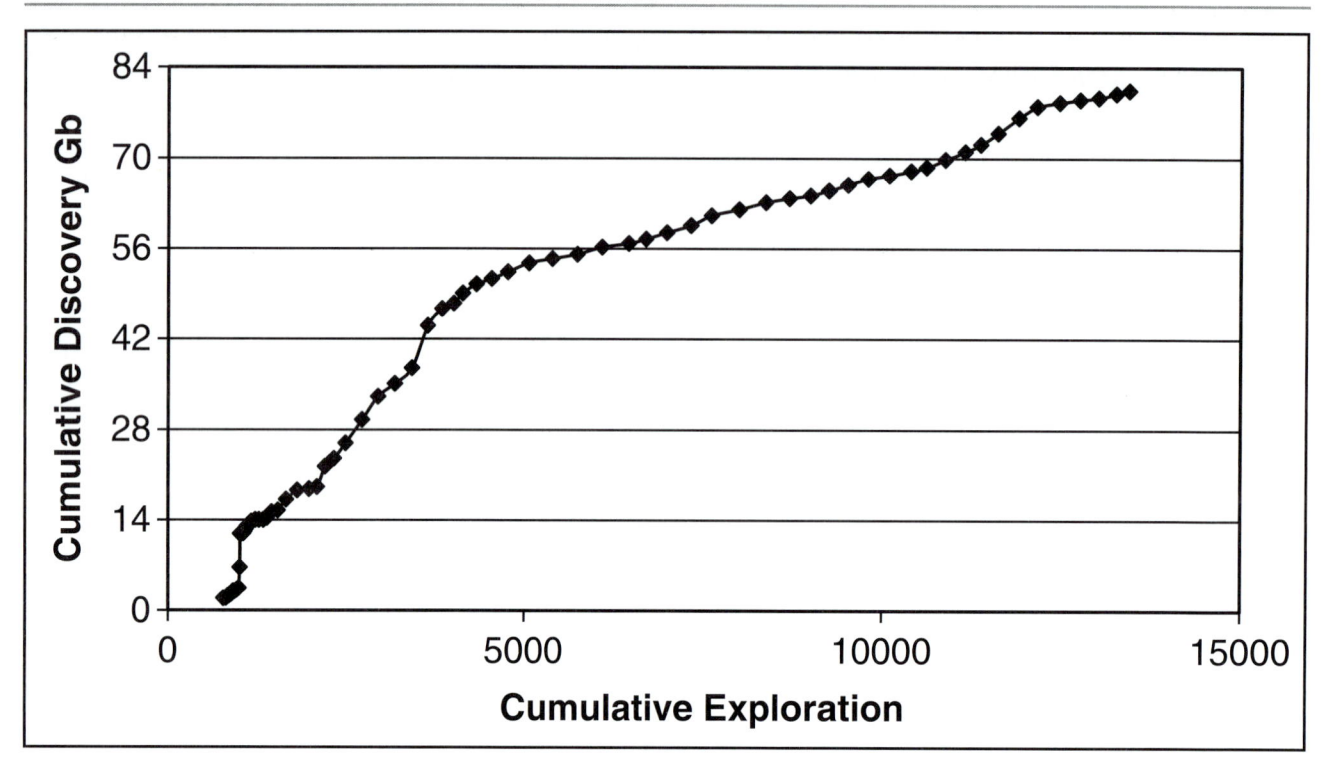

Fig. 26.3 Asia Pacific discovery trend

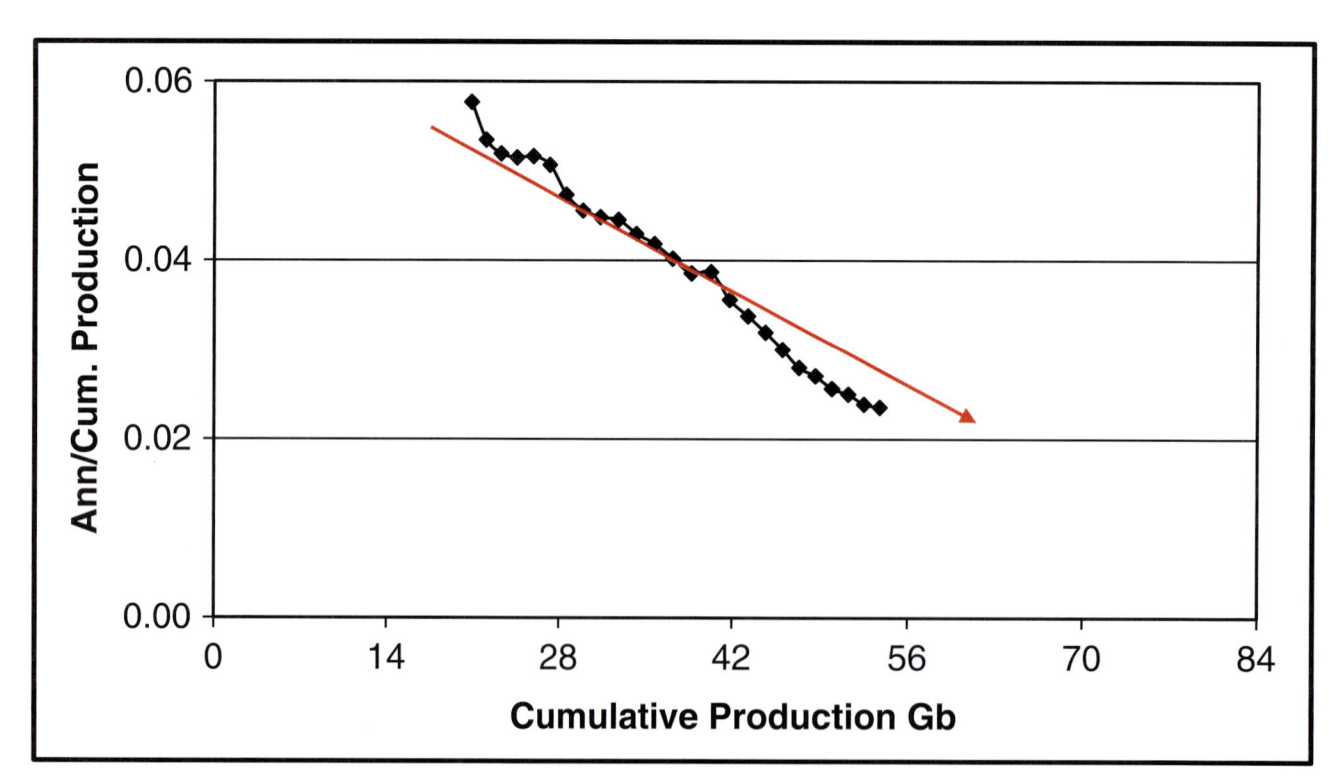

Fig. 26.4 Asia Pacific derivative logistic

26 Asia Pacific Region

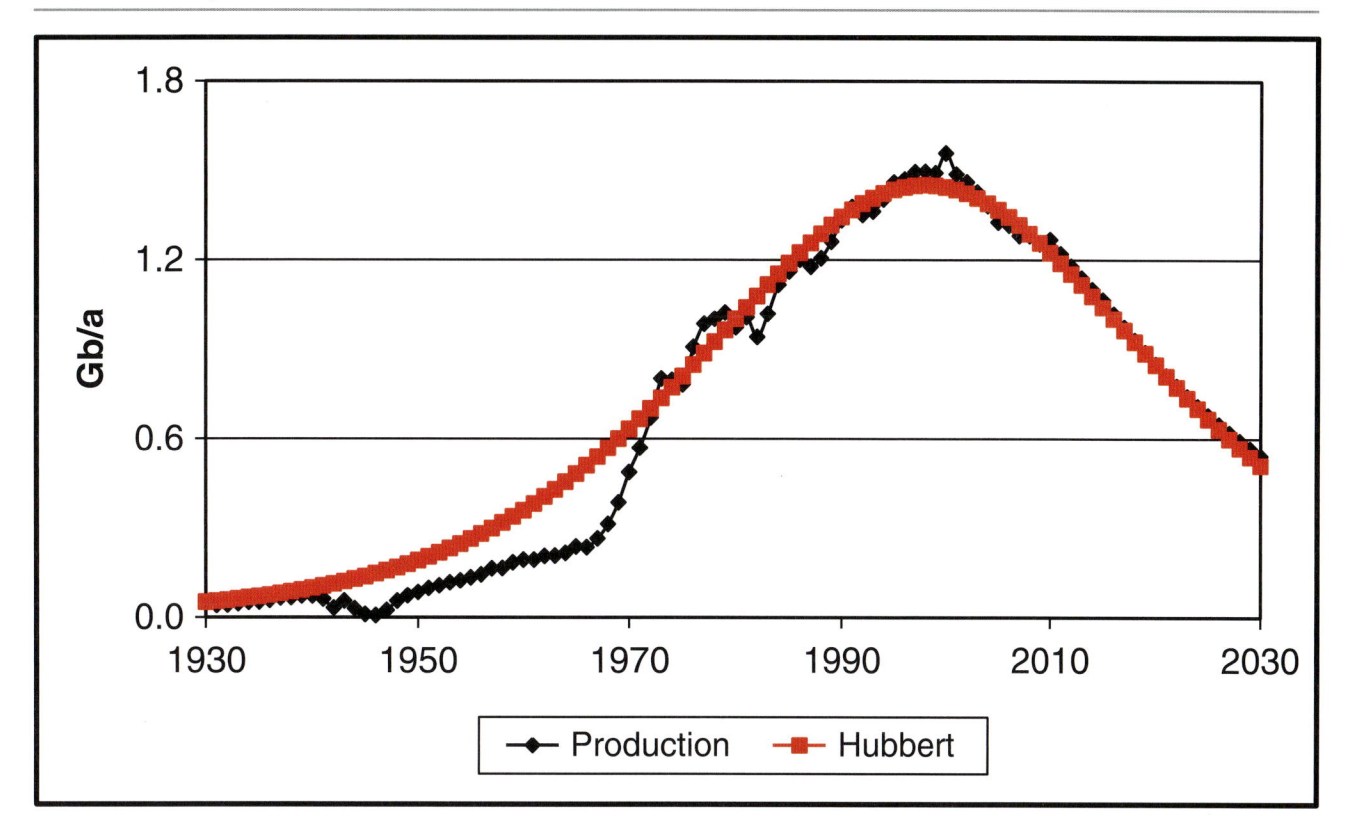

Fig. 26.5 Asia Pacific production: actual and theoretical

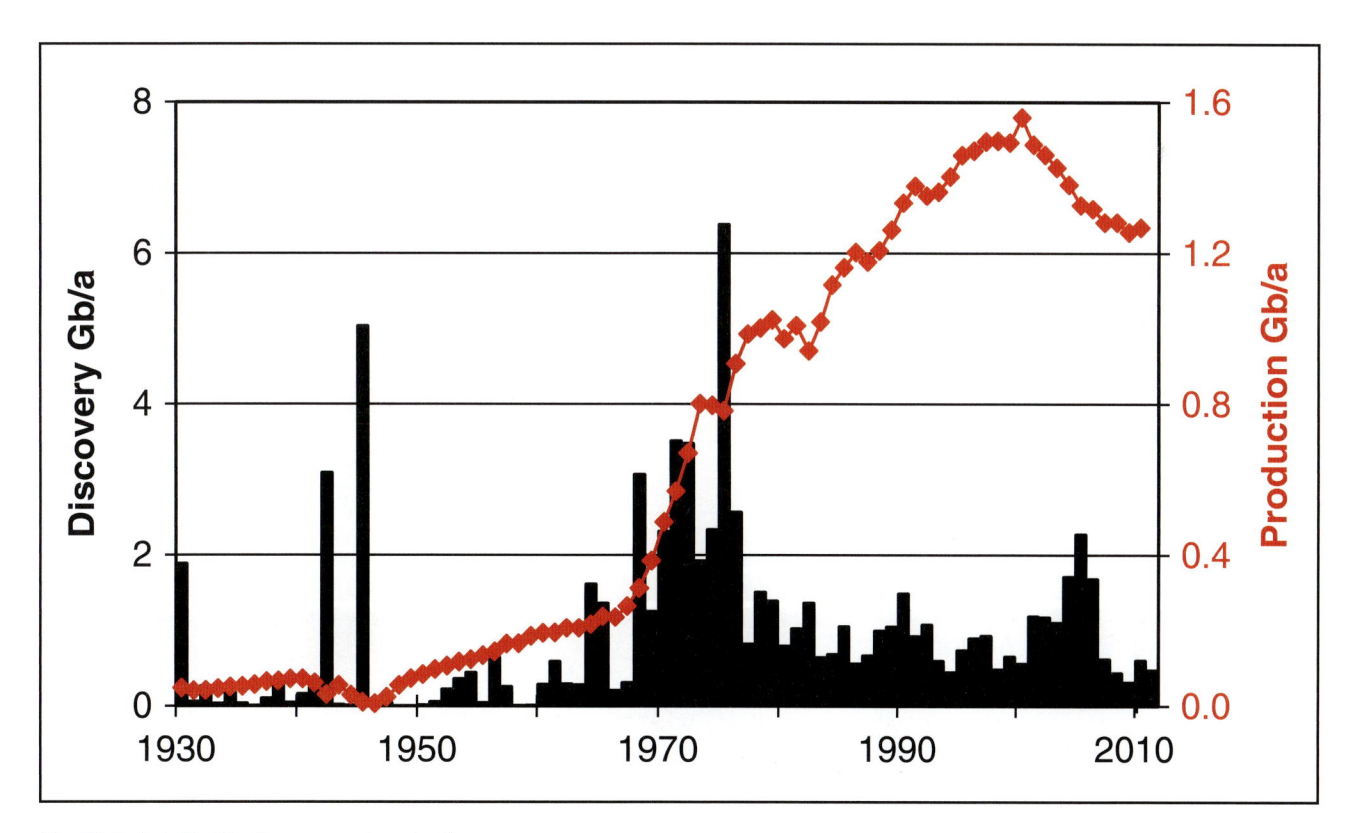

Fig. 26.6 Asia Pacific discovery and production

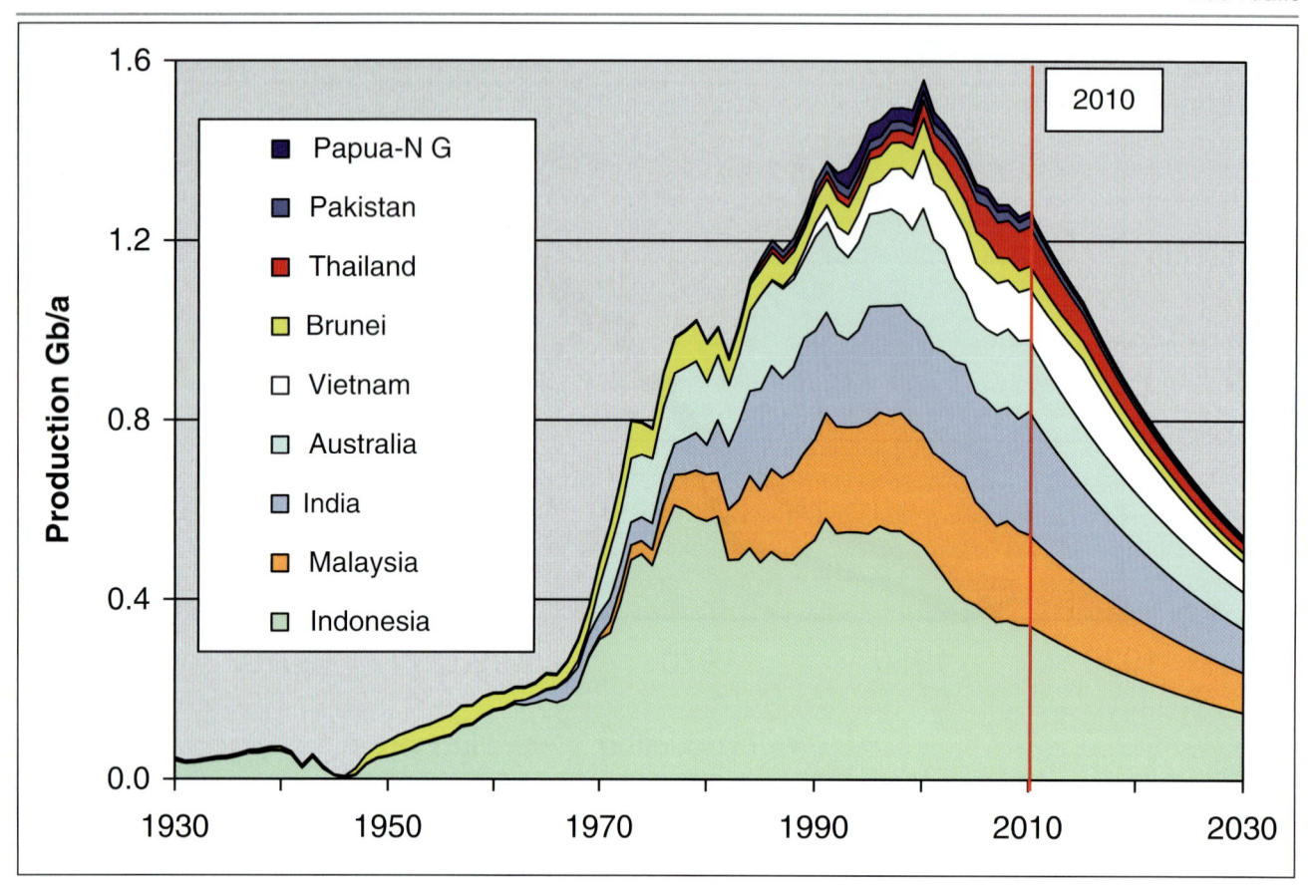

Fig. 26.7 Asia Pacific oil production

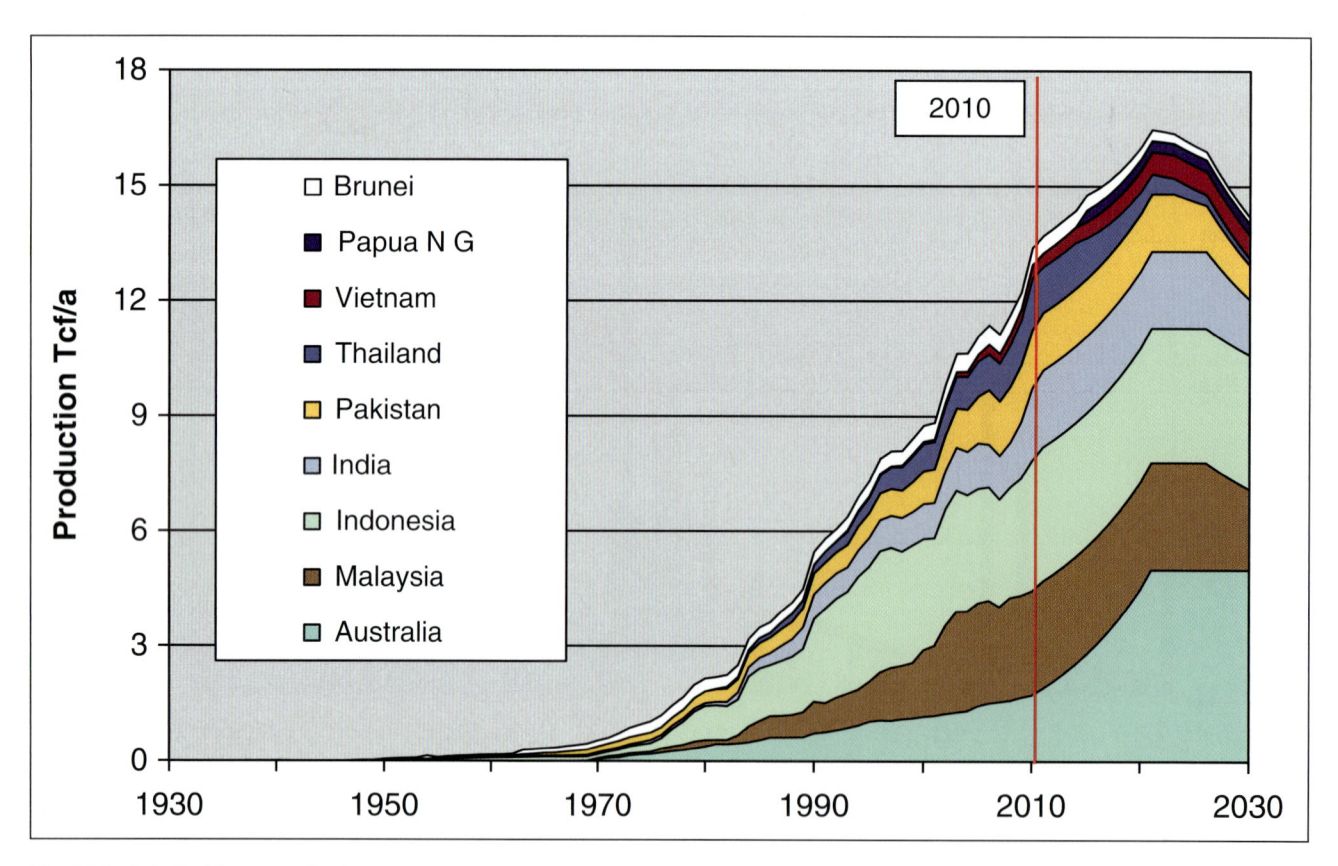

Fig. 26.8 Asia Pacific gas production

Table 26.4 Asia-Pacific: Oil Resource Base

		PRODUCTION TO 2100															
ASIA PACIFIC		Regular Conventional Oil															2010
		KNOWN FIELDS									Revised 05/10/2010						
		Present		Past			Reported Reserves				Future	Total	NEW	ALL	TOTAL	DEPLETION		PEAK	
		kb/d	Gb/a		5yr	Disc	Average	Deductions		%	Known	Found	FIELDS	FUTURE		Current	Mid-		
Country	Region	2010	2010	Gb	Trend	2010		Static	Other	Disc.	Gb	Gb	Gb	Gb	Gb	Rate	Point	Disc	Prod
Australia	G	436	0.16	7.19	0%	0.16	3.81	0.00	0.00	94%	4.09	11.28	0.72	4.81	12.0	3.2%	2003	1967	2000
Brunei	G	136	0.05	3.48	–6%	0.04	1.10	0.20	0.00	98%	0.92	4.40	0.10	1.02	4.5	4.6%	1990	1929	1979
India	G	752	0.27	7.88	2%	0.03	6.03	1.03	0.00	98%	4.86	12.74	0.26	5.12	13.0	5.1%	2005	1974	2010
Indonesia	G	943	0.34	23.86	–1%	0.11	4.14	0.69	0.00	97%	7.32	31.19	0.81	8.14	32.0	4.1%	1993	1944	1977
Malaysia	G	554	0.20	7.14	–2%	0.03	4.90	0.64	0.00	98%	4.62	11.76	0.24	4.86	12.0	4.0%	2004	1971	2004
Pakistan	G	61	0.02	0.65	–2%	0.00	0.32	0.00	0.00	97%	0.31	0.97	0.03	0.35	1.0	6.0%	2003	1984	2006
Papua-New Guinea	G	30	0.01	0.46	–6%	0.01	0.12	0.04	0.00	97%	0.51	0.97	0.03	0.54	1.0	2.0%	2013	1987	1993
Thailand	G	242	0.09	1.02	4%	0.04	0.43	0.00	0.00	89%	1.23	2.00	0.25	1.23	2.3	6.7%	2011	1981	2011
Vietnam	G	318	0.11	2.08	–2%	0.05	2.72	0.58	0.00	94%	3.08	5.16	0.34	3.42	5.5	3.2%	2016	1975	2015
ASIA PACIFIC	G	3472	1.27	53.76	–1%	0.48	23.47	3.17	0.00	0%	26.71	80.46	2.79	29.49	83.3	4.1%	2001	1974	2000

Table 26.5 Asia-Pacific: Gas Resource Base

		Production to 2100																
ASIA PACIFIC		Conventional Gas																2010
		KNOWN FIELDS										Revised 15/11/2011						
		Present			Past		Discovery		Reserves		FUTURE	TOTAL	NEW	ALL	TOTAL	DEPLETION		PEAKS	
		Tcf/a	FIP	Gboe/a		5yr	Tcf	%	Reported	Deduct Non-Con	FINDS	FOUND	FIELDS	FUTURE		Current	% Mid-		
Country	Region	2010	2010	2010	Tcf	Trend	2010	Disc.	Average		Tcf	Tcf	Tcf	Tcf	Tcf	Rate	Point	Expl Disc Prod	
Australia	G	1.73	0.00	0.31	32.38	0.03	8.95	83%	145.38	33.60	150.09	182.48	37.52	187.62	220	0.91%	15% 2029	1985 1971 2030	
Brunei	G	0.44	0.00	0.08	15.96	–0.03	0.18	98%	16.13	0.00	8.59	24.55	0.45	9.04	25	4.64%	64% 2002	1975 1963 2006	
India	G	1.88	0.00	0.34	24.13	0.14	3.60	86%	47.76	0.80	44.69	68.83	11.17	55.87	80	3.26%	30% 2003	1991 1976 2023	
Indonesia	G	3.41	0.00	0.61	77.98	0.03	2.65	90%	131.52	7.00	120.71	198.70	21.30	142.02	220	2.34%	35% 2021	1983 1973 2020	
Malaysia	G	2.72	0.00	0.49	42.50	0.00	0.98	93%	105.53	1.30	47.25	116.75	8.25	82.50	125	3.19%	34% 2017	1970 1970 2017	
Pakistan	G	1.51	0.00	0.27	26.19	0.01	1.09	88%	37.86	0.00	31.04	57.24	7.76	38.81	65	3.75%	40% 2014	2003 1952 2014	
Papua-New Guinea	G	0.00	0.00	0.00	0.08	–0.04	0.80	90%	20.85	0.00	17.93	18.01	1.99	19.92	20	0.02%	0% 2030+	1990 1987 2030+	
Thailand	G	1.36	0.00	0.25	15.68	0.09	0.10	98%	14.55	0.00	13.60	29.28	0.72	14.32	30	8.70%	52% 2008	1983 1973 2010	
Vietnam	G	0.33	0.00	0.06	2.29	0.07	0.16	95%	14.2	0.00	21.58	23.86	1.14	22.71	25	1.43%	9% 2030	1996 1995 2030	
ASIA PASIFIC	G	13.39	13.39	2.41	2377.20	0.48	18.51	89%	533.71	50.70	482.49	719.69	90.31	572.80	810	2.28%	29% 2023	1990 1973 2021	

Table 26.6 Asia-Pacific: Oil Production Summary

Regular Conventional Oil Production					
Mb/d	2000	2005	2010	2020	2030
Australia	0.72	0.45	0.44	0.31	0.23
Brunei	0.19	0.19	0.14	0.08	0.05
India	0.65	0.66	0.75	0.45	0.26
Indonesia	1.43	1.07	0.94	0.62	0.41
Malaysia	0.69	0.63	0.55	0.37	0.25
Pakistan	0.05	0.07	0.06	0.03	0.02
Papua-New Guinea	0.07	0.04	0.03	0.03	0.02
Thailand	0.11	0.18	0.24	0.12	0.06
Vietnam	0.36	0.35	0.32	0.31	0.18
ASIA PACIFIC	4.27	3.63	3.47	2.33	1.49

Table 26.7 Asia-Pacific: Gas Production Summary

Gas Production					
Tcf/a	2000	2005	2010	2020	2030
Australia	1.17	1.45	1.73	4.49	5.00
Brunei	0.42	0.47	0.44	0.30	0.17
India	0.91	1.19	1.88	2.00	1.45
Indonesia	2.90	2.98	3.41	3.50	3.50
Malaysia	1.73	2.68	2.72	2.80	2.10
Pakistan	0.86	1.21	1.51	1.50	0.89
Papua-New Guinea	0.00	0.00	0.00	0.30	0.30
Thailand	0.71	0.93	1.36	0.59	0.18
Vietnam	0.05	0.17	0.33	0.54	0.60
ASIA-PACIFIC	8.76	11.09	13.39	16.02	14.19

Table 26.8 Asia-Pacific : Oil and Gas Production and Consumption, Population and Density

ASIA PACIFIC Country Major	\multicolumn{3}{c	}{OIL}			GAS				POPULATION & AREA					
	Production	Consumption	Trade	Production	Consumption	Trade	Population	Growth Factor	\multicolumn{3}{c	}{Area km²}			Density
		p/capita	(+)		p/capita	(+)			Onshore	Off	Total	
	Mb/a	Mb/a	b/a	Mb/a	Gcf/a	kcf/a	Gcf/a	M		\multicolumn{3}{c	}{M}			
Australia	159	351	15.7	−192	1730	1141	50.9	589	22.4	7.33	7.77	2.56	10.33	3
Brunei	50	6	15.5	43	440	105	335	335	0.4	40.0	0.01	0	0.01	69
India	275	1161	1.0	−887	1883	2277	−394	−394	1188.9	5.2	3.3	0.41	3.71	360
Indonesia	344	472	2	−127	3406	1460	1946	1946	235.5	5.8	1.91	1.38	3.29	123
Malaysia	202	205	7.1	−2.6	2718	1145	1573	1573	28.9	14.5	0.33	0.22	0.55	87
Pakistan	22	150	0.8	−127.5	1513	1400	113	113	184.8	9.3	0.8	0.01	0.81	231
Papua	11	12	1.8	−1	4	4	0	0	6.8	9.9	0.46	1.2	1.66	15
Thailand	88	361	5.3	−272	1364	1592	−228	−228	68.1	9.7	0.52	0.27	0.79	132
Vietnam	115	117	1.3	−2	330	290	40	40	88.9	8.9	0.33	0.36	0.69	267
Sub–Total	1266	2834	1.6	−1567	13388	9414	5.2	3974	1825	5.8	15.1	6.4	21.5	120.85
Minor														
Afghanistan	0.00	1.75	0.06	−2	1	1	0.0	0	29.1	6.4	0.65			45
Bangladesh	1.83	35.77	0.22	−34	711	711	4.3	0	164.4	14.9	0.14			1142
Bhutan	0.00	0.37	0.52	0	0	0	0.0	0	0.7	9.0	0.05			15
Cambodia	0.00	11.68	0.77	−12	0	0	0.0	−3545	15.1	8.5	0.18			83
Japan	1.87	1625.15	12.76	−1683	174	3718	29.2	0	127.4	2.8	0.38			337
Laos	0.00	1.10	0.17	−1	0	0	0.0	0	6.4	0.4	0.24			27
Mongolia	0.00	6.21	2.22	−6	0	0	0.0	0	2.8	3.3	1.57			2
Nepal	0.00	7.30	0.26	−7	0	0	0.0	5	2.8	9.3	0.15			190
N.Zealand	20.09	54.65	12.42	−35	152	147	33.5	20	4.4	4.7	0.27			16
Philippines	8.83	113.15	1.20	−104	121	101	1.1	−297	94	11.1	0.3			313
Singapore	0.00	394.2	77.29	−394	0	296.65	58.2	−1484	5.1	15	0			8239
S.Korea	0.00	821.78	16.81	−822	30	1514.6	31.0	0	48.9	7.0	0.1			493
Sri Lanka	0.00	33.58	1.62	−34	0	0	0.0	−525	20.7	10.1	0.07			316
Taiwan	0.2	365.73	15.76	−365	11	535	23.1	−249	4	11.5	0.04			641
Other	0.00	30.00	7.50	−30	3778	4027	1006.7			0	0.25			16
Sub–Total	33	3502	6.1	−3470	4978	11052	19.2	−6074	574	16	4			144
Asia Pacific All	1299	6336	2.64	−5038	18366	20466	8.5	−2100	2399	7	19			126

RUSSIA
HUNGARY
UKRAINE
KAZAKHSTAN
ROMANIA
CROATIA
UZBEKISTAN
ALBANIA
AZERBAIJAN
TURKMENISTAN
CHINA

Part IV

Eurasia

Albania

Table 27.1 Albania regional totals (data through 2010)

	Production					Peak Dates			Area	
	Amount		Rate				Oil	Gas	'000 km²	
	Gb	Tcf	Date	Mb/a	Gcf/a	Discovery	1928	1977	Onshore	Offshore
PAST	0.5	0.49	2000	2	2	Production	1983	1982	30	2
FUTURE	0.3	0.76	2005	3	2	Exploration	1987		Population	
Known	0.2	0.65	2010	4	2	Consumption	Mb/a	Gcf/a	1900	1.0
Yet-to-Find	0.05	0.11	2020	4	5	2010	12	1	2010	3.2
DISCOVERED	0.70	1.14	2030	3	5		b/a	kcf/a	Growth	3.2
TOTAL	0.75	1.25	Trade	−8	1	Per capita	3.8	0.34	Density	111

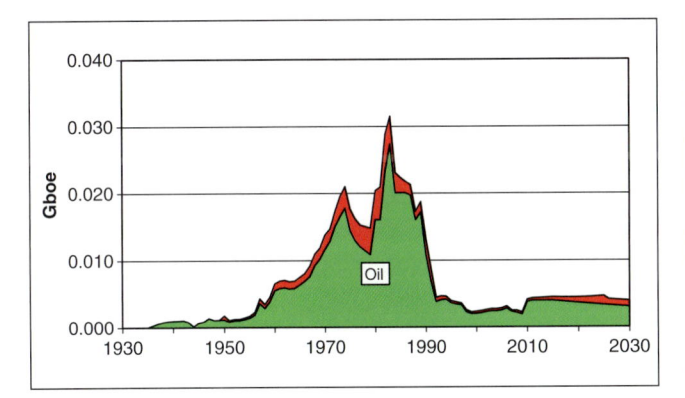

Fig. 27.1 Albania oil and gas production 1930 to 2030

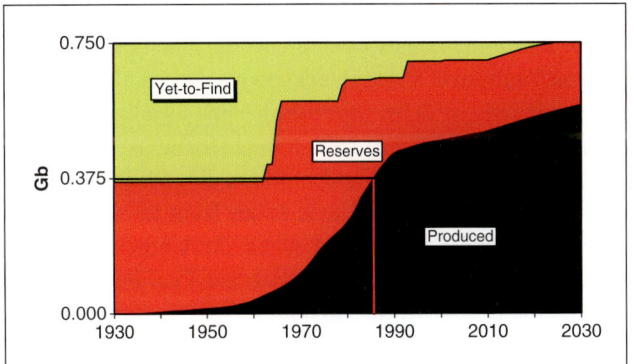

Fig. 27.2 Albania status of oil depletion

Essential Features

Albania is a small mountainous country on the eastern border of the Adriatic Sea. The Dinaric Alps in the north rise to almost 3,000 m, forming rugged forested country, while to the south follows a gentler terrain supporting the main centres of population. An indented coastline is flanked by islands. The country supports a population of 3.2 million.

Geology and Prime Petroleum Systems

The Dinaric Alps form a branch of the main Alpine chain, being dominated by Mesozoic limestones. The frontal ranges are thrust towards the Adriatic, giving some foreland Tertiary and Mesozoic basins, one of which is locally oil-bearing.

Exploration and Discovery

Oil seepages attracted early attention to the prospects of Albania, where the first borehole was sunk in 1918, making a small discovery. That was followed in 1928 by larger finds at Patos-Marinza and Kucove, yielding some 285 Mb from Upper Tertiary strata at shallow depth. Exploration was effectively then suspended during the Second World War, but resumed at a low level afterwards, yielding a small number of modest oil and gas finds. Efforts were made to explore the offshore in the 1990s, which included the drilling of a few deep boreholes to investigate the Lower Tertiary at depths of as much as 5,000 m. One such borehole was even located in as much as 712 m of water. But the results so far have been disappointing.

Production and Consumption

Oil production commenced in 1929 and rose slowly over the ensuing years, before climbing after the Second World War to reach an overall peak of 75 kb/d in 1983. It declined steeply from 1990 onwards to no more than 6 kb/d in 1998 before rising to 11 kb/d in 2010. It is expected to remain at about this level to 2017 before declining gently, being by than 70% depleted.

Gas production commenced in 1950 and rose to a peak of 30 Gcf a year in 1982, before declining steeply to the present levels. It is expected to increase slightly in the future.

Oil consumption is running at 12 Mb/a, most of which is imported. Gas consumption exceeds production by a small factor, with the balance being imported.

The Oil Age in Perspective

Albania is an ancient country that was well developed in Greek and Roman times when it covered much of what later became Yugoslavia, being then known as Illyria. Christianity came to it early, with the first bishop being installed in 58 AD. When the Roman Empire split, Albania found itself a prominent member of the Eastern Empire of Byzantium, but was later invaded by various Slavic tribes from the north. The Byzantine links survived and when the Christian Church split in 1054, Albania came under the Patriarch of what became the Greek Orthodox Church. Various further Slavic invasions followed, eventually causing many Albanians to emigrate especially to Greece and the Greek isles.

A new conflict arose in the fifteenth century when the country was attacked and eventually subjugated by the Ottoman Turks, who insisted that the people convert from Christianity to Islam, imposing a feudal system under Turkish landlords.

Opposition to Turkish rule grew over the centuries until the country finally won its independence in 1912. The ensuing First World War plunged it into a new chaos, and it barely survived the competing claims of its neighbours in the division of the Balkans in the Versailles Peace Treaty that followed the war.

A new leader came to prominence thereafter, eventually declaring himself to be King Zog, the First, whose reign lasted until the outbreak of the Second World War in 1939, when the country was invaded by Italy. In 1941, the German Army took over control in connection with its Balkan campaign, but the Allies encouraged Resistance movements to oppose the occupation. As in other occupied countries, the Communists took a lead in these movements, in part exploiting the deep-seated divisions between peasant and landlord. At the end of the war, Enver Hoxha, the Communist Resistance Leader, came to power. He declared the country to be the People's Republic of Albania in 1946, after which it became a full-blown Communist State, establishing collective farms and nationalising industry. It also secured external support in turn from Yugoslavia, the Soviet Union and China. But the transformation was not an easy task because of deep-seated blood feuds, tribal and factional disputes and certain age-old traditions, including the confinement of women to the home in accord with Islamic practice. It became a draconian regime, oppressing dissidents and closing churches of all denominations, according to the atheistic principles of Communism.

Hoxha died in 1985, and the country began to move towards the West, a process that was accelerated by the fall of the Soviet Government of Russia. A new dispute however then emerged in relation to the former Albanian province of Kosovo. It had become a semi-autonomous region of Serbia, which now moved to persecute its inhabitants, who were predominantly Albanian and Muslim. This prompted a reaction with the formation of a new resistance movement to oppose the Serbs, which was in turn forcefully suppressed by the Serbs under their leader, the late Slobodan Milosevic.

The North Atlantic Treaty Organisation (NATO) was formed in 1949 as a defensive pact in the early days of the Cold War, whereby the signatory countries would come to the defence of any member that was attacked. The rules of engagement were subsequently modified, first allowing NATO to intervene if any member was perceived to be threatened, and later if its *vital interests* were deemed to be at risk. This was fairly broadly interpreted when it mounted a bombing campaign in March 1999, known as the Kosovo War. The territory was later occupied by military forces, including the establishment of a major US Base, known as Camp Bondsteel, located on the route of a proposed major oil pipeline to bring Caspian oil to the Adriatic coast of Albania for export.

It is evident from the foregoing that Albania has had a tortured history, being so-to-speak at a boundary between fundamental political, ethnic and religious forces. But ironically it may be relatively well placed to face the Second Half of the Age of Oil. It has some indigenous oil and gas which were subject to low depletion rates due to the politico-economic situation, such that production may even rise in the years ahead. It has remained a predominantly agricultural country with relatively low energy demands. If the new pipeline from the Caspian is constructed, it may find itself in the powerful role of a transit country.

27 Albania

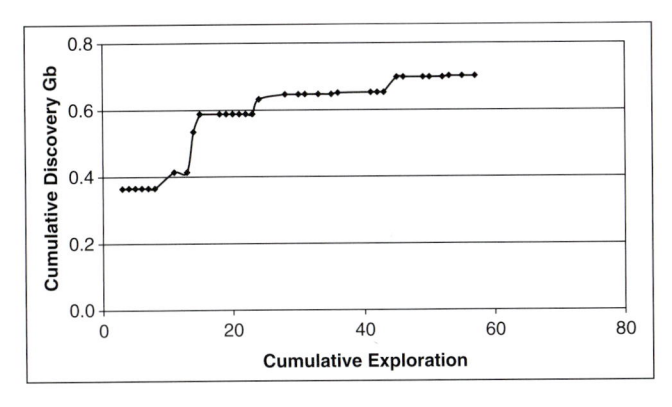

Fig. 27.3 Albania discovery trend

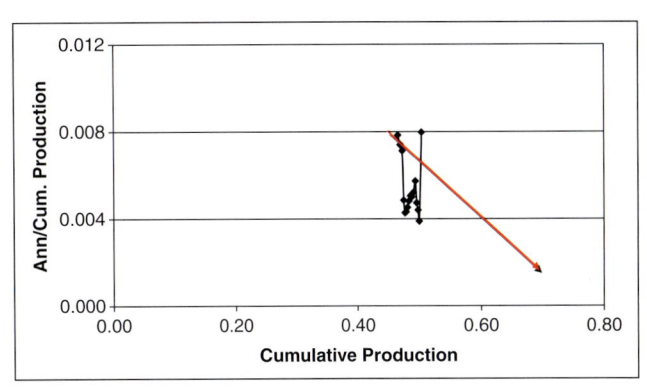

Fig. 27.4 Albania derivative logistic

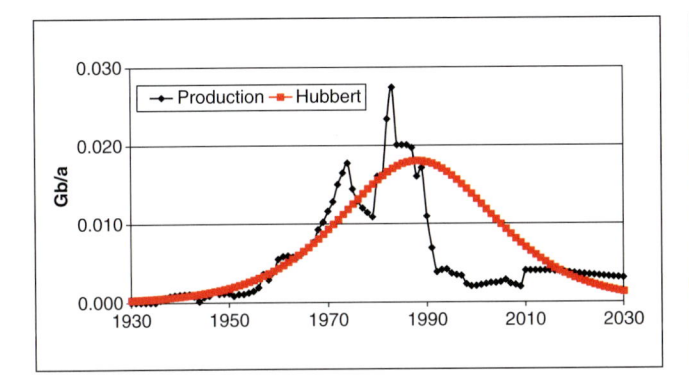

Fig. 27.5 Albania oil production: actual and theoretical

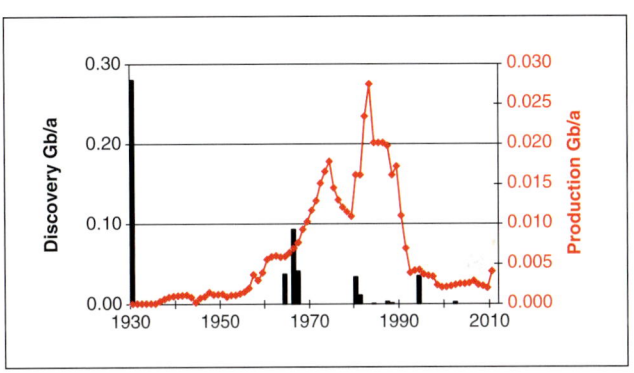

Fig. 27.6 Albania discovery and production

Azerbaijan

Table 28.1 Azerbaijan regional totals (data through 2010)

Production to 2100						Peak Dates			Area	
Amount			Rate				Oil	Gas	'000 km²	
	Gb	Tcf	Date	Mb/a	Gcf/a	Discovery	1871	1999	Onshore	Offshore
PAST	10	16	2000	102	487	Production	2014	2030	90	75
FUTURE	14	54	2005	158	206	Exploration	1953		Population	
Known	12	43	2010	378	926	Consumption	Mb/a	Gcf/a	1900	1.0
Yet-to-Find	1.4	11	2020	313	1350	2010	38	350	2010	9.2
DISCOVERED	23	59	2030	231	1350		b/a	kcf/a	Growth	8.1
TOTAL	24	70	Trade	+340	+576	Per capita	4.1	38	Density	106

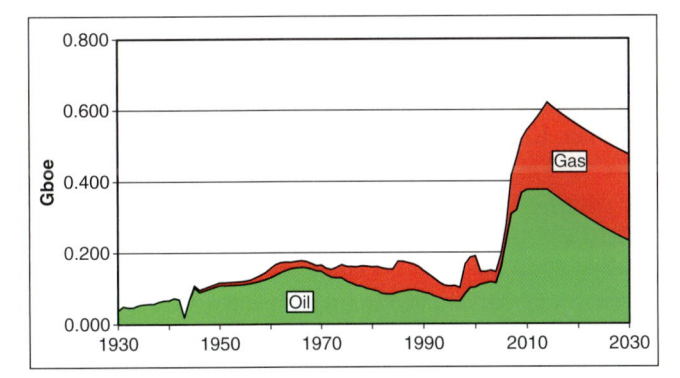

Fig. 28.1 Azerbaijan oil and gas production 1930 to 2030

Fig. 28.2 Azerbaijan status of oil depletion

Essential Features

Azerbaijan covers an area of about 90,000 km² on the western shore of the Caspian Sea, including the oil-bearing Apsheron Peninsula. It is bordered by Iran to the south, Russia to the north, and Armenia and Georgia to the west. A semi-autonomous enclave of Christian Armenians, known as Nogorno-Karabakh, lies in the west of the country, claiming yet-to-be recognised independence. About one-third of the country is made up of fertile lowlands, while the rest comprises the Caucasus Mountains, rising to almost 4,500 m. It enjoys a dry sub-tropical climate of cold winters and hot summers, moderated by altitude. The lowlands are cultivated, partly with the help of extensive irrigation canals. The present population amounts to about 9.2 million, mainly belonging to the *Shi'ia* branch of Islam.

Geology and Prime Petroleum Systems

In geological terms, the country straddles the palaeo-delta of the Volga River, comprising a sequence of much-faulted Tertiary sands and clays, giving multiple reservoirs. The interbedded clays in the lower part of the sequence provide rather lean, gas-prone source-rocks. Gas seepages have given rise to mud-volcanoes, being large mounds of mud brought to the surface by gas seepages, which sometimes catch fire. To the south and east, the sequences are buried beneath a Pliocene fill in the foredeep of the Elburtz

Mountains of Iran, where the source-rocks are over-matured, yielding gas-condensate.

Exploration and Discovery

Azerbaijan boasts some of the world's oldest giant fields, including Balakhany-Sanbunchi with about 2.5 Gb (1871), Bibi-Eybat with slightly less (1873), and Surakhany with almost 1 Gb (1904). A second phase of exploration during the last half of the twentieth century delivered a further batch of generally smaller discoveries onshore and in the near offshore. The entry of foreign companies after the fall of the Soviets brought in two already identified major offshore fields, Azeri-Chirag-Guneshli and Shakh Deniz, but the results have failed to live up to early expectations. About 530 exploration boreholes have now been drilled on what is quite a small prospective area, finding some 23 Gb of oil and 59 Tcf of gas.

Production and Consumption

Oil production is currently running at about 1,035 kb/d, following the opening of the export pipeline to Ceyhan in Turkey. It is expected to remain at this level until about 2015, when the resource will be about 50 % depleted, before declining at about 3 % a year. The country, with relatively low oil consumption, is in a position to remain a major exporter for many years to come, the uncertain political environment allowing.

It is reported that gas consumption exceeds production, even though its endowment has been depleted by only 40 %, which suggests that the statistics are unreliable.

The Oil Age in Perspective

Azerbaijan has lain throughout its history at the cross-roads between Russia, Iran and Turkey. Originally populated by nomadic Turkic tribes having links with Iran, it came briefly under Christian influences based in Armenia, before reverting to the *Shi'ia* branch of the Muslim world.

Russia conquered much of the territory during the early nineteenth century, taking the Apsheron Peninsula where surface seepages of oil had long been known. Baku, the capital, soon developed into an important oil centre, attracting immigrant labour to work in the oilfields. They were developed mainly by the Nobel Brothers of Sweden and Shell Oil in co-operation with the Rothschild Bank, and by 1900, Baku was providing more than half of the World's supply of oil. Operating conditions were appalling, and formed a natural breeding ground for revolution and revolt. The interior of the country was a no-mans-land of divergent ethnic and religious groups, given to banditry and conflict. The Christian factions generally prospered from the oil wealth while the Muslims were substantially disenfranchised by lack of property rights. The situation gave rise to much resentment, amongst not only the workers themselves but various intellectuals who supported their cause. A particularly bloody uprising in 1905 brought to prominence Joseb Dzhugashvili, later adopting the name of Josef Stalin, who was an oil workers' leader, in a movement that culminated in the Bolshevik Revolution of 1917, changing the World in many ways.

An Azerbaijani State was declared in 1918 by nationalists, backed by the Turkish Army, but that was followed by a massacre of predominantly Christian Armenians. A British Army briefly occupied the territory after the war, but put up little resistance when the Red Army marched into Baku on January 15th, 1920. It then became a fully-fledged member of the USSR, remaining under the close control of Moscow, which no doubt was particularly interested in retaining access to its oil.

The long-standing conflict with the Christian Armenians in the Nagorno-Karabakh enclave erupted again in 1988 and continued for several years with great loss of life. Political instability ensued until 1993, when Haydar Aliyev, a former KGB Official and Communist leader, came to power and managed to survive various pressures and abortive coups.

In 1998, he was granted an enthusiastic welcome in Washington on signing oil concessions with eight Western oil consortia. Much attention was devoted to the means of exporting oil from this landlocked territory. The shortest route through Iran was rejected on political grounds. An alternative proposal for a line through Georgia to the Black Sea was also considered. But it involved either the risks of increased tanker traffic through the environmentally sensitive Bosphorus, or trans-shipment in Bulgaria to a second pipeline to the Adriatic passing through Kosovo, where a large US military base, named Bondsteel, was established for possibly not unrelated reasons. But finally it was decided to choose the most costly route to Ceyhan on the Mediterranean coast of Turkey. This 42-in. line, which is 1,760 km in length, is now in operation under the management of BP, having cost some $4 billion. The pipeline passes through mountainous country, which is occupied by disaffected people with a long tradition of violence, based on deep-seated ethnic and religious conflicts. Their passions have no doubt been further inflamed by recent events in Iraq.

In 2003, the Presidency passed to Ilham Aliyev, the son of the previous strongman, in a process seen as less than democratic by some elements in the country.

The new oil wealth flowing into the country may stimulate jealousies both internally and in relation to neighbouring countries, suggesting continued turbulence into the Second Half of the Age of Oil.

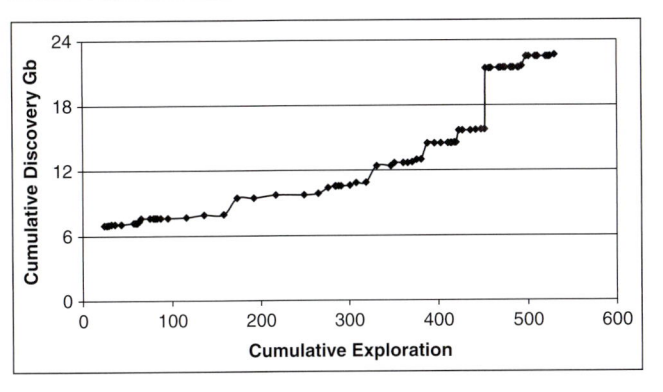

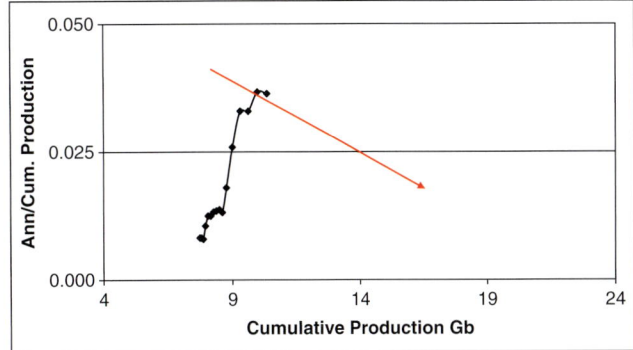

Fig. 28.3 Azerbaijan discovery trend

Fig. 28.4 Azerbaijan derivative logistic

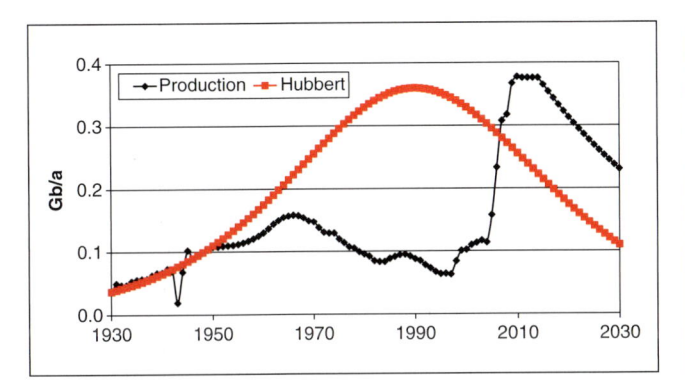

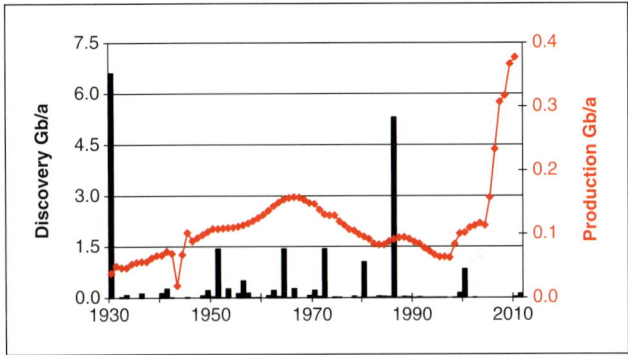

Fig. 28.5 Azerbaijan production: actual and theoretical

Fig. 28.6 Azerbaijan discovery and production

China

Table 29.1 China regional totals (data through 2010)

Production					Peak Dates			Area	
Amount			Rate			Oil	Gas	'000 km²	
	Gb	Tcf	Date	Mb/a	Gcf/a	Discovery	1960	2000	Onshore	Offshore
PAST	39	43	2000	1186	962	Production	2010	2030	9,610	400
FUTURE	31	157	2005	1317	1763	Exploration	2003		Population	
Known	24	141	2010	1488	3334	*Consumption*	Mb/a	Gcf/a	1900	350
Yet-to-Find	6.1	16	2020	925	5000	2010	3354	3768	2010	1,346
DISCOVERED	64	184	2030	575	5000		b/a	kcf/a	Growth	3.7
TOTAL	70	200	*Trade*	−1866	−434	Per capita	2.5	2.8	Density	140

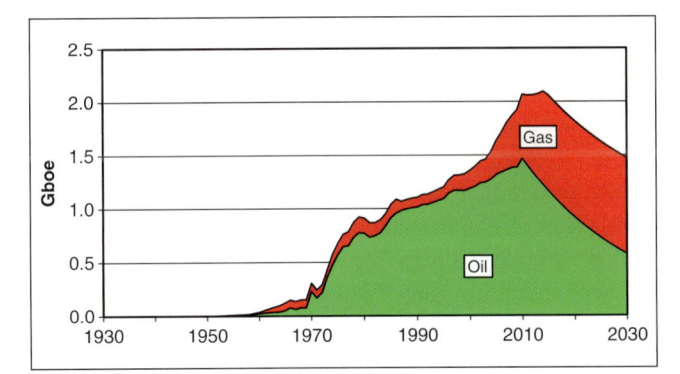

Fig. 29.1 China oil and gas production 1930 to 2030

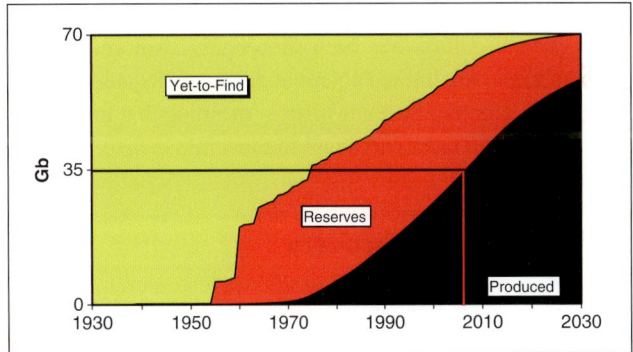

Fig. 29.2 China status of oil depletion

Essential Features

China covers an area of about 10 million km², being surpassed in size only by Russia and Canada. It supports a population of about 1.3 billion people, amounting to one-fifth of the World's total. They are made up of several different tribal and ethnic groups, speaking different languages. Accordingly, it is a land of great diversity. Highlands, flanking the Himalayas in the west and rising to an altitude of as much as 4,000 m, give way to dissected mountainous county, which in turn passes into the borderlands of the South China Sea. Its mountainous character helps explain the diversity of its people and the fact that it remained isolated from the rest of the World for much of its history.

The country lies at the junction of two climatic systems. The Siberian system to the north delivers arid deserts with an extreme range of temperature, while the Pacific to the southeast brings rain and more moderate and fertile conditions. Three major river systems, including the Yangtze, flow eastwards through the country to provide fertile valleys, where most of the people are concentrated. Shanghai is the largest city with 23 million inhabitants, followed by the capital, Beijing, with 20 million.

Geology and Prime Petroleum Systems

In geotectonic terms, China lies between the Siberian Shield and the Pacific Oceanic Plate. It is broadly divided into two provinces. In the remote west behind the Himalayas, lie several compressional basins, including the large Tarim Basin, which has a deep prospective Palaeozoic sequence, whereas to the east lie a series of Mesozoic and Tertiary basins, bounded by a major shear zone. They in turn give way to a large continental shelf bordering several back-arc basins at the edge of the Pacific Plate.

C.J. Campbell, *Campbell's Atlas of Oil and Gas Depletion*,
DOI 10.1007/978-1-4614-3576-1_29, © Colin J. Campbell and Alexander Wöstmann 2013

The principal source-rocks occur in the Cretaceous and in the somewhat lean, lake-bed deposits of the early Tertiary that give waxy crudes causing production problems. The Sagliao Basin in the northeast contains the giant Daqing Field, found in 1959. It is adjoined by the North China basins holding almost half of the country's known oil. One of them has given the giant Shengli Field, which was found in 1984 with about 1 Gb. An offshore extension into the Gulf of Bohai has yielded some recent discoveries, including the giant Jidong Nanpu found in 2006. Early reports speak of it holding 2 Gb, but it would not be surprising if they prove to be subject to downward revision in the future.

The continental shelf was opened to Western companies in the 1980s but has disappointed, save for a sizeable gas field, near Heinan Island, and the recent giant field in the Gulf of Bohai. Much misplaced media interest has been directed at the Spratley Islands, north of Borneo, whose ownership is in dispute with several neighbouring countries, including China. As so often the case, areas that are for some reason closed to exploration are deemed to be floating on oil.

In general, China has been thoroughly explored, but is found to have only limited potential due to inadequate source-rocks and a complex tectonic history, meaning that individual fields are for the most part small to moderate in size.

Exploration and Discovery

Desultory exploration commenced early in the last Century, but did not take off in earnest until after the Second World War when relative political stability was imposed by the Communists. Discovery peaked in 1960 with the giant Daqing Field, at which point some 300 exploration boreholes had been drilled. Exploration was stepped up subsequently with the drilling of some 2,000 more boreholes, but the results have been discouraging. It seems that a total of about 64 Gb have been discovered, of which 39 Gb have been produced. Future discovery is here estimated at about 6 Gb. The gas situation is rather more promising, with indicated reserves of about 140 Tcf. Attention is now turning to the remote Tarim Basin, which, according to reports, is promising.

Production and Consumption

Oil production commenced in 1939 and rose gradually to 1964. It increased markedly thereafter, passing 1 Mb/d in 1973 to reach an overall peak at 4 Mb/d in 2010. At that point, some 56% of the estimated resource had been produced, suggesting that it is set to decline in the years ahead at the current Depletion Rate of about 5% a year. In earlier years, production operations were not blessed with the most advanced technology, but those limitations were partly compensated for by close drilling.

Gas production has followed a similar path to reach current production of about 3.3 Tcf/a.

China's oil use has been growing rapidly to stand at about 3,354 Mb/a today. It means that the country has to import about half of its needs, being second only to the United States in total consumption. Close to 15 million new cars have taken to the roads over the past 12 months, and the demand for fuel obviously grows in parallel. Imports have thus risen to 1,866 Mb/a and are set to grow still higher as indigenous production declines, which explains why Chinese companies are scouring the world for exploration rights. The country's growing dependence on imported energy threatens its economic prosperity, which may slump in the future.

Gas consumption stands at about 3.8 Tcf/a, which is set to expand as new pipelines are constructed.

The Oil Age in Perspective

Archaeological research shows that some of the earliest members of *Homo sapiens* had managed to reach China. The long history, with its many cultural flowerings, has spanned at least 4,000 years, as empires and dynasties waxed and waned, but need not concern us here, save to mention the importance of various Mongol invasions from the northwest.

By the early nineteenth century, the country was being exposed to Western influences, principally related to the opium and tea trades. The British East India Company was growing opium in Bengal which it exported to China, partly to pay for the rising imports of tea from that country. It led to the First Opium War in 1840 when Britain sent a military expedition to China, resulting both in the secession of Hong Kong, which became a British colony, and the opening of five ports to foreign trade. Furthermore, a certain rivalry developed between Britain, France, Russia and later Germany, to bring China into their spheres of influence to foster trade. There were also conflicts due to opposition to the arrival of Christian missionaries and resultant clashes with the large Muslim community of the western territories. In summary, a long period of turmoil, revolt and foreign intervention ensued, limiting economic development, and at times causing great suffering and loss of life.

China supported Britain and France during the First World War, sequestering German and Austrian assets, which opened the door for penetration by Japan, then an ally of Britain and France. The inter-war years saw the emergence of competing Nationalist and Communist movements, backed by a new intellectual class, opposed to the great disparities of wealth and influence, as well as foreign commercial intrusion. Japan sought to strengthen its grip on China by taking Manchuria and other territories, but triggered resistance that led to the emergence of Communist armies which fought their way east in what became known as the Long March. Mao Zedong, and other Communist leaders who later came to prominence, led these endeavours, being opposed by the Nationalists under General Chiang Kai-Shek. Civil war erupted notwithstanding the Japanese threat, but by 1937, the rival factions had patched up an uneasy national co-operation in the face of a full scale war with Japan. Chiang Kai-Shek emerged as the national leader, supported by his impressive wife, who evidently wore the trousers. But his opposition to the Communists remained not far beneath the surface.

Japan had early victories, partly with the help of air power, taking territories in the east of the country, including Nanking, but Chinese resistance was unbroken. The outbreak of the Second World War brought military and financial support from the Soviet Union and the United States. The defeat of the Japanese by the British Army in Burma led to more direct foreign military intervention, with the US Air Force taking part in operations over mainland China. Indeed, Roosevelt, the US President, sought unsuccessfully to place General Stillwell in charge of all Chinese forces. Meanwhile, Communist-led resistance built a political grip behind the Japanese lines, such that, by the end of the war, it was in de facto control of large sections of the country. By 1949, it had cemented its power, declaring The People's Republic of China as a full-blown Communist State. Land reform swept the country, bringing hope to the suffering peasantry.

The Nationalists managed to hold on to the Chinese island of Taiwan, which with the US support has remained independent. Comparable conflicts were also affecting the adjoining territory of Korea, where war broke out in 1950 between a strongly Communist north, which was backed by China and the Soviet Union, and a Nationalist south supported by the United States and its allies. It ended in stalemate in 1953.

The 1960s saw a certain retreat from doctrinaire Communism with the downsizing of State farms and industries, although Mao Zedong tried to maintain his vision with the Cultural Revolution. The Red Guards controlled daily life on almost military lines. But Mao Zedong suffered a serious stroke in 1972 and died four years later.

Deng Xiaoping emerged as a successor with a reforming mission to move China into a more capitalist and international direction, which gathered momentum after the fall of the Soviets in 1990, and his own death in 1997. This paved the way for Britain to return Hong Kong to China in 1997, on the expiry of the lease, signed in 1842.

The rampant industrialisation and commercialisation of China in recent years is evidently not sustainable in energy terms. Its oil industry was developed by State-controlled entities, which were probably able to conduct efficient exploration in a scientific manner, freed of the commercial distortions of the West. The truth is that China is not a particularly prospective territory, as explained above.

In the face of these limited resources, the country embarked on a massive hydroelectric scheme, known as the Three Gorges Scheme. The dam, costing $24 billion, is a mile wide and 170 m high. It will provide 18,000 MW, or 10% of China's electricity, but at a cost of flooding precious farm land and displacing two million people. The food supply for the immense population is failing with the depletion of the aquifers and the advance of deserts.

The Communists did introduce draconian birth control measures, but famine now becomes the more likely mechanism by which to match population to the limited carrying capacity of the largely barren and mountainous lands. Overgrazing and dustbowls are rendering large tracts of the interior uninhabitable (see Brown L.R., 2003, *Plan B*, W.W. Norton).

The worldwide energy crisis that unfolds during the Second Half of the Oil Age is likely to have a devastating effect on China, and especially on its new aspirations to become an industrial dynamo.

In fact the country appears to be well aware of its energy predicament. Its companies have sought to secure stakes in Middle East oil, especially in Iran and Iraq, and it has also signed a contract to exploit Saudi Arabia's gas. It is involved in a heavy oil project in Venezuela, and there are ambitious pipeline projects under consideration to secure oil and gas from Central Asia and Siberia. The country does, however, have substantial coal reserves, on which it will increasingly have to rely, possibly with adverse environmental consequences.

China's long-suffering people have known a history of strife and turmoil and have endured much poverty in a land incapable of supporting their numbers. The looming energy crisis means that the recent economic miracle can be no more than short-lived. The long walk back to the paddy fields is likely to be fraught with tension and turmoil by a disillusioned people. It will test the iron grip of the government to avoid the break-up of the country into tribal regions under warlords, especially as food production suffers from the deteriorating natural environment. Urban conditions may become exceedingly difficult, adding to the pressure for emigration.

The growing critical dependence on Middle East imports puts China on a geopolitical collision course with the United States. It is noteworthy that China holds substantial dollar reserves giving it a certain control over the US economy, carrying the threat of unloading its holdings which would devalue the dollar further.

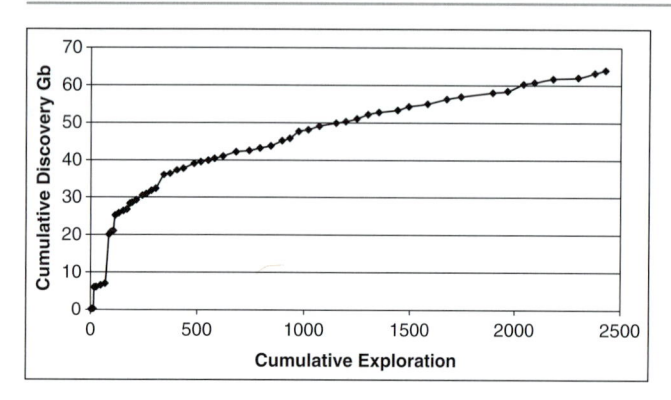

Fig. 29.3 China discovery trend

Fig. 29.4 China derivative logistic

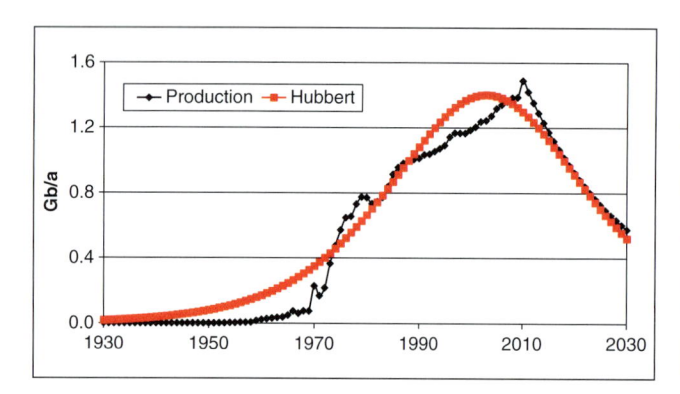

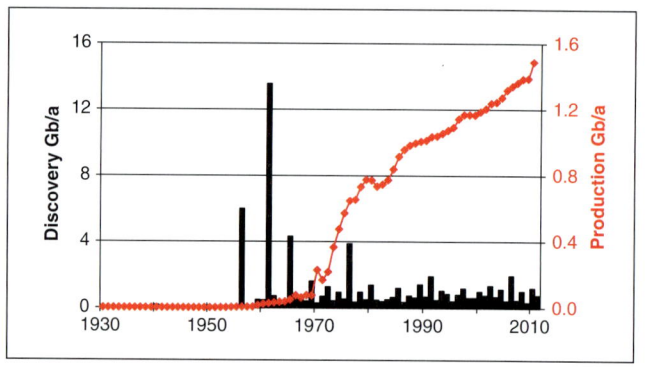

Fig. 29.5 China oil production: actual and theoretical

Fig. 29.6 China discovery and production

Croatia

Table 30.1 Croatia regional totals (data through 2010)

Production to 2100						Peak Dates				Area	
Amount		Rate				Oil	Gas	'000 km²	
	Gb	Tcf	Date	Mb/a	Gcf/a	Discovery	1957	1974	Onshore	Offshore
PAST	0.5	1.5	2000	8.8	59	Production	1988	2011	60	33.2
FUTURE	0.5	1.5	2005	6.8	54	Exploration	1985		Population	
Known	0.3	1.2	2010	5.3	67	*Consumption*	Mb/a	Gcf/a	1900	3
Yet-to-Find	0.1	0.3	2020	4.7	65	2010	36	100	2010	4.4
DISCOVERED	0.9	2.7	2030	4.2	28		b/a	kcf/a	Growth	4.5
TOTAL	1	3	*Trade*	−30	−33	Per capita	8.1	23	Density	73

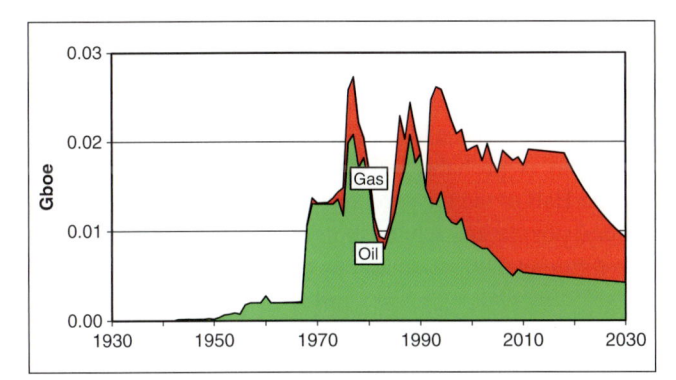

Fig. 30.1 Croatia oil and gas production 1930 to 2030

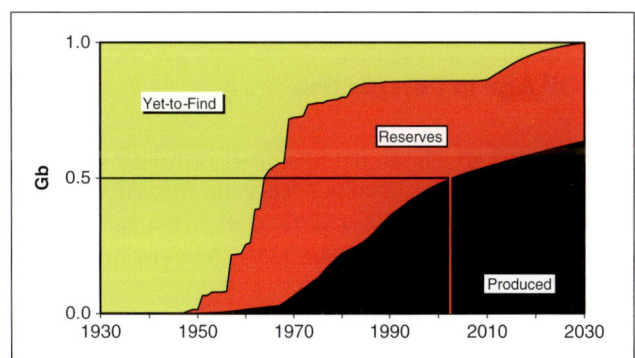

Fig. 30.2 Croatia status of oil depletion

Essential Features

Croatia is a crescent-shaped country in the north-western part of the Balkans, flanking the Adriatic Sea, having been previously part of Yugoslavia. The Dinaric Alps border an indented coastline with many islands and give way eastwards to rich agricultural plains which are drained by the Sava River flowing eastwards to the Black Sea. It supports a population of 4.4 million.

Geology and Prime Petroleum Systems

The Dinaric Alps are built largely of Mesozoic carbonates, including dolomites, which give a karst terrain, characterised by crags and deep gorges. The western part of the Pannonian Basin, which consists mainly of late Tertiary sediments overlying the frontal thrust-belt of the Dinaric Alps, extends into Croatian territory. Important sinistral transcurrent faults have cut the basin, offsetting structural trends. Triassic source-rocks, which are widespread in the Adriatic, have locally charged a range of reservoirs, including Mesozoic and older rocks in the thrust-belt, but the prime source seems to have been Miocene clays, which have charged overlying reservoirs with oil and gas.

Exploration and Discovery

Oil was produced in hand-dug wells in the vicinity of seepages as early as 1856. The first recorded borehole was drilled in 1884, but exploration subsequently lapsed until after the Second World War. It then resumed at a modest level with no more than a few boreholes being drilled annually thereafter, giving a peak of 17 boreholes in 1985.

A minor discovery was made in 1930 followed by a number of modest finds of oil and gas during the late 1950s and 1960s. The largest was the Struzec Field, found in 1957 with some 126 Mb of oil and 76 Mcf of gas, but discovery subsequently dwindled. In total, about 860 Mb of oil have been found, with limited potential for new finds.

Production and Consumption

The production of oil commenced in 1943 at a low rate. It then rose slowly to 29 kb/d by 1968, before reaching a peak at 57 kb/d in 1977 and subsequently declining to 15 kb/d in 2010.

Significant gas production commenced in 1969 and is expected to rise to a peak of 77 Mcf in 2011 and to remain at about that level until terminal decline sets in 2019 at about 8% a year.

Oil consumption stands at 36 Mb/a, meaning that the country relies on imports. It also has to import gas.

The Oil Age in Perspective

Croatia was part of the territory of Illyria during Roman times (as already described in connection with Albania) but was invaded by the Croats, a Slavic tribe, in the seventh century. It became a monarchy in 925 before entering into a form of union with Hungary in 1102. It succeeded in resisting the Ottoman Turks, who were expanding through the Balkans, by cementing its relations with Hungary, which itself was absorbed into the Austro-Hungarian Empire under the Habsburg kings. As a result, it became a predominantly Catholic country.

The defeat of the Austrians in the First World War was followed in 1929 by the establishment of the Kingdom of Yugoslavia, which included Croatia, but much internal conflict remained, especially between the Croats and neighbouring Serbs.

Germany invaded the country in 1941 during the Second World War and encouraged an extreme fascist group, the Ustase, to form an independent State, but it eventually fell to the Communist-led Resistance movement of Marshall Tito, becoming the People's Republic of Croatia, one of the six constituent republics of Yugoslavia.

Renewed Serbo-Croat conflict followed the death of Marshall Tito in 1980, leading to a full scale war, which was not resolved until 1995, when the Croatian army launched a successful campaign, in which thousands lost their lives. A degree of stability and democratic progress has followed, such that Croatia is expected to join the European Union.

The country has had a difficult political history in recent years, but is probably relatively well placed to face the Second Half of the Oil Age, having a modest primarily agricultural population, which should be able to find a sustainable future in their mountainous land.

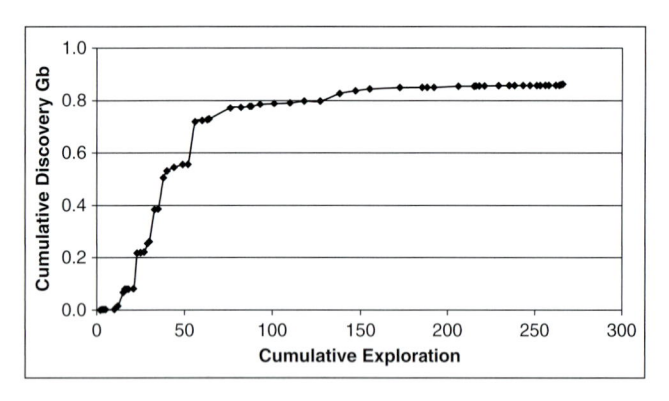

Fig. 30.3 Croatia discovery trend

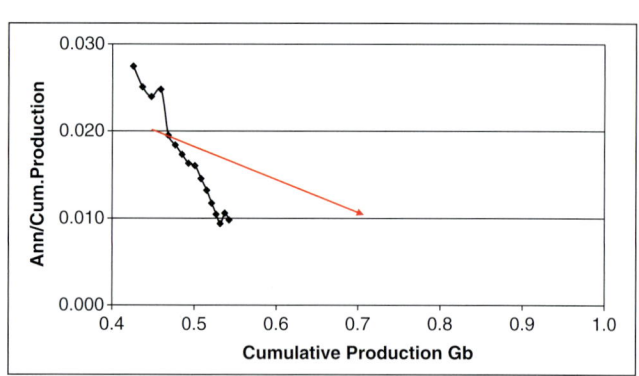

Fig. 30.4 Croatia derivative logistic

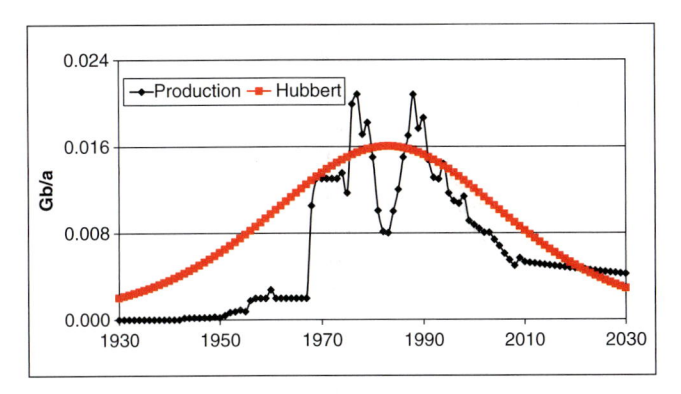

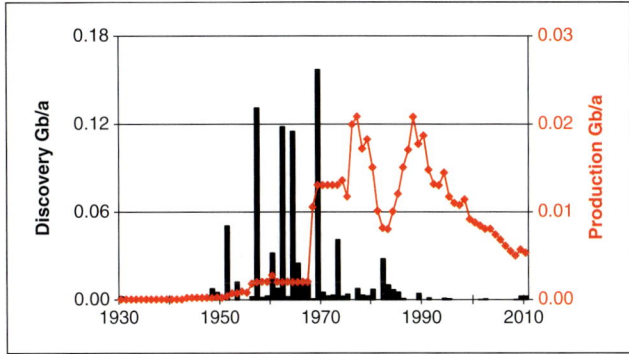

Fig. 30.5 Croatia production: actual and theoretical

Fig. 30.6 Croatia discovery and production

Hungary

Table 31.1 Hungary regional totals (data through 2010)

	Production					Peak Dates			Area	
	Amount		Rate				Oil	Gas	'000 km²	
	Gb	Tcf	Date	Mb/a	Gcf/a	Discovery	1965	1965	Onshore	Offshore
PAST	0.7	7.7	2000	10	113	Production	1979	1985	90	0
FUTURE	0.3	3.3	2005	7	108	Exploration	1964		*Population*	
Known	0.2	3.0	2010	5	101	*Consumption*	Mb/a	Gcf/a	1900	7.5
Yet-to-Find	0.1	0.3	2020	4	75	2010	53	426	2010	10
DISCOVERED	0.9	10.7	2030	4	55		b/a	kcf/a	Growth	1.3
TOTAL	1.0	11	Trade	−48	−325	Per capita	5.3	43	Density	111

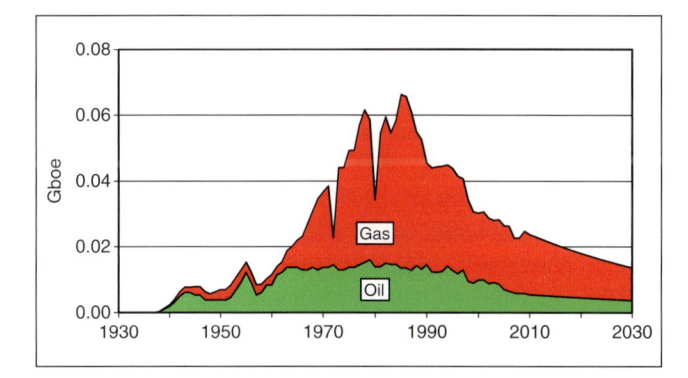

Fig. 31.1 Hungary oil and gas production 1930 to 2030

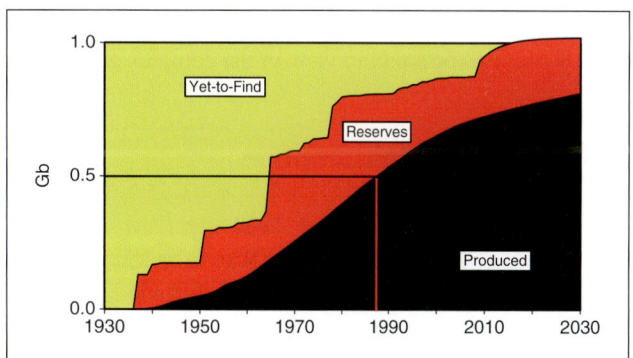

Fig. 31.2 Hungary status of oil depletion

Essential Features

Hungary lies in the heartland of Eastern Europe, having common frontiers with Austria, Slovenia and Croatia to the west; Slovakia to the North; Serbia to the south; and the Ukraine and Romania to the east. It consists mainly of extensive plains around 200 m above sea-level, which are flanked to the north by the junction of the Alpine and Carpathian mountain chains. A large lake, known as Lake Balaton, lies in the western part of the country near the capital, Budapest, which sits on the great Danube River flowing southward through the plains. The population of almost ten million are mainly descended from the Magyars from Asia, with gypsies comprising about 2%.

Geology and Prime Petroleum Systems

Hungary lies in the centre of the Pannonian Basin, already described in connection with Croatia. A major sinistral transcurrent fault divides it into two blocks: the Pelso Block to the north and Tisza Block to the south, which are in turn flanked by branches of the Carpathian orogenic belt. Three separate but similar petroleum systems have developed in the western, central and eastern parts of the country, yielding as many as 500, generally small, oil and gas fields.

The first discovery was the Budafa Field in the western province, which was found in 1937 with 45 Mb. It was followed by a larger nearby find in 1951 with 151 Mb in Miocene as well as Cretaceous and Triassic carbonates. The central province contains a large number of mainly small oil and gas finds occurring in complex structures related to the transcurrent fault system, producing from a variety of reservoirs, which are mainly, but not exclusively, of late Tertiary age. The gas is notably rich in carbon dioxide, derived from the underlying carbonates. The principal source-rock appears to be of Miocene age, although older secondary sources, principally in the Jurassic, have been identified.

Exploration and Discovery

The first recorded exploration borehole was drilled in 1913, and drilling continued at a low level until after the Second World War when it increased markedly. The peak of exploration came in 1964 when as many as 49 boreholes were sunk. Discovery followed a parallel pattern with numerous small finds of oil and gas in what is evidently a dispersed habitat due to the structural complexity. The peak of discovery was in 1965 when 340 Mb of oil and 3 Tcf of gas were found. It has since dwindled to very low levels.

There is some scope for future discovery but the size of finds is likely to be small in view of the complex geology.

Production and Consumption

Oil production commenced in 1937 at a low level, but increased after the Second World War to reach a peak in 1979 at 44 kb/d. It has since sunk to 14 Mb/d and is expected to continue to decline at a relatively low depletion rate of 2% a year, reflecting the fact that it comes from a large number of ageing small fields, approaching exhaustion. Gas production rose roughly in parallel to peak in 1985 at 293 Mcf/a and has also declined steeply since then. The country accordingly has to import most of the oil and gas it requires.

The Oil Age in Perspective

The frontier of the Roman Empire was the Danube River, which flows through Hungary. The country later bore the brunt of invasions from the east, starting with the Magyars in 896, to be followed by the Mongols in 1241, who massacred about half of the indigenous population. That was succeeded in the next century by a national flowering under King Louis the Great (1342–1382) whose dominion expanded over neighbouring areas. It was in turn eclipsed by the invasion of Ottoman Turks, leading Hungary to enter into an alliance with the Austrian Hapsburg kings, which was cemented in a union of the two countries in 1867. Hungary found itself dragged into the First World War, as a result of the union, but began to move towards independence, being in part inspired by the Bolshevik Revolution in 1917. The war devastated the economy and caused great suffering.

A Communist rising followed the war, which delayed a settlement with Allied powers until 1920 when new frontiers were defined at the Treaty of Trianon. It cost Hungary 68% of its land and 58% of its population, causing much resentment, which led to continued tensions and difficulties during the inter-war years. The Great Depression of 1929 caused an economic and financial crisis, leading to the emergence of a Fascist leader, Miklos Horthy, who sought to ally Hungary with Germany and Italy.

Attempts to remain neutral during the early stages of the Second World War failed in 1941, when Hungary declared war on the Soviet Union, believing that Germany was set to win the war. It was in due course occupied by German troops, but eventually faced defeat in 1945, being occupied by Soviet troops.

The Peace Treaty of 1947 imposed reparations and a return to the frontiers set in 1920. The Communist Party came to power in the following year, declaring the People's Republic of Hungary, which became a full member of the Communist bloc. Industry was nationalised and land distributed to collective farms. But an anti-Communist revolution broke out in 1956, backed in part by the Catholic Church, only to be suppressed by Soviet military force. The country then moved in a more liberal direction, with the last Soviet troops leaving in 1991, following the collapse of the Soviet Government of Russia.

Hungary has since joined the western camp, becoming a member of NATO in 1999 and the European Union in 2006, but it faces continued tensions and economic difficulties having long roots in its tortured past. These obstacles may emerge again in the difficult transition to the Second Half of the Age of Oil, but, generally speaking, the country is not badly placed having a modest population and ample agricultural land.

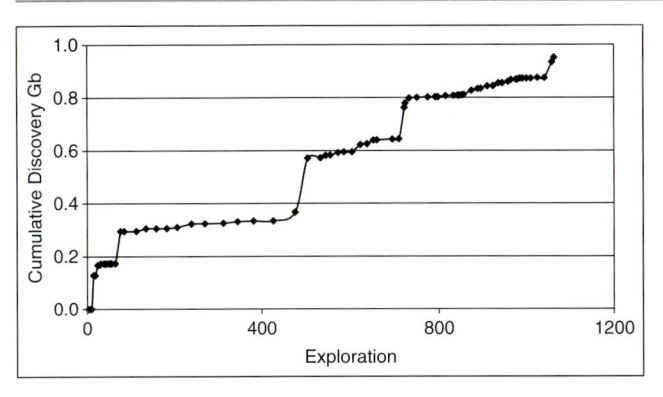

Fig. 31.3 Hungary discovery trend

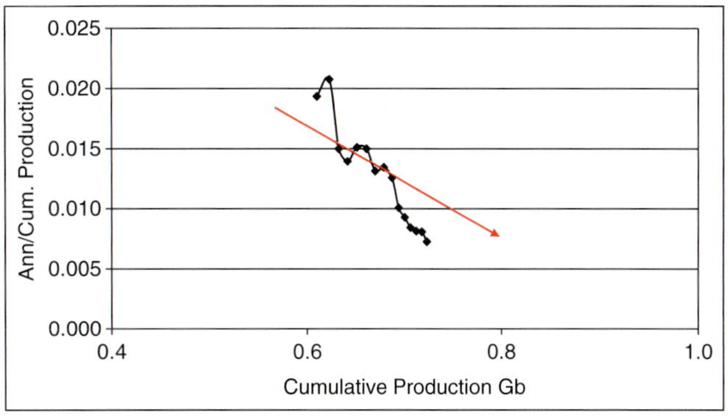

Fig. 31.4 Hungary derivative logistic

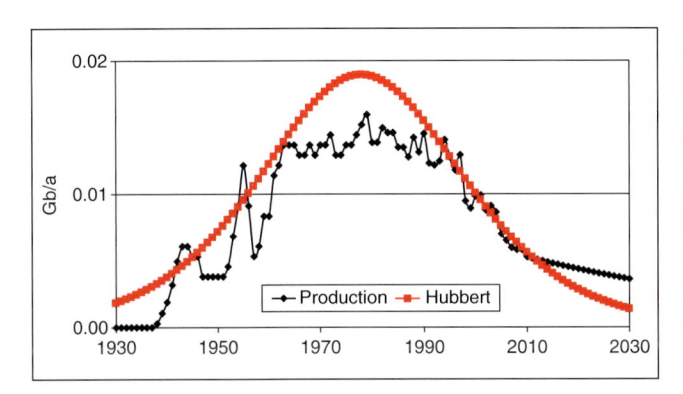

Fig. 31.5 Hungary production: actual and theoretical

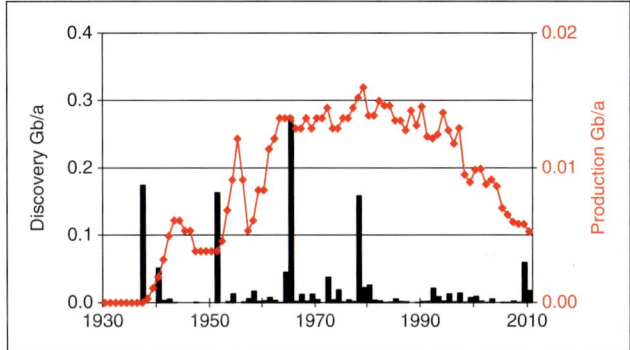

Fig. 31.6 Hungary discovery and production

Kazakhstan

Table 32.1 Kazakhstan regional totals (data through 2010)

Production						Peak Dates			Area	
Amount			Rate				Oil	Gas	'000 km²	
	Gb	Tcf	Date	Mb/a	Gcf/a	Discovery	2000	1979	Onshore	Offshore
PAST	11	14	2000	262	314	Production	2018	2027	2,730	90
FUTURE	34	111	2005	470	882	Exploration	1988		Population	
Known	29	100	2010	557	1312	Consumption	Mb/a	Gcf/a	1900	3.5
Yet-to-Find	5.2	11	2020	750	3402	2010	91	303	2010	16.6
DISCOVERED	40	114	2030	679	3600		b/a	kcf/a	Growth	4.4
TOTAL	45	125	Trade	+466	+1009	Per capita	5.5	18	Density	6

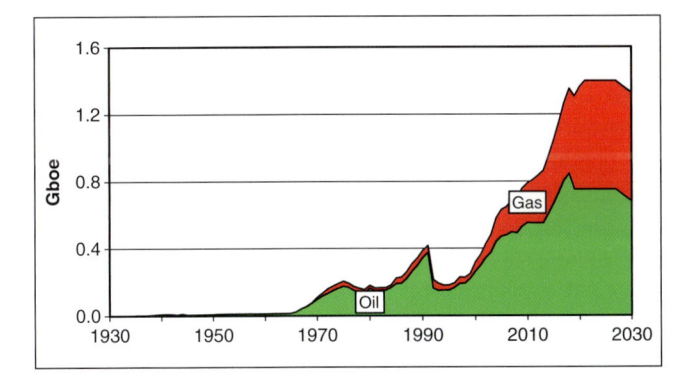

Fig. 32.1 Kazakhstan oil and gas production 1930 to 2030

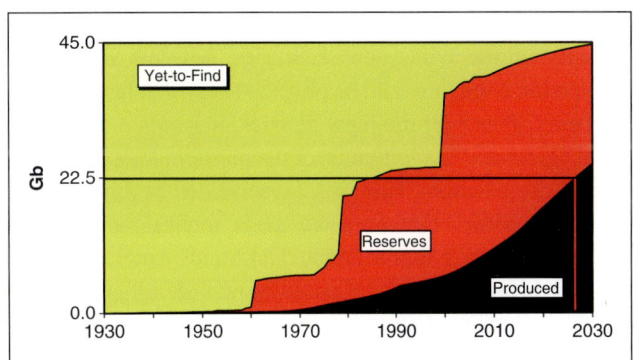

Fig. 32.2 Kazakhstan status of oil depletion

Essential Features

Kazakhstan covers an area of some 2.7 million km², having common frontiers with Russia to the north and China to east. Much of the country comprises low-lying plains, including some extensive deserts, bordering the inland Caspian and Aral Seas, but to the south and east are high mountain ranges, associated with the Himalayan chain. The highest peak rises to as much as 7,000 m. The country suffers from an extreme climate of hot summers and cold winters, with generally low rainfall. The Aral Sea suffered an environmental crisis in 1986 when rivers flowing into it were diverted for irrigation, causing it to be so severely polluted and saline that crops could barely grow in the vicinity. The country, which supports a population of 16.6 million, was a member of the Soviet Union until 1991.

Geology and Prime Petroleum Systems

In geological terms, the country is dominated by two petroleum systems. The most important is the Pre-Caspian Basin in the northwest. The sedimentary column is extremely thick ranging from 8 to 20 km. The margins were folded and faulted during the late Palaeozoic. Silurian source-rocks have charged Carboniferous carbonate reservoirs, commonly lying beneath thick deposits of salt, which form an excellent petroleum seal. To the south, lies a less prolific Jurassic trend, probably relying mainly on Upper Jurassic source-rocks.

C.J. Campbell, *Campbell's Atlas of Oil and Gas Depletion*,
DOI 10.1007/978-1-4614-3576-1_32, © Colin J. Campbell and Alexander Wöstmann 2013

Exploration and Discovery

Exploration in Kazakhstan has had a long history with the first recorded borehole being drilled in 1908. It continued at a modest level until the 1950s, when drilling increased to about 20 boreholes a year, before surging during the 1980s and 1990s. The peak came in 1988 when as many as 81 boreholes were drilled, but has since declined markedly.

A few modest discoveries were made prior to the Second World War, but the first major finds were in 1960 and 1961 when 4.6 Gb was found, followed by 15.5 Gb in 1979. This includes the important Tengis Field, to the northeast of the Caspian, which contains some 6 Gb of high-sulphur oil in a Carboniferous reef at a depth of almost 4,000 m.

Russia, having abundant conventional onshore supplies, had no particular reason to explore the offshore Caspian, which became of great interest to western oil companies after the fall of the Soviet Government. Ownership of the Caspian proved legally contentious. If it were deemed a lake, its mineral rights under international law should be the common property of the contiguous countries, whereas if it were regarded as a sea, it would be physically divided between the countries. Whatever the legal niceties, it seems that a pragmatic agreement has left much of the northern Caspian under Kazakh jurisdiction. This was perceived to be highly prospective territory that attracted great interest, with some reports suggesting that it might rival Saudi Arabia. A huge offshore prospect, named Kashagan, was identified, attracting the attention of the veteran New York promoter, Jack Grynberg. He persuaded Mr Nazarbayev, the President of Kazakhstan, to grant rights to a consortium of major oil companies (led by BP) in return for an over-riding royalty, which he may possibly have shared.

Operating conditions are appalling as shallow water impedes the entry of drilling equipment and a frightful wind coats everything in ice during the winter. If that was not enough, the area lies in the breeding grounds of the sturgeon, an important source of caviar for the Russian market. But eventually boreholes were drilled at either end of the prospect at enormous cost, finding some 9–15 Gb. By most standards, this would be a fine discovery, but BP, Statoil and later, British Gas, pulled out. The operation was thus turned over to Agip, which left poor Mr. Grynberg to sue for his over-riding royalty.

Probably, the structure is a huge platform containing discrete reefal reservoirs, separated by rocks lacking sufficient porosity and permeability to be effective. Other structures in the vicinity were later tested successfully, suggesting that overall Kazakhstan has considerable potential. The Government is now in conflict with the companies over the development of the Kashagan Field, which has been subject to delay. It is subject to a Production-Sharing contract, meaning that the higher the allocated corporate costs, the lower is the Government's revenue, providing fertile ground for dispute.

Insufficient is known about the country to make a reliable assessment but the indications are that about 40 Gb have been discovered, of which only 11 Gb have been produced. Future discovery is here assessed at about 5.2 Gb, giving a rounded total of 45 Gb. With such substantial reserves, the country has little incentive to explore for more in the short term.

Approximately 114 Tcf of gas have been found, of which only 14 Tcf have been produced, meaning that there is a substantial export potential, with Russia or China being the obvious main customers.

Production and Consumption

Little is known regarding the oil production prior to the Second World War, but it had reached a modest 16 kb/d by 1945, growing slowly to 1966 when it increased steeply to reach 1.5 Mb/d in 2010. If the above estimates are about correct, it might be reasonable to model production rising to about 2 Mb/d by 2016 with the opening of new pipelines, followed by a plateau to the onset of decline around 2027. The options for the new pipelines are to route them through Russia; southwards through Turkmenistan and Iran to the Persian Gulf; eastwards to China; or westwards to the Black Sea, and thence onward by tanker or pipeline across Eastern Europe through Kosovo to a terminal on the Adriatic. All options carry potentially grave geopolitical constraints suggesting that production will not in fact rise as modelled on the resource base. It is easy to understand why BP, Statoil and British Gas have pulled out.

Since consumption stands at no more than about 91 Mb/a, the country is in a position to be an increasingly important exporter, barring the development of an unfavourable geopolitical climate. It likewise has a considerable export capacity for gas.

The Oil Age in Perspective

Little is known about the early history of Kazakhstan prior to its incorporation in the great Mongol Empire in the fifteenth century. The Kazakhs, being primarily a nomadic people, later developed into a tribal society, led by their respective leaders, known as Khans. Russia began to encroach on their territory in the early eighteenth century, imposing a form of protectorate. It then became subject to massive waves of immigration, which saw the arrival of Russians, Slavs, Jews, Germans and others, as well as the imposition of the Russian language. The indigenous people were driven from their lands and, in many cases, made destitute. Various unsuccessful uprisings against Russian rule erupted. The Bolshevik Army of Russia defeated the so-called White Russians in 1919–1920 which led to the establishment of a Kazakh State in 1925. It became a full member of the Soviet Union in 1936.

The country was not directly affected by the Second World War although it faced the arrival of many refugees. In 1953, the Soviet Government embarked on an ambitious plan to convert the grasslands to grain production.

The current President, Nursultan Nazarbayev, who was born in 1940 in humble circumstances, rose to prominence in the Communist regime, being appointed President in 1990 before steering the country to independence a year later on the fall of the Soviet Government. He was re-elected President in 1999. An old-time Communist strongman by background, he is generally regarded as a pragmatic leader, successfully controlling the electoral system. He was returned in recent elections, granting himself certain powers for life.

Independence prompted many Russians to leave, such that about half the population is now of Kazakh origins. They belong to the Muslim faith, although that was partially suppressed during the Communist era. The non-Kazakh sectors of the population are concentrated in the cities, including Almaty, the capital.

Kazakhstan is in an anomalous position thanks to its remote landlocked location and its history as a former member of the Soviet Union. It evidently has great potential as an exporter of oil and gas, but is likely to face intractable political and geopolitical pressures, which can only grow as alternative world supplies decline in the years ahead.

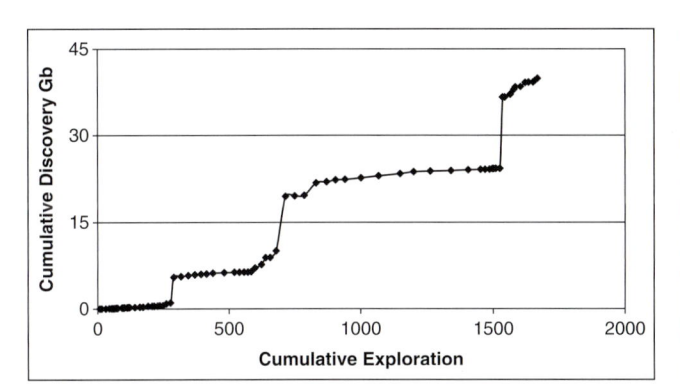

Fig. 32.3 Kazakhstan discovery trend

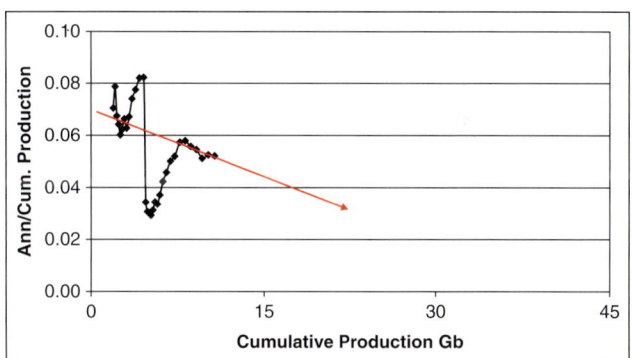

Fig. 32.4 Kazakhstan derivative logistic

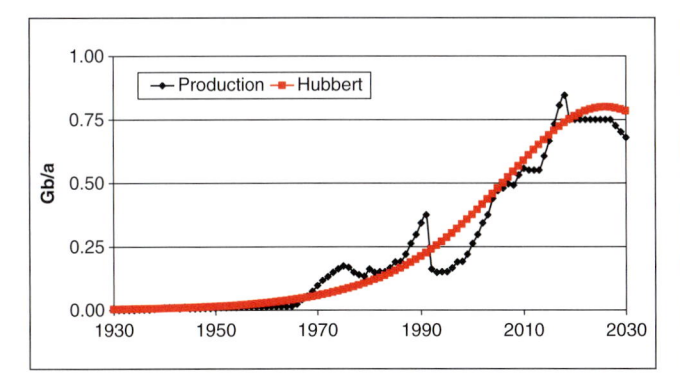

Fig. 32.5 Kazakhstan oil production: actual and theoretical

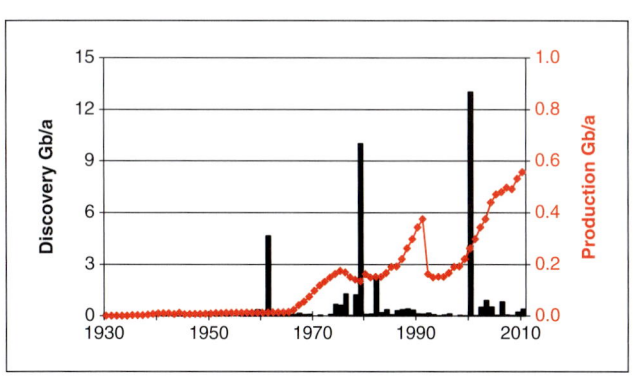

Fig. 32.6 Kazakhstan discovery and production

Romania

Table 33.1 Romania regional totals (data through 2010)

Production					Peak Dates			Area	
Amount			Rate			Oil	Gas	'000 km²	
	G5b	Tcf	Date	Mb/a	Gcf/a	Discovery	1890	1954	Onshore	Offshore
PAST	5.6	45	2000	44	488	Production	1976	1984	240	10
FUTURE	1.4	2.3	2005	38	413	Exploration	1969		Population	
Known	1.3	2.1	2010	32	374	Consumption	Mb/a	Gcf/a	1900	7
Yet-to-Find	0.1	0.2	2020	26	89	2010	72	455	2010	21.4
Discovered	6.9	47	2030	21	26		b/a	kcf/a	Growth	3.0
TOTAL	7.0	48	Trade	−39	−81	Per capita	3.3	21	Density	89

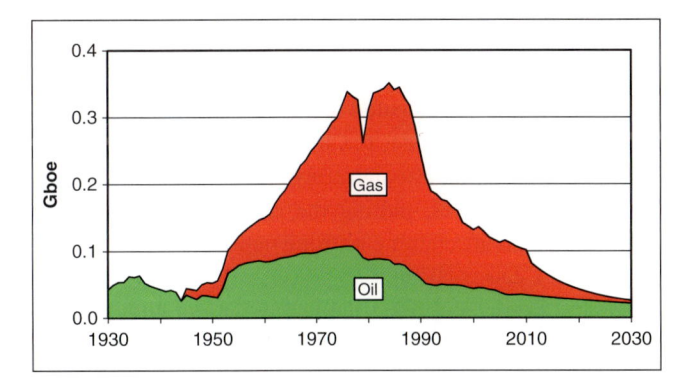

Fig. 33.1 Romania oil and gas production 1930 to 2030

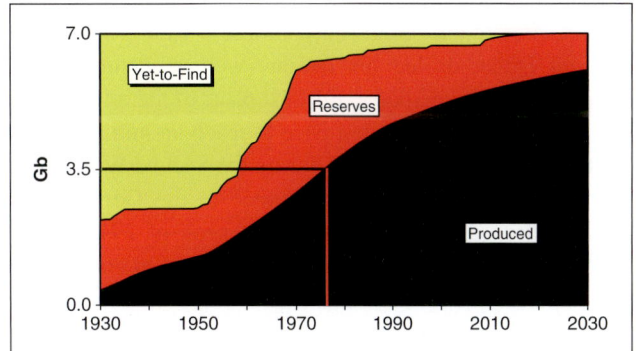

Fig. 33.2 Romania status of oil depletion

Essential Features

The topography of Romania is dominated by a great arc of the Carpathian Mountains that swings through the centre of the country. The highest peak rises to 2,500 m. The passes through the mountains exercised a certain strategic influence in earlier years, being the gateways for migrants from the east. The mountains are flanked by coastal plains to the east, bordering the Black Sea, and to the west by an interior basin of rolling country, known as Transylvania, which is in turn flanked by the Apuseni Massif. The great Danube River forms the frontier with Bulgaria to the south. The country supports a population of nearly 22 million.

Geology and Prime Petroleum Systems

The Pannonian Basin, which has already been described in connection with Hungary and Croatia, extends over western Romania. It is mainly of late Tertiary age and overlies a complex tectonic collision zone, responsible for the Carpathian Mountains.

The Transylvanian Basin to the west is a moderately folded, oil-bearing, late Tertiary Basin. It is bordered by Cenozoic volcanics, associated with obducted oceanic crust in front of the Carpathian chain, which itself comprises a set of thrust sheets. To the east is the Moesian Platform where source rocks of various ages from the Mid-Devonian to the Miocene have charged intervening and overlying reservoirs with oil and gas.

Exploration and Discovery

Romania has had a long oil history with the first recorded discovery being made in 1835 from hand-dug wells in the vicinity of oil seepages. Conventional exploration followed with a number of sizeable finds of oil and gas being made around the turn of the last Century. The peak of oil discovery was in 1890 when 760 Mb were found. Another surge followed the Second World War before decline set in around 1985. Offshore exploration in the Black Sea commenced in 1977 and has delivered a few modest discoveries. A total of some 600 exploration boreholes have now been drilled, delivering 6.9 Gb of oil.

Some 47 Tcf of gas have also been found, with the peak coming in 1954 when 5.2 Tcf were found.

Production and Consumption

Although Romania's oil endowment is comparatively modest, it had an important place in history, being one of the world's major producers in the early years of the last century. In fact, production commenced in 1857, two years before Colonel Drake drilled his famous well in Pennsylvania, which is widely regarded as the birthplace of the oil industry. Production grew at a modest pace to reach an initial peak in 1936 at 174 kb/d. It dwindled during the Second World War, despite its great strategic importance to the German war effort, before climbing to an overall peak of 294 kb/d in 1976. It has since fallen to 88 kb/d and is set to continue to decline at a low depletion rate of 2.1% a year, reflecting the tail end of a large number of small onshore fields.

Gas production commenced after the Second World War and rose to a peak of 1.5 Tcf/a in 1984 before declining to its present level of 374 Gcf/a.

Oil consumption stands at 72 Mb/a, meaning that the country imports more than half of its needs. It also has to import much of its gas requirement, reportedly standing at almost 81 Gcf a year.

The Oil Age in Perspective

Romania has had a very long history having been settled 2,000 years before Christ by immigrants from the east who mingled with Neolithic stock. After centuries of resistance, it fell to the Roman Empire in AD 101, becoming the province of Dacia. That was followed by successive invasions from the east by Huns, Goths, Slavs and others, who arrived over the rest of the first millennium. Christianity under the Eastern Patriarch was adopted towards the end of this period.

During the Middle Ages, the various principalities that constituted the country fell to the Empire of the Ottoman Turks, while retaining a degree of independence as vassal States, paying tribute. That chapter was followed during the eighteenth and nineteenth centuries by growing Russian influence, including an invasion in 1848, as well as various moves to greater unification and independence. The country began to participate in alliances with the powers of Western Europe, but found itself in growing conflict with its neighbour, the Austro-Hungarian Empire under the Hapsburg kings.

In 1916, Romania entered the First World War on the side of Britain and France but suffered badly when it became isolated and cut off from military supplies, due to the Bolshevik Revolution of 1917. It nevertheless benefited greatly from the ensuing Versailles Peace Treaty when new frontiers were drawn expanding its territory, such that as much as 30% of its population were of non-Romanian origins.

Conflicts between authoritarian and democratic rule marked the inter-war years, exacerbated by the First Great Depression of 1929–1930. A military dictatorship arose during the Second World War making an alliance with Germany. Romanian troops participated in the invasion of Russia. The country then changed sides towards the end of the war before being occupied by the Russian army in 1945. The Communist Party came to power, putting the country firmly into the Soviet bloc, albeit retaining a strong national identity.

Nicolae Ceaucescu came to power in 1967, when he was a popular figure, and ruled the country in an authoritarian style for 22 years, resisting Soviet pressure, being amply supported by his wife who took a leading role in political affairs. He tried to revolutionise economic life in the country under strict Communist ideology, but the experiment failed, causing great hardship. A popular rising accompanied by a military coup finally ended the regime in 1989. The Ceaucescu's were forced to flee the capital, being eventually arrested and shot after a summary military trial.

The country then moved into the western camp, with various elections and changes of government, becoming a full member of the European Union in 2007, but tensions between the various regions and ethnic groups probably simmer beneath the surface. It may be imagined that they will resurface during the Second Half of the Age of Oil as the different regions are forced to develop, or recover, greater independence to live within their own means. The population is not however excessive, so that the rich soils and benign climate can support the people into the foreseeable future.

33 Romania

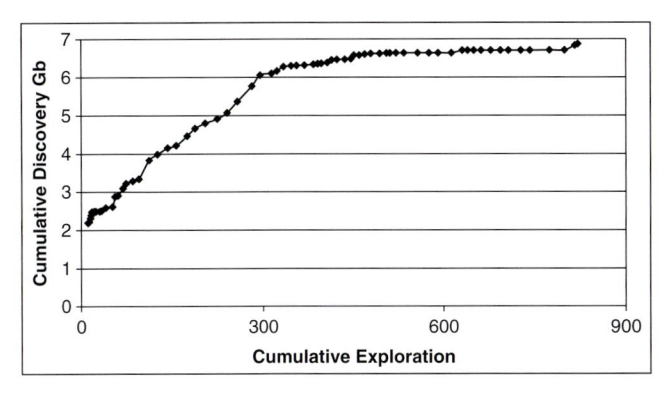

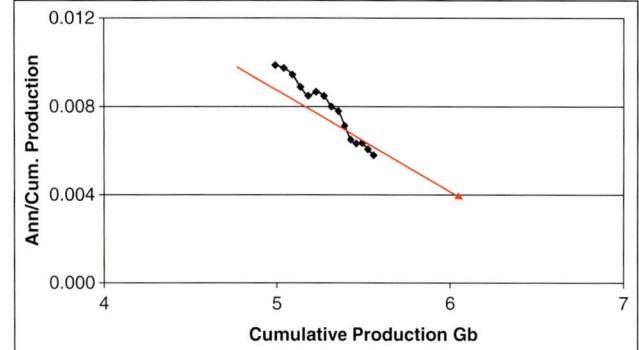

Fig. 33.3 Romania discovery trend

Fig. 33.4 Romania derivative logistic

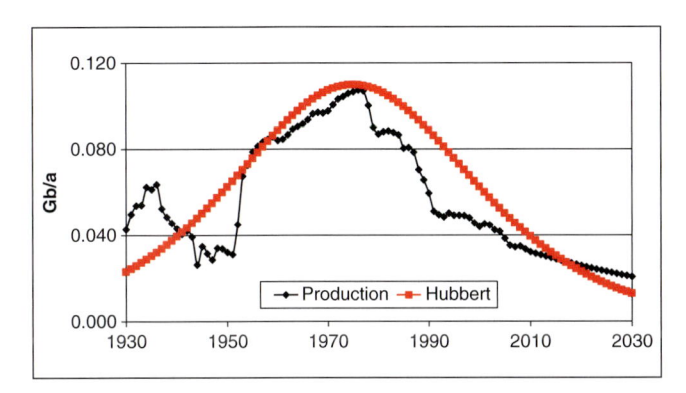

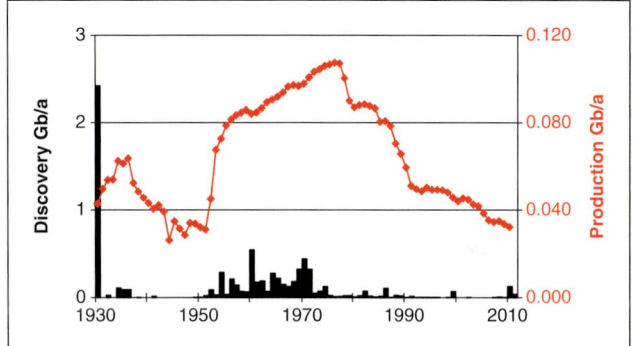

Fig. 33.5 Romania production: actual and theoretical

Fig. 33.6 Romania discovery and production

Russia

Table 34.1 Russia regional totals (data through 2010)

Production to 2100					Peak Dates			Area		
Amount		Rate				Oil	Gas	'000 km²	
	Gb	Tcf	Date	Mb/a	Gcf/a	Discovery	1960	1966	Onshore	Offshore
PAST	150	631	2000	2,365	18,568	Production	1983	2025	17,140	1,400
FUTURE	80	869	2005	3,070	20,361	Exploration	1988		Population	
Known	64	608	2010	3,178	20,033	*Consumption*	Mb/a	Gcf/a	1900	90
Yet-to-Find	16	261	2020	2,153	25,000	2010	1072	17495	2010	142.8
DISCOVERED	214	1239	2030	1,458	17,769		b/a	kcf/a	Growth	1.6
TOTAL	230	1500	*Trade*	+2,106	+2,538	Per capita	7.5	123	Density	8

Applies to Regular Conventional Oil and Gas only, excluding extensive Polar regions.

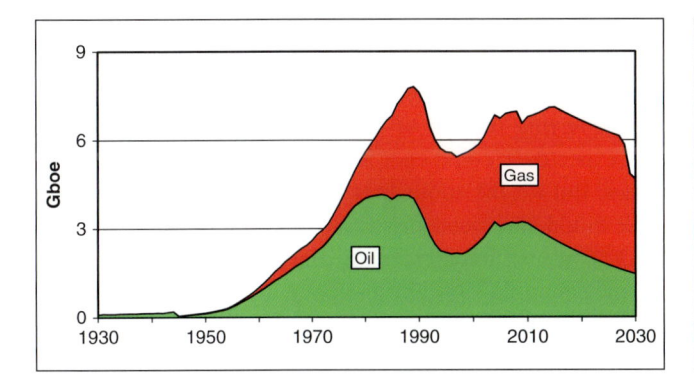

Fig. 34.1 Russia oil and gas production 1930 to 2030

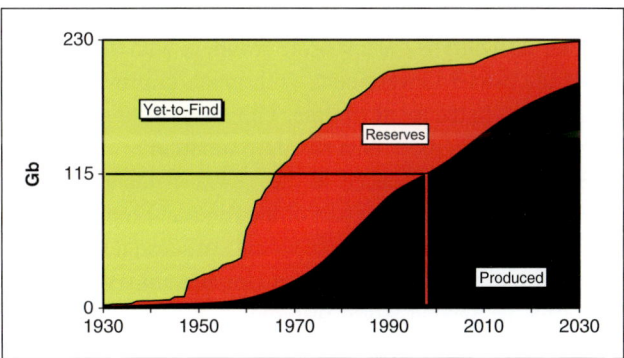

Fig. 34.2 Russia status of oil depletion

Essential Features

Russia is the world's largest country covering an area of 17 million km² (including the Arctic regions), which is almost double the size of the United States. It supports a population of about 143 million, being rather sparsely populated. The country, which has an extreme continental climate, can be divided into six main physiographic regions, described generally from west to east. First, is the Russian Plain, which is a glaciated terrain of lakes and rivers, being drained principally by the Volga River that flows south into the Caspian Sea. Second, are the northward-trending Ural Mountains, which is an ancient chain rising to no more than 2,000 m and cut by accessible passes. Third, are the huge plains of West Siberia, which are drained by the Ob and Yenisey Rivers, flowing northward to the Arctic. Fourth is the Central Siberian Plateau, covering extensive tracts at an altitude of 300–700 m, being flanked to the south and east by mountainous country, and including Lake Baikal, the world's largest lake, covering 31,500 km². Fifth, are the mainly mountainous Pacific borderlands, including the peninsulas of Sakhalin and Kamchatka, which flank the Sea of Okhotsk. Sixth are the Arctic Seas and islands, some of which are of a substantial size.

Geology and Prime Petroleum Systems

Much of the eastern part of the country is underlain by the Siberian Shield, composed of ancient non-prospective rocks. A Permian tectonic plate boundary gave rise to the Urals Mountain.

C.J. Campbell, *Campbell's Atlas of Oil and Gas Depletion*,
DOI 10.1007/978-1-4614-3576-1_34, © Colin J. Campbell and Alexander Wöstmann 2013

This large territory has a large number of sedimentary basins, from which we may recognise six prime petroleum systems.
- The Western basins between the Barents and Caspian Seas with Silurian source-rocks, including the Volga-Ural (Pre-Caspian) basin, and the North Caucasus.
- The West Siberian basins with Jurassic source-rocks.
- The Arctic domain, which is predominantly gas-prone due to the deep burial of source-rocks under the weight of fluctuating ice-caps in the geological past, including the Timan-Pechora trend, which runs offshore into the Arctic Ocean.
- The locally productive Tertiary deltaic basins of the Pacific margin, especially at Sakhalin.

Exploration and Discovery

It is important to note that *Regular Conventional Oil and Gas* excludes Polar Oil by definition, which poses a particular difficulty in the case of Russia, because the available database is insufficient to apply the boundary accurately. Another difficulty is the changing frontiers with the rise and fall of the Soviet Government, with much information confused or lost in the mists of time. The assessment given here is accordingly no more than an approximation

Exploration commenced in the 1840s in the vicinity of Baku on the Caspian, then part of the Russian Empire, where hand-dug wells were sunk in the vicinity of seepages. It lapsed during the early years of the Soviet Union, until it was revitalised after the Second World War. In fact, the Soviet explorers proved to be highly efficient, being able to apply scientific methods, free of commercial constraints. They even had the luxury of being allowed to drill boreholes for geological information. They pioneered the geochemical breakthrough that defined source-rocks and generating belts.

Drilling commenced prior to the Second World War at a very low level, but picked up in the 1950s and 1960s with more than 200 exploration boreholes being drilled in most years, reaching an overall peak in 1988 when 464 were drilled.

Discovery in sub-Arctic Russia peaked around 1960 and was followed by the corresponding peak of production in 1987. Exactly how much was found is hard to know, because the Soviet classification of reserves ignored commercial constraints. It is normal to equate the reserve categories $A+B+C_1$ under the Soviet classification with the so-called *Proved Reserves* of the West, but decline curve analysis shows that reserves of most Russian fields, reported on that basis, have to be reduced by about 30% to obtain realistic estimates.

It is clear that the reserve estimates of around 60 Gb as reported by the *Oil & Gas Journal* are far too low. Exactly how far is difficult to know, but we tentatively favour a figure of about 64 Gb. It gives a fairly low depletion rate of 3.8%, which is one argument against higher estimates. If we add to this 30 Gb of Polar oil, together with substantial deposits of heavy oil in Eastern Siberia and NGL from gas fields, which are here excluded from *Regular Oil* by definition, we could approach the total of 100–120 Gb, as has been reported by Russian companies.

Production and Consumption

Early oil production is unsure and also confused because it does not distinguish the different regions of the former Soviet Union, but it is estimated that about one billion barrels had been produced by 1937, at which date production had risen to about 350 kb/d. It fell steeply during the latter years of the Second World War but rose thereafter passing a low of 1 Mb/d in 1956 to reach an overall peak of 11.36 Mb/d in 1983. It then collapsed with the fall of the Soviet Government to 5.9 Mb/d in 1998, before recovering to 8.7 Mb/d in 2010, at which point 65% of the assessed endowment had been produced, if we exclude, by definition, Polar production. In part, the recent rise was making good the production that would have already been secured but for the dislocations accompanying the fall of the Soviet regime. Production is now expected to commence its terminal decline at almost 4% a year. But for the anomalous fall in production, the overall peak would have been passed in the 1990s.

Oil consumption is currently running at 1,072 Mb/a, making the country a substantial exporter of 2,106 Mb/a. But on present trends and assessments, export capacity will fall to zero by around 2015 or even sooner if domestic consumption should increase faster than expected.

In earlier years, associated gas must have been substantially flared, but after the Second World War it began to be exploited as a fuel for heating and electricity generation. Production in sub-Arctic Russia grew steadily to peak in 1991 at 22 Tcf a year, but has since fallen to about half that amount, as Arctic supplies rose to a dominant position. It should be noted therefore that the figures above indicate the import of Arctic gas to meet Russia requirements reflecting this study's differentiation between Arctic and sub-Arctic gas production. It is difficult to forecast the future but it is here assumed that new sub-Arctic production will indeed be brought on stream to reach a plateau at about 25 Tcf/a by 2015, at the indicated midpoint of depletion.

The Oil Age in Perspective

The country of Russia was occupied over its long history by Slavs, Huns and others, migrating from the plains of Mongolia. The western part of the country came under the control of the Varanginians, who may have been related to the Vikings, establishing a trade route from the Baltic to the Black Sea. The orthodox Christianity of Byzantium was adopted around 1,000 AD.

Later, in their turn, came Mongol and Tartar invaders, but they were generally assimilated in a growing number of principalities and petty kingdoms that were developing in Western Russia, including Muscovy on the site of the present capital. Ivan the Terrible began to expand Muscovite influence in the sixteenth century, being largely the pawn of his noblemen. He espoused European influences, being responsible for the construction of the Kremlin with the help of Italian craftsmen.

The Romanov Dynasty followed and continued in power to the twentieth century. Peter the Great (1689–1725) consolidated power, settling disputes with Turkey to the south, and Sweden and Poland to the west, which paved the way for the expansion of a new Russian Empire. His main achievement was the establishment of a competent administration and an improved educational system, as well as the founding of St. Petersburg on the shores of the Baltic, which gave Russia an outlet for world trade.

The next luminary was Catherine the Great, the German widow of an ineffectual Czar, who came to power in 1763 after a coup d'etat, organised by her lover, Count Orlov. Her reign was marked by both amorous and territorial conquests. A general state of tension developed after her death in 1796 with various wars against Turkey and the European powers, which resulted in the invasion by Napoleon who was, however, defeated at the gates of Moscow in 1812.

The Czars faced great difficulties in administering their vast territories and relied heavily on the nobility who owned and controlled an under-class of serfs to work their lands. The nineteenth century also saw the development of industry, mining and railways, with the emergence of a culture of prosperous capitalists, miners and industrial workers, effectively drawing the curtains on the earlier, essentially feudal, environment.

The Crimean War of 1853–1856 found Russia in conflict with Britain, France and Turkey, who were opposing the threat of Russian expansion into the Middle East, whose importance then lay, not in its oil, but in its strategic position facing the British Empire. Defeat led the reigning Czar to move towards the liberation of the serfs, which was naturally resisted by the nobility. It was not altogether welcomed by the liberated Serfs who found themselves having to borrow money. Progress was slow and sowed the seeds of revolution, in some cases encouraged by sympathetic intellectuals. Russia's large Jewish population was mistrusted by both the officials and the serfs alike, who were possibly reacting to the hidden pressures of usury. Waves of anti-Semitic pogroms swept the country, forcing many Jews to emigrate.

Russia's eyes turned eastward during the early years of the twentieth century where, in company with Britain, France and Germany, it sought to capture the markets of China and Japan, exploiting also the conflicts between those countries. The trans-Siberian railway was constructed. But a surprise attack by the Japanese in 1904 led to war, in which Russia suffered several defeats that in turn stimulated more domestic unrest. In the following year, disgruntled workers demonstrated in St. Petersburg, with the intent of delivering a petition to Czar Nicholas, but they were brutally cut down in a massacre that became known as Bloody Sunday. The Bolshevik Movement, amongst others, gained strength, pressing for reform.

Meanwhile in Western Europe, a newly united industrial Germany was challenging the mercantile empires of Britain and France, who reacted by signing a complex set of alliances, including a pact of mutual assistance between France and Russia. The catalyst for the outbreak of the ensuing world war in 1914 was a move by Serbia, whose Slav population was supported by Russia, to secede from the Austrian Empire. With the outbreak of hostilities, a Russian army marched into East Prussia, but was repulsed. The privations of war exacerbated the tensions at home, which erupted in February 1917 in a spontaneous popular outburst against the Government that was soon exploited by the Bolsheviks who proposed a Soviet administration. Several of these leaders, including Lenin and Trotsky, were Jewish, who were perhaps resentful of previous anti-Semitic oppression, becoming advocates of Zionism. A civil war followed in 1918 between the so-called Red and White armies. Czar Nicholas and his family were arrested and murdered, and an oil workers' leader from Baku, later to be known as Joseph Stalin, came into prominence, eventually taking control of the Government after Lenin's death in 1928.

The inter-war years saw Russia, leading the Union of Soviet Socialist Republics (USSR), develop largely in isolation, with all strands of its economy placed under State ownership and control. Stalin proved to have an iron hand, suppressing any hint of opposition by ruthless means. Millions of people lost their lives. Even so, the Soviet experiment appealed to many intellectuals in other countries, inspiring elements within the socialist movement, seeking a milder variant.

The Second World War was essentially an extension of the first, and after an initial alliance with Germany under a non-aggression pact, Russia again joined with Britain and France. After initial reverses which brought the German army to the gates of Moscow, the Red Army began to advance, and in 1945 raised its *Hammer and Sickle* banner over the ruins of Berlin. Russia had suffered grievously in the war and was not about to give up the territories it had conquered in East Europe, where puppet Communist regimes were established. The British and French empires were extinguished by the war, leaving the United States and the Soviet Union to glower at each other for the next 40 years in what became known as the Cold War. That in turn ended in 1991 when a moderate Communist leader, Mikhail Gorbachev, was ousted by the late Boris Yeltsin. A new capitalism, complete with robber barons, came to Russia, leaving many Russians to look back with a certain nostalgia for the old days when they knew where they stood. The Soviet Empire was dismembered, with many of its component parts becoming independent countries facing their own internal conflicts. Foreign oil companies flocked to Russia, and new entrepreneurs in Russia itself took positions with capitalistic verve.

Vladimir Putin succeeded Boris Yeltsin as President in 2000 and was re-elected in 2004 for a second term. He comes from a humble Communist background, his grandfather having been no less than the personal cook to both Stalin and Lenin, but he advanced rapidly to graduate in Law at Leningrad University in 1975. He later joined the Intelligence Service, before returning to St. Petersburg, where he worked for the city administration, becoming active in politics. In 1997, he submitted a doctorial thesis at the university entitled *The Strategic Planning of Regional Resources*, which may be of great significance, suggesting that he is fully aware of the depletion and the geopolitical significance of Russia's oil and gas.

Since cementing his power as President, Putin has successfully recovered State control of the country's oil and gas industries, extinguishing some of the new local companies, including Yukos, whose founder finds himself in jail for tax evasion. Some major companies, such as BP and Shell, continue to operate, but face a diminished role, more akin to that of a contractor than the proprietor of national resources.

Mr. Putin was succeeded by Mr. Medvedev in early 2008, who had run Gasprom, Russia's State Gas Company, although Mr. Putin remained in an influential role as Prime Minister and recently won re-election as President. It remains to be seen what policy a future government will adopt, but looking ahead, it seems that Russia is relatively well placed to face the Second Half of the Age of Oil. It has substantial oil and gas resources, mainly under State control, which it will likely try to preserve for the benefit of its own citizens. At first, this may be challenged by, especially, Europe as a hostile act, but gradually, as the true nature of global depletion becomes appreciated, people may understand the logic. The rouble may emerge as a strong well-managed currency, relative to the dollar and euro which weaken, as the market upon which they were built declines in parallel with global oil supply.

Urgent steps will have to be taken to reduce the massive waste of energy in Russia, especially for domestic heating, which was provided virtually for free under the Soviet regime, but otherwise the country could probably learn to live within the bounds of its natural resources, having been spared the excesses of industrialisation and affluence found in the West. Its birth rate has already fallen markedly, reducing the pressures.

In international affairs, it is likely that Russia will remain largely aloof, maintaining what could be called the Cool War. It may well come to the help of the Middle East in recovering from the impact of the Anglo-American invasion, especially in the realm of oil production, and it may help supply China with much needed energy. So far as Europe is concerned, there seems to be a special relationship with Germany.

There arises a certain irony whereby countries and people who suffered during the first half of the Age of Oil may do better during the second half as the Planet reverts to a more sustainable structure.

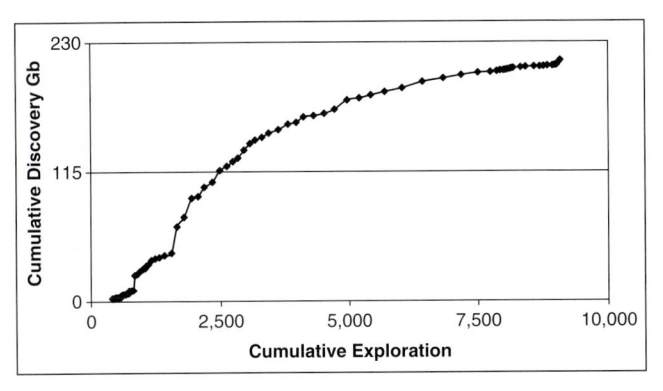

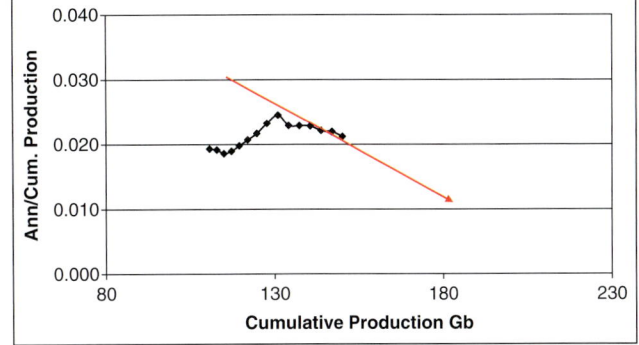

Fig. 34.3 Russia discovery trend

Fig. 34.4 Russia derivative logistic

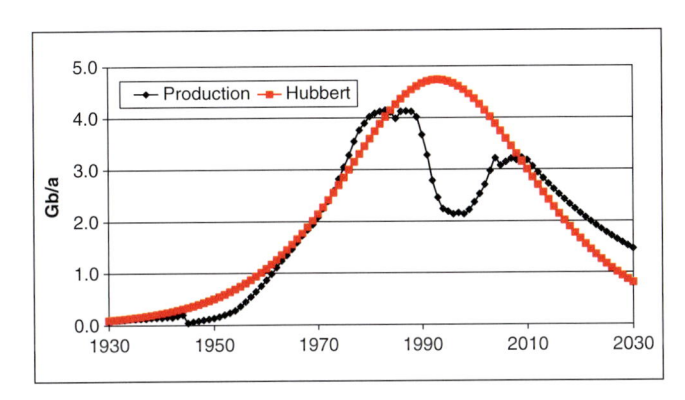

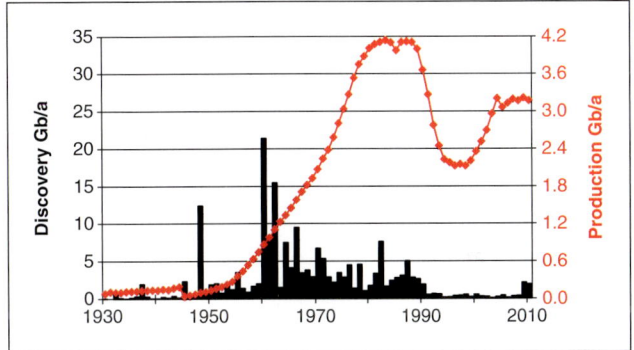

Fig. 34.5 Russia oil production: actual and theoretical

Fig. 34.6 Russia discovery and production

Turkmenistan

Table 35.1 Turkmenistan regional totals (data through 2010)

Production						Peak Dates			Area	
Amount			Rate				Oil	Gas	'000 km²	
	Gb	Tcf	Date	Mb/a	Gcf/a	Discovery	1956	1973	Onshore	Offshore
PAST	3.5	78	2000	52	1,642	Production	1973	2030	490	75
FUTURE	2.0	297	2005	67	2,225	Exploration	1986		Population	
Known	1.6	223	2010	65	1,600	*Consumption*	Mb/a	Gcf/a	1900	0.9
Yet-to-Find	0.4	74	2020	47	4,149	2010	43	720	2010	5.1
Discovered	5.0	301	2030	35	4,500		b/a	kcf/a	Growth	6
TOTAL	5.5	375	*Trade*	+22	+880	Per capita	8.5	141	Density	10

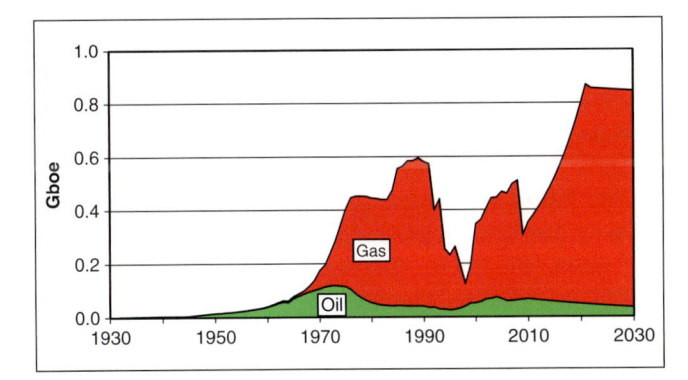

Fig. 35.1 Turkmenistan oil and gas production 1930 to 2030

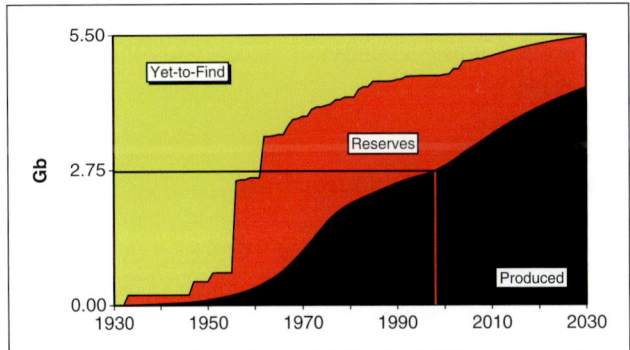

Fig. 35.2 Turkmenistan status of oil depletion

Essential Features

Turkmenistan is a remote country adjoining the eastern shore of the Caspian Sea, being bordered by Iran, Afghanistan and Uzbekistan. It is mainly covered by deserts, although the foothills of the Himalayan Mountains extend into the southeastern parts of the country. A curious shallow saline lagoon, known as Kara-Bogaz Gol, covers an area of 12,000 km² on the shore of the Caspian. Extreme evaporation in the hot climate makes it the world's largest deposit of salt.

The population, which numbers about five million, is composed largely of Turkmen who were a nomadic tribal people, although there are in addition a substantial number of immigrants from Russia and elsewhere who live in the few cities.

Geology and Prime Petroleum Systems

The deserts of southeastern Turkmenistan overlie the Amu Darya Basin which is filled by a thick sequence of Mesozoic and Tertiary sediments. The prime source-rocks are Lower–Middle Jurassic coals, which have given rise to substantial amounts of gas in a process akin to that responsible for the southern North Sea gas fields. In addition, Upper Jurassic basinal black shales provide a limited source of oil in the east-central part of the basin. The oil and gas from these sources have migrated upwards to charge reservoirs, primarily in Upper Jurassic carbonates and Lower Cretaceous sandstones, but also to a lesser extent the overlying Paleocene reservoirs. Thick Upper Jurassic salt deposits are an important element influencing the migration of gas and acting as a seal.

Another system is represented by the paleo-delta of the Volga River that occurs in coastal regions and offshore in the Caspian, where Upper Tertiary sands have been charged by oil and gas from Lower Tertiary sources.

Exploration and Discovery

The first recorded exploration borehole was drilled as long ago as 1882, but there was little subsequent activity until after the Second World War. Exploration was then stepped up to a peak in 1990 when 33 boreholes were sunk. Drilling declined with the fall of the Soviet Government but has since recovered somewhat. Minor gas discoveries were made in the early 1950s and were followed by giant finds during the 1960s and 1970s, giving an overall peak of discovery in 1973 when 57 Tcf of gas was found. Modest oil discoveries were made in parallel, with the peak coming in 1956 when 1.8 Gb were found.

Exploration was extended offshore into the Caspian during the 1960s where a number of modest finds of oil and gas were made.

Production and Consumption

Oil production commenced in the 1930s at a low level, but increased after the Second World War, thanks in part to the construction of pipelines, and reached a peak in 1973 at 324 kb/d. It then declined to a low in 1995 before recovering to its present level of 178 kb/d. It is now set to decline at about 3% a year. Oil consumption stands at 43 Mb a year making the country a modest exporter for the next few years.

Gas production commenced in 1945 rising to pass 2 Tcf/a by 1977, set primarily by pipeline capacity, although also falling in the late 1990s for political reasons. Gas consumption stands at 720 Gcf a year, making the country a substantial exporter, partly to Iran.

The Oil Age in Perspective

Archaeologists have unearthed Bronze Age remains in Turkmenistan, but little is known of its early history. It was invaded in the eleventh century by various Turkic tribes from Central Asia but remained isolated and essentially tribal until it was brought into the Russian Empire in the nineteenth century. That move was opposed by the people who occasionally rose in rebellion.

The Turkmens, who are predominantly Muslim, were enthusiastic participants in the 1916 Russian Revolution and were involved in the ensuing civil war that preceded the establishment of the Soviet Government. The country formally became a member of the Soviet Union in 1924 when its national identity was defined. A degree of Europeanization followed with the suppression of local customs and traditions, but met with resistance.

Turkmenistan declared its independence in 1991, following the fall of the Soviets, but the President, Saparmurad Niyazov, a classic Soviet strongman, remained in power until his death in 2006. The succession of Gurbanguly Berdimuhammedow, who is allegedly an illegitimate son of Niyazov, remains in dispute.

The country will probably revert to nomadic tribalism during the Second Half of the Oil Age, rediscovering a sustainable life style that worked for centuries in the past. Many of the recent immigrants, primarily city dwellers, might return to their origins in Russia and elsewhere.

35 Turkmenistan

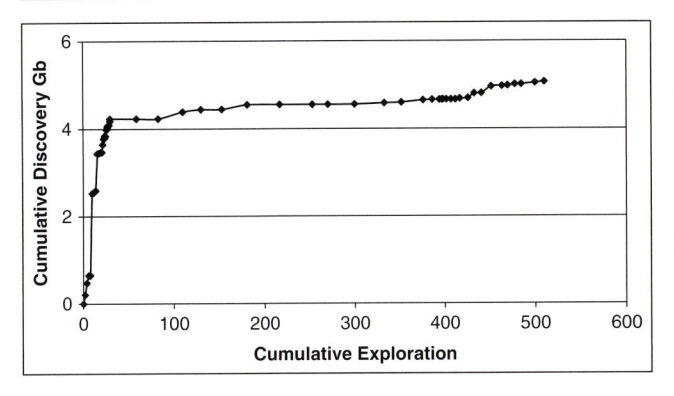

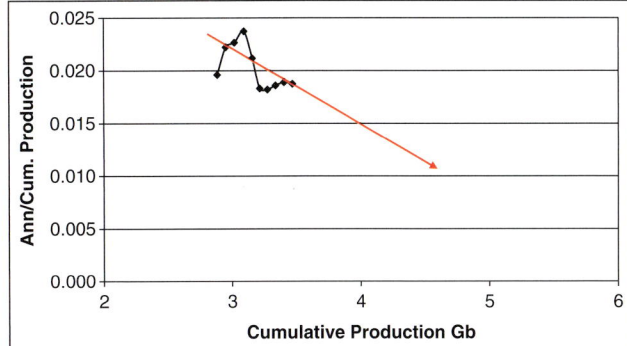

Fig. 35.3 Turkmenistan discovery trend

Fig. 35.4 Turkmenistan derivative logistic

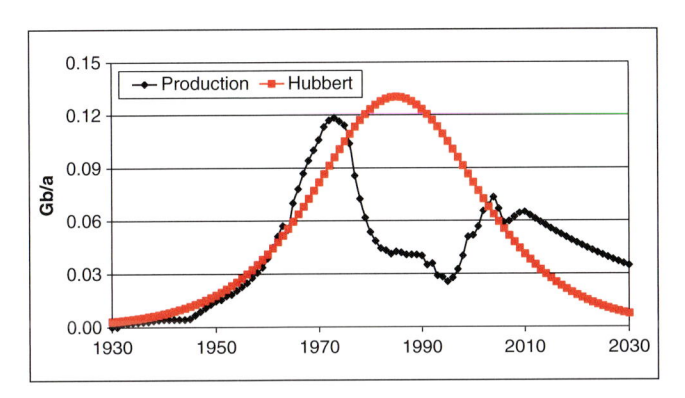

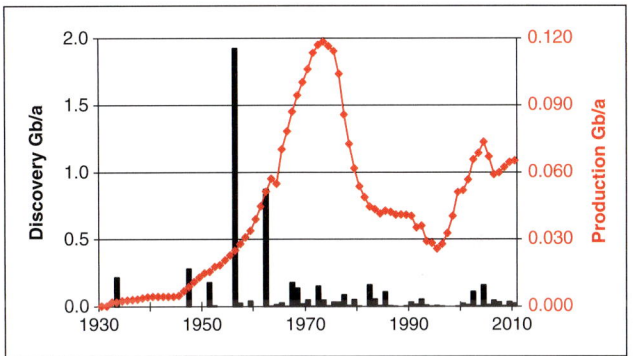

Fig. 35.5 Turkmenistan production: actual and theoretical

Fig. 35.6 Turkmenistan discovery and production

Ukraine

Table 36.1 Ukraine regional totals (data through 2010)

	Production					Peak dates			Area	
	Amount		Rate				Oil	Gas	'000 km²	
	Gb	Tcf	Date	Mb/a	Gcf/a	Discovery	1962	1950	Onshore	Offshore
PAST	2.9	64	2000	27	636	Production	1970	1975	610	35
FUTURE	1.6	36	2005	33	685	Exploration	2000		Population	
Known	1.6	32	2010	26	684	*Consumption*	Mb/a	Gcf/a	1900	15
Yet-to-Find	0.1	3.6	2020	22	669	2010	108	1,877	2010	45.7
DISCOVERED	4.4	96	2030	19	531		b/a	kcf/a	Growth	3.2
TOTAL	4.5	100	*Trade*	−82	−1,193	Per capita	2.4	41	Density	75

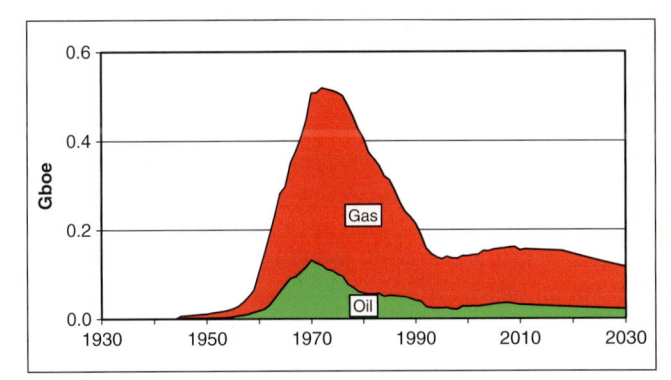

Fig. 36.1 Ukraine oil and gas Production 1930 to 2030

Fig. 36.2 Ukraine status of oil depletion

Essential Features

The Ukraine consists mainly of plains which are flanked by the foothills of the Carpathians to the west, and the Black Sea and the Sea of Azov to the south. The latter are separated by the Crimea, which is a partly mountainous virtual island, connected to the mainland by a low-lying narrow isthmus. The country is drained by the Dnieper and Donets Rivers, with associated lakes and tributaries, flowing south into the Black Sea. It is known for its rich soils and agricultural lands and also has substantial iron and coal deposits, supporting a large iron and steel industry. The country has a population of nearly 46 million.

Geology and Prime Petroleum Systems

A rift system, known as the Dnieper-Donets Basin, separates Precambrian massifs to the northeast and southwest. It is filled with over 10,000 m of Devonian and Carboniferous strata, capped by up to 4,000 m of Mesozoic and Tertiary rocks. Important deposits of salt are found in the Permian and Late Devonian sequences. There is also a thick development of Devonian volcanic rocks, which cut part of the rift.

The prime source-rocks are in the Devonian and Lower Carboniferous sequence, with the overlying Permian salt representing an important seal. The deeper coals may also have been a source of gas. The primary reservoirs are in the

C.J. Campbell, *Campbell's Atlas of Oil and Gas Depletion*,
DOI 10.1007/978-1-4614-3576-1_36, © Colin J. Campbell and Alexander Wöstmann 2013

Upper Carboniferous to Permian sequence. It is evident that this is primarily a gas province.

Exploration and Discovery

Records of the early exploration in Ukraine are lost in the mists of history, but some successful drilling was evidently conducted prior to the Second World War. It was followed in 1950 by a major gas discovery, known as the Shebelinka Field, holding 22 Tcf. But the main thrust of exploration came in the 1960s and 1970s when ten to twenty boreholes a year were drilled, yielding another major gas find in 1968 with 10 Tcf. Efforts declined on the fall of the Soviets but picked up again in the late 1990s, with the overall peak of drilling coming in 2000 when 27 boreholes were sunk. A large number of small to modest oil discoveries have been made. The two largest are the Dolyna Field, found in 1957 with about 500 Mb, and the Lelyaky Field, found in 1962 with about 400 Mb.

Overall, it is evident that this is a mature gas-prone province, with deeply buried petroleum systems. A few modest finds may yet be made, but the scope is limited.

Production and Consumption

Oil production built up after the Second World War to reach a modest peak of 361 kb/d in 1970, since when it has declined to about 71 kb/d, being set to continue to decline at a relatively low Depletion Rate of about 2 %. This may imply that operations are not being conducted in the most modern way, which ironically means that more is left in the ground for the future. Consumption stands at 108 Mb/a, of which 82 Mb/a are imported.

Gas production also built up after the Second World War to reach a plateau in the 5–6 Tcf a year range during the 1970s, since when it has declined to about 700 Gcf a year. Gas consumption stands at about 2 Tcf/a. The extra supplies needed are imported from Russia which has recently caused some political dispute.

The Oil Age in Perspective

A number of Greek colonies were established along the northern shore of the Black Sea in the sixth and seventh centuries BC, which were later absorbed into the Roman Empire. The area was in turn subject to successive invasions of Goths from the Baltic regions, Huns, Magyars, Slavs and others during the first millennium. The frontiers have changed many times over a turbulent history. A form of national identity began to emerge in the tenth century under the so-called Kievan Empire, which extended from the Black Sea to the Baltic, including what is now Moscow. It was built by the Varanginian merchant-warriors, who may have been relatives of the Vikings from Scandinavia, and later adopted Orthodox Christianity.

During the Middle Ages, the country fell to Mongol invasions under Ghengis Khan, before becoming part of a new Polish-Lithuanian Empire in a development which was accompanied by religious conflict between the Roman and Eastern branches of the Catholic Church.

The famous Cossacks originated on the steppes of the Ukraine evolving from hunters into cavalry, given to occasional revolts and rebellions. One such rebellion led to the division of the Ukraine into Polish and Russian spheres of influence. The Polish sector, known as Galicia, later fell to the Austrians. Disputes and conflicts continued until the end of the eighteenth century when much of the territory was fully absorbed into the Russian Empire, being designated as a province. Landlords, many of Polish origins, remained in dominant positions, controlling large numbers of serfs, while a substantial Jewish community was subject to varying degrees of oppression. Coal mining led to industrialisation and the growth of industrial workers and miners adding a new political force.

The First World War helped define the frontiers, leading to what became the Soviet Ukraine, which was duly incorporated into the Soviet Union in 1922. The country then enjoyed, if that is the word, a period of rapid industrialisation, but pressures for greater independence built during the 1930s, being brutally suppressed by Moscow in what is known as the Great Terror. Crop failures and Soviet collectivisation led to serious famines in the 1930s, in which millions died.

The Ukraine was invaded in 1941 by German troops, who were at first welcomed as liberators, but soon showed themselves to be even more brutal. Some 600,000 Jews were executed and 2.5 million Ukrainians transported to Germany as slave labour, prompting the growth of various underground Resistance movements. The defeat of the German army at Stalingrad in 1943 turned the tables, allowing the Soviet army to march westward liberating the Ukraine one year later. The Paris Peace Treaty of 1947 finally defined the frontiers of the present Ukraine. The country had suffered terribly in the war with the loss of some five million people. But peace was little reprieve as a new epoch of Soviet terror ensued, leading to yet another famine, during which, ironically, food was exported. Conditions improved under Khrushchev, who succeeded Stalin in 1953, as he had a more sympathetic feeling for the Ukraine, having spent some of his early years there.

In 1986 came the Chernobyl disaster when a nuclear power plant in the western Ukraine exploded, causing radioactive contamination over a wide area, which has been responsible for the premature death of some 9,000 people.

At last in 1990, the moderate policies of Gorbachev allowed the Ukraine to declare its full independence, but that led to an economic recession including rampant inflation. The situation began to improve around the end of the century, although new political conflicts have arisen with the appearance of the so-called Orange Movement opposing the authoritarian style of the Government. There is speculation that it has had capitalist backing from overseas, possibly in part related to the European Union's ambitions for eastward expansion. NATO too is expressing an interest in the Ukraine.

One of Russia's principal export gas pipelines passes through the Ukraine, giving rise to recent tensions over transit fees and the pricing of imports for the Ukrainian domestic market. They led to temporary interruptions in the supply to Europe in 2006, causing much concern. Another point of tension is the Pivdenne-Brody oil pipeline, built by the Ukraine to transport Caspian oil to Europe and the Baltic, but it has yet to be put into operation.

The Ukraine has evidently had one of the more turbulent of histories, which is probably ultimately due to the fact that its extensive plains of rich agricultural land lack natural boundaries. Furthermore, the country lay on the path of various massive migrations out of Asia, which were driven, it may be supposed, by climate change. Its location on the Black Sea also gives it a strategic importance. This history is relevant when looking ahead to ask how it will fare during the Second Half of the Age of Oil. While it could evidently support its own population well enough, the omens are that it will again become victim to new waves of immigrants and expansive regimes attracted by its agricultural potential and perhaps its considerable coal deposits.

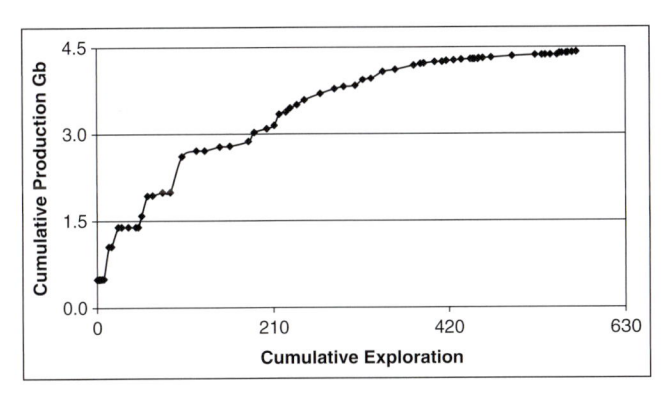

Fig. 36.3 Ukraine discovery trend

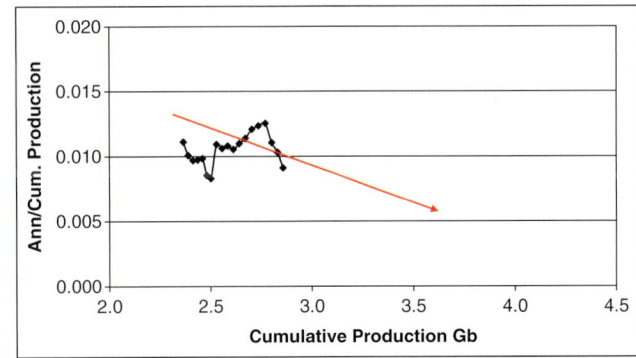

Fig. 36.4 Ukraine derivative logistic

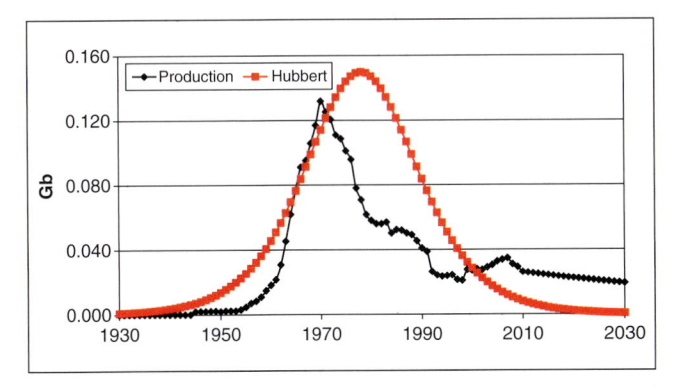

Fig. 36.5 Ukraine production: actual and theoretical

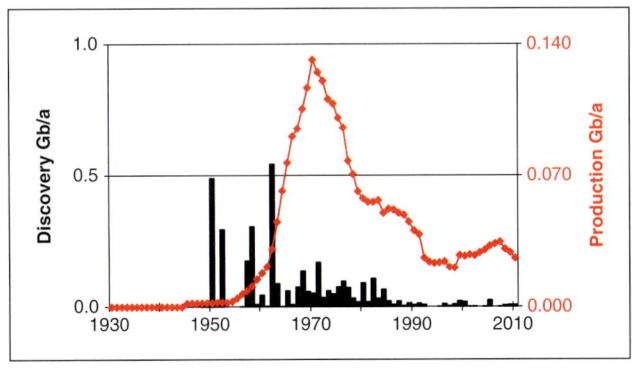

Fig. 36.6 Ukraine discovery and production

Uzbekistan

Table 37.1 Uzbekistan regional totals (data through 2010)

Production						Peak Dates			Area	
Amount			Rate				Oil	Gas	'000 km²	
	Gb	Tcf	Date	Mb/a	Gcf/a	Discovery	1985	1974	Onshore	Offshore
PAST	1.0	68	2000	33	1,992	Production	1996	2015	450	20
FUTURE	1.9	62	2005	25	2,108	Exploration	1991		Population	
Known	1.4	53	2010	14	2,123	Consumption	Mb/a	Gcf/a	1990	5
Yet-to-Find	0.6	9.3	2020	22	2,200	2010	53	1614	2010	28.5
DISCOVERED	2.4	121	2030	34	1,288		b/a	kcf/a	Growth	5.5
TOTAL	3	130	Trade	−39	+509	Per capita	1.8	57	Density	63

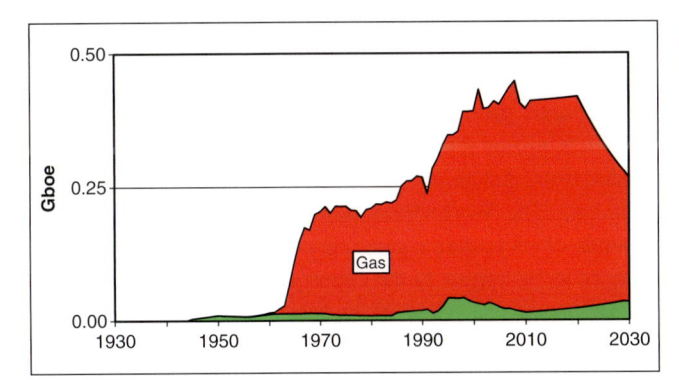

Fig. 37.1 Uzbekistan oil and gas production 1930 to 2030

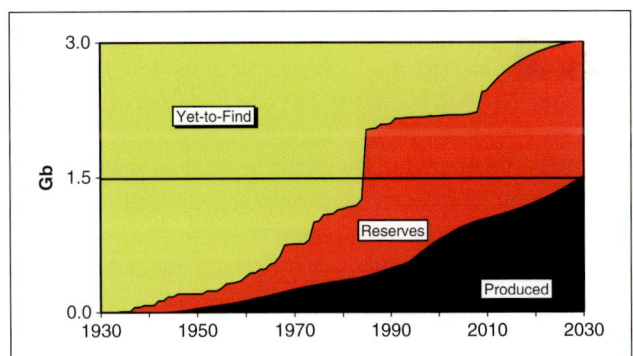

Fig. 37.2 Uzbekistan status of oil depletion

Essential Features

Uzbekistan is a remote, arid, land-locked country to the east of the Caspian, being sandwiched between Turkmenistan to the south and Kazakhstan to the north. The Aral Sea in the northern part of the country is a saline inland sea lacking an outlet. It is drying up and has become heavily polluted by pesticides and nutrients from agricultural activities in the watershed of the rivers that drain into it. Mountainous country develops to the southeast, with the highest peak rising to 4,300 m. The intervening valleys, including especially the Fergana Valley, are fertile, being cultivated with the help of irrigation canals. These valleys hold most of the population as well as the major towns, including the historic city of Tashkent. The population of 28.5 million is predominantly made up of devout Sunni Muslims. The economy is dominated by the production of cotton and gold. The country also possesses one of the largest military establishments in Central Asia with some 65,000 troops under arms.

Geology and Prime Petroleum Systems

The northern margin of the Amu Darya Basin of Turkmenistan extends into Uzbek territory and has yielded a number of oil and gas fields. The primary source-rock is provided by Lower Jurassic gas-prone coals, which have charged overlying sandstones primarily with gas, although also locally with oil.

Another productive basin develops around the Aral Sea where several complex petroleum systems associated with Triassic rifts systems have been identified.

Exploration and Discovery

Although a few exploration boreholes had been drilled earlier, the main thrust followed the Second World War when some 10 and 20 boreholes a year were drilled making some modest discoveries. A total of some 2.4 Gb of oil have been found, with 1985 being the peak discovery year, when the Kokdumalak Field, containing 470 Mb and 6 Tcf of gas, was found. Gas discovery followed a similar path, giving a total discovery of 121 Tcf, being dominated by the Shorton Field with 22 Tcf, which was found in 1974.

Production and Consumption

Oil production prior to 1990 was at a modest level, not exceeding 50 kb/d, but has since risen to a peak of 115 kb/d in 1996 before declining slightly. It is expected to rise slightly in the future in view of the anomalously low depletion rate of 0.7%. Gas production reached a first plateau of about 1 Tcf a year in 1969, before rising to a second one at about double that amount in 2001, which no doubt was set by pipeline capacity and domestic demand. It will likely remain at about this level until around 2020 before terminal decline sets in.

Oil consumption stands at 53 Mb/a, making the country a modest importer. Gas consumption is running at 1,614 Gcf/a, giving scope for exports.

The Oil Age in Perspective

Uzbekistan has been inhabited since the earliest days. In the first millennium, it had an important role on the famous Silk Road between Europe and the East, falling at times under the Parthan and Sassanid empires of Persia. The towns of Samarkand and Bukhara have been famous throughout history. The country was subject to massive waves of immigration during the Middle Ages from Siberia, including the Mongols under Ghengis Khan. Tamerlane became a great national leader during the fourteenth century. He succeeded in defeating the Ottoman Turks, and even reached the Middle East. The name Uzbek is probably derived from Oz Beg, a Muslim leader of the Golden Horde. The country had risen to be one of the most powerful in Central Asia by the late eighteenth century.

It then fell to Russian conquest in the late nineteenth century, being incorporated into the province of Turkmenistan, before becoming a member of the Soviet Union in 1925, which relied on it heavily for cotton and grain. It gained independence in 1991 on the fall of the Soviets, but remains under the authoritarian rule of President Karimov, who vigorously suppresses any moves of opposition, especially from Islamic groups. In fact in 2001, he provided facilities for a US military base from which to attack neighboring Afghanistan, but this was closed in 2005 following a local massacre. The World Bank declines to make loans, on the grounds of its human rights record, but this may, ironically, turn out to be hugely to the country's advantage.

Looking ahead, Uzbekistan has fertile lands sufficient to support its people, who have lived in substantial isolation for much of their history. It also has modest gas resources to provide an adequate energy supply for the next few years. The present political situation is evidently difficult, but the Uzbeks can perhaps look forward to the future with a degree of optimism.

37 Uzbekistan

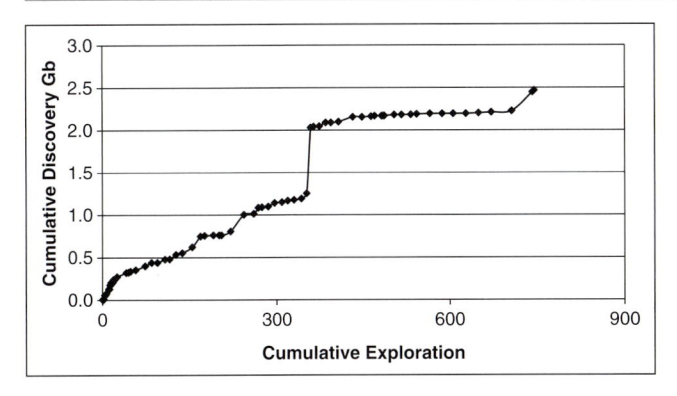

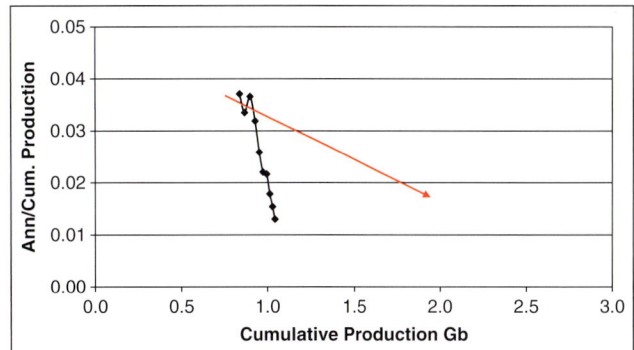

Fig. 37.3 Uzbekistan discovery trend

Fig. 37.4 Uzbekistan derivative logistic

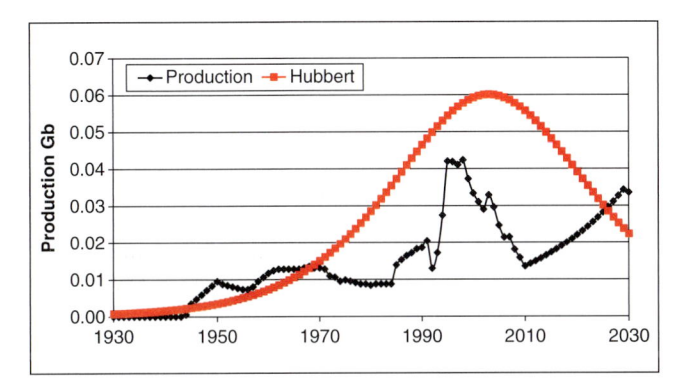

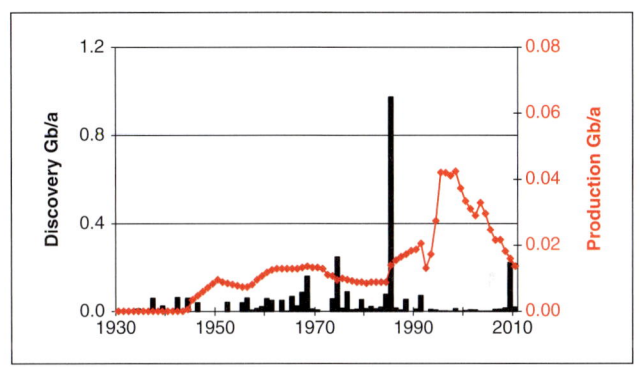

Fig. 37.5 Uzbekistan production: actual and theoretical

Fig. 37.6 Uzbekistan discovery and production

Eurasia Region

Table 38.1 Eurasia regional totals (data through 2010)

	Production					Peak Dates			Area	
	Amount		Rate				Oil	Gas	'000 km²	
	Gb	Tcf	Date	Mb/a	Gcf/a	Discovery	1960	1966	Onshore	Offshore
PAST	225	970	2000	4092	25262	Production	1988	2021	31530	2140
FUTURE	167	1593	2005	5194	28806	Exploration	1988		Population	
Known	136	1207	2010	5752	30556	Consumption	Mb/a	Gcf/a	1900	475
Yet-to-Find	30	386	2020	4271	42003	2010	5490	29611	2010	1633
DISCOVERED	362	2176	2030	3061	34152		b/a	kcf/a	Growth	3.4
TOTAL	392	2563	Trade	+314	+1196	Per capita	3.1	17	Density	52

Note: Data refer to main producing countries only, **except consumption and trade data** which include other countries and exclude Non-Conventional oil and gas in Polar Regions.

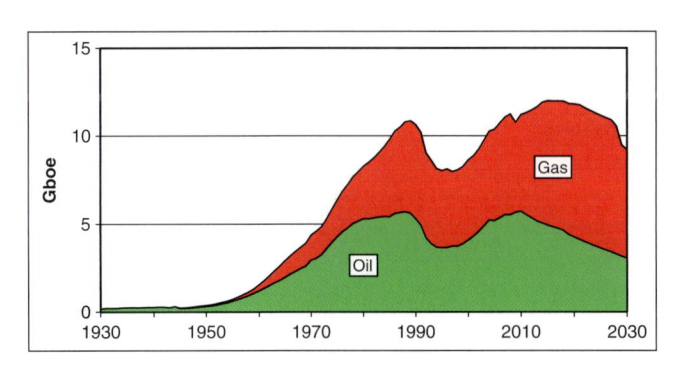

Fig. 38.1 Eurasia oil and gas production 1930 to 2030

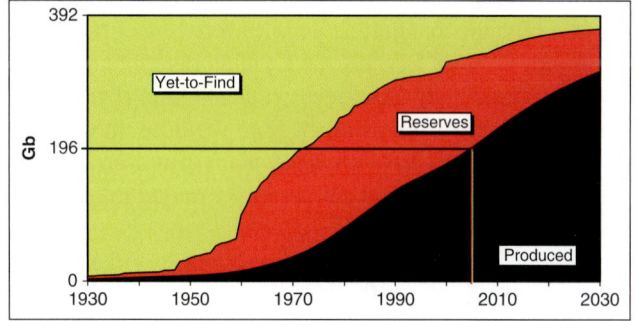

Fig. 38.2 Eurasia status of oil depletion

The Eurasia Region is defined for this purpose as the former Communist bloc of the Soviet Union, Eastern Europe and China. These countries had a common economic and political environment affecting the pattern of oil and gas development. It is noteworthy, however, that the discovery pattern is substantially the same as for the other regions because it is imposed by Nature.

Table 38.2 Other countries

Afghanistan	Estonia	Moldova
Armenia	*Georgia	Mongolia
*Belarus	*Krygistan	*Poland
Bosnia	Latvia	*Serbia
*Bulgaria	*Lithuania	*Slovenia
Czech Rep.	Macedonia	*Tajikstan
(*) = with minor production		

In addition to the eleven countries with significant oil and gas production, which have been considered in detail, are a number of other countries that fall within the region, as listed in the table. Those indicated (*) have minor amounts of oil and/or gas which are evaluated with other minor producers in Chapter 11, as the amounts are too small to model meaningfully. They are however consumers of oil and gas.

The region has an estimated total endowment of 392 Gb of producible *Regular Conventional* oil, amounting to 20% of the World's total. It also has an endowment of about 2,563 Tcf of gas, principally in Russia. It is important to note that the Polar Regions of Russia have substantial gas deposits, which are classed by definition as *Non-Conventional*, being discussed in Chapter 12.

The current overall depletion rate of *Regular Conventional Oil* is a modest 3% a year reflecting the fact that production is mainly onshore and in many cases in mature fields. It is set to decline at about this rate, falling from 16 Mb/d in 2010 to about 8 Mb/d by 2030. The population for the entire region, including the additional countries listed in the table, is 1.7 billion, or 26% of the world's total, with a population density of 52/km². This, in large measure, reflects the over-populated nation of China. Total oil consumption for the main producing countries stands at 4,932 Mb/a, meaning that the Region is a net exporter of 820 Mb/a.

It has an endowment of about 2,176 Tcf of gas, of which 1,593 Tcf remain, including the estimate of what is left to be found. Production stands at 30 Tcf/a year and will probably peak around 2020 at 42 Tcf/a, before declining slightly. Reported consumption for main producing countries stands at 27 Tcf/a, meaning that about 3 Tcf/a is exported.

In geographic terms, it is a region of great diversity ranging from the polar wastes to arid deserts and the foothills of the Himalaya Range in the south. Several great inland seas and saline lakes are present in the form of the Caspian and Aral Sea as well as Lake Baikal. The climate also varies greatly across the region, but is generally extreme.

The early history of the region was dominated by massive migrations out of Siberia, presumably reflecting adverse climate changes, as people were forced to spread south into Western Europe, the Caspian region and China. The migrations may have been responsible in part for the continuing strife and conflict that seem to have been endemic. Frontiers have changed many times as empires waxed and waned, and as communities sought to establish their separate identities.

The Russian Empire extended over much of the region, even taking parts of what had been China. It was succeeded, after the 1917 Revolution, by what amounted to a Communist Empire of the Soviet Union which extended its reach into Eastern Europe after the Second World War. Many of the countries then became effectively vassal states under authoritarian administrations although their national identities and priorities were never quite extinguished.

The collapse of the Soviet Union in the early 1990s led to the emergence of new administrations and national entities, many remaining under authoritarian governments and facing tensions of various kinds. China remained under Communist government but adopted capitalist practices with a huge expansion of industry and manufacturing, as the world began to use its cheap labour force.

In energy terms, Russia emerges supreme with a rich endowment of both oil and gas that it now seeks to preserve to meet national needs, causing new tensions with Western Europe, which demands access. China, for its part, faces a desperate situation as its oil and gas decline, deserts encroach and aquifers deplete.

Looking ahead to the Second Half of the Age of Oil, the greatest tensions are likely to arise in over-populated China, whose new found economic prosperity is built on fragile foundations. Russia will likely emerge relatively well, having less far to fall. The countries of Eastern Europe will sail in the same boat as their western neighbours, while Central Asia will probably revert to its tribal structure.

Confidence and Reliability Ranking

The table lists the standard deviation of the range of published data on reserves. A relative confidence ranking in the validity of the underlying assessment is also given: the lower the Surprise Factor, the less the chance for revision. In general terms, it can be said that the region is characterised by unreliable data.

Table 38.3 Eurasia: Range of Reported Reserves and Scope for Surprise

	Standard Deviation Public Reserve Data		Surprise Factor	
	Oil	Gas	Oil	Gas
Albania	0	0.04	7	9
Azerbaijan	0.81	9.15	3	4
China	2.51	12.06	5	7
Croatia	0	0.08	8	8
Hungary	0	1.42	2	2
Kazakhstan	5.67	11.17	7	8
Romania	0.06	10.94	2	2
Russia	9.0	68.46	6	8
Turkmenistan	0.57	105.02	8	10
Ukraine	0.2	35.71	6	7
Uzbekistan	0.31	2.70	8	10
			Scale 1–10	

Albania

Albania has not been thoroughly explored although it is an old oil province. There is scope for surprise oil discoveries, but probably small ones. The gas situation is even less sure.

Azerbaijan

Azerbaijan has had a very long oil history and has been thoroughly explored so there is little scope for surprises. However, the accuracy of the reported reserves is open to doubt.

China

China has been thoroughly explored but it is a large territory which may hold some surprises. The validity of the data is open to question.

Croatia

Croatia probably holds scope for relatively small surprises.

Hungary

Hungary has had a long oil history and there is confidence in the assessment.

Kazakhstan

Kazakhstan is relatively unexplored and so offers plenty of scope for surprises, especially in deep sub-salt plays, but the disappointing results of the giant Kashagan prospect, once hailed as rivalling the Middle East, are grounds for caution.

Romania

Romania, like Hungary, has had a long oil history and there is confidence in the assessment.

Russia

Although Russia covers a huge area, it has been systematically explored under the Soviets, so that all the major basins and most of the giant fields within them have been identified. The overall assessment is believed to be valid in terms of order of magnitude, but there are serious doubts about the accuracy of the reported reserves, as illustrated by the high standard deviation. It should be noted that Polar oil and gas are treated as *Non-Conventional,* as defined herein, but there are difficulties in allocating the resources of the two domains.

Turkmenistan and Uzbekistan

These two countries share essentially the same basin, which has a modest potential especially for gas and has generally been under-explored, so that no great confidence can be placed on the assessment.

Ukraine

The Ukraine has a modest potential, and the assessment can be classed as no more than moderately reliable.

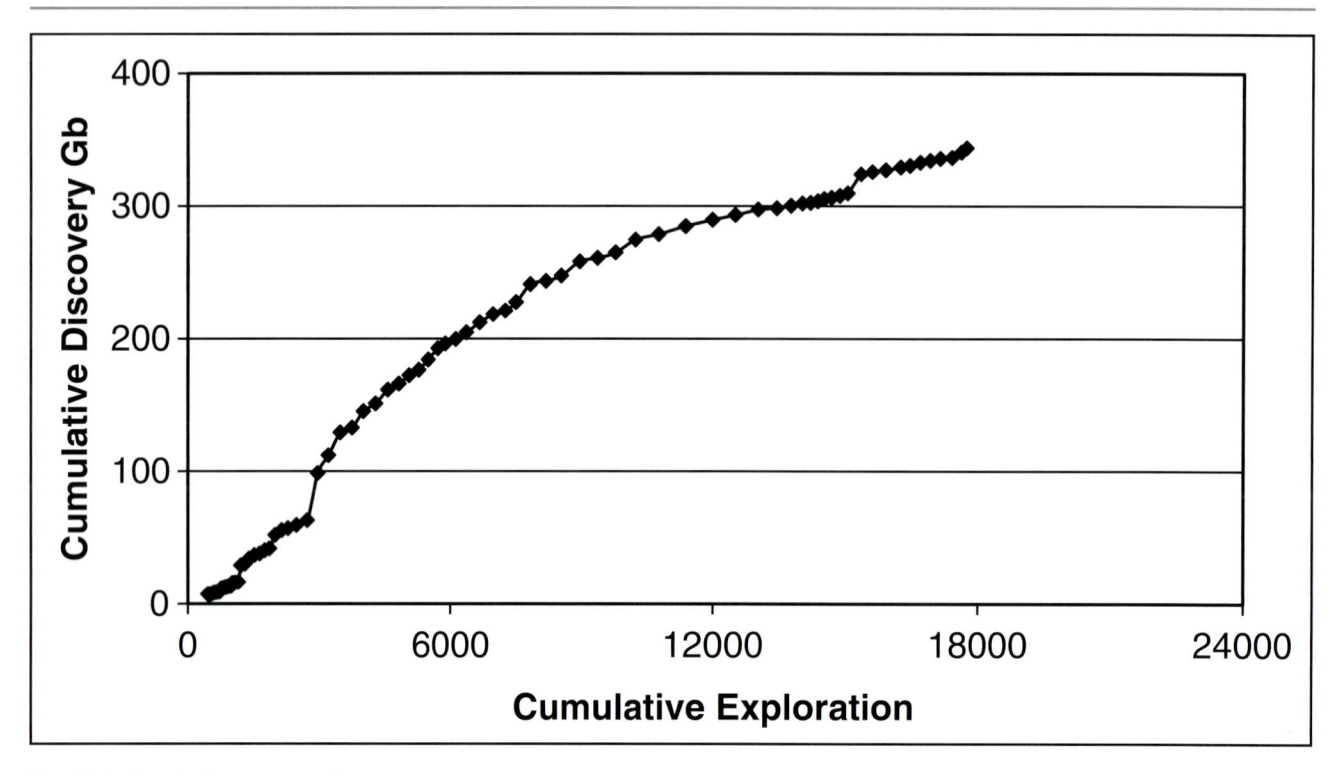

Fig. 38.3 Eurasia discovery trend

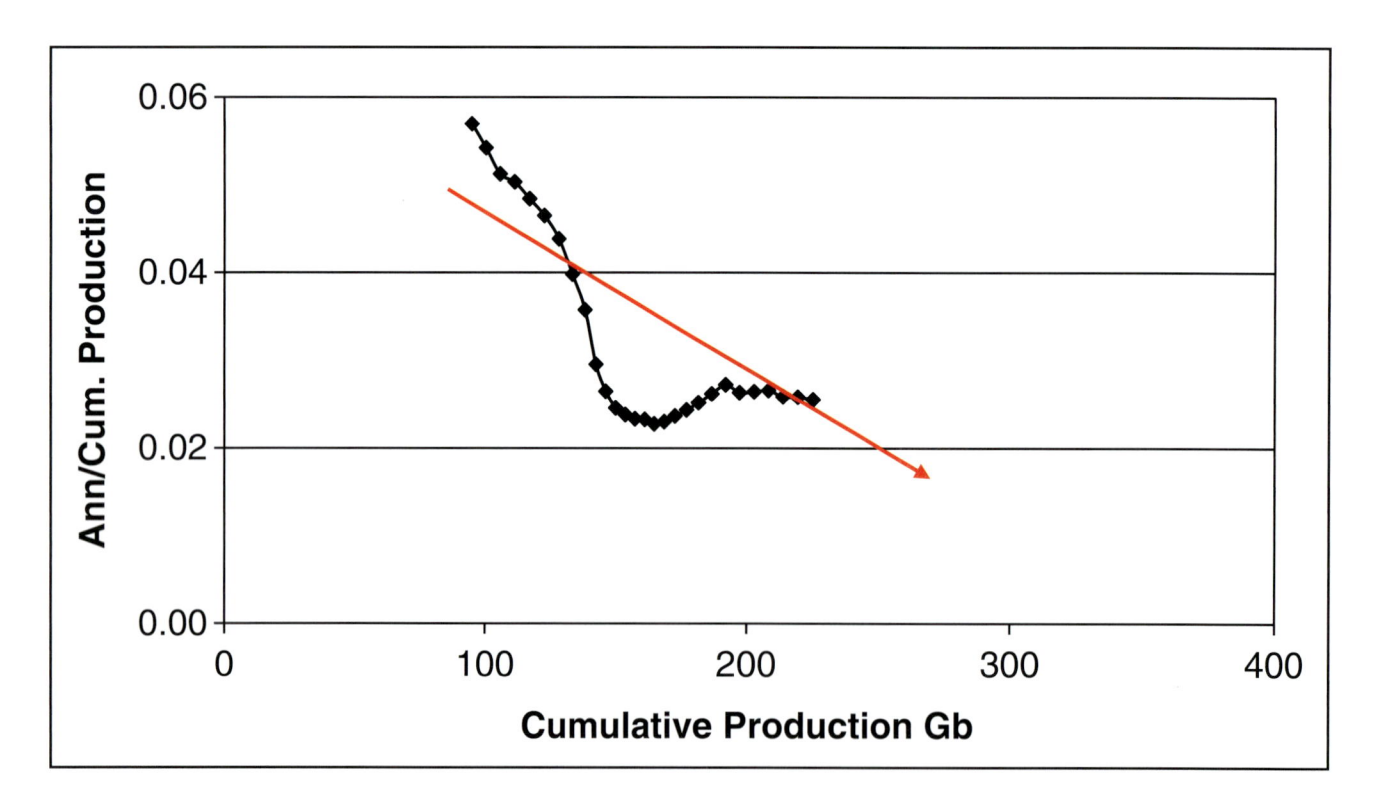

Fig. 38.4 Eurasia derivative logistic

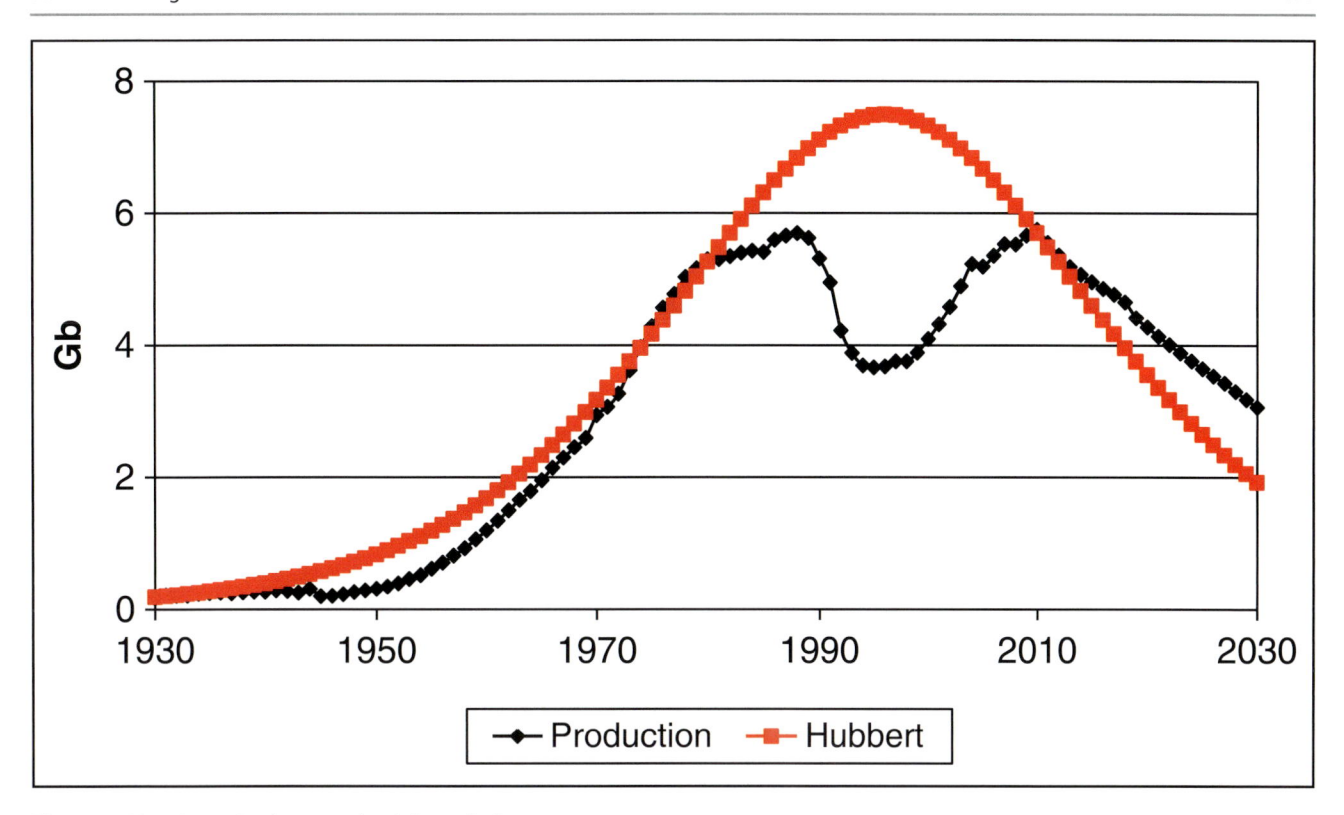

Fig. 38.5 Eurasia production: actual and theoretical

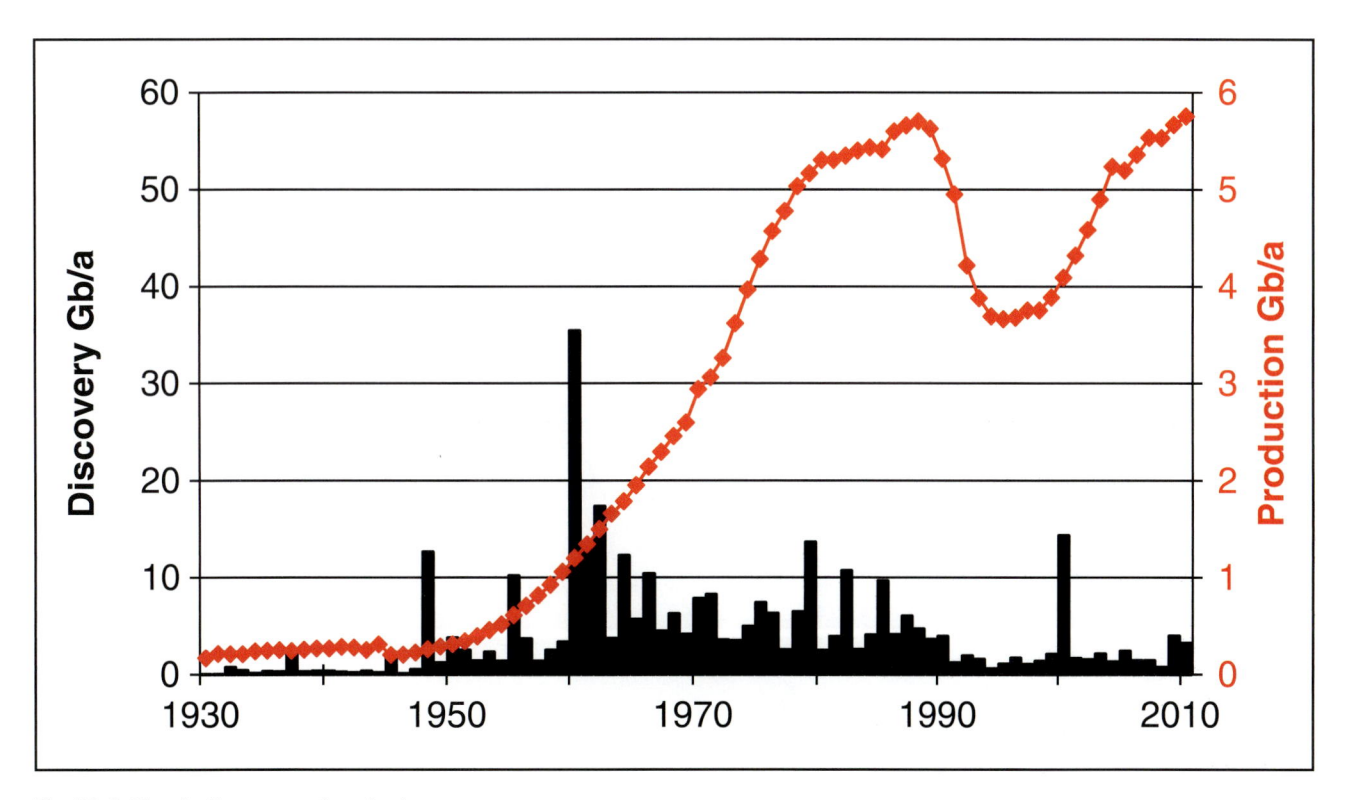

Fig. 38.6 Eurasia discovery and production

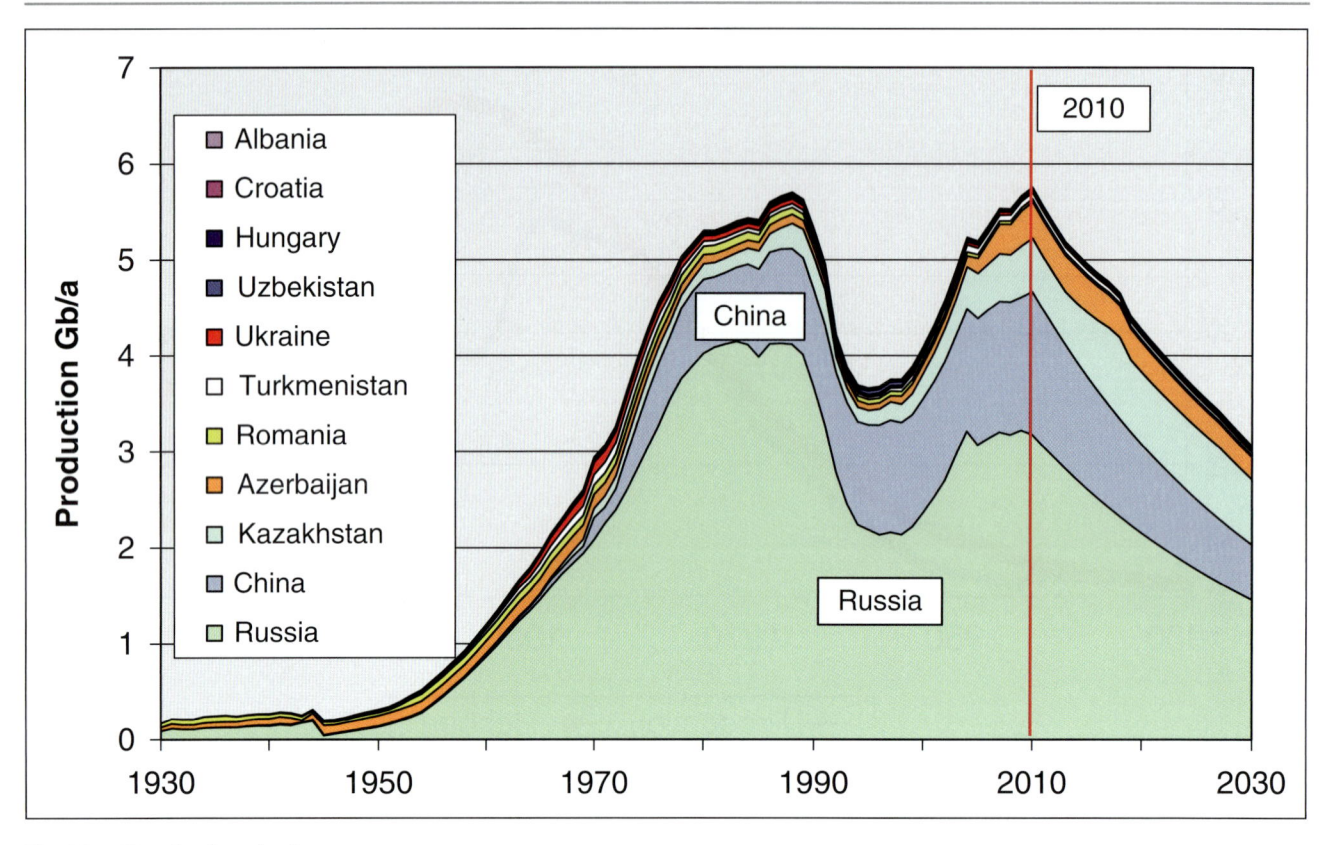

Fig. 38.7 Eurasia oil production

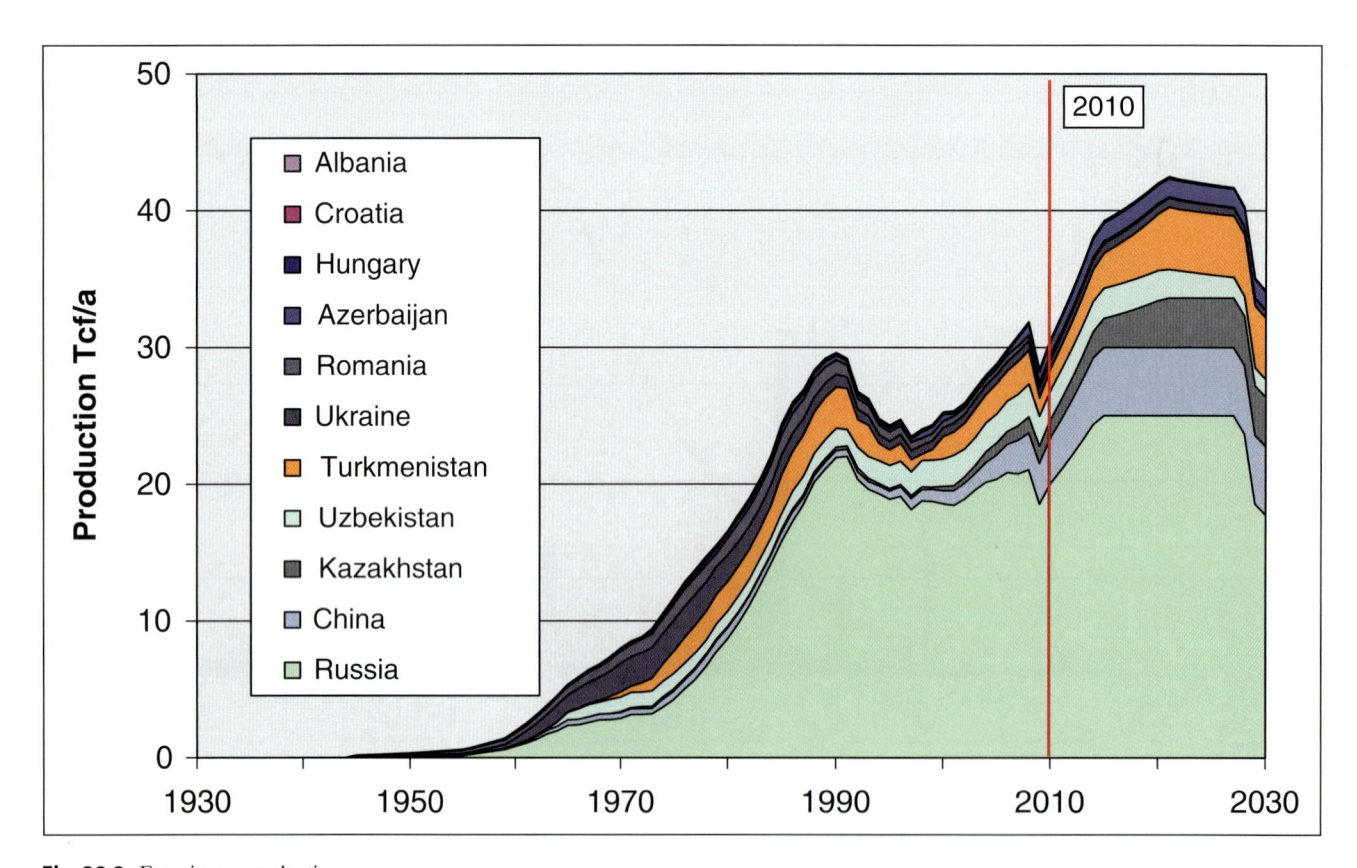

Fig. 38.8 Eurasia gas production

38 Eurasia Region

Table 38.4 Eurasia: Oil Resource Base

		\multicolumn{18}{c	}{PRODUCTION TO 2100}
EURASIA		\multicolumn{15}{c	}{Regular Conventional Oil}			2010
		\multicolumn{8}{c	}{KNOWN FIELDS}					Revised				05/10/2011
	Region	Present		Past			Reported Reserves			Future	Total	NEW	ALL		DEPLETION		PEAK	
		kb/d	Gb/a	Gb	5yr Trend	Disc 2010	Average	Deductions Static	Other	% Disc.	Known Gb	Found Gb	FIELDS Gb	FUTURE Gb	TOTAL Gb	Rate	Mid-Point	Disc	Prod
Country		2010	2010																
Russia	B	8707	3.18	149.99	0%	2.0	69.84	−12.30	−45.00	93%	64.01	214.00	16.00	80.01	230.0	3.8%	1998	1960	1987
China	B	406	1.49	39.45	2%	0.73	17.91	0.00	0.00	91%	24.44	63.89	6.11	30.55	70.0	4.6%	2005	1960	2010
Kazakhstan	B	1525	0.56	10.68	3%	0.3	34.91	0.00	0.00	89%	29.17	39.85	5.15	34.32	45.0	1.6%	2028	2000	2025
Azerbaijan	B	1035	0.38	10.377	12%	0.12	6.60	−0.50	0.00	94%	12.27	22.64	1.36	13.63	24.0	2.7%	2014	1871	2014
Romania	B	88	0.03	5.56	−2%	0.03	0.56	0.00	0.00	98%	1.29	6.86	0.14	1.44	7.0	2.2%	1976	1890	1976
Turkmenistan	B	178	0.07	3.47	2%	0.01	0.88	0.00	0.00	93%	1.62	5.09	0.41	2.03	5.5	3.1%	1999	1956	1973
Ukraine	B	71	0.03	2.86	−5%	0.01	0.51	−0.21	0.00	98%	1.56	4.42	0.08	1.64	4.5	1.6%	1989	1962	1970
Uzbekistan	B	37	0.01	1.04	−7%	0.02	0.75	−0.17	0.00	80%	1.96	2.41	0.59	1.96	3.0	0.7%	2029	1985	2029
Hungary	B	14	0.01	0.72	−4%	0.01	0.02	0.00	0.00	93%	0.21	0.93	0.07	0.28	1.0	1.9%	1987	1965	1979
Croatia	B	15	0.01	0.54	−3%	0.00	0.07	0.00	0.00	86%	0.32	0.86	0.14	0.46	1.0	1.2%	2003	1957	1988
Albania	B	0	0.00	0.50	8%	0.00	0.20	−0.01	0.00	93%	0.20	0.70	0.05	0.25	0.75	1.6%	1986	1928	1983
EURASIA	B	15758	5.75	225.20	1%	3.39	123.36	−1328	45.00	92%	136.45	361.65	30.10	166.55	391.8	3.3%	2005	1960	1988

Table 38.5 Eurasia: Gas Resource Base

		\multicolumn{18}{c	}{PRODUCTION TO 2100}
EURASIA		\multicolumn{17}{c	}{Regular Conventional Gas}				2010
		\multicolumn{9}{c	}{KNOWN FIELDS}						Revised			15/11/2011		
	Region	Present			Past		Discovery		Reserves		FUTURE	TOTAL	FUTURE	ALL	TOTAL	DEPLETION			PEAKS		
		Tcf/a	FIP	Gboe/a	5yr	Tcf	%	Reported Average	Deduct Non-Con	KNOWN Tcf	FOUND Tcf	FINDS Tcf	FUTURE Tcf	Tcf	Current Rate	%	Mid-Point	Expl	Disc	Prod
Country		2010	2010	2010	Tcf	Trend	2010	Disc.													
Russia	B	20.03	0.03	3.61	631	−1%	31.06	83%	1590.49	−991.90	608.25	1239	261	868.9	1500	2.25%	42%	2014	1988	1966	2014
Turkmenistan	B	1.60	0.30	0.29	78	−6%	0.40	80%	215.28	0.00	222.71	301	74	296.9	375	0.54%	21%	2030	1986	1973	2030
China	B	3.33	0.73	0.60	43	12%	1.93	93%	94.75	0.00	141.06	184	16	156.7	200	2.08%	22%	2022	2003	200	2022
Uzbekistan	B	2.12	0.02	0.38	68	−1%	0.30	92%	62.13	0.00	52.68	121	9	62.0	130	3.31%	52%	2009	1991	1974	2015
Kazakhstan	B	1.31	0.92	0.24	14	8%	0.85	91%	72.17	0.00	99.85	114	11	110.9	125	1.17%	11%	2027	1988	1979	2027
Ukraine	B	0.68	−0.02	0.12	64	0%	0.31	96%	36.69	0.00	32.13	96	4	35.7	100	1.88%	64%	1989	2000	1950	1975
Azerbaijan	B	0.93	0.35	0.17	16	57%	0.89	85%	37.92	0.00	43.17	59	11	54.0	70	1.69%	23%	2030	1953	1999	2030
Romania	B	0.37	−0.01	0.07	45	−3%	0.06	100%	12.20	0.00	2.09	47	0	2.3	48	13.86%	95%	1982	1969	1954	1985
Hungary	B	0.10	0.01	0.02	7.7	−1%	0.08	97%	1.04	0.00	2.97	10.7	0	3.30	11	2.99%	70%	1992	1964	1965	1985
Croatia	B	0.07	0.00	0.01	1.5	0%	0.00	90%	1.00	0.00	1.19	2.7	0.3	1.48	3	4.31%	51%	2014	1985	1974	2014
Albania	B	0.002	0.000	0.00	0.5	0%	0.00	91%	0.06	0.00	0.65	1.1	0.1	0.76	1.25	0.23%	39%	2030	1987	1977	1982
EURASIA	B	30.56	2	5.50	970	0%	35.89	85%	2123.71	−992	1207	2176	386	15934	2563	1.88%	38%	2020	1988	1966	2020

Table 38.6 Eurasia: Oil Production Summary

Regular Conventional Oil Production					
Mb/d	2000	2005	2010	2020	2030
Russia	6.45	9.04	8.71	5.90	4.00
China	3.25	3.61	4.08	2.53	1.57
Kazakhstan	0.72	0.99	1.53	2.05	1.86
Azerbaijan	0.28	0.43	1.03	0.86	0.63
Romania	0.12	0.11	0.09	0.07	0.06
Turkmenistan	0.14	0.18	0.18	0.13	0.09
Ukraine	0.07	0.09	0.07	0.06	0.05
Uzbekistan	0.09	0.07	0.04	0.06	0.09
Hungary	0.03	0.02	0.01	0.01	0.01
Croatia	0.02	0.02	0.01	0.01	0.01
Albenia	0.01	0.01	0.01	0.01	0.01
EURASIA	11.21	14.57	15.76	11.70	8.39

Table 38.7 Eurasia: Gas Production Summary

Gas Production					
Tcf/a	2000	2005	2010	2020	2030
Russia	18.6	20.4	20.0	25.0	17.8
Turkmenistan	1.64	2.22	1.60	4.15	4.50
China	0.96	1.76	3.33	5.00	5.00
Uzbekistan	1.99	2.11	2.12	2.20	1.29
Kazakhstan	0.16	0.78	1.26	3.09	3.60
Ukraine	0.64	0.69	0.68	0.67	0.53
Azerbaijan	0.49	0.21	0.93	1.35	1.35
Romania	0.49	0.41	0.37	0.09	0.03
Hungary	0.11	0.11	0.10	0.08	0.06
Croatia	0.06	0.05	0.07	0.06	0.03
Albania	0.00	0.00	0.00	0.00	0.01
EURASIA	25.3	28.8	30.6	42	34.2

38 Eurasia Region

Table 38.8 Eurasia: Oil and Gas Production and Consumption, Population and Density

EURASIA	PRODUCTION AND CONSUMPTION							POPULATION & AREA		
	OIL				GAS						
	Production	Consumption	Trade	Production	Consumption	Trade	Population	Growth	Area	Density
			p/capita	(+)		p/capita	(+)		Factor	Km²	
Major	Mb/a	Mb/a	b/a	Mb/a	Gcf/a			M		M	
Russia	318	1072	7.5	2106	20033	17495	122.5	2538	142.8	1.6	17.14	8
Albania	4	12	3.8	−8	2	1	0.3	1.0	3.2	3.2	0.03	111
Azerbaijan	378	38	4.1	340	926	350	38.0	576	9.2	8.1	0.09	106
China	1488	3354	2.5	−1866	3334	3768	2.8	−434	1345.9	3.7	9.61	140
Croatia	5	36	8.1	−30	67	100	22.7	−33.00	4.4	4.5	0.06	78
Hungary	5	53	5.3	−48	101	426	42.6	−325	10	1.3	0.09	107
Kazakhstan	557											
Romania	32	72	3.3	−39	374	455	21.3	−81	21.4	3.0	0.24	89
Turmenistan	65	43	8.5	22	1600	720	141.2	880	5.1	6.0	0.49	10
Ukraine	26	108	2.4	−82	684	1877	41.1	−1193	45.7	3.2	0.61	75
Uzbekistan	14	53	1.8	−39	2123	1614	56.6	509	28.5	5.5	0.45	63
Sub-total	**5752**	**4932**	**3.0**	**820**	**30556**	**27109**	**16.6**	**3447**	**1633**	**3.4**	**31.53**	**52**
Other												
Armenia	0.00	18.98	6.12	−19	0.0	61.1	19.7	−61.1	3.1		0.03	
Belarus	10.95	59.50	6.26	−49	7.7	770.6	81.1	−762.9	9.5		0.21	
Bosnia/Herzeg.	0.00	10.40	2.74	−10	0.0	7.4	2.0	−7.4	3.8		0.05	
Bulgaria	0.37	33.22	4.43	−33	0.4	76.6	10.2	−76.3	7.5		0.11	
Czech Rep.	1.28	71.44	6.80	−70	7.2	327.7	31.2	−320.6	10.5		0.08	
Estonia	0.00	11.32	8.71	−11	0.0	24.8	19.0	−24.8	1.3		0.05	
Georgia	0.37	4.75	1.10	−4	0.0	58.3	13.5	−58.3	4.3		0.07	
Krygistan	0.37	5.84	1.04	−5	0.4	16.3	2.9	−15.9	5.6		0.20	
Latvia	0.00	14.97	6.80	−15	0.0	53.7	24.4	−53.7	2.2		0.06	
Lithuania	0.73	24.46	7.64	−24	0.0	109.5	34.2	−109.5	3.2		0.07	
Macedonia	0.00	6.94	3.30	−7	0.0	2.8	1.3	−2.8	2.1		0.03	
Moldova	0.00	7.30	1.78	−7	0.0	76.6	18.7	−76.6	4.1		0.03	
Mongolia	0.00	6.21	2.22	−6	0.0	0.0	0.0	0.0	2.8		1.57	
Poland	4.65	206.03	5.39	−201	214.9	607.3	15.9	−392.5	38.2		0.37	
Serbia	3.65	32.12	4.40	−28	15.2	79.8	10.9	−64.6	7.3		0.09	
Slovakia	0.07	30.59	14.57	−31	3.7	221.1	105.3	−217.4	2.1		0.02	
Tajikstan	0.08	14.60	1.95	−15	1.4	8.0	1.1	−6.5	7.5		0.14	
Other	30.20	0.00	0.00	0	0.0	0.0	0.0	0.0	0.0		0.00	
Sub-Total	**52.7**	**558.67**	**85**	**−506**	**250.8**	**2501.6**	**21.7**	**−2251**	**115.10**		**3.16**	
EURASIA	**5804**	**5490**	**3.1**	**314**	**30806.8**	**29611**	**17**	**1196**	**1747.90**		**34.7**	

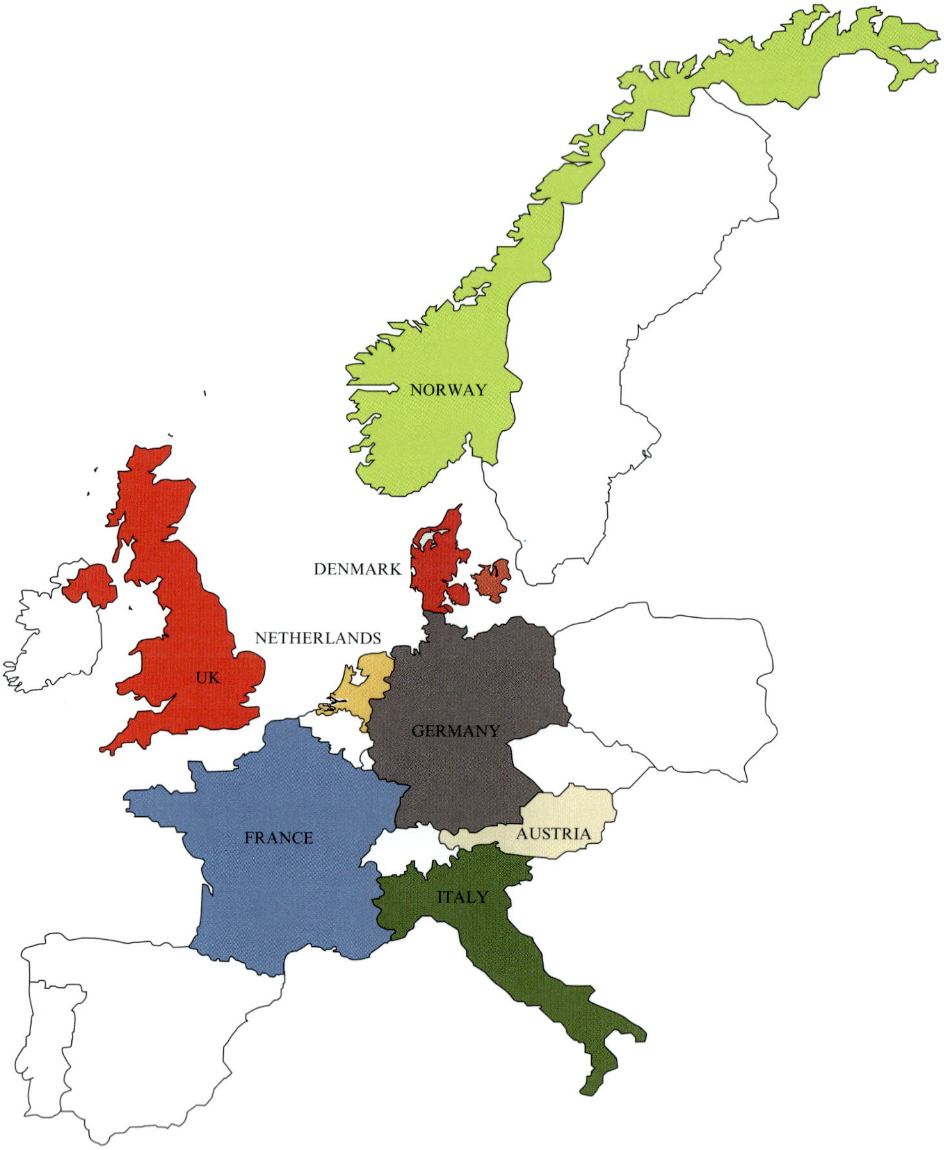
NORWAY
DENMARK
NETHERLANDS
UK
GERMANY
FRANCE
AUSTRIA
ITALY

Part V

Europe

Austria

Table 39.1 Austria regional totals (data through 2010)

Production to 2100						Peak Dates			Area	
Amount			Rate				Oil	Gas	'000 km²	
	Gb	Tcf	Date	Mb/a	Gcf/a	Discovery	1949	1949	Onshore	Offshore
PAST	0.83	3.4	2000	7	64	Production	1955	1975	80	0
FUTURE	0.12	1.6	2005	6	58	Exploration	1975		*Population*	
Known	0.09	1.6	2010	6	61	*Consumption*	Mb/a	Gcf/a	1900	6.0
Yet-to-Find	0.02	0.1	2020	4	44	2010	101	335	2010	8.4
DISCOVERED	0.93	4.9	2030	2	30		b/a	Gcf/a	Growth	1.4
TOTAL	0.95	5	*Trade*	−95	−274	Per capita	12.1	40	Density	100

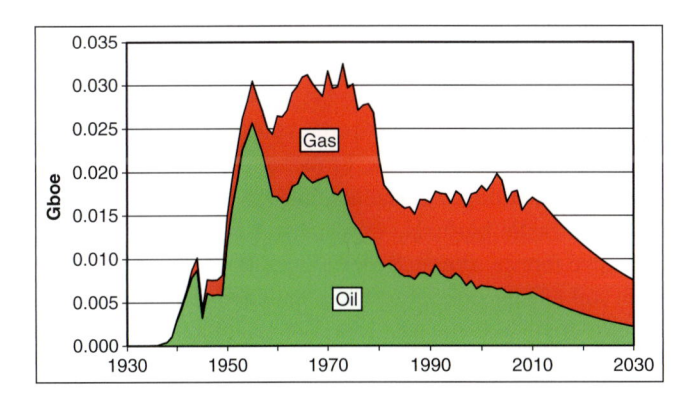

Fig. 39.1 Austria oil and gas production 1930 to 2030

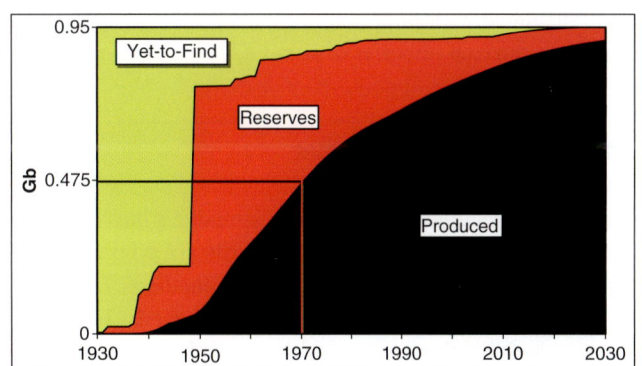

Fig. 39.2 Austria status of oil depletion

Essential Features

The western part of Austria is made up of impressive Alpine Ranges, rising to 3,800 m, while to the east lie foothills and extensive plains. The country is drained by the Danube River and its tributaries, flowing eastwards to the Black Sea. It supports a population of 8.4 million.

Geology and Prime Petroleum Systems

The Alpine orogenic belt is flanked to the east by the Vienna Basin containing oil and gas fields. The oil was generated primarily in Middle to Upper Miocene clays, which were deposited in a relatively small restricted basin. It migrated during the Pliocene to charge reservoirs in complex and faulted structures, partly forming the Alpine foothills. A localised subsidiary system has also been identified in the Lower Tertiary.

Exploration and Discovery

Exploration drilling commenced in 1891, but lapsed for many years before resuming at a modest scale before and during the Second World War when it yielded a number of small discoveries. The breakthrough came in 1949 with the discovery of the giant Matzen Field, holding some 565 Mb of oil. Exploration has continued since, but has provided no more than minor finds despite the drilling of almost 860 exploration boreholes, meaning that most viable prospects have already been investigated. In total, about 1 Gb have been discovered, and the scope for future discovery in this very mature area is therefore severely limited. In addition, some 4.9 Tcf of gas have been discovered, of which about 1.6 Tcf remain.

Production and Consumption

Oil production commenced in 1933 and reached a peak at 70 kb/d in 1955, much being flush production from the Matzen Field. It thereafter declined gently to 17 kb/d in 2010. Future production is expected to continue to decline at the present depletion rate of about 5 % a year. Austria consumes about 100 Mb of oil a year, meaning that it has to import 93 % of its needs.

Gas production commenced in the late 1930s, rising to a peak of 88 Gcf/a in 1975, followed by a gentle decline to about 61 Gcf/a in 2010. It is expected to remain at about this level to 2012 before terminal decline sets in. Being an onshore area, even low levels of production can be economically sustained. Consumption is running at 335 Gcf/a making the country an importer of about 82 % of its needs.

The Oil Age in Perspective

Austria was occupied by Romans, Germanic tribes and others, before falling to Charlemagne in 796 AD, who gave it a form of national identity. After various tribulations, the country came together in the thirteenth century under the Hapsburg kings, who built their power partly through expedient marriages, such that, by the fifteenth century, their dominion extended over Spain, Hungary, the Netherlands and much of the Balkans. Vienna became one of the great capitals of Europe, enjoying a cultural pre-eminence with grand architecture and the music of Mozart.

Austria played an important part in the events leading to the First World War, as its Serbian province with a Slav population was seeking independence, being in part backed by Russia. The conflict triggered various alliances, including those between France, Russia and Britain, as well as one between Germany and Austria itself, and led to war in 1914. The entry of the United States in 1917 ended the stalemate, bringing defeat to Austria and Germany.

The Austrian empire then disintegrated with the secession of various territories, including Hungary and its Balkan provinces, leading to the declaration of a new Republic in 1918. This was duly confirmed by the Peace Treaty, giving it rights to about one-eighth of the previous Austro-Hungarian kingdom. Internal conflict resumed in the 1930s, with the assassination of the Chancellor being followed by a peaceful union with Germany in 1938, one year before the resumption of hostilities in the Second World War, in which it again faced eventual defeat. The victorious Allies recognised the Republic, which became a full member of the European Union in 1995.

Austria has always found itself at the cross-roads between eastern and western Europe. In facing the future, it has the advantage of strong regional administrations and is also well advanced with biofuels, especially wood-chips from its extensive forests, although they have suffered to some extent from soil and air pollution. The population has increased by a factor of 1.4 over the past century to eight million, giving a density of 99/km^2, but the country does not seem to be too badly placed to find a self-sustainable future during the Second Half of the Age of Oil.

39 Austria

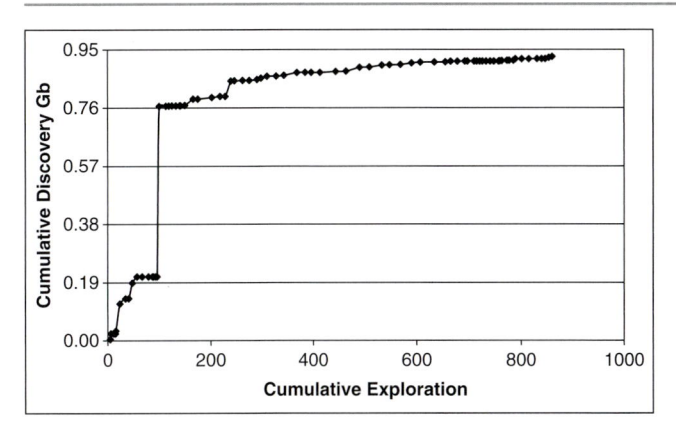

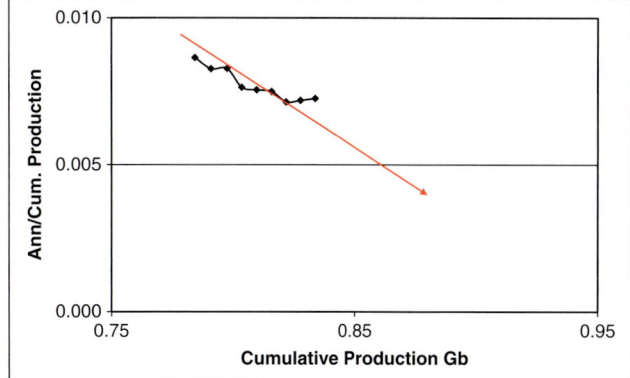

Fig. 39.3 Austria discovery trend

Fig. 39.4 Austria derivative logistic

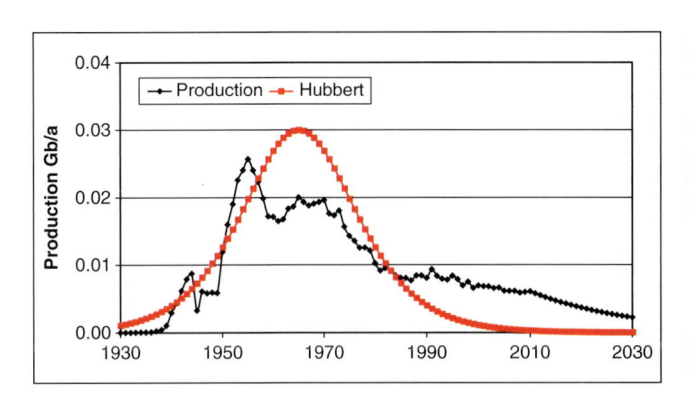

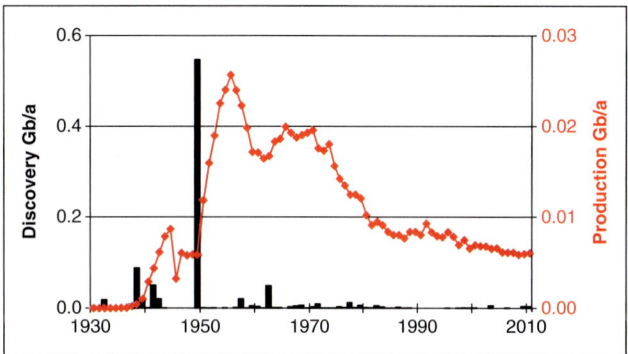

Fig. 39.5 Austria production actual and theoretical

Fig. 39.6 Austria discovery and production

Denmark

Table 40.1 Denmark regional totals (data through 2010)

Production to 2100						Peak Dates			Area	
Amount			Rate				Oil	Gas	'000 km²	
	Gb	Tcf	Date	Mb/a	Gcf/a	Discovery	1971	1968	Onshore	Offshore
PAST	2.3	6.7	2000	132	422	Production	2004	2007	40	110
FUTURE	1.2	2.3	2005	138	376	Exploration	1985		Population	
Known	1.0	2.1	2010	90	289	*Consumption*	Mb/a	Gcf/a	1900	2.5
Yet-to-Find	0.2	0.2	2020	44	89	2010	61	175	2010	5.6
DISCOVERED	3.3	8.8	2030	22	29		b/a	kcf/a	Growth	2.2
TOTAL	3.5	9.0	*Trade*	+29	+114	Per capita	11	31	Density	129

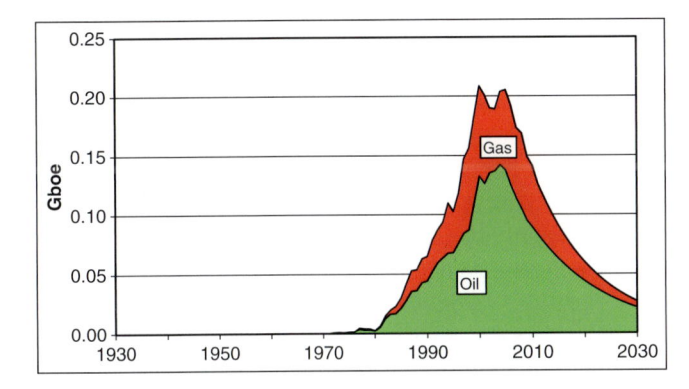

Fig. 40.1 Denmark oil & gas production 1930 to 2030

Fig. 40.2 Denmark status of oil depletion

Essential Features

Denmark is a low-lying peninsula between the North Sea and the seaways leading to the Baltic to the east. An archipelago of large islands crosses the Kattegat seaway, with Copenhagen, the capital, being located on the easternmost island, Sjaeland, only 20 km from the Swedish coast. The country supports a population of 5.6 million.

Geology and Prime Petroleum Systems

Much of Denmark is underlain by substantially non-prospective rocks, but the southern limit of the northern North Sea rift system with its prolific oil and gas extends into Danish waters. It contains rich Upper Jurassic source-rocks which have locally charged overlying reservoirs in Cretaceous Chalk and lower Tertiary sandstones. The Chalk reservoirs are particularly difficult, having low permeability. Slumping during deposition gave rise to local areas where high porosity was preserved but the low permeability results in low recovery rates, which are nevertheless susceptible to advanced technological methods. The reservoirs are also subject to formation collapse as the oil is removed, leading to subsidence of the sea-floor above.

Exploration and Discovery

Exploration drilling commenced in 1936, and grew after the war, when it was rewarded by a modest discovery in 1966, which in fact drew early attention to the oil potential of the northern North Sea. In 1971, came the discovery of the

C.J. Campbell, *Campbell's Atlas of Oil and Gas Depletion*,
DOI 10.1007/978-1-4614-3576-1_40, © Colin J. Campbell and Alexander Wöstmann 2013

giant offshore Dan Field with about 750 Mb and 0.9 Gcf of gas, being produced from the Cretaceous chalk. A total of about 190 exploration boreholes have now been drilled, yielding some 3.3 Gb of oil and 8.8 Tcf of gas. Exploration drilling peaked in 1985 when 14 boreholes were sunk, but has since dwindled to no more than two or three a year. Exploration drilling is expected to end around 2020 by which time all viable prospects will have been tested.

Production and Consumption

Oil production commenced in 1972 and reached a peak at 389 kb/d in 2004, before falling to 246 kb/d in 2010. It is now set to decline at a depletion rate of 7 % a year. This is a relatively high rate, but reflects the advanced technology of offshore North Sea operations, which can drain the reservoirs quickly. Oil consumption is running at 61 Mb a year, making the country a declining net exporter over the next 10 years until internal demand overtakes production.

Gas production commenced in 1977, rising to a peak of 376 Gcf/a in 2005, since when it has fallen to 289 Gcf/a and is expected to continue to decline at about 10 % a year, being now 75 % depleted. Consumption is running at 175 Gcf/a, making the country a net exporter of about 114 Gcf/a year.

The Oil Age in Perspective

Archaeological remains show that Denmark was inhabited by Bronze and Iron Age peoples, before coming to prominence in the Viking Age during the first millennium, when its sailor-kings led foreign expeditions of pillage, conquest and settlement. Both Iceland and Greenland became Danish territories.

By the eleventh century, it had established dominion over the rest of Scandinavia before being consumed in civil wars and internal power struggles. Later attempts to forge a union with Norway and Sweden met with only brief success before also foundering in conflict. During the eighteenth century, the country sought to emancipate its peasants reducing the power of the landlords. It also adopted a policy of neutrality in European wars, which antagonised Britain, when it sought to blockade European ports during the Napoleonic Wars. That led to a naval attack in 1801, which destroyed the Danish fleet, prompting the country to become a staunch ally of Napoleon. Tensions arose later over the southern provinces of Schleswig and Holstein, which sought greater autonomy, prompting Bismarck to successfully invade in 1864 and incorporate them into what became Germany.

Denmark managed to remain neutral in the First World War and recovered Schleswig in the ensuing Peace Treaty. Although primarily an agricultural country, its shipping and fishing industries prospered in the inter-war years, which came to an end with an unprovoked German invasion in April 1940. It was occupied for the duration of the war, but supported an active resistance movement.

Post-war economic recovery was relatively rapid, and the country became an early and enthusiastic member of the European Economic Community, the predecessor of the European Union. It remains a monarchy although generally favouring mildly socialist forms of government.

Denmark has led Europe in the development of wind energy and the adoption of intelligent energy-saving policies, including the provision of rental bicycles in cities, and communal heating systems. Its population remains at a modest 5.5 million, despite having doubled over the First Half of the Oil Age, and it does not face any particular minority tensions. It is accordingly relatively well placed to face the post-peak world.

40 Denmark

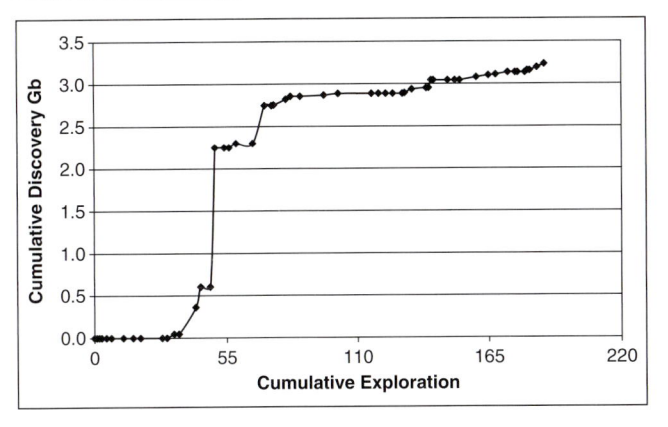

Fig. 40.3 Denmark discovery trend

Fig. 40.4 Denmark derivative logistic

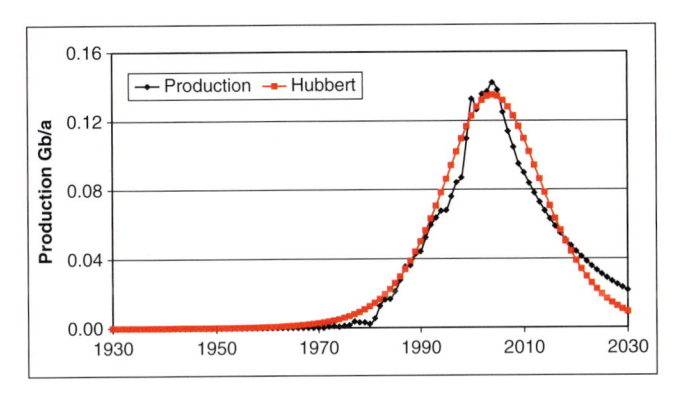

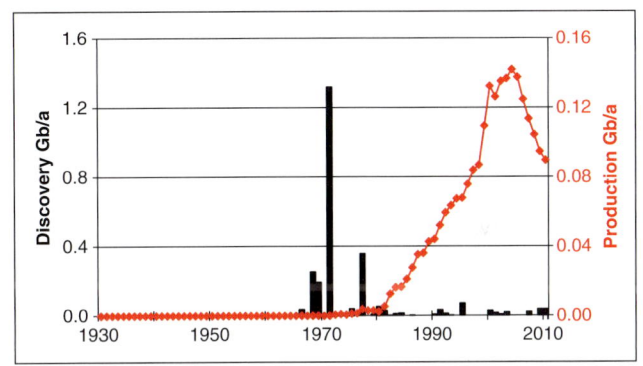

Fig. 40.5 Denmark production: actual and theoretical

Fig. 40.6 Denmark discovery and production

France

Table 41.1 France regional totals (data through 2010)

Production to 2100					Peak Dates			Area	
Amount		Rate				Oil	Gas	'000 km²	
	Gb	Tcf	Date	Mb/a	Gcf/a	Discovery	1954	1949	Onshore	Offshore
PAST	0.79	11.4	2000	11	66	Production	1988	1978	554	177
FUTURE	0.16	0.4	2005	8	63	Exploration	1959		*Population*	
Known	0.13	0.2	2010	7	44	*Consumption*	Mb/a	Gcf/a	1900	38
Yet-to-Find	0.03	0.16	2020	4	16	2010	679	1,699	2010	63
DISCOVERED	0.92	11.6	2030	3	6		b/a	kcf/a	Growth	1.6
TOTAL	0.95	12.0	*Trade*	−673	−1655	Per capita	10.8	27	Density	114

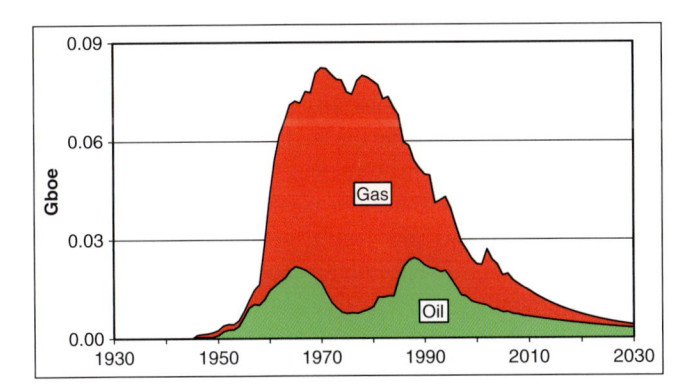

Fig. 41.1 France oil and gas production 1930 to 2030

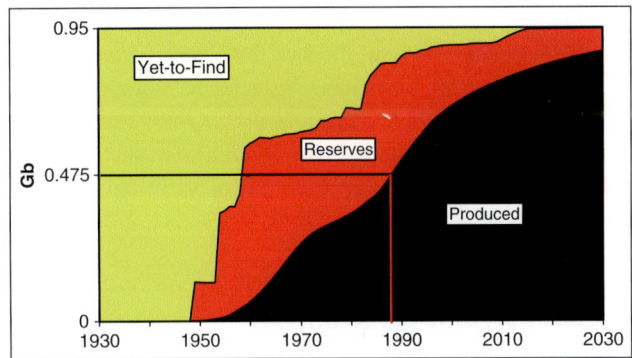

Fig. 41.2 France status of oil depletion

Essential Features

France, the largest country in Western Europe, is made up of beautiful rolling farm-lands, flanked by the mountains ranges of the Alps to the east, the Pyrenees to the south and the Vosges to the northeast. The remains of older mountain belts form both the Massif Central, in the middle of the country, and the rocky peninsula of Normandy projecting into the Atlantic. The Seine and Garonne Rivers drain the western part of the country into the Atlantic, while the Rhone, in the east, flows southward into the Mediterranean. France also owns the island of Corsica in the Mediterranean. The country supports a population of 63 million.

Geology and Prime Petroleum Systems

France exhibits a wide range of geology, including two principal petroleum systems forming respectively the Aquitaine and Paris Basins. The Aquitaine Basin contains Upper Jurassic source-rocks that have charged overlying reservoirs with both oil and, where deeply buried, gas which has provided a major gas field at Lacq. The basin also has a small offshore extension. The Paris Basin relies on lean Lower Jurassic source-rocks, which have given a number of modest fields.

There are a number of other minor systems not deserving any particular mention, save for a tar-deposit at Pechelbronn

C.J. Campbell, *Campbell's Atlas of Oil and Gas Depletion*,
DOI 10.1007/978-1-4614-3576-1_41, © Colin J. Campbell and Alexander Wöstmann 2013

in the Alsace region in the northeast of the country, which is of historical interest, being the scene of possibly the world's first oil well, drilled in 1813. It was also here that Conrad Schlumberger developed the first electric well-logging technology which transformed the oil industry worldwide by making it possible to identify the oil-bearing strata penetrated in a borehole. A total of 24 Mb was produced at Pechelbronn between 1934 and 1964.

The source-rocks offer potential for shale-gas but so far developments have been limited being widely opposed for environmental reasons.

Exploration and Discovery

Modern exploration commenced in the 1920s when a few boreholes were sunk, yielding a minor discovery. It resumed in earnest after the Second World War to reach a peak in 1959 when as many as 126 exploration boreholes were drilled, mainly in the Aquitaine Basin. Exploration drilling dwindled during the 1960s before building to a second peak of 75 boreholes in 1986, since when it as fallen to an annual average less than five over the past few years.

Only two major finds have been made, comprising the Lacq Gas field in 1949 with about 9 Tcf, and Parentis Field in 1954, with 240 Mb of oil, both of which lie in the Aquitaine Basin. The discoveries in the Paris Basin are relatively modest in size.

Production and Consumption

Apart from Pechelbronn, described above, oil production commenced at a very low level in 1928, not rising significantly until the 1960s. It reached a peak in 1988 at 67 kb/d before declining to 18 kb/d in 2010. It is now set to continue to decline at the current Depletion Rate of about 4 % a year. The country consumes 679 Mb/a of oil, almost all of which is imported.

Gas production commenced in 1946 and rose to a plateau of about 375 Gcf/a in the years between 1970 and 1981. This was followed by a decline to around 44 Gcf a year by 2010, being 97 % depleted. Future production is set to decline at about 10 % a year to exhaustion in the not distant future. Consumption stands at around 1,700 Gcf a year, most of which is imported.

The Oil Age in Perspective

Various Celtic tribes occupied France prior to the Roman conquest by Julius Caesar in 50 BC, giving the Roman language and administration to the country. When that empire collapsed, France was subject to invasions by Goths, Vandals, Burgundians and Franks from the east and north, as well as Vikings, who settled in Normandy, and later went on to successfully invade England in 1066. The next few centuries were marked by internal and external conflicts between the various kingdoms and duchies that developed in the country. Additional tensions between the Catholic and Protestant faiths arose, but a unified kingdom gradually came into being.

The country developed imperial aspirations in the eighteenth century, taking Louisiana and Quebec in North America, as well as parts of India, but eventually lost them, mainly to Britain. Conditions were not altogether happy at home as intellectuals and the people at large grew tired of the elitist, essentially feudal, system that had evolved under the monarchy. These pressures built, leading to the famous French Revolution of 1789, which culminated in the execution of the king and other leading aristocrats in a movement that was to become the inspiration for socialism throughout Europe.

Weak government ensued with corresponding internal conflicts putting the country's defences at risk. A reaction saw the emergence of a Corsican general, Napoleon Bonaparte, who came to power in 1798, imposing a new authoritarian regime. It embarked on a policy of successful foreign conquest to become the leading power in continental Europe. One positive feature was the Napoleonic Code, devised to reform the European legal system and introduce, amongst other things, the metric system of measurement. But it was a short-lived empire, facing military defeat in Russia and finally against British and German armies at the Battle of Waterloo in 1815, whereupon Napoleon was forced to abdicate, dying in exile six years later.

The country briefly reverted to a monarchy before facing another revolution in 1848, when a nephew of Napoleon returned to power, making an alliance with the country's old enemy, Britain, to fight the Crimean War against Russia.

In later years, it began to feel threatened by the growing power of German unity, and on a trivial pretext declared war in 1870, only to be roundly defeated by the Prussians, who demanded large financial reparations and took the border territories of Alsace-Lorraine with their rich iron and coal deposits.

Although internal political tensions remained, the Third Republic became firmly established, and began to foster alliances with Russia, Italy and eventually Britain intending to counter the perceived threat of German power. Meanwhile, imperial ambitions had returned as reflected by the acquisition of territories in North Africa, including Algeria, West Africa, Syria and parts of Indo-China.

Renewed tensions erupted in 1911 when Germany sent a warship to Morocco to frustrate efforts to bring that country into the French Empire. An escalation of these pressures occurred in 1914 when a separatist movement in Austria received support from Russia. Germany came to Austria's aid, declaring war on Russia on August 1st and on France, three days later. This was partly a strategic military move to avoid having to fight on two fronts, being based on the erroneous assumption that a lightening strike against France would be successful. These seemingly superficial events, which triggered the First World War, may in fact reflect deeper economic and financial conflicts, as Germany sought to conquer new world markets and put its currency on the financial map. The war was greeted in France with patriotic fervour, but an initial campaign to re-take Alsace-Lorraine was frustrated when Germany marched south through neutral Belgium and Luxembourg. This move instigated what became a gruesome war of attrition on the killing fields of Flanders, where 1.4 million Frenchmen lost their lives.

Victory finally came in 1918, but peace found a devastated country, burdened by heavy war debt, especially to the United States. Elections threw up a series of generally weak governments, which followed a policy of appeasement in the face of renewed German threats during the 1930s, before the resumption of hostilities in 1939 in what became the Second World War. Despite impressive fortifications along the frontier, it failed to resist the German onslaught, and the country capitulated one year into the war. A pro-Fascist government, known as the Vichy Government, with a new constitution was formed to seek accommodation with Germany, which was expected to emerge victorious from the war. It did not however enjoy wide popular support, and General de Gaulle, who had escaped to England, formed a Government in Exile, whilst a growing resistance movement fought the German occupation. Anglo-American forces liberated the country in 1944, and General de Gaulle received a tumultuous welcome when he returned to form a government.

The post-war years saw the country facing economic and political problems at home, with the Communists achieving a strong following. The overseas empire was progressively lost, being accompanied by wars in Algeria and Vietnam. De Gaulle remained in power under the so-called Fifth Republic, ruling in an autocratic manner until 1969, only to dies one year later. His rule was followed by a succession of governments, which were at first inclined towards socialism but have progressively moved to the right. France was a founder member of what became the European Union, being also an active member of NATO during the Cold War, but has since somewhat distanced itself from the globalist policies of the United States. A large number of Algerians moved to France in preference to remaining in their newly independent homeland. They form a less than fully integrated community, which occasionally gives rise to friction and tension.

Despite the political travails outlined above, France has recorded great achievements in the cultural domain, with the flourishing of philosophers, writers, artists and scientists, many of world renown. In economic terms, it is primarily an agricultural country, with its vineyards producing some of the world's most famous vintages, but it also has a strong industrial base built originally on the coal mines of the north. It has long had a strong financial centre, with French banks playing an important role in international finance.

At first, it was able to supplement its limited domestic oil and gas resources by drawing on the rich deposits of Algeria, but when that territory was lost, de Gaulle perceptively recognised the country's vulnerability and turned to nuclear energy, supplemented locally by tidal power.

Looking ahead, France appears to be relatively well placed to face the *Second Half of the Age of Oil*. The population has grown by a factor of no more than 1.6 over the First Half of the Oil Age to 63 million, and is now dwindling as its birth-rate declines. Its rich agricultural land and benign climate should be sufficient to feed its people in the years ahead. It has come to terms with a strong nuclear industry, not sharing the environmental concerns widely expressed in other countries. An effective and workable form of government has evolved, and the country no longer faces any particular threats from its neighbours, avoiding contentious international engagements, although recently joining the US and Britain in a military attack upon the government of Libya.

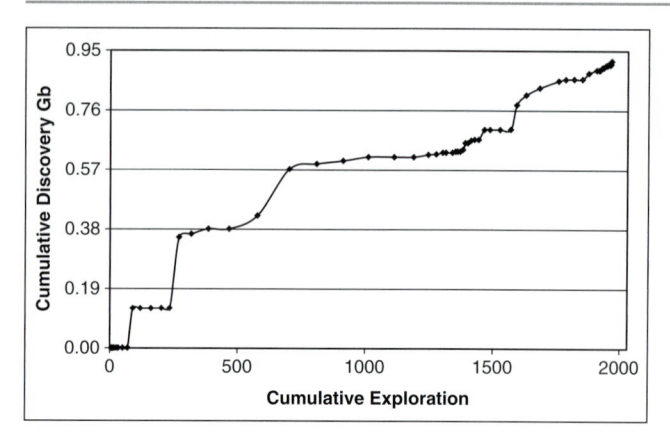

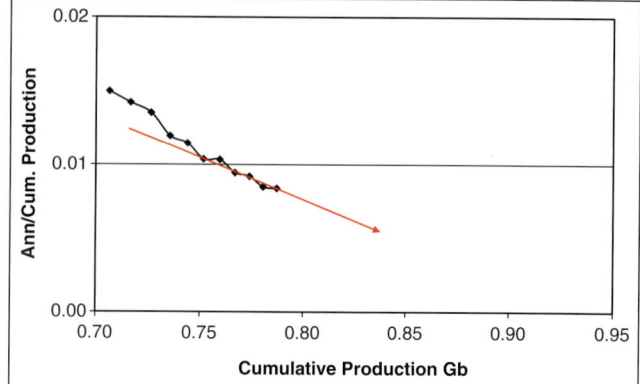

Fig. 41.3 France discovery trend

Fig. 41.4 France derivative logistic

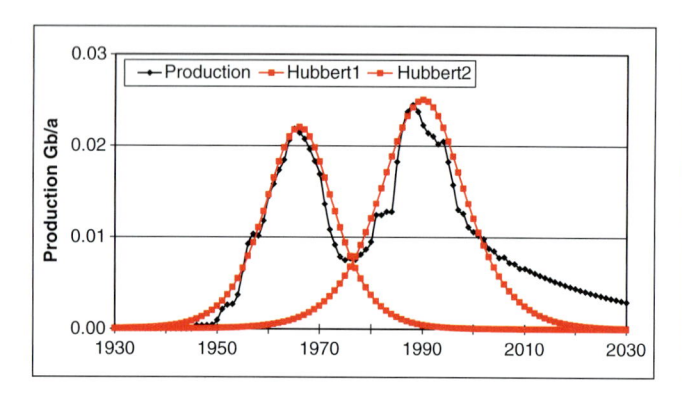

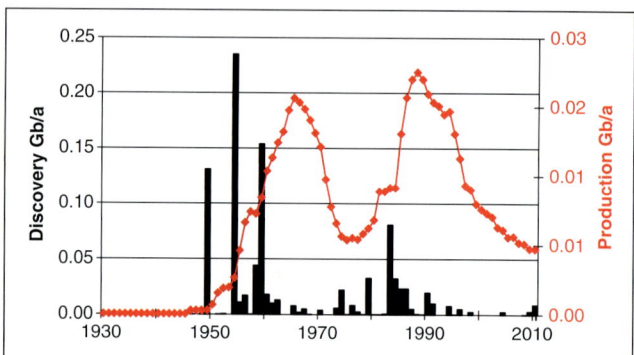

Fig. 41.5 France production: actual and theoretical

Fig. 41.6 France discovery and production

Germany

Table 42.1 Germany regional totals (data through 2010)

	Production to 2100					Peak Dates			Area	
	Amount		Rate				Oil	Gas	'000 km²	
	Gb	Tcf	Date	Mb/a	Gcf/a	Discovery	1949	1969	Onshore	Offshore
PAST	1.96	35	2000	17	836	Production	1967	1987	358	40
FUTURE	0.54	13	2005	16	750	Exploration	1958		Population	
Known	0.43	11	2010	10	464	*Consumption*	Mb/a	Gcf/a	1900	55
Yet-to-Find	0.11	1.3	2020	9	324	2010	911	3,437	2010	81.8
DISCOVERED	2.39	47	2030	7	226		b/a	kcf/a	Growth	1.5
TOTAL	2.50	48	*Trade*	−900	−2,973	Per capita	11.1	42	Density	228

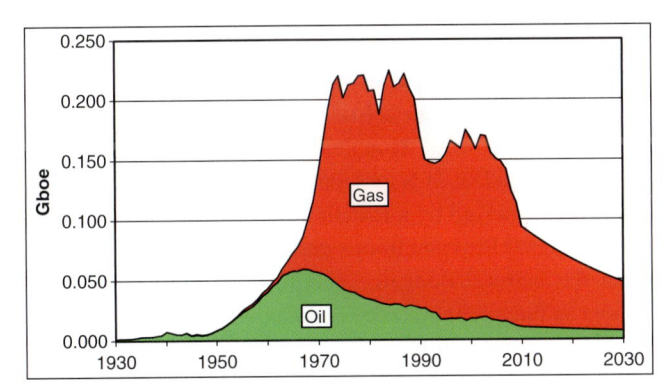

Fig. 42.1 Germany oil and gas production 1930 to 2030

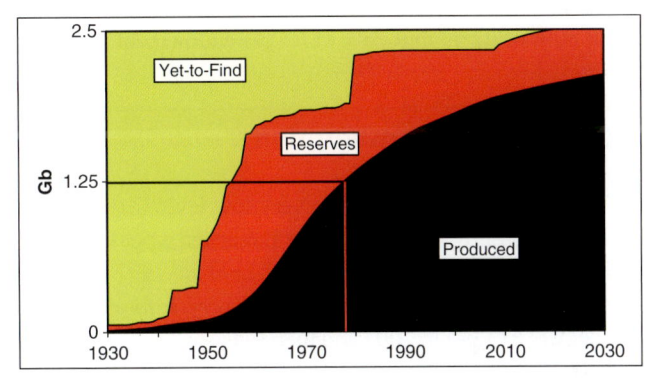

Fig. 42.2 Germany status of oil depletion

Essential Features

Germany, which supports the largest economy in Europe, is made up of three physiographic regions, comprising central uplands, including the Harz Mountains, which separate extensive plains in the north from hilly country in the south, bordering the Vosges and Alpine mountain chains. The mighty Rhine River drains the southern part of the country, while the Elbe and Weser rivers flow into the North Sea. The south-eastern part of the country is drained by the headwaters of the Danube, flowing eastward eventually to the Black Sea. The country, which enjoys a temperate climate, supports extensive forests, leaving about one-third of it under agriculture. With almost 82 million inhabitants, it is the most populous country in Europe.

Geology and Prime Petroleum Systems

Germany has a diverse geology including six minor petroleum systems. By far the most important is the Carboniferous gas system that extends eastwards from the Netherlands across the northern part of the country into Poland. Deeply buried coal deposits have been converted to gas under a process akin to coking, and have charged overlying reservoirs mainly in Permian desert sandstones. A seal of salt above is an important additional factor.

The other systems rely primarily on lean Lower Jurassic and Lower Cretaceous source-rocks which have generated oil, including finds in the North Sea and in the Rhine valley. A Permian oil source-rock is also responsible for small fields in the northeast.

C.J. Campbell, *Campbell's Atlas of Oil and Gas Depletion*,
DOI 10.1007/978-1-4614-3576-1_42, © Colin J. Campbell and Alexander Wöstmann 2013

Germany is an example of an intensely explored mature area that has yielded a number of different relatively small petroleum systems, based on lean source-rocks, which are locally valuable but too small to have much global impact.

Exploration and Discovery

Although a few boreholes had been drilled earlier, serious exploration commenced in 1934, when as many as 22 boreholes were sunk, and it continued at a modest level over subsequent years. A new chapter opened in the 1950s and 1960s, reaching a peak in 1958 when 143 boreholes were drilled. A number of modest oil discoveries and a few significant gas finds were made. Most of the discoveries were onshore, but a useful gas field, with 2.7 Tcf, was found in the southern North Sea in 1965. Exploration has since declined markedly with no more than an average of about five boreholes a year being drilled over the past decade. It is evident that the scope for new discovery is severely limited. The country has thus reached an advanced state of depletion, having produced about 80% of its oil and 75% of its gas endowment.

Production and Consumption

Oil production commenced at a very low level in 1887, not rising significantly until the 1950s. It reached a peak in 1967 at 162 kb/d before declining to 47 kb/d in 1997 when it briefly recovered, following a new discovery. It rose to a second peak at 51 kb/d in 2003 and then sank to 28 kb/d in 2010. It is now expected to resume its overall decline at the current Depletion Rate of almost 2% a year. Consumption is running at 911 Mb/a year, almost all of which is imported.

Gas production commenced in 1944, rising to a plateau of about 1 Tcf/a from 1974 to 1989, which was followed by a decline to around 464 Gcf/a by 2010, when 75% had been depleted. It is now expected to continue its terminal decline at about 6% a year. Consumption stands at 3,437 Gcf/a, of which around 85% is imported.

The Oil Age in Perspective

Bronze and Iron Age peoples evolved into various tribes that successfully resisted the Roman armies, with the great Rhine River being a natural frontier. The country was later subject to invasions from the east with the arrival of Huns and Goths. The Rhine froze in 410 AD allowing Alaric the Goth to march south and sack Rome, effectively bringing the Roman Empire to an end. The Saxons from northwest Germany successfully invaded and colonized the British Isles during fourth and fifth centuries.

Various kingdoms arose thereafter being brought together to some extent by Charlemagne around 800 AD, but they later went their separate ways, with Hannover, Saxony, Prussia and Bavaria becoming prominent. Christianity was gradually adopted, leading to the so-called Holy Roman Empire which sought to weld together ecclesiastic and secular power. It met with only partial success, being followed by the schism between the Protestant and Catholic faiths, led by Martin Luther in the sixteenth century. A pandemic of 1348–1349, known as the Black Death, decimated the country, but there was nevertheless a rise of weaving and other crafts under important trade guilds. A trading association, known as the Hanseatic League, helped develop external trade, especially in the north.

By a quirk of history, the grand-daughter of King James I of England married George, the son of the Elector of Hannover, who became the King of England in 1714, siring the present British royal family, which diplomatically changed its name from Saxe-Coburg-Gotha to Windsor during the First World War.

The French Revolution of 1789 and the subsequent rise of Napoleon were welcomed by some sections of German society, including the growing number of industrial workers. This put pressure on the aristocracy, which at first sought an accommodation with Napoleon but later faced military conflict. The defeat of Napoleon, at the Battle of Waterloo in 1815, was achieved by a British army with the decisive support of 45,000 Prussian troops under General Blücher.

In the cultural domain, such famous philosophers and writers as Nietzsche, Goethe and Hegel, found patronage in the various royal courts.

The Industrial Revolution of the eighteenth century saw Britain, France and, to a lesser extent, Russia evolving into imperial powers seeking overseas markets, leaving a fragmented Germany somewhat isolated and disadvantaged. Prussia, which had become a leading power under its Chancellor, Bismarck, sought to rectify the situation by unifying the country. The first step was the removal of trade tariffs, which led to an epoch of rapid economic growth, including the expansion of the railways. France, for her part, began to feel threatened by this new European economic power, and in 1870 declared a war, known as the Franco-Prussian War, only to be roundly defeated. The peace terms required France both to cede Alsace-Lorraine, with its rich deposits of iron and coal deposits, to Germany and pay reparations of five billion francs. A new German empire was effectively born in this way although it lacked overseas territories. The economic boom was however short-lived, as Germany was not spared the effects of a worldwide Depression in 1873, prompting widespread emigration. Its response was the creation of large corporate empires to control the iron and steel industry, banking and agriculture, which enjoyed protectionist government policies. Some modest overseas territories in East Africa, Southwest Africa, China and the Asia-Pacific region were secured. It also had a particular interest in the Middle East, which was underpinned by a plan to build a railway from Berlin to Baghdad as a strategic weapon to rival the maritime power of the British Navy.

Germany forged an alliance with Austria, which was facing not only internal separatist movements from its Slav minorities but threats from Russia that was in conflict with the Turkish Ottoman Empire occupying adjoining parts of southeast Europe. Various alliances were forged between the European powers to try to keep these growing conflicts under control. There were, in addition, internal tensions as underprivileged industrial workers, who courted socialist ideals, threatened the power of the elite landlords and industrialists. Streaks of anti-Semitism arose in Germany, as elsewhere, as people began to identify what they perceived to be sinister banking forces, largely controlled by Jewish families.

The delicate balance of power collapsed with the outbreak of the First World War in 1914 triggered by the assassination of an Austrian Archduke in Serbia. It degenerated into an appalling war of attrition, thanks in part to advances in military technology and the development of railways which were able to continually supply the frontlines with reinforcements. The entry of the United States in 1917 heralded the surrender of Germany in the following year, although its troops were still undefeated on foreign soil. President Wilson had proposed *Peace but no Victory*, but the eventual peace treaty attributed guilt for the war on Germany and imposed severe reparations, causing great hardship, including starvation. This led to a resurgence of Socialist and Communist movements, now taking inspiration from the Bolshevik Revolution of 1917 in Russia, which put pressure on the industrial and banking elite. This indirectly led them to support a new political party of so-called *National Socialists*, which promoted the recovery of national strength and grandeur through authoritarian rule, having reason to be disillusioned with the democratic process. The Great Depression of 1929 cast its shadow nowhere more heavily than on war-torn Germany, putting six million people out of work and causing rampant inflation, paving the way for the National Socialists, under Adolf Hitler, to come to power in 1933. As part of its efforts to bring a new dignity to the German people, the Government promoted the idea that Germans were racially superior. This theory was in part inspired by the science of eugenics developed in Sweden, which claimed that superior humans could be bred in the same way as can racehorses. It had the effect of prompting renewed anti-Semitism in Germany. Other groups within the citizenry, such as homosexuals, gypsies and those with certain maladies were also considered inferior.

The Third Reich (Third Empire) was born and indeed began to make substantial progress, stabilising the country and creating employment, partly through impressive road building programmes. Workers were given the opportunity to buy a Volkswagen (literally the *People's Car*) as a symbol of the success of the new regime. It also had ambitions to secure an eastward empire. This took tangible form in the late 1930s with the annexation of Austria, parts of Czechoslovakia and Lithuania, where there were long-established German communities. These actions did not meet any serious opposition from the British Government, which might indeed have welcomed any moves against Communist Russia as it was facing Socialist pressures from its own working class. The regime also received strong support from American banks, investment houses and oil companies, who admired the decisive management of the economy.

Later, the Soviet Union came to an accommodation for the division of Poland and the establishment of the different spheres of influence in Eastern Europe. This led to the German invasion of Poland, which upset the balance of power, prompting Britain to declare what became the Second World War in September 1939. Russia took the opportunity to garrison several Baltic States, facing stiff opposition and initial defeat in Finland, which led Germany to underestimate Russia's military strength and risk invading it in June 1941. Meanwhile, American oil companies, especially Texaco, continued to supply Germany with critical fuel until the United States entered the war at the end of the year.

The early years of the war moved in Germany's favour with the fall of France and military successes, which put German armies at the gates of Moscow and Cairo, but its fortunes began to turn with entry of the United States in December 1941. There followed a time of great suffering: the Jews were persecuted and sent to labour camps where millions suffered torture, deprivation and death; the civilian population was indiscriminately bombed; and the retreating troops suffered in the frozen wastes of Russia. Defeat came in 1945, whereupon the leaders of the regime either committed suicide or were executed.

The country was then divided under British, American and Russian occupations. The former allies now found themselves glaring at each other under the so-called Cold War, which left Germany divided into a Communist east and a Capitalist west. The latter installed a democratic government and was given every economic and financial help to form a bastion against Communist influence, becoming the Federal Republic, while the east, known as the Democratic Republic, became part of the Communist bloc.

The fall of the Soviet Government paved the way for reunification in 1990. The country nevertheless retains a strong degree of regional government and feeling; reflecting deeper historical divisions. Some of the inhabitants of Bavaria may think of themselves to be Bavarians first and Germans second.

West Germany was a founder member of what has become the European Union, and enjoyed great economic prosperity in the post-war years. While it accepted a large number of immigrant workers, especially from Turkey, who live mainly in Berlin, it does not face any particular internal ethnic tensions. That said, the former East Germany remains somewhat economically depressed, with certain disaffected elements in society entertaining degrees of resentment. Many have been forced to emigrate westward.

Looking ahead, the country may expect a difficult future. The population has grown by a factor of 1.5 to 82 million over the past century. Its limited oil and gas resources are now heavily depleted; its coal industry has long been in decline. Substantial lignite deposits, which may assume more importance in the future, remain, although presenting environmental challenges. The country has already chosen to close its nuclear industry for environmental reasons, and perhaps because it foresees the worldwide decline of uranium supplies. In other words, its electricity supply depends largely on imports, principally from Russia, which gives that country a decisive control over its destiny. On the positive side, it has made impressive progress with solar energy, especially in the south, and wind energy in the north. It is also well placed to revert to a more regional form of government fostering local communities and markets to find new sustainable lifestyles. In the absence of massive immigration, its population may fall naturally back to a supportable level without undue tension.

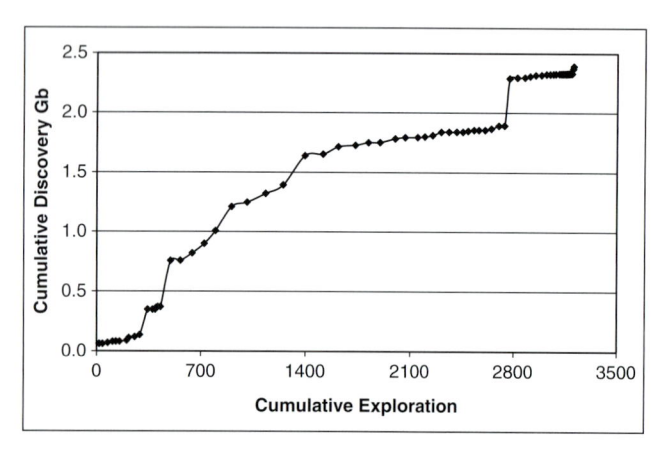

Fig. 42.3 Germany discovery trend

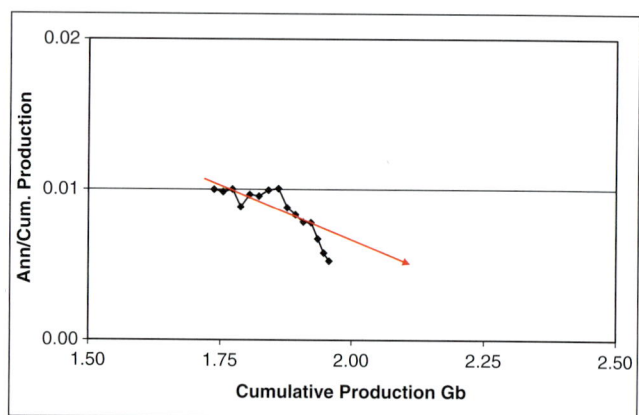

Fig. 42.4 Germany derivative logistic

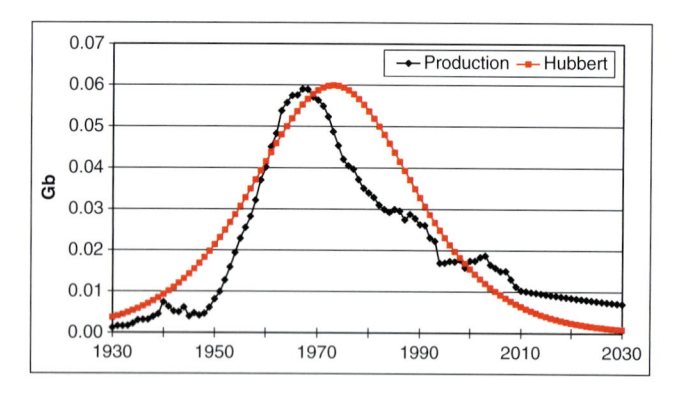

Fig. 42.5 Germany production: actual and theoretical

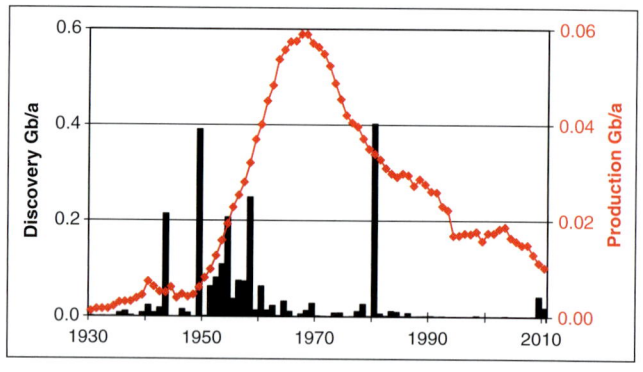

Fig. 42.6 Germany discovery and production

Italy

Table 43.1 Italy regional totals (data through 2010)

Production to 2100					Peak Dates			Area	
Amount			Rate			Oil	Gas	'000 km²	
	Gb	Tcf	Date	Mb/a	Gcf/a	Discovery	1989	1968	Onshore	Offshore
PAST	1.16	26	2000	33	587	Production	1997	1994	302	140
FUTURE	0.84	6.3	2005	42	426	Exploration	1962		*Population*	
Known	0.71	5.7	2010	35	293	*Consumption*	Mb/a	Gcf/a	1900	32
Yet-to-Find	0.13	0.6	2020	23	186	2010	558	2,930	2010	60.8
DISCOVERED	1.87	31	2030	15	118		b/a	kcf/a	Growth	1.9
TOTAL	2.0	32	*Trade*	−523	−2,637	Per capita	9.17	48	Density	201

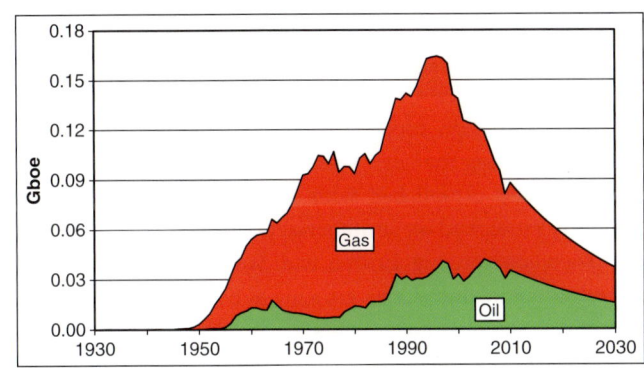

Fig. 43.1 Italy oil and gas production 1930 to 2030

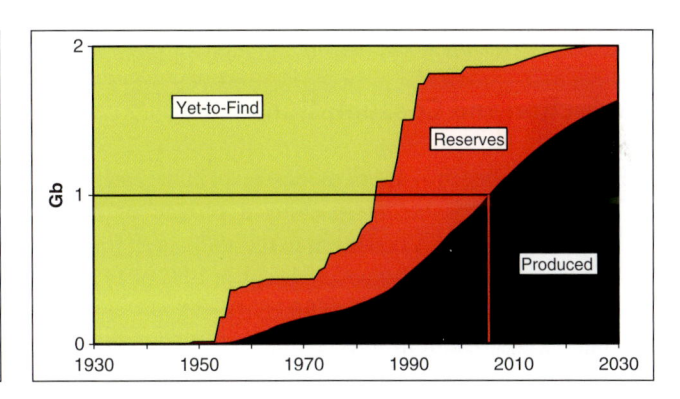

Fig. 43.2 Italy status of oil depletion

Essential Features

Italy forms a mountainous peninsula extending into the Mediterranean, also having jurisdiction over the islands of Sicily and Sardinia. The Apennine mountain range, which forms the backbone of the country, rises to over 3,000 m, and is flanked to the east by a low-lying coastal strip, bordering the Adriatic Sea. The Po Valley, which flows eastward across the northern part of the country, separates the Apennines from the frontal ranges of the Alps. The country supports a population of almost 61 million.

Geology and Prime Petroleum Systems

The prime petroleum system relies on the relatively lean source-rocks in the Triassic *Taormina Formation*, which has charged overlying Jurassic carbonates in Sicily, offshore in the Adriatic and along the eastern seaboard. Tertiary gas-prone shales within the Po Valley have also fed overlying sandstones, and indeed are still active, partially recharging them. A third, more complex, system lies at depth within the Apennines themselves, which is made up of a pile of slumped and thrusted masses, which have been uplifted from the margins of the Tyrrhenian Sea to the west as a result of a tectonic-plate collision.

Exploration and Discovery

Although desultory exploration had commenced in the early years of the last century, it did not take off in earnest until after the Second World War, rising rapidly to a peak in 1962, when as many as 102 exploration boreholes were drilled. It thereafter declined gradually to the late 1980s, before falling precipitously, such that less than ten boreholes a year have been drilled over the recent past. Evidently, the stock of untested prospects has been seriously depleted, although surprises can always come at depth within the complex structures of the Apennines and its foothills. Offshore exploration commenced in the 1960s in the Adriatic and off Sicily, but it too has declined steeply in recent years.

The first recorded discovery was in 1860, but nothing significant was found for almost 100 years. Landmarks were the Ragusa Field in Sicily which came in with 165 Mb in 1954, followed by the Villa Fortuna Field with 250 Mb in 1984, being in turn succeeded by a few more of comparable size over the succeeding years. A total of 1.9 Gb of oil has been found.

Gas exploration has been rather more successful with numerous small to modest discoveries. An important find was the Porto Corsini Field in 1961 with 1.75 Tcf. An offshore extension and several other fields of comparable size were added over the next few years, before discovery dwindled. In total about 31 Tcf of gas have been found.

Production and Consumption

Oil production began at a very modest level in the early years of the last century, before reaching a plateau of around 35 kb/d in the 1960s. A downturn in the 1970s was followed by an irregular increase to an overall peak in 1997 of 112 kb/d. Production has since declined to 96 kb/d and looks set to fall at almost 4% a year, mirroring the preceding decline in discovery. The country consumes 558 Mb a year, of which it has to import over 90%.

Gas production commenced in the 1930s before rising in the post-war years to an irregular plateau at about 550 Gcf/a from 1970 to 2000 before declining. The overall peak was in 1994 at 729 Gcf/a. With 80% of the endowment already gone, future production is expected to decline at its current depletion rate of 4%. Gas consumption stands at 2,930 Gcf/a meaning that it has to import almost 90% of its needs, partly through pipelines from North Africa.

The Oil Age in Perspective

The Etruscan civilization reached its peak in the sixth century BC, and was followed during the early years of the first millennium by the great Roman Empire, which at its peak held dominion over a region stretching from Britain to the Middle East. Its legal system has had a lasting influence on these countries, some of whose languages also have Latin roots. Christianity was at first opposed and suppressed before being accepted in the fourth century, when Rome became the seat of the Catholic Popes.

The Empire fell in 470 AD to invading Goths and other tribes from the north, and the country disintegrated into a number of city states, whose fortunes waxed and waned through the Middle Ages. Venice was a major port for imports from the eastern Mediterranean, while Florence became a cultural centre supporting no less than the great artists, Leonardo da Vinci and Michelangelo, as well as the astronomer Galileo, who, to the dismay of the Church, discovered that the Earth rotated round the Sun. Florentine bankers were also famous, playing a decisive role in European mediaeval history, as they funded their favoured political powers at home and abroad.

The country was later invaded on several occasions by France, Spain and other countries, with Napoleon crowning himself as King of Italy in 1805. That prompted moves to unification, led by a fiery young patriot, named Giuseppe Mazzini. He was however defeated by the Austrians who appropriated the northern province of Lombardy. Another patriot, by the name of Garibaldi, then emerged, only to be in turn defeated by the French in 1867. An additional factor during these times was the existence of the temporal Papal States, owned by the Catholic Popes, who at times were able to mobilize foreign support to protect their properties. But eventually during the latter years of the nineteenth century, a Kingdom was established with wide popular support, and promptly embarked on an imperial policy, securing colonies in Africa, comprising Eritrea and Somaliland. Even so, a degree of social unrest was growing at home, with the development of socialist movements, especially in the industrial north. Anarchists made an appearance, assassinating the King in 1900.

War with the Ottomans of Turkey broke out in 1911 over the rights of Italian settlers in Libya, which was duly annexed. Italy managed to remain neutral during the first year for the First World War, but was persuaded to join Britain and France in return for the offer of the Austrian province of South Tirol, in the event of eventual victory. However, the Austrians successfully invaded in 1917, facing largely disillusioned Italian troops, but were later repulsed in the last days of the war at the cost some 600,000 Italian lives.

Italy was accordingly present in the victorious carve-up at the Treaty of Versailles, securing South Tirol and the Croatian port of Trieste. Its position in Libya was also confirmed, but the war had drained the country giving rise to many social problems over the ensuing years.

A new leader, Benito Mussolini, appeared on the scene in 1922 as founder of the Fascist Party which sought to combine nationalism with socialism, gaining an absolute majority in elections two years later. He resumed imperial ambitions, largely to accommodate surplus population beyond that supportable by the economy at home, mounting a war against Abyssinia. He also laid claims to Djibouti, Tunis, Corsica and even Nice on the French mainland, and signed an alliance with Germany in 1939 following an invasion of Albania. At first, Italy preferred neutrality, but the fall of France prompted Mussolini, who by now had secured dictatorial powers, to declare war on Britain and France on the assumption that Germany was set to emerge victorious. It was a mistaken policy as defeat followed defeat in North Africa over the ensuing years, with Anglo-American forces landing in Sicily in 1943 followed by a successful assault upon the mainland. Mussolini and his mistress ended their days in 1945, hanging from lampposts in Milan, as communist Resistance fighters gained the upper hand.

The post-war epoch brought Italy a period of economic prosperity and political stability, enhanced when it joined the European Union, supplying a President by the name of Romano Prodi.

Its population has increased by a factor of 1.8 over the past century to reach 61 million, giving a density of 200/km^2, but it is now declining through a natural fall in birth rate. The country receives a number of largely illegal immigrants who cross the Mediterranean and Adriatic in small boats, but it has no fraught ethnic minority. The recent tensions in North Africa have greatly increased the flood of illegal immigrants some of whom lose their lives as their boats run aground or sink. Recent elections have marked a swing to the right with calls for a new regionalism and stricter immigration controls. Italy was not spared the financial crisis affecting the European Union. It cost Silvio Berlusconi, the President, his job as the country faces a chapter of austerity.

The consumption of oil and gas is clearly excessive, but the country can probably face the future with a degree of equanimity in its sunny, smiling land, also tapping geothermal energy from its volcanoes.

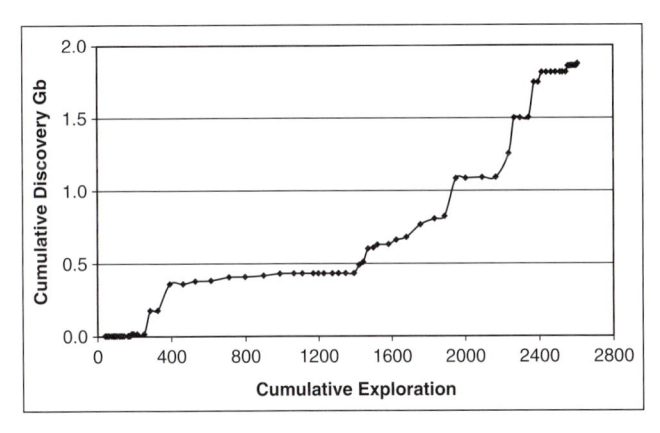

Fig. 43.3 Italy discovery trend

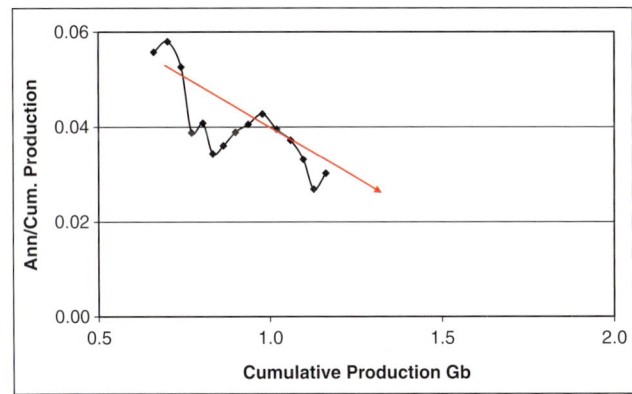

Fig. 43.4 Italy derivative logistic

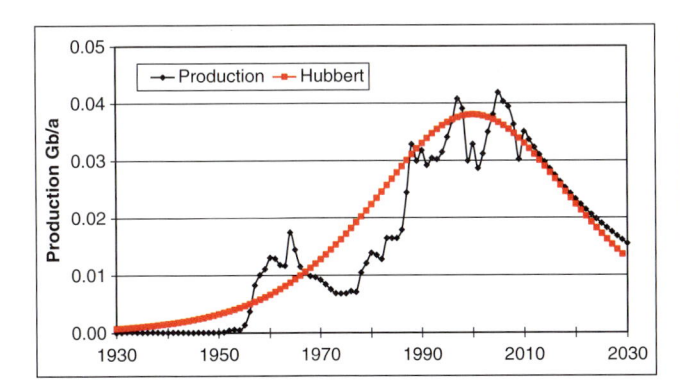

Fig. 43.5 Italy production: actual and theoretical

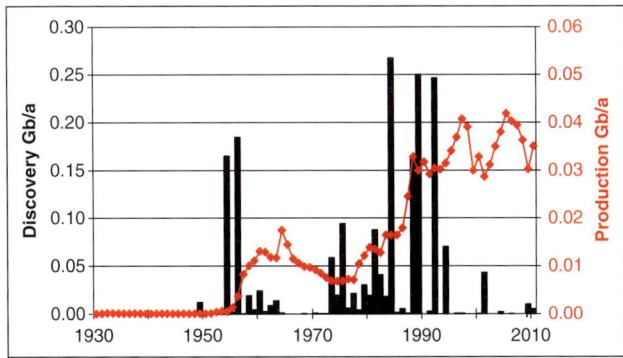

Fig. 43.6 Italy discovery and production

Netherlands

Table 44.1 Netherlands regional totals (data through 2010)

Production to 2010					Peak Dates				Area	
Amount		Rate				Oil	Gas	'000 km²	
	Gb	Tcf	Date	Mb/a	Gcf/a	Discovery	1943	1959	Onshore	Offshore
PAST	0.91	118	2000	11	2,564	Production	1986	1976	34	40
FUTURE	0.44	47	2005	12	2,776	Exploration	1985		*Population*	
Known	0.42	45	2010	7	3,131	*Consumption*	Mb/a	Gcf/a	1900	4.6
Yet-to-Find	0.02	2.4	2020	6	1,646	2010	368	1,938	2010	16.7
DISCOVERED	1.33	163	2030	5	865		b/a	kcf/a	Growth	3.6
TOTAL	1.35	165	*Trade*	−361	+1,193	Per capita	22	116	Density	491

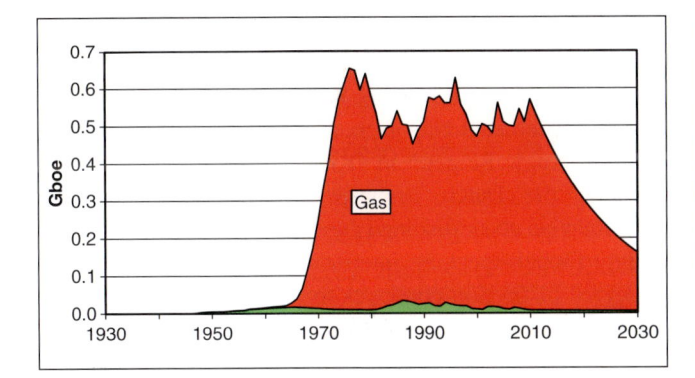

Fig. 44.1 Netherlands oil and gas production 1930 to 2030

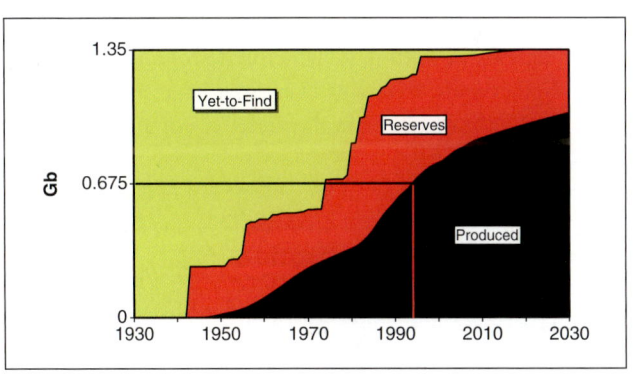

Fig. 44.2 Netherlands status of oil depletion

Essential Features

The Netherlands is a relatively small country of 41,000 km² on the northwest coast of Europe. It lies on the delta of the River Rhine, bordering Germany and Belgium. A large, partly reclaimed, inland sea, known as the Zuider Zee, extends in from the North Sea, while a string of islands runs parallel with the coast. Most of the country is low-lying, with about one-fifth being below sea-level behind a complex system of dykes, some of which have been in existence since the Middle Ages. The delta is subsiding and the sea-level is rising, making the country increasingly vulnerable to floods: one flood in 1953, which was caused by storms and high tides, claimed the lives of over 2,000 people. Even so, the country supports a population of nearly 17 million.

Geology and Prime Petroleum Systems

The geology of the Netherlands is largely obscured below recent deposits, so what is known comes from drilling operations. The prime petroleum system comprises deeply buried Carboniferous coals that have been converted to gas by a process of natural coking. The gas migrated upwards to fill Permian, and, to a lesser degree, Triassic sandstone reservoirs. Late Permian salt deposits commonly form an important seal.

A much less important oil system is also present in the form of lean Lower Jurassic source-rocks which have locally charged overlying Jurassic and Cretaceous reservoirs.

C.J. Campbell, *Campbell's Atlas of Oil and Gas Depletion*,
DOI 10.1007/978-1-4614-3576-1_44, © Colin J. Campbell and Alexander Wöstmann 2013

Exploration and Discovery

A few boreholes were sunk in the early years of the last century, but the main impetus came in 1937 when Shell exhibited an operating drilling rig at an industrial fair. To everyone's surprise, it encountered some unexpected indications of oil. This was followed up during the War when as many as 15 boreholes were drilled in 1943 under the German occupation.

Another surprise came in 1957, when a weekend communications failure led to the unintentional deepening of a well near Groningen in the north of the country which encountered a very large gas accumulation in Permian desert sandstones, which had not hitherto been considered to be even remotely prospective. The field was eventually found to contain almost 100 Tcf of gas, making it the largest in Europe.

It prompted a surge of subsequent drilling as the trend was tracked offshore across the southern North Sea to reach a peak in 1985 when 48 boreholes were sunk. A total of almost 1,300 exploration boreholes have now been drilled, meaning that almost all the prospects, even for very small accumulations of oil or gas, which are economically viable in the Netherlands, have now been tested.

The effort has resulted in the discovery of 163 Tcf of gas and 1.33 Gb of oil.

Production and Consumption

Oil production commenced in 1945 and reached an initial peak of 45 kb/d in 1965. It then declined to 25 kb/d in 1980 before rising to an overall peak of 93 kb/d in 1986. It has since fallen to 20 kb/d and is expected to continue to decline at its current Depletion Rate of about 1.6 % a year.

We may note in passing that different databases report a wide range of production figures, possibly due to differing treatment of gas liquids. The EIA data have been used herein. The country consumes about 368 Mb/a of oil, of which 361 Mb/a are imported.

Gas production commenced in 1950 and rose steadily to 3.6 Tcf/a by 1976. It has declined since then on a gentle slope to 3.1 Tcf/a in 2010. As much as 70 % has now been depleted, suggesting that a terminal decline at about 5 % a year follows. The Netherlands adopted a wise and cautious policy in depleting its gas slowly in order to make it last as long as possible. Gas consumption stands at 1,938 Gcf/a, allowing the country to export 1,193 Gcf/a, albeit on a declining trend.

The Oil Age in Perspective

Several archaeological sites confirm settlement in the earliest of times. The Rhine formed the northern limit of the Roman Empire, with the so-called Low Countries, including what are now the Netherlands, lying on the border. The shifting sand dunes, channels and swamps of the delta were settled by a warlike tribe, known to the Romans as Belgae, who were in turn subject to Viking raids and settlement.

The Middle Ages saw the development of city States and principalities, often in conflict with each other, as well as facing the successive external pressures of the Frankish Empire, the Holy Roman Empire, the Burgundian dukes, the Hapsburg dynasties of Spain and Austria, and finally Napoleon. Religious divisions also contributed to the political evolution, leaving the country predominantly Protestant.

The province of Flanders became a centre of weaving, securing its wool from England. Trade and banking flourished during the seventeenth and eighteenth centuries, based partly on a seafaring tradition and the country's geographical location at the mouth of the Rhine River, which was an important trading route. There was also a cultural flowering, highlighted by the famous painters Rembrandt, van Gogh, Frans Hals and others.

The modern State effectively came into being in 1814 when a monarchy under King William I was established following the defeat of Napoleon, although it saw the secession of what is now Belgium in 1830. Language was a divisive element: the Dutch language with its Germanic and Scandinavian roots distinguishes the coastal areas from the mainly French-speaking interior of what is now Belgium. A degree of friction continues to simmer between the two communities.

Overseas territories, notably the Dutch East Indies (now Indonesia) and South Africa, were acquired, partly settled and developed in the nineteenth and twentieth centuries. South Africa was lost to Britain in the Boer War of 1899–1902, while Indonesia gained its independence after the Second World War. Amsterdam became an important financial centre.

The Netherlands was neutral during the First World War, and tried to be so again in the Second, but was invaded by Germany in 1940. Queen Wilhelmina escaped to England where she presided over a government in exile. The post-war epoch saw gradual economic recovery, with the country playing a central part in the European Union and NATO.

The population has grown from about 4.6 million in 1900 to 17 million today, making it a very crowded place with a density of almost 400/km^2, of whom about 5 % are Muslim immigrants, many from the former colony of Indonesia, and later from Morocco and Turkey.

The country is not well placed to face the Second Half of the Age of Oil: indeed much of it may disappear altogether if global warming causes the sea-level to rise much higher. Even to-day, much energy is consumed pumping water from areas that are already below sea-level.

The Dutch, who have already rejected a proposed EU Constitution in a referendum, may move towards devolution, even rediscovering the benefits of a local currency by which to better manage their affairs. The turning point to a new realism may come if and when the Government forbids the export of its remaining gas, having belatedly come to recognise that it needs to conserve this critical resource for local use, in order to ameliorate the severe tensions of the post-peak transition.

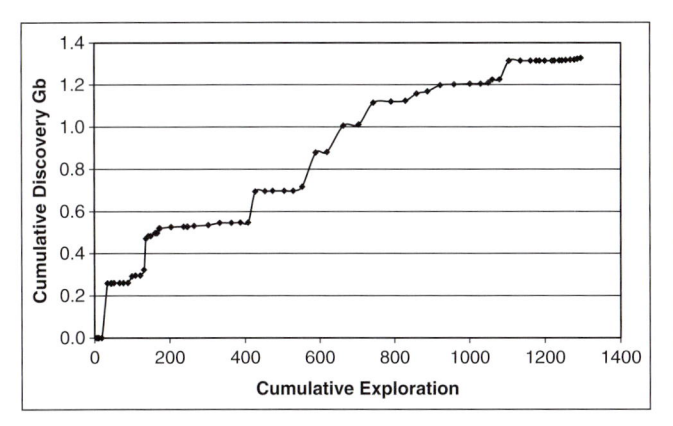

Fig. 44.3 Netherlands discovery trend

Fig. 44.4 Netherlands derivative logistic

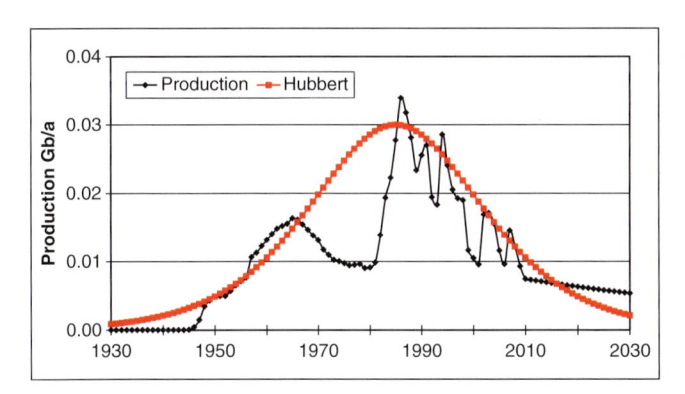

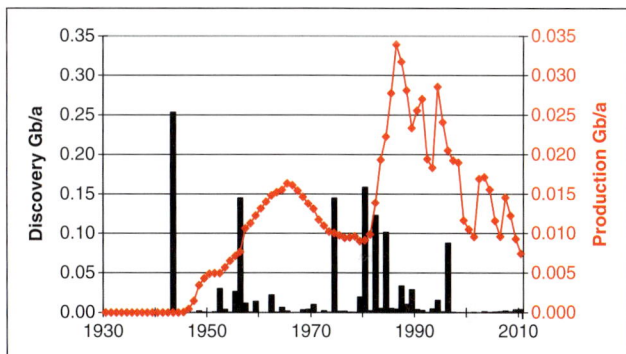

Fig. 44.5 Netherlands production: actual and theoretical

Fig. 44.6 Netherlands discovery and production

Norway

Table 45.1 Norway regional totals (data through 2010)

	Production to 2100					Peak Dates			Area	
Amount			Rate				Oil	Gas	'000 km2	
	Gb	Tcf	Date	Mb/a	Gcf/a	Discovery	1979	1979	Onshore	Offshore
PAST	23.5	79	2000	1,176	3,188	Production	2001	2018	325	2,300
FUTURE	10.6	91	2005	985	4,619	Exploration	2009		*Population*	
Known	9.0	87	2010	682	5,253	*Consumption*	Mb/a	Gcf/a	1900	2.25
Yet-to-Find	1.6	4.6	2020	365	4,264	2010	81	203	2010	4.9
DISCOVERED	32.4	165	2030	195	1,581		b/a	kcf/a	Growth	2.1
TOTAL	34	170	Trade	+602	+5,050	Per capita	16.5	41	Density	15

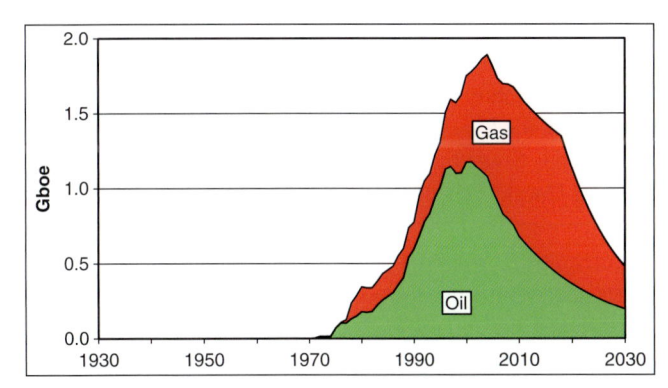

Fig. 45.1 Norway oil and gas production 1930 to 2030

Fig. 45.2 Norway status of oil depletion

Essential Features

Norway forms the mountainous margin of Scandinavia. The coastline is cut by deep fjords flanked by a string of offshore islands, which shelter the inland waters. To the southeast, a deep inlet from the Skagarrak leads to Oslo, the capital. Norway also exercises jurisdiction of the Arctic islands of Svalbard (Spitzbergen) and Jan Mayen. The country measures some 1,700 km in length, reaching far into the Arctic Circle: from southern Norway to its North Cape is as far as it is to Rome. It should be noted that oil and gas within the Polar Region and in water deeper than 500 m do not qualify as *Regular Conventional* by definition and are excluded from this assessment, being considered in Chap. 12. The country with 4.9 million inhabitants is sparsely populated.

Geology and Petroleum Systems

Norway's mainland is composed mainly of Palaeozoic rocks of the ancient Caledonian orogeny, but younger partly oil-bearing sediments form an extensive continental shelf covering the Norwegian and Barents Seas. The prime petroleum system comprises a Jurassic rift in the North Sea that formed as the Atlantic opened, when the continents moved apart. It contains rich late Jurassic source-rocks which have charged primarily underlying sandstone reservoirs with oil in rotated fault-blocks, although also locally feeding overlying Cretaceous chalk and Tertiary reservoirs. The source-rocks become gas-prone along the eastern margin, and have also entered the gas-window where deeply buried to the west. Comparable systems are also present in the

Haltenbanken area off mid-Norway and along the western margin of the Barents Sea. A secondary system within the same belt is provided by lean Middle Jurassic sources associated with coal.

Other systems develop in the huge Barents Sea to the north, where they have generally suffered from substantial vertical movements of the crust under the weight of alternating ice-caps in the geological past. Jurassic, Triassic and even older source-rocks are locally present but have generally been depressed into the gas-window over geological time. Seal integrity has also been damaged by the vertical movements, leading to the escape and re-migration of oil and gas.

Exploration and Discovery

Exploration commenced in 1962 and grew at a comparatively modest pace under strict government control to reach a peak in 1997 when 37 boreholes were sunk. Exploration drilling then declined before surging in 2006 to pass an all-time peak of 46 in 2009 with exploration for ever smaller prospects being driven by high oil prices.

One of the early finds was the giant Ekofisk Field, which came in as a surprise in 1969. In this area, Upper Jurassic source-rocks at peak generation underlie salt-induced structures, containing Chalk deposits, which just here contain adequate porosity to hold oil, thanks to having been deposited as slumps. The reservoirs are difficult to produce, due to low permeability, but various enhanced recovery techniques have lifted recovery to about 30 %. A surprise late discovery from a stratigraphic trap in the North Sea has recently added about 1 Gb of oil.

The peak of discovery came in 1979, when prime Jurassic prospects yielded a number of giant fields in the North Sea, including Statfjord and Oseberg, as well as the massive Troll gas field to the northeast. Haltenbanken off mid-Norway came in during the early 1980s delivering a number of fields in a similar setting, including Heidrun, the North Sea's last giant with about 1 Gb of oil, found in 1985. A late major gas discovery was Ormen Lange with 14 Tcf of gas, which was found off mid-Norway in 1997 in water depths of almost 1,000 m.

Production and Consumption

Oil production commenced in 1971 and grew steadily to peak at 3.2 Mb/d in 2001 before declining to 1.9 Mb/d in 2010, being now set to fall at the current depletion rate of about 6 % a year. This is a relatively high rate and reflects the high efficiency of offshore operations in Norway. The country with its small population consumes only 81 Mb/a, leaving over 90 % for export.

Gas production commenced in 1971, and grew to a plateau, set by pipeline capacity of around 1 Tcf/a, which lasted until 1991. It then increased again towards a new plateau in the 4,000–5,000 Gcf/a range, made possible by the construction of a new export pipeline, and that may last until around 2027. Consumption is minimal at 203 Gcf/a, meaning that most is available for export, including that from the deepwater and Polar areas.

The Oil Age in Perspective

Norway was already occupied some 14,000 years ago by hunters emanating from Europe. Later, came more settled communities who fished the lakes and fjords, sustaining themselves with difficult agriculture in a cold climate. They were isolated communities under petty kings and warlords. The Viking era, with an advanced culture, followed during the first millennium, when warriors in longboats headed south to colonise and trade, as well as to rape and pillage. The Norwegian Vikings went westwards to Iceland, Greenland, the Shetlands and Ireland, where they established settlements. Some may even have landed in North America.

King Harald I succeeded in unifying the Kingdom in the ninth century, but dissent amongst his successors led to fragmentation, with the country falling at different times under the control of the Danish and Swedish kings until 1297, when the three countries were unified, with Norway becoming a province of Denmark. The Black Death pandemic decimated the population during the fourteenth century, thanks to a particularly active rat, called *ratus norvegicus,* which was later well known to seamen and infested the New World.

Denmark, which had been an ally of France in the Napoleonic wars, was forced on defeat to cede Norway to the Swedish king in 1814. This was opposed by the Norwegian people, who wished for independence, and various conflicts and disputes with Sweden occupied the ninth century as a growing wave of nationalism built momentum. It was in part stimulated by a cultural flowering, as exemplified by the famous author, Ibsen, and the rediscovery of the ancient Norwegian language and folklore. Finally in 1905, an independent kingdom was declared with the crown being offered to Prince Carl of Denmark, who became Haakon VII. The First World War soon followed but the Scandinavian countries, including Norway, managed to maintain their neutrality.

The inter-war years saw the gradual development of fisheries, canning and shipping. Norway's great hydroelectric potential was tapped, being used particularly to refine aluminium and produce synthetic agricultural nutrients in a development of great significance for Europe. Even so, life was hard, leading to emigration to the New World, as well as a growing spirit of egalitarian co-operation at home.

Norway was victim of an unprovoked German invasion on 9th April 1940. It was forced to surrender after a short struggle, but not before the curator of a museum on the mouth of the Oslo fjord had managed to cause the sinking of a battleship by firing an ancient canon. The King escaped to England to establish a government in exile, while a puppet Nazi regime under Vidkum Quisling was established in Oslo. An active Resistance movement throughout the country contributed to the eventual liberation on 8 May 1945, but was unable to prevent the retreating German troops from destroying several towns in the north under their *scorched earth* policy.

Post-war reconstruction was built on the already well-entrenched co-operative spirit, with virtually all aspects of national life being placed under heavy government control. The shipping industry was rebuilt, partly with generous tax treatment, giving rise to various dynasties. They conquered world markets with capitalistic drive on the high seas, but changed their coats in home waters to become unostentatious and responsible patrons of their communities. Socialist governments, built more on co-operation than envy, dominated the post-war epoch.

If anyone in Norway thought about oil at this time, they pictured the sands of Arabia, little imagining that the stormy waters of the North Sea might one day give them a key oil position. Few noticed the first hint coming from the discovery of the giant Groningen gas field in the Netherlands. It in turn attracted attention to the adjoining waters of the southern North Sea, which was soon to be rewarded with a string of gas fields extending into British waters.

Not long afterwards the European office of Phillips Petroleum of Bartlesville, Oklahoma, turned its eyes north to wonder what the northern North Sea might offer, opening talks for exploration with Norway. At that time, jurisdiction extended only 3 miles from the shore, so the countries bordering the North Sea had to decide how to divide it. At first, ever-fair Norway opposed the notion of a median-line on the grounds that it would give a disproportionate share of the mineral resources to the coastal States at the expense of the inland countries. Britain, with fewer scruples, pressed for a median-line solution, eventually winning the support of Norway. In fact, Norway is bounded by a deep trench, which would have deprived it of the prospective tracts if water-depth alone had been taken into account. By this thin thread hung the train of events which would eventually deliver untold wealth to Norway, making it one of the world's largest exporters of oil and gas.

The first concessions (licences) were awarded in 1968 covering the southern part of the shelf, and led to the surprise discovery of the giant Ekofisk Field, owing its presence to the remarkable combination of geological circumstances, described above. Norway became an oil nation.

The next milestone came when Shell discovered the Brent Field in 1971 in the British sector of the northern North Sea, as improved seismic technology led to the identification of Jurassic troughs beneath the younger sediments. The field lay on a structural trend extending into Norwegian waters, where a huge structure was soon identified, yielding the Statfjord oilfield in 1973, which remains to this day the largest in the North Sea, with over 3.5 Gb of oil.

Norway reeled at the prospect of unimaginable wealth, and soon began to re-examine its oil policy, rightly fearing that oil might undermine its carefully balanced economy and society. To that point, the concessions had been granted on the basis of a normal royalty and corporation tax, but now the country moved to toughen its terms while respecting, in its ever-honest fashion, the rights already granted.

Britain had already created a State Oil Company under its then socialist government, which set an easy precedent for Norway to follow. Den Norske Stats Oljeselskap (or Statoil) was established under what at first sight seemed a highly advantageous arrangement, whereby it would hold a mandatory 50 % in all concessions, with its exploration costs being met by the foreign companies. It even retained the right to increase its share to as much as 85 % in the event of success. A special oil tax was also introduced. The world price of oil was soaring at the time in response to the oil shocks, and the industry accepted these outrageous terms, not wishing to be left out of what was rightly perceived to be one of the world's last great oil provinces. The Norwegians earned the sobriquet of being *blue-eyed Arabs*. But all was not what it seemed, for the companies' ever-ingenious tax lawyers soon found that they could take the cost of carrying Statoil as a charge against their taxable income. So, at the end of the day, it was the long-suffering Norwegian taxpayer who met the cost of the creation of the State Company, which started to burn up national wealth at a prodigious rate. It now employs more than 11,000 well-paid people.

In addition to Statoil were two other strong national companies, Norsk Hydro and Saga, which later merged and have now been absorbed into Statoil.

The Norwegian Petroleum Directorate was established to run exploration, deciding which companies would work together as groups, which prospects would be drilled and how many commitment wells were to be imposed, effectively treating the foreign companies as if they were contractors. But the companies did not object as the cost of all of this was taken as a charge against taxable income under terms that meant they were effectively spending *20 cent dollars*, enjoying a colossal unseen subsidy.

At first, the Government moved with admirable caution so as to accommodate the new industry into the economy. New licensing was delayed until 1979, when a number of prime prospects were awarded, yielding a string of giant discoveries to the north and east of Statfjord. But with the passage of time, the early caution was abandoned as the country succumbed to the political pressures of new Norwegian rig owners and contractors, who sought rapid expansion, and the people at large began to develop an unquenchable thirst for wealth; a departure from the attitude of their somewhat Spartan, God-fearing antecedents.

In short, the bulk of Norway's oil had already been found by the early 1980s, and what followed has been little more than a mopping up operation to find and produce ever smaller accumulations, apart from a surprise large late find in 2011. This unwelcome reality is however countered by optimists who continue to believe in exploration, drawing attention to the vast size of the Norwegian shelf, and dream that technological progress might extract more oil from known fields. Some improvement in recovery has indeed been achieved in the difficult chalk reservoirs, for which there was plenty of scope as very low recovery factors of below 20 % were at first assumed. For a brief moment, it seemed that Statfjord could recover as much as 70 %, setting a precedent for other similar fields. It was later realised that its complex east flank also held large amounts of oil-in-place which was in fact replenishing the reservoirs, returning the recovery factor to about 45 %. Some of the more recent small fields have given disappointing results, as companies were forced to make optimistic assessments to justify development at all. The licensing terms have been progressively ameliorated to match the dwindling oil prospects, and to keep the exploration business alive.

The golden days of Norwegian oil are accordingly coming to an end, but a new chapter of gas production is opening, calling again for clear thinking Government policy.

Some hopes have been expressed for new discovery in the deepwater Atlantic Margin that flanks Norway and the British Isles. The probability, however, is that the critical Upper Jurassic source-rocks are, at best, only locally present, and, even where present, too deeply buried to yield oil, save in some freak occurrences where re-migration from earlier accumulations has occurred. The province may, however, have considerable gas potential, although it will be very costly.

Norway's control of European gas supply carries geopolitical risks as it is never easy to be a rich man in a crowd of beggars, and Europe may put pressure on Norway to deplete its gas rapidly in order to counter the stranglehold of Russian supply policies.

With hindsight and a realisation of inevitable depletion, it might have been a better policy for Norway to have used its State Company to develop its oil and gas much more slowly, having said goodbye to the foreign companies after thanking them for their pioneering contribution. Although the country has invested $250 billion in an oil fund to try to save something for the future, oil and gas in the ground might have proved to be a much better asset. Statoil, evidently perceiving the limitations of the homeland, was allowed to move overseas on the well-known principle that *distant fields are greener*. It has been an expensive experiment with little to show for the investment.

In general, Norway's politics have moved to the right in recent years, eclipsing the long record of Socialist Government. Its electoral system however tends to give rise to coalition governments in which small parties may find themselves with a disproportionate power.

Norway was invited to join the European Union in 1994, but wisely declined after a referendum. The farmers and fishermen feared for the subsidies, and the country's oil wealth has enabled it to stand aside. Nevertheless, it voluntarily complies with much European legislation, not wanting to find itself too isolated. It has been an enthusiastic member of NATO, having had a common frontier with the former Soviet Union, and it has contributed greatly to various UN peace-keeping missions. The wealth of the country and its fair-thinking people allowed a relatively massive scale of immigration. A reaction prompted to a gruesome act in 2011 when an eccentric activist set off a bomb in Oslo and killed a large number of young members of the Labour Party attending a festival on a nearby island.

Norway with its small population and large reserves of remaining oil and gas is well placed to face the Second Half of the Age of Oil. It has a tradition of closely knit communities who can be expected to pull together in facing the future, after the wave of extreme affluence that sweeps the country has subsided.

45 Norway

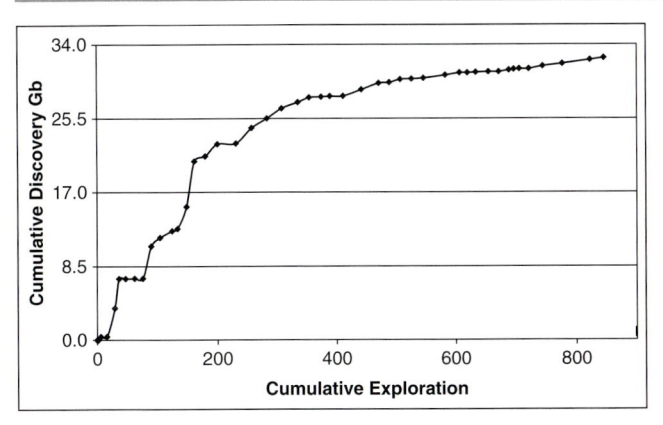

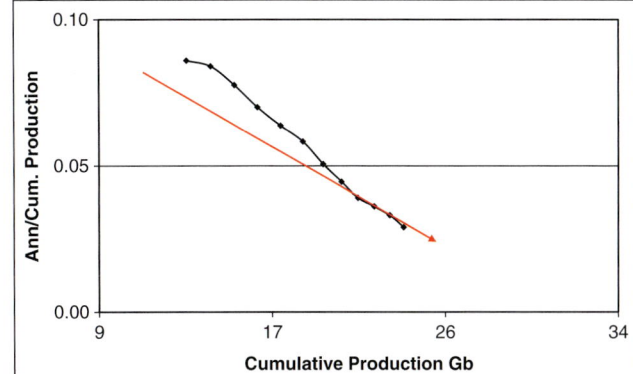

Fig. 45.3 Norway discovery trend

Fig. 45.4 Norway derivative logistic

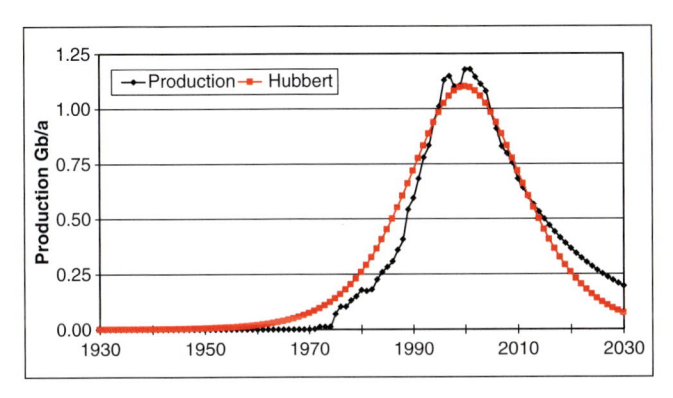

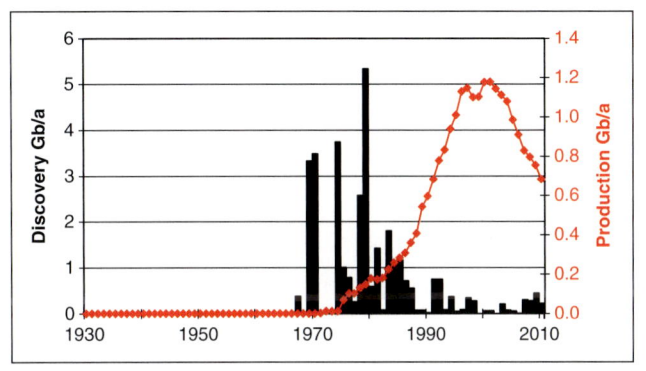

Fig. 45.5 Norway production: actual and theoretical

Fig. 45.6 Norway discovery and production

United Kingdom

Table 46.1 United Kingdom regional totals (data through 2010)

Production to 2010					Peak Dates				Area		
Amount			Rate				Oil	Gas	'000 km²	
	Gb	Tcf	Date	Mb/a	Gcf/a	Discovery	1974	1966	Onshore	Offshore
PAST	24.7	94	2000	830	4124	Production	1999	2000	246	55
FUTURE	7.3	26	2005	602	3379	Exploration	1990		Population	
Known	6.6	23	2010	450	2171	Consumption	Mb/a	Gcf/a	1900	38
Yet-to-Find	0.7	2.6	2020	265	963	2010	592	3,329	2010	62.7
DISCOVERED	31.3	117	2030	156	428		b/a	kcf/a	Growth	1.6
TOTAL	32	120	Trade	−142	−1158	Per capita	9.4	53	Density	255

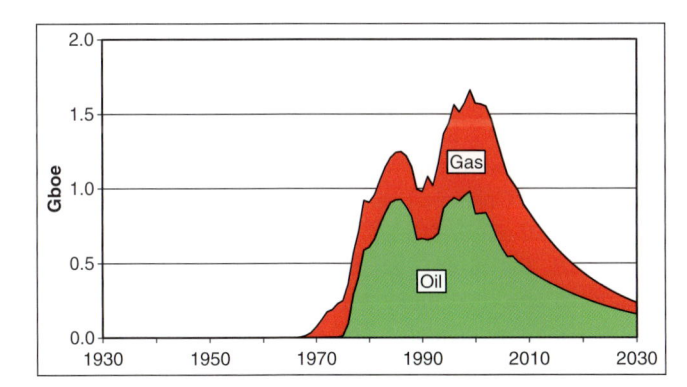

Fig. 46.1 United Kingdom oil and gas production 1930 to 2030

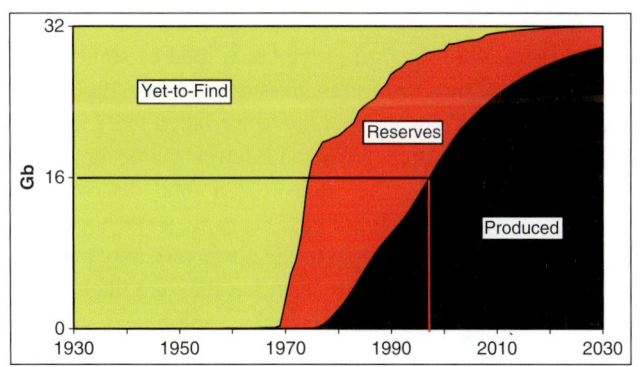

Fig. 46.2 United Kingdom status of oil depletion

Essential Features

The United Kingdom (or the United Kingdom of Great Britain and Northern Ireland to give it its full name) comprises the island of Britain itself and an enclave in the north of Ireland. Britain itself is made up of England, Scotland and Wales, which are discrete regions with long histories that are rediscovering their identity. Mountainous country develops to the north and west, but for the most part it is made up of rolling temperate farmland, on which is superimposed spreading urban and suburban settlement. It is flanked to the north by a series of islands, including the Hebrides and Orkneys. The southeast is drained mainly by the River Thames and its tributaries flowing into the southern North Sea, while the Clyde flows westwards through Scotland in the north. The country supports a population of almost 63 million, of whom about 10 % are relatively recent immigrants, many coming from its former Empire.

Geology and Prime Petroleum Systems

The ancient Caledonian mountain chain forms the foundation of the island's geology, but was broken up during the opening of the North Atlantic, being flanked by younger strata. Britain was one of the birthplaces of the science of geology, lending its place-names for geological epochs throughout the world: the Devonian, for example, being named after the county of Devon.

There are two prime petroleum systems. The first comprises a system of rifts in the northern North Sea of which approximately half lies in British waters. As described in connection with Norway, it contains prolific Upper Jurassic oil and gas source-rocks that have charged over- and underlying reservoirs. These source-rocks become gas-prone where deeply buried. The second system straddles the southern North Sea comprising deeply buried coal measures that have charged overlying Permian desert sands with gas, in a natural process akin to coking (see also Netherlands). Overlying salt deposits form an important seal in some areas. A number of other systems exist, both onshore and offshore, which involve lean Lower Jurassic and Carboniferous sources, but are relatively insignificant.

Exploration and Discovery

The first recorded exploration borehole was drilled in 1895. Exploration continued at a modest rate until the end of the Second World War by which time 118 boreholes had been drilled, making a few minor onshore discoveries. The main surge followed in the 1960s, as the southern North Sea gas province was developed, to be followed a few years later by the northern North Sea oil province. A total of some 3,450 boreholes have now been sunk, passing a peak in 1990 when 168 were drilled. The number has since fallen to 27 in 2010 reflecting the fact that fewer and fewer viable prospects remain to be tested. They delivered a total discovery of 31.3 Gb of oil and 117 Tcf of gas. There are some hopes that the so-called Atlantic Margin to the northwest, may prove to be a new and productive province, but they are likely to be dashed due to limited source-rock development and the adverse affects of large vertical movements of the crust, although some almost freak discoveries have been made involving remigration from earlier accumulations.

Production and Consumption

Oil production commenced in 1919 but remained at an insignificant level until around 1970 when it grew rapidly from the North Sea fields. It reached an early peak of 2.5 Mb/d in 1986, but then suffered an anomalous decline due to a serious accident at the Piper Field in the North Sea, before recovering to an overall peak of 2.7 Mb/d in 1999. It has since declined to 1.2 Mb/d, and is set to continue to decline at the current Depletion Rate of about 6 % a year, reaching virtual exhaustion by around 2050. Consumption stands at 592 Mb/a, which translates to a per capita usage of almost 10 b/a. It follows that the demand for imports is set to rise steeply.

Gas production commenced in the mid-1960s, as the southern North Sea was brought in, being later supplemented from the northern North Sea. Production peaked in 2000 at 4,124 Gcf/a, before falling to 2,171 Gcf/a in 2010, being now set to decline at about 8 % a year. Consumption, much used for electricity generation, stands at 3,329 Gcf/a, meaning that, if this trend continues, the demand for imports from ever more distant sources will rise steeply.

The Oil Age in Perspective

The United Kingdom had a strong Neolithic culture, long before falling to the Romans in 55 BC. That occupation lasted only a few centuries, but left an indelible mark. It was followed by the *dark ages* of Viking and Saxon incursions, culminating in the arrival of recycled Danish Vikings from Normandy in 1066 in what proved to be the last successful military invasion to-date. Christianity had arrived some 300 years earlier and spread through the country. The Normans gradually exerted their control, but Wales was not finally incorporated into the Kingdom until 1536, with Scotland holding out even longer until 1603, when King James VI of Scotland became King James I of England. Ireland too was invaded by Norman barons and gradually subjugated. However, conflicts continued, being exacerbated by religious tensions between the Catholic and Protestant persuasions, to erupt into a civil war in 1642 between the King and his Parliament. The King lost and was executed, paving the way for a so-called Commonwealth under Oliver Cromwell, who mounted a repressive campaign on Catholic Ireland.

In 1801, the diverse people and interests of the British Isles were absorbed under a new monarchy becoming the United Kingdom. Seafarers had stimulated trade and exploration throughout the world, paving the way for the British Empire which at its peak, in the reign of Queen Victoria, became the premier world power, benefitting from the use of its pound sterling for world trade. Great achievements were recorded in the fields of science, literature and culture.

The Industrial Revolution began on home territory in the eighteenth century at first with mills powered by water to make cloth for export to its colonial markets. The wealth, so created, led to the rapid growth of capitalism, banking, usury, investment and a financial economy. Self-sufficient peasants left the land to become wage-earners, consumers and taxpayers, many living in gruesome industrial slums. Mechanisation based on iron and steel took many directions. Iron smelting made new demands for energy, coming first from firewood and later from coal. The coal was at first collected from beaches and outcropping seams before mining commenced. The pits were deepened into regular mines, and minerals, including tin in Cornwall, were also mined. But the mines were subject to flooding which led to the development of steam-driven pumps. The pumps evolved into steam engines that were later used to power transport. Sail gave way to steam and a rail network was developed.

Britain successfully resisted and eventually defeated an epoch of French expansion under Napoleon. Later, during the nineteenth century, it found itself increasingly threatened by a newly united Germany that was overtaking it in industrial prowess. Germany, however, lacked the benefit of pound sterling, which by now was the world trading currency, delivering a handsome hidden tribute to the banks in the City of London, controlled by a few well-known names. Pressures, caused mainly by this conflict, eventually led to two world wars during the twentieth century. Although victorious in military terms, a weakened Britain voluntarily gave up most of its once splendid Empire that had mainly brought order and a fair administration to its territories. It half-heartedly joined a newly united European community preferring to retain its particular financial links with the United States. This carried advantages in the new, global, economic and military hegemony of the United States.

Massive immigration from the former Empire followed the Second World War. The population has expanded by a factor of 1.6 since 1900 to its present level of 62 million, with the immigrants and their descendants making up a growing percentage.

The United Kingdom has had a long oil history, both within its own territory, and through the early prominence of its oil companies in the Middle East, Mexico and Venezuela. BP was the flagship with major holdings in Iran, Iraq and Kuwait, while Shell, an Anglo-Dutch enterprise, had a strong position in the Western Hemisphere. BP was once almost a national oil company with a 51 % government shareholding, and corresponding responsibilities, but was later privatised.

The UK's brief oil age is now in decline, in no small way due to a governmental policy to pump and sell the resource as fast as possible, especially during the 1990s. The major companies are withdrawing to be replaced by smaller firms, mopping up satellite fields and step-outs, as well as scavenging tail end production from ageing offshore platforms.

It is difficult to imagine how the UK will cope with the Second Half of the Age of Oil. Failure by the government to recognise natural depletion, and to be more far-sighted in its policies, has left the country unprepared, being now forced to re-develop nuclear power, unpopular as it is. Re-commissioning old abandoned coal mines will prove difficult and costly. The growing contribution of solar, wind, wave and tidal power, including that from bio-fuels, is useful, indeed vital, but insufficient to meet the country's energy needs with its current population, at least in the manner they are accustomed to live.

The country with its important global banking influence was seriously affected by the financial and economic crash of the recent past. A new coalition Conservative/Liberal Government moves to reduce the deficit and cut State expenditures and encourage localism, but meets opposition and may well trigger major industrial disputes and indeed riots. In 2011, together with the US "leading from behind" and France, it mounted a military operation against the government of oil-rich Libya with the blessing of the United Nations.

In the unfolding circumstances, the United Kingdom may become less united. Already Scotland and Wales are gaining greater control of their destinies with their own legislatures, and various ethnic groups rediscover their identity for survival. Pressures against further European integration are likely to mount, as the consequences of its outdated economic and financial principles become more evident. Perhaps the best hope is that Europe, including Britain, should rediscover the Treaty of Maastricht which encourages regionalism under the slogan that no decision should be taken at any level higher than it need be.

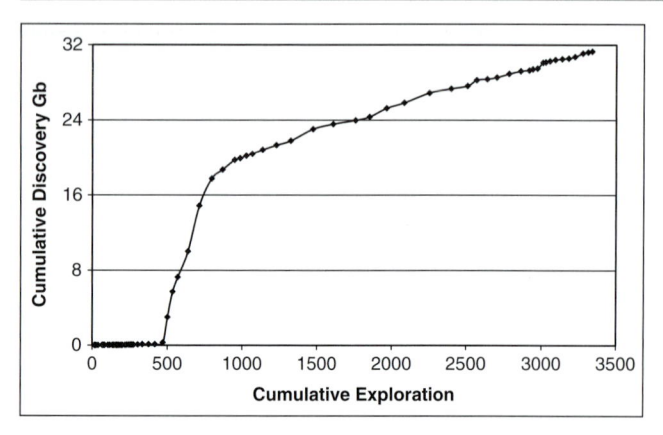

Fig. 46.3 United Kingdom discovery trend

Fig. 46.4 United Kingdom derivative logistic

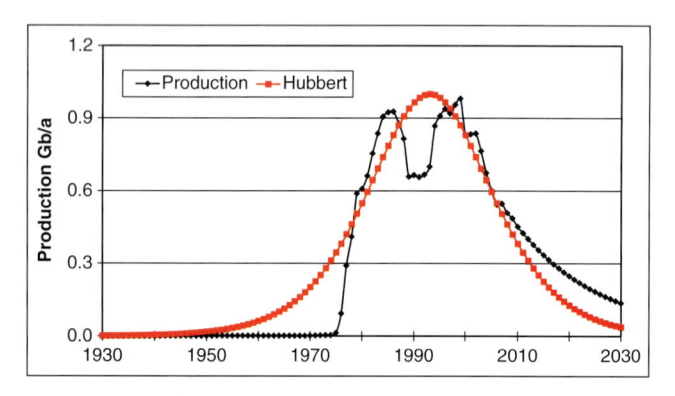

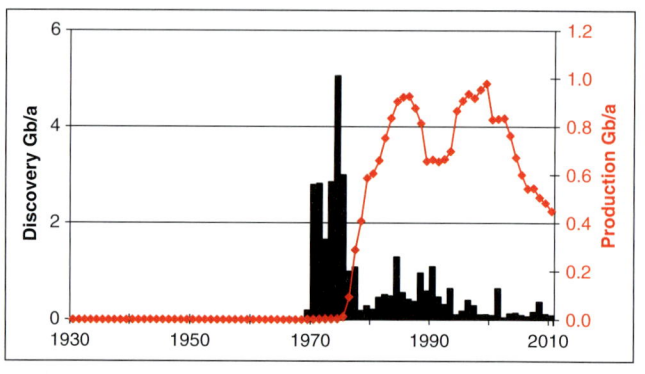

Fig. 46.5 United Kingdom production: actual and theoretical

Fig. 46.6 United Kingdom discovery and production

Europe Region

Table 47.1 Europe regional totals (data through 2010)

Production to 2100						Peak Dates			Area	
Amount		Rate					Oil	Gas	'000 km²	
	Gb	Tcf	Date	Mb/a	Gcf/a	Discovery	1974	1959	Onshore	Offshore
PAST	56	373	2000	2217	11851	Production	1999	2004	1950	2860
FUTURE	21	188	2005	1807	12447	Exploration	1985		Population	
Known	18	176	2010	1288	11706	Consumption	Mb/a	Gcf/a	1900	178
Yet-to-Find	2.9	12	2020	702	7531	2010	4701	16902	2010	304
DISCOVERED	74	549	2030	385	3282		b/a	kcf/a	Growth	1.7
TOTAL	77	561	Trade	−3411	−5180	Per capita	11	41	Density	156

*Note: Data refer only to main producing countries, **except for Consumption and trade data** which include minor countries.*

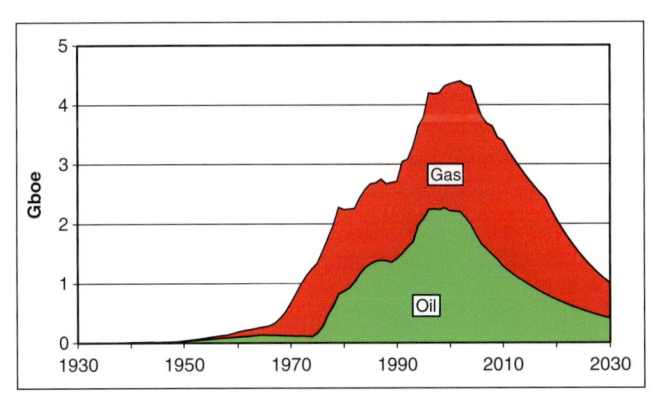

Fig. 47.1 Europe oil and gas production 1930 to 2030

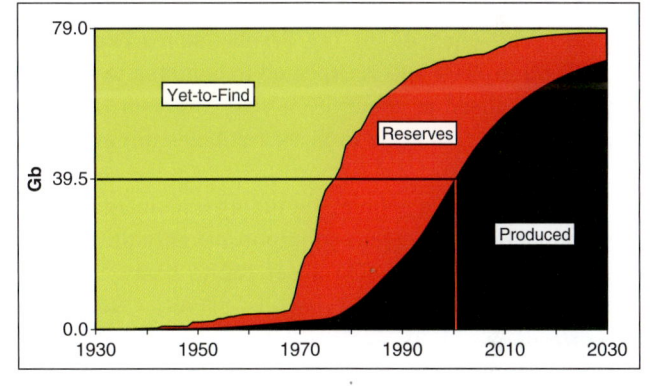

Fig. 47.2 Europe status of oil depletion

The Europe Region is defined for this purpose as the countries of Western Europe, excluding the former Communist bloc of Eastern Europe, because they lived under a very different environment over much of the past 50 years. The table above refers the main oil and gas producers, with the remaining other countries being listed below. Those indicated (*) have minor amounts of oil and/or gas which are included in the assessment of other minor producers in the world in Chapter 11, as the amounts are too small to model meaningfully. Some of them are however major consumers of oil and gas.

Table 47.2 Europe: Minor countries in the Region (*with minor oil/gas production)

Andorra	*Ireland	San Marino
Belgium	Liechtenstein	*Spain
Channel Islands	Luxembourg	*Switzerland
Finland	Malta	*Sweden
*Greece	Monaco	*minor oil and/or gas
Iceland	Portugal	

In oil terms, the Europe Region is dominated by Britain and Norway which share the greater part of the North Sea, the largest new petroleum province to be discovered since the Second World War, yet holding only enough to supply the world for under three years. The region has a total endowment of 77 Gb of oil, amounting to only about 4% of the World's total. It also has had substantial gas deposits, amounting to some 561 Tcf, but they are now almost 70% depleted.

Although oil and gas had been found locally in the nineteenth century, the major oil finds were made during a brief epoch in the 1970s when the giant fields of the North Sea were brought in, giving an overall peak of discovery in 1974. This delivered a corresponding peak of oil production in 1999, 25 years later, which is a fairly normal time-lag. The current overall depletion rate is relatively high at 5.7% a year, reflecting both the great advances in offshore technology and the market pressures to produce as fast as possible, thus accelerating depletion. Production is set to continue to decline at this rate, falling from 3.5 Mb/d in 2010 to about 1 Mb/d by 2030.

The gas situation has been dominated by two major finds: the Groningen Field in 1959 in the Netherlands and the Troll Field in 1979 in Norway, both of which opened important North Sea trends and contain the bulk of the Region's endowment of 561 Tcf. Production reached a peak of 13 Tcf/a in 2004 and is expected to continue to decline in the years ahead at about 6% a year, although Britain's share collapses faster due to market policies driving rapid depletion.

The Region as a whole, including the insignificant producing countries, needs to import roughly 66% of its oil and 26% of its gas. On a per capita basis, average consumption is almost 11 barrels of oil a year and 41 kcf/a of gas. It is worth noting in passing that the Netherlands has an anomalous consumption rate of 22 barrels of oil per annum, which is almost double the regional average. It may be a misleading statistic distorted by the production and export of refined product from Dutch refineries.

Countries with Minor Production

Some of the countries listed in the above table, such as Spain and Portugal, which are minor in oil production terms, have had a major role in history, possessing important empires, whereas others such as Andorra, Liechtenstein and San Marino, are small communities that by historical accident managed to maintain their independence and identity. This may prove a blessing in the future, as raw necessity may force people to adopt more local structures by which to adapt better to the issue of oil depletion: indeed Sweden has already announced a policy of weaning itself of oil dependence by 2020. In summary, this group of countries produces 1.9 Mb/a of oil and 15 Gcf/a of gas but consumes 1,350 Mb/a of oil and 2.8 Tcf/a of gas.

The Oil Age in Perspective

An attempt has been made in the preceding pages to review each of the main oil producing countries from an historical and political perspective, to see how they are placed to face the future. For practical purposes, the Oil Age covers the twentieth and twenty-first centuries, meaning that we are now about half-way through it.

The dominant theme, through the ages, seems to have been conflict both between different tribes or nations, as well as between different classes of people. There seems to have always been a class of elite who lorded their privileged position over the masses. In earlier years, this was the relationship between landlord and peasant, or monarch and subject, but in later years was translated to that between an industrial-financial elite and their workers. While the wars and conflicts are highlighted in the history books, it would be a mistake to assume that people failed to live happily in harmony with their circumstances whatever they were.

In addition were religious disputes between Catholics and Protestants, which were probably more about earthly power than argument over the nature of divinity. Despite these sundry tensions, the region did make great achievements in the realm of culture, producing examples of fine architecture, as well as excelling in the fields of science, philosophy, literature, art and music.

The Industrial Revolution, fuelled by coal, and later by oil-based energy, opened a new chapter of growing mercantilism and urban living; the population began to exceed the fertility of the land in the nineteenth century which led to massive emigration to North America and Australasia. Even so, the population has doubled over the First Half of the Age of Oil, in part due to a massive influx of immigrants, attracted by economic prosperity. The growing population was accompanied by a fall in the number of agricultural workers, as oil-driven mechanisation and the application of synthetic nutrients and pesticides, which have actually damaged the soil, made many farm workers redundant. Much of Europe's food is now imported.

The First Half of the Oil Age has now ended, delivering massive amounts of cheap energy, but it was far from being a Golden Age as the prosperity has been accompanied by a certain loss of dignity and an explosion of crime and violence. The Second Half now dawns, and there is a hope therefore that a new benign age may evolve as oil declines, such that people start to again live in harmony with themselves, their neighbours and above all the natural environment. The transition however threatens to be a difficult time as the large urban populations are very dependent on oil and gas supply. Riots have already broken out in several countries accompanying a financial collapse, putting exceptional strains on Greece, Ireland, Portugal, Spain and Italy.

Confidence and Reliability Ranking

The table below lists the standard deviation of the range of published data on reserves. A relative confidence ranking in the validity of the underlying assessment is also given: the lower the Surprise Factor the less the chance for revision.

Table 47.3 Europe: Range of Reported Reserves and Scope for Surprise

	Standard Deviation Public Reserve Data		Surprise Factor	
	Oil	Gas	Oil	Gas
Austria	0.02	0.28	1	2
Denmark	0.19	0.88	1	2
France	0.01	0.01	1	2
Germany	0.08	1.45	0	4
Italy	0.25	0.53	2	3
Netherlands	0.09	4.69	0	1
Norway	0.49	13.39	2	4
UK	0.83	5.76	1	2
Range in published data			Scale 1–10	

Austria

Austria has been thoroughly explored under an efficient regime. It is a mature, relatively small province at an advanced stage of depletion. The data quality is relatively good although official statistics may err on the side of conservatism, and there is a wide range for gas data.

Denmark

Denmark has also been thoroughly explored under an efficient regime. Peak oil has arrived prior to depletion midpoint and the depletion rate is high, suggesting that the estimates might be too low, although that might be due the dominance of the giant Dan Field. The estimated gas potential too might be under-evaluated.

France

France is a mature, thoroughly explored country, but it is a large country with complex geology, so there remains the possibility of some small new surprise discoveries, especially of gas. The data seems to be consistently reported.

Germany

The former West Germany has been thoroughly explored leaving little scope for surprises, but the former East Germany may have more gas potential than recognised.

Italy

There are some uncertainties about the reliability of data from Italy. It has been thoroughly explored, but there may be some scope for surprises in the complex structure of the deep Apennine thrust belts.

Netherlands

The Netherlands have been very thoroughly explored in an efficient regime, leaving negligible scope for surprise.

Norway

Norway has been very thoroughly explored, and its national database is outstanding in its quality. There has been a large late surprise in 2011. The Atlantic Margin offers some scope for positive surprise, especially for gas, although the chances are not rated highly here. The extensive Arctic tracts of the Barents Sea do not qualify as *Regular Conventional Oil and Gas*, as herein defined, but the public databases do not make this distinction. These areas are here rated to be primarily gas-prone, subject to remigration, giving hints of encouragement that tend to disappoint.

United Kingdom

The United Kingdom has been very thoroughly explored in an open market environment giving every possible incentive to discover and deplete the reserves of oil and gas at the maximum rate possible. There is accordingly very little remaining scope for surprises. The last remaining possibilities for a surprise discovery are in the deepwater Atlantic Margin, which is probably gas-prone.

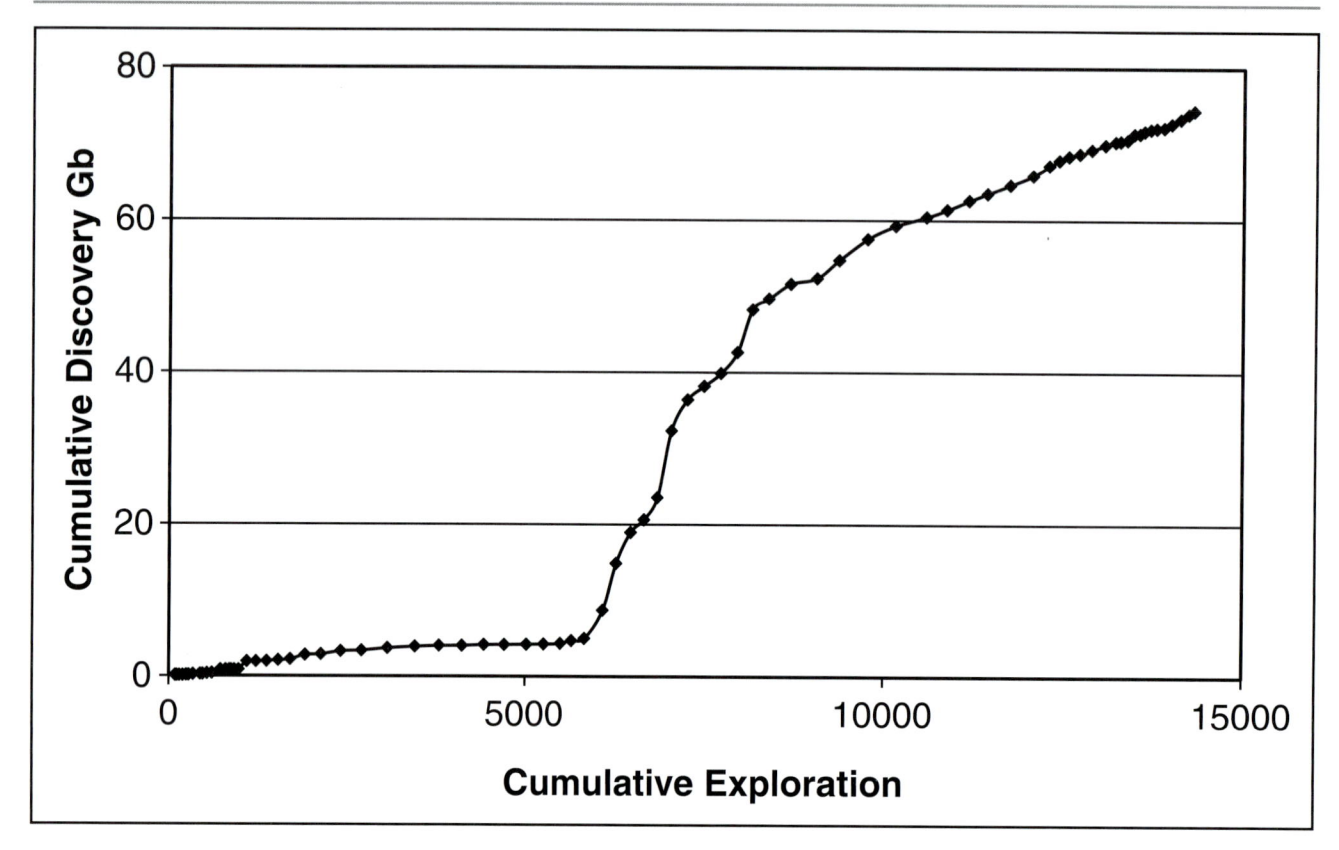

Fig. 47.3 Europe discovery trend

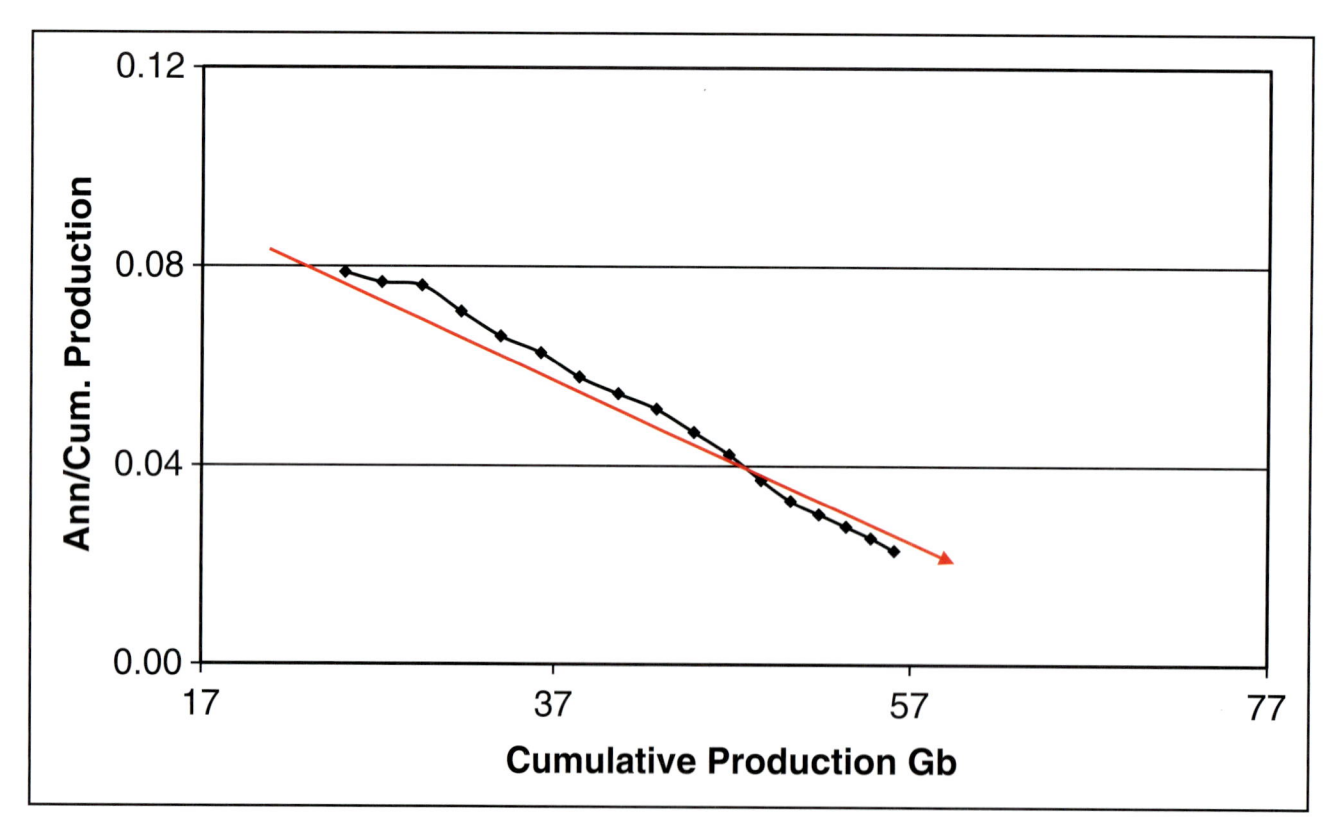

Fig. 47.4 Europe derivative logistic

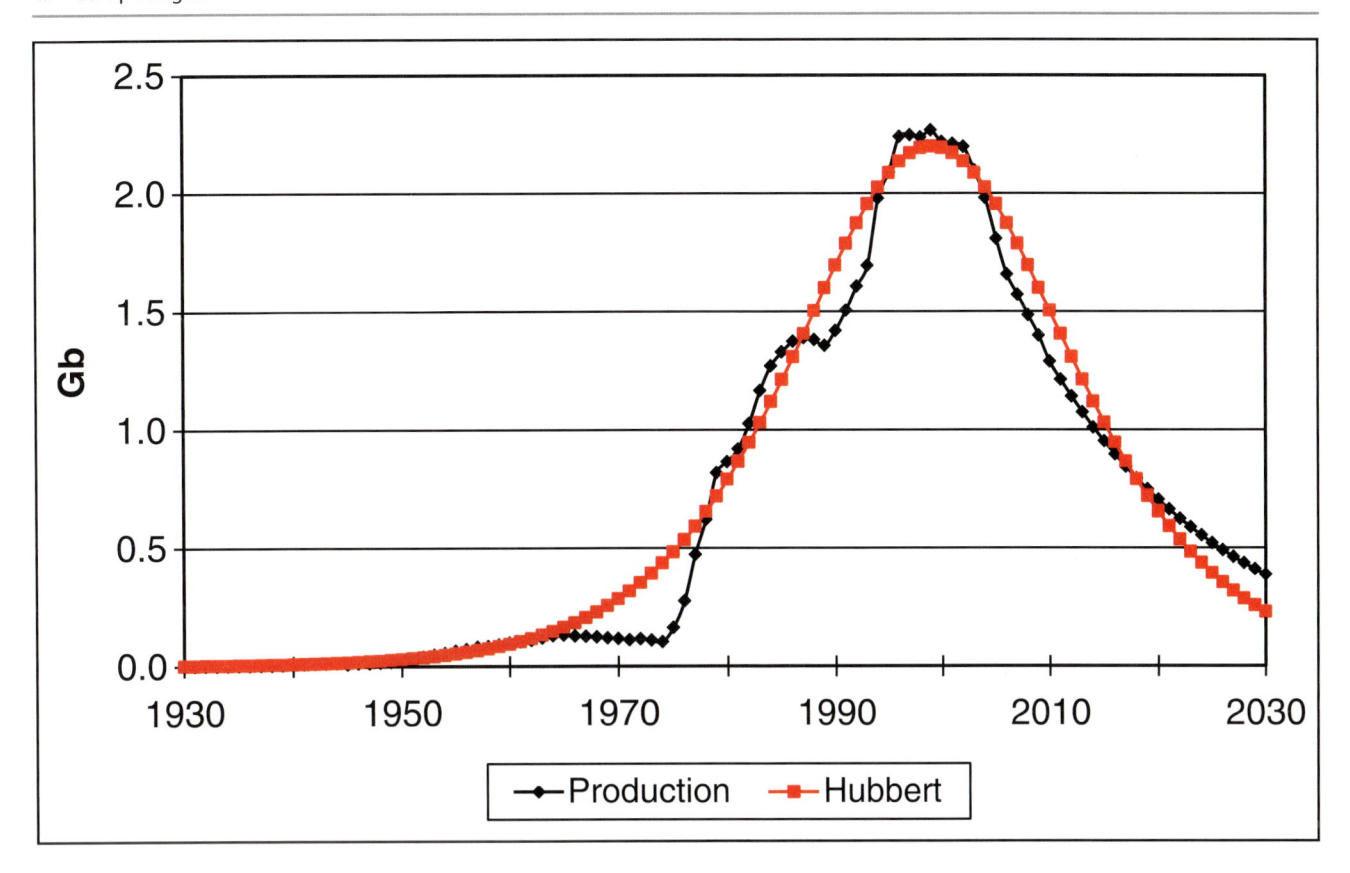

Fig. 47.5 Europe production: actual and theoretical

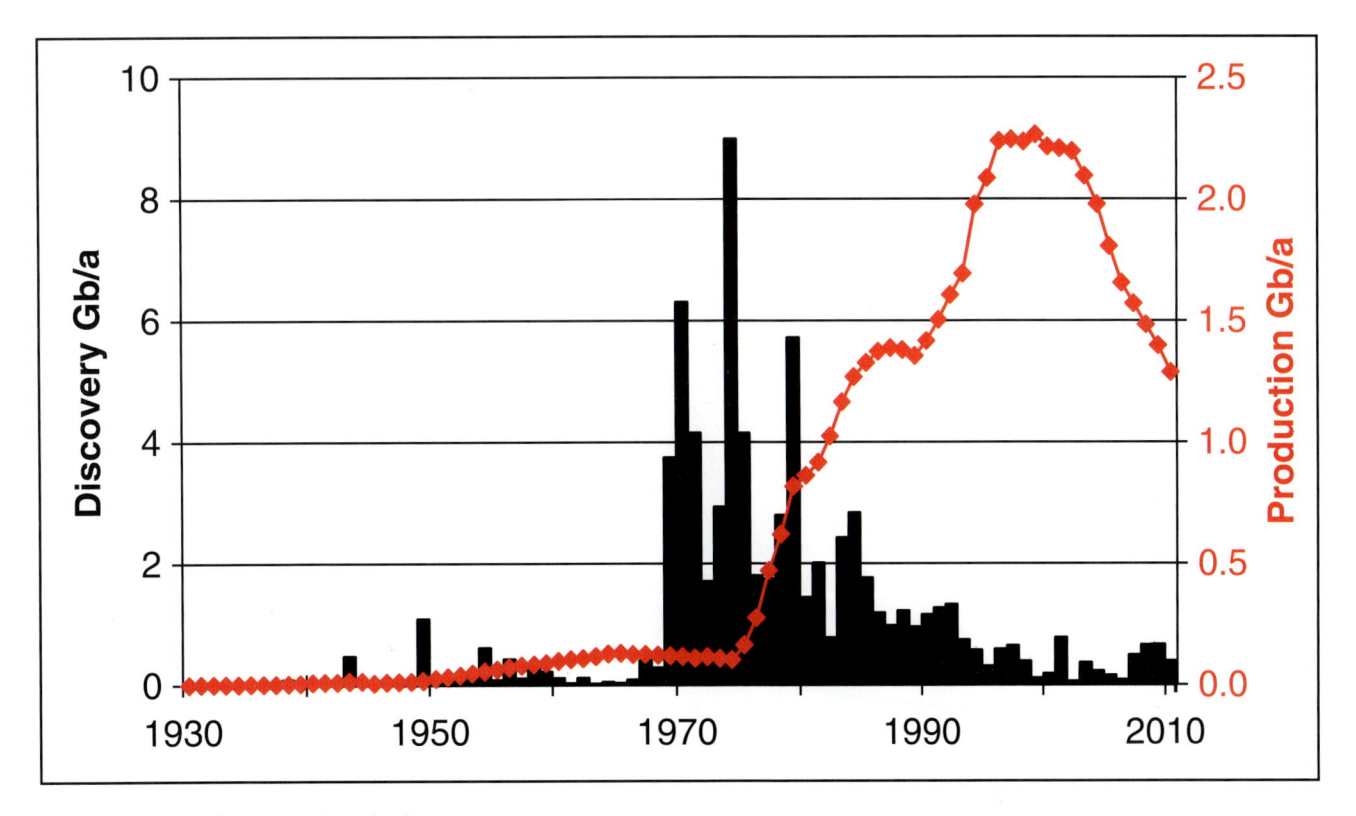

Fig. 47.6 Europe discovery and production

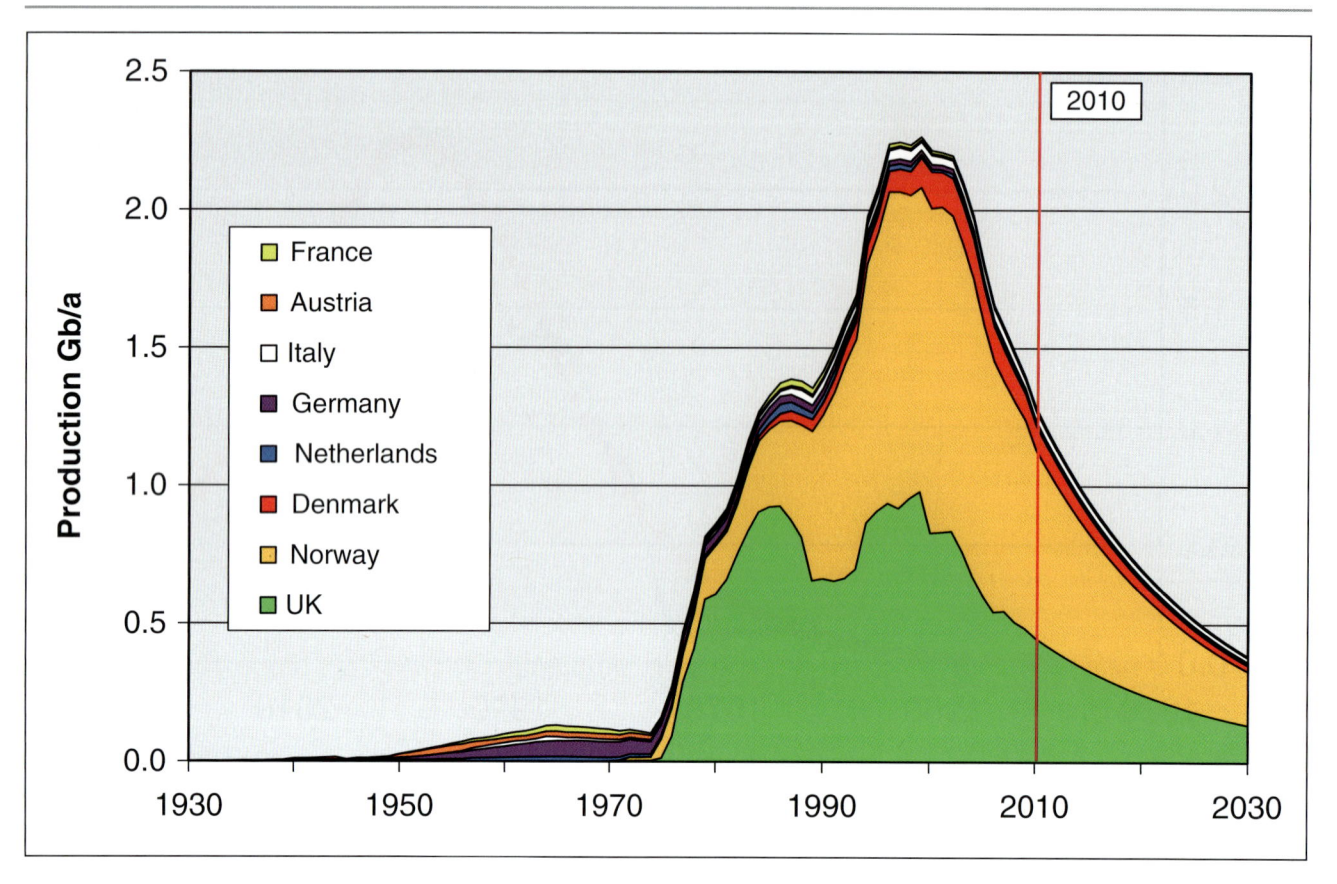

Fig. 47.7 Europe oil production

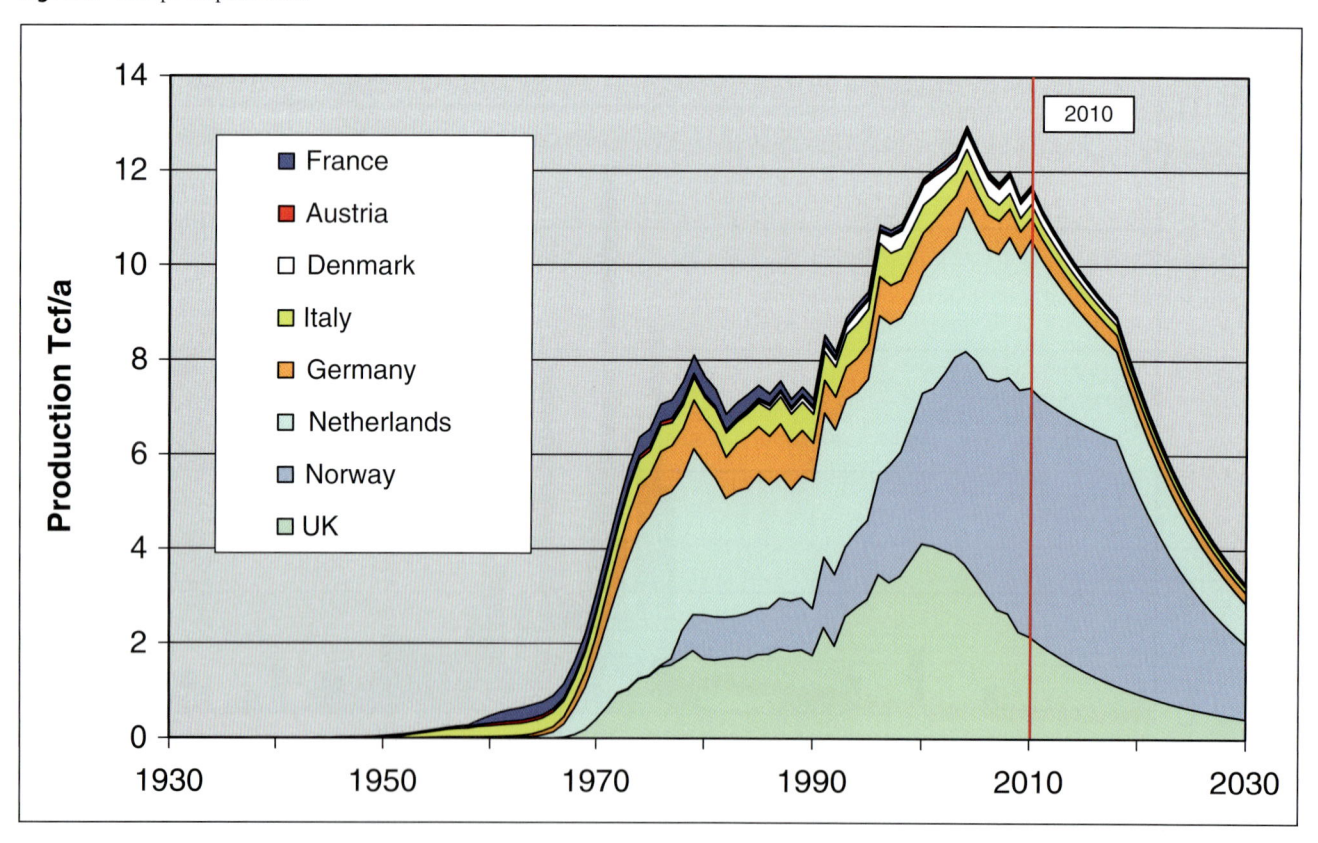

Fig. 47.8 Europe gas production

47 Europe Region

Table 47.4 Europe: Oil Resource Base

		PRODUCTION TO 2100														
EUROPE		**Regular Conventional Oil**														2010
		KNOWN FIELDS											Revised: 05.10.2011			
	Region	Present		Past		Discovery		Reserves		FUTURE KNOWN Gb	TOTAL FOUND Gb	FUTURE FINDS Gb	ALL FUTURE Gb	TOTAL Gb	DEPLETION		PEAK DATES	
Country		kb/d 2010	Gb/a 2010	Gb	5yr trend	Gb 2010	% Disc.	Reported Average	Deducted Non-Con						Current Rate	Mid-Point	Disc	Prod
Norway	F	1869	0.68	23.45	−5%	0.22	95%	6.38	0	8.97	32.42	1.58	10.55	34.2	6.10%	2002	1979	2001
UK	F	1233	0.45	24.71	−3%	0.09	98%	3.32	0	6.56	31.27	0.73	7.29	32.0	5.80%	1998	1974	1999
Denmark	F	246	0.09	2.28	−6%	0.04	93%	0.97	0	0.97	3.26	0.24	1.22	3.5	6.90%	2005	1971	2004
Germany	F	28	0.01	1.96	−6%	0.02	96%	0.23	0	0.43	2.39	0.11	0.54	2.5	1.90%	1978	1949	1967
Italy	F	96	0.04	1.61	−3%	0.00	94%	0.72	0	0.71	1.87	0.13	0.84	2.0	4.00%	2005	1989	1997
Netherlands	F	20	0.01	0.91	−5%	0.00	98%	0.17	0	0.42	1.33	0.02	0.44	1.4	1.60%	1992	1943	1986
France	F	18	0.01	0.79	−3%	0.01	97%	0.10	0	0.13	0.92	0.03	0.16	1.0	3.90%	1987	1954	1988
Austria	F	7	0.01	0.83	0%	0.00	98%	0.07	0	0.09	0.93	0.02	0.12	1.0	4.90%	1970	1949	1955
Europe	F	3528	1.29	56.08	−4%	0.39	96%	11.87	0	18.30	74.38	2.87	21.17	77	5.70%	1999	1974	1999

Table 47.5 Europe: Gas Resource Base

		PRODUCTION TO 2100																
EUROPE		**Regular Conventional Gas**																2010
		KNOWN FIELDS											Revised					15/11/2011
	Region	Present			Past		Discovery		Reserves		FUTURE KNOWN	TOTAL FOUND	FUTURE FINDS	ALL FUTURE	TOTAL	DEPLETION			PEAKS
		Tcf/a	FIP	Gboe/a	5yr Tcf Trend		Tcf	%	Reported Average	Deduct Non-Con						Current Rate	%	Mid-point	
Country		2010				2010	Disc.				Tcf	Tcf	Tcf	Tcf	Tcf				Expl Disc Prod
Norway	F	5.25	1.55	0.95	78.55	2%	0.64	97%	81.75	0	86.88	165.43	4.57	91.45	170	5.43%	46%	2012	1997 1979 2018
UK	F	2.17	0.12	0.39	94.35	−6%	0.32	98%	12.56	0	23.09	117.43	2.57	25.65	120	7.80%	79%	1999	1990 1966 2000
Denmark	F	0.29	0.03	0.05	6.71	−5%	0.04	97%	2.53	0	2.06	8.77	0.23	2.29	9	11.20%	75%	2004	1985 1968 2000
Germany	F	0.46	−2.39	0.08	35.33	−7%	0.22	97%	4.88	0	11.40	46.73	1.27	12.67	48	3.54%	74%	1995	1958 1969 1987
Italy	F	0.29	0.02	0.05	25.72	−5%	0.02	98%	2.70	0	5.65	31.37	0.63	6.28	32	4.46%	80%	1992	1962 1968 1994
Netherlands	F	3.13	0.50	0.56	117.87	3%	0.40	99%	45.90	0	44.78	162.64	2.36	47.13	165	6.23%	71%	1997	1985 1959 1976
France	F	0.04	0.00	0.01	11.35	−6%	0.04	99%	0.25	0	0.24	11.59	0.16	0.40	12	9.87%	97%	1977	1959 1949 1978
Austria	F	0.06	0.01	0.01	3.36	−1%	0.01	98%	0.75	0	1.56	4.92	0.08	1.64	5	3.57%	67%	1996	1975 1949 1975
EUROPE	F	11.71	−0.16	2.11	373.23	0%	1.69	98%	151.32	0	175.66	548.89	11.86	187.52	561	5.88%	67%	2004	1986 1959 2004

Table 47.6 Europe: Oil Production Summary

Regular Conventional Oil Production					
Mb/d	2000	2005	2010	2020	2030
Norway	3.22	2.70	1.87	1.00	0.53
UK	2.28	1.65	1.23	0.68	0.37
Denmark	0.36	0.38	0.25	0.12	0.06
Germany	0.05	0.04	0.03	0.02	0.02
Italy	0.09	0.11	0.10	0.06	0.04
Netherlands	0.03	0.03	0.02	0.02	0.01
France	0.03	0.02	0.02	0.01	0.01
Austria	0.02	0.02	0.02	0.01	0.01
EUROPE	6.07	4.95	3.53	1.92	1.06

Table 47.7 Europe: Gas Production Summary

Gas production						Total
Tcf/a	2000	2005	2010	2020	2030	Tcf
Norway	3.188	4.619	5.253	4.264	1.581	170
UK	4.12	3.38	2.17	0.96	0.43	120
Denmark	0.42	0.38	0.29	0.09	0.03	9
Germany	0.84	0.75	0.46	0.32	0.23	48
Italy	0.59	0.43	0.29	0.19	0.12	32
Netherlands	2.56	2.78	3.13	1.65	0.87	165
France	0.07	0.06	0.04	0.02	0.01	12
Austria	0.06	0.06	0.06	0.04	0.03	5
EUROPE	11.85	12.45	11.71	7.53	3.28	561

Table 47.8 Europe : Oil and Gas Production and Consumption, Population and Density

EUROPE	OIL Production	OIL Consumption p/capita		OIL Trade (+)	GAS Production	GAS Consumption p/capita		GAS Trade (+)	Population	Growth	Area Mkm² Onshore	Area Mkm² Off	Area Mkm² Total	Density
Major	Mb/a	Mb/a	b/a	Mb/a	Gcf/a	Gcf/a	kcf/a	Gcf/a	M	Factor	Onshore	Off	Total	
Norway	682.29	81	16.5	602	5253	203	41	5050	4.9	2.1	0.33	2.30	2.63	15
UK	450.11	592	9.4	−142	2171	3329	53	−1158	62.7	1.6	0.25	0.06	0.30	256
Denmark	89.80	61	10.9	29	289	175	31	114	5.6	2.2	0.04	0.11	0.15	130
Germany	10.27	911	11.1	−900	464	3437	42	−2973	81.8	1.5	0.36	0.04	0.39	229
Italy	35.04	558	9.2	−523	293	2930	48	−2637	60.8	1.9	0.3	0.14	0.44	202
Netherlands	7.44	368	22.0	−360	3131	1938	116	1193	16.7	3.6	0.03	0.04	0.07	557
France	6.60	679	10.8	−673	44	1699	27	−1655	63.0	1.6	0.05	0.18	0.73	115
Austria	6.05	101	12.1	−95	61	335	40	−274	8.4	1.4	0.08	0.00	0.08	100
Sub-total	1288	3351	11	−2063	11706	14047	46	−2341	303.9	1.7	1.94	2.86	4.80	157
Minor														
Belgium/Lux		227.26	20.66	−227	0.00	716	65	−716	11	1.5			0.031	355
Greece	0.7	135.54	11.99	−135	0.04	135	12	−134	11.3	4.2			0.132	86
Finland		79.34	14.69	−79	0.00	166	31	−166	5.4	1.9			0.338	16
Iceland		6.36	21.2	−6	0.00	0	0	0	0.3				0.103	3
Ireland		58.29	12.67	−58	13.63	201	44	−188	4.6	0.8			0.07	66
Portugal	1.3	101.27	9.46	−100	0.00	182	17	−182	10.7	1.8			0.092	116
Spain		525.07	11.37	−525	1.80	1265	27	−1263	46.2	2.1			0.505	91
Sweden		128.15	13.63	−128	0.00	59	6	−59	9.4	1.7			0.45	21
Switzerland		88.57	11.21	−89	0.00	130	16	−130	7.9	1.8			0.041	193
Sub-total	1.9	1350	12.6	−1348	15.47	2855	27	−2839	106.8	1.8			1.762	61
Europe	1290	4701	11	−3411	11721	16902	41	−5180	411	1.7			6.6	63

MEXICO
TRINIDAD
VENEZUELA
COLOMBIA
ECUADOR
PERU
BRASIL
BOLIVIA
CHILE
ARGENTINA

Part VI

Latin America

Argentina

Table 48.1 Argentina regional totals (data through 2010)

Production to 2100					Peak Dates			Area	
Amount		Rate				Oil	Gas	'000 km²	
	Gb	Tcf	Date	Mb/a	Gcf/a	Discovery	1962	1977	Onshore	Offshore
PAST	10.4	43	2000	278	1,585	Production	1998	2004	2790	1000
FUTURE	4.7	27	2005	257	1,822	Exploration	1985		*Population*	
Known	4.0	22	2010	234	1,663	*Consumption*	Mb/a	Gcf/a	1900	5
Yet-to-Find	0.7	5.5	2020	143	1,029	2010	226	1,529	2010	40.5
DISCOVERED	14.3	65	2030	88	493		b/a	kcf/a	Growth	7.8
TOTAL	15.0	70	Trade	+8	+134	Per capita	5.6	38	Density	15

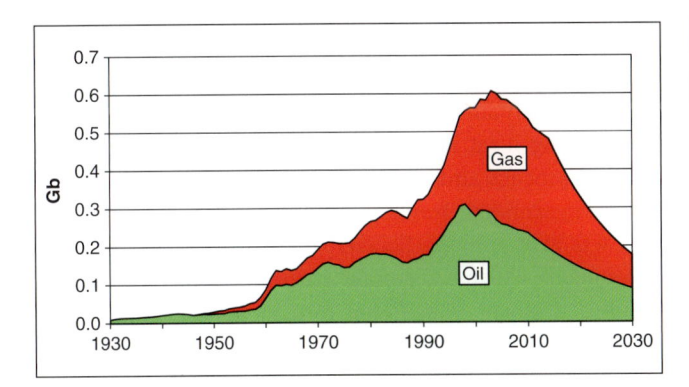

Fig. 48.1 Argentina oil and gas production 1930 to 2030

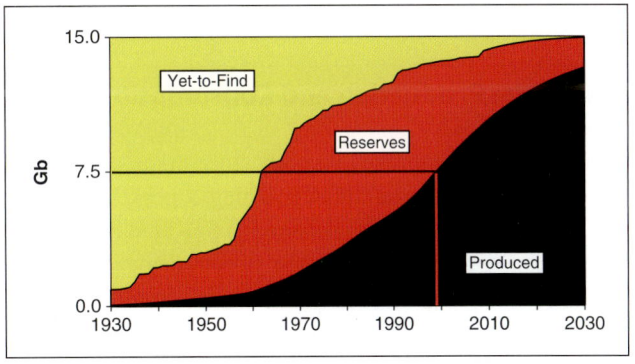

Fig. 48.2 Argentina status of oil depletion

Essential Features

Argentina covers an area of 2.8 million km² at the southern end of the South American continent, and supports a population of about 41 million. The impressive Andes Mountains form the western boundary with Chile, while to the east lie extensive plains, known as the *Pampas*. The Straits of Magellan separate the mainland from Tierra del Fuego in the extreme south. The climate is mainly temperate, which made it an attractive destination for European immigrants. The country is flanked to the east by a very large continental shelf bordering the Malvinas (Falkland Islands), which it endeavoured to repossess in 1982 by military means, although its oil prospects appear limited, despite some recent discoveries. Argentina also lays claim to various other islands and sections of Antarctica.

Geology and Prime Petroleum Systems

Argentina lies outside the most prolific zones of oil generation of South America, but has nevertheless supported a modest oil industry for many years. There are four discrete oil provinces: the Mendoza and Neuquen Basins in the sub-Andean province of the north; Comadoro-Rivadavia on the Atlantic Coast and the Austral Basin in the extreme south. Jurassic and Cretaceous source-rocks have charged

Cretaceous and Tertiary reservoirs. The Mendoza Basin has some additional Palaeozoic potential. The Neuquen Basin has substantial potential for production of oil and gas by artificial fracking.

Petroleum Exploration and Discovery

Exploration commenced in 1907 and continued at a modest pace until the 1930s when it picked up to some degree. The prime chapter followed in the 1980s with a peak of drilling in 1985 when 134 boreholes were drilled. Exploration has fallen since, but still continues at a fair level, averaging almost 50 boreholes a year until 2010 when only 13 were reported. Most of the activity has been onshore, although the continental shelf off the east coast has received some attention, yielding a few modest finds. Exxon took a strong early position, followed in the 1960s by Amoco. A State company, YFP, has been active for many years partly in partnership with various privately owned national companies.

Overall the peak of oil discovery was in 1962. A total of 14.3 Gb of oil and 65 Tcf of gas have been found.

Petroleum Production and Consumption

Oil production has grown gradually since the first field was placed on production in 1908, reaching a peak in 1998 at 847 kb/d, since when it has declined to 641 kb/d and is now expected to continue to fall at the current depletion rate of 4.8% a year. Oil consumption is running at 226 Mb/a making the country a small exporter.

Gas production commenced in 1947 and has also grown gradually to reach a plateau at about 1,800 Gcf/a year, which will likely continue until around 2014, when 70% will have been depleted before terminal decline sets in at about 7% a year. Gas consumption stands at 1,529 Gcf/a making the country a small exporter, primarily to Chile.

The Oil Age in Perspective

Some 300,000 hunters and fishermen are thought to have lived in Argentina prior to the arrival of the first European explorers from Spain. The latter were led by Ferdinand Magellan, who made his famous voyage in 1520 passing into the Pacific through the straits now bearing his name. He was followed six years later by Sabastian Cabot, who explored the eastern seaboard. Settlement commenced in 1536 after another expedition from Spain. It was not as vigorous as in Mexico or Peru where gold and silver acted as the lure, but several farming colonies established themselves. Buenos Aires developed as a port in the late eighteenth century as trade with Europe increased, but faced a certain threat from Brazil.

Moves to independence began in the early nineteenth century when the Spanish motherland was preoccupied with the Napoleonic wars. Buenos Aires declared a form of independence in 1810, but conflict with royalists continued. Full independence for what was described as the United Provinces of the Rio de la Plata came six years later, but internal disputes later led to the secession of what are now Uruguay, Paraguay and eastern Bolivia. As in other South American countries, Argentina faced the eternal conflict between centralist and federalist forms of government.

Large-scale immigration from Europe commenced in the 1860s. It brought economic expansion, including the construction of railways, which helped unify the country. British capital flowed in during the early years of the twentieth century, which saw the establishment of successful cattle ranching, and was accompanied by a degree of political stability. The population had grown to eight million by the outbreak of the First World War, but new conflicts began to emerge between the land-owning and industrial elite and the people at large. Financial mismanagement led to defaults on foreign loans and new uncertainties.

In 1933, Britain entered into a special trade arrangement, guaranteeing meat imports and removing tariffs on wheat, which in practice almost brought the country into the British Empire in economic terms.

It was neutral in the Second World War under a military government, which had certain national socialist (or Fascist) leanings. Colonel Juan Peron rose to power in the 1940s bringing in social reform with welfare legislation and greater political freedom, having the support of the trade unions. He became a charismatic and popular socialist dictator, ably assisted by his beautiful wife, Evita, who succeeded in describing herself as one of the "shirtless" (*descamisadas*) whilst haranguing the workers in a fur coat.

Even so, he eventually lost control in 1955 and went into exile having failed to curb the growing influence of the military and the Church on Argentine politics. Subsequent conflict and financial mismanagement paved the way for his return, now with his third wife, Isabel, but he had only one year in office before he died in 1974. Isabel succeeded him as the first lady President, but a combination of high oil prices, associated with the First Oil Shock, and an outbreak of foot-and-mouth disease caused an economic recession, leading to yet another military government two years later.

The new regime under General Videla ruled with an iron hand, closing Congress and banning trade unions. It initiated a reign of terror in which some 15,000 members of the opposition lost their lives, often after torture and imprisonment. He was succeeded in 1981 by General Galtieri, who in the following year endeavoured to retrieve the Malvinas Islands from Britain, to be robustly repulsed by Mrs. Thatcher.

A form of democratic government then returned and has persisted to the present day, but the country has been plagued by recurring economic crises, culminating in a collapse of the currency in 2003. As in many countries, democracy was in large measure a cloak for various forms of exploitation by the privileged elite, which if excessive prompted military takeover.

We may conclude from this long story of alternating political and economic difficulties that Argentina, which is a rich country of farmland, has truly been a victim of globalism. Its wealth flowed to foreign investors and local elites, who no doubt promptly exported their capital, leaving the people to suffer, complain and periodically revolt.

Looking ahead, the country seems to be relatively well placed to face the Second Half of the Oil Age. Its population density is not excessive, and it should be able to feed itself: its steaks being famous. The indigenous people are almost extinct and therefore pose no ethnic conflict. The European immigrants, many coming from Italy, seem to have integrated well to live in harmony. There must be considerable potential for, especially, wind and tidal energy, capable of meeting much of the country's future energy needs after its oil and gas are depleted.

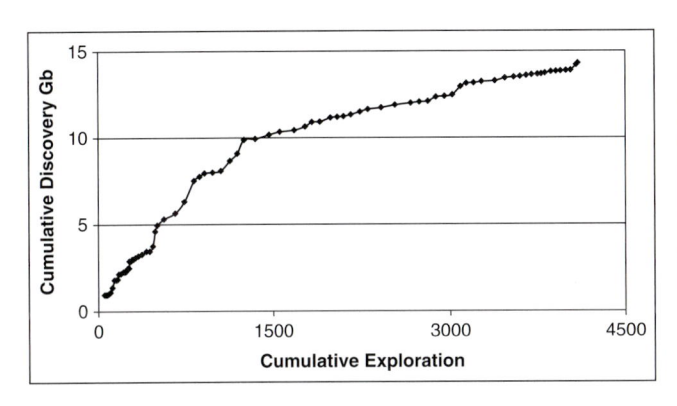

Fig. 48.3 Argentina discovery trend

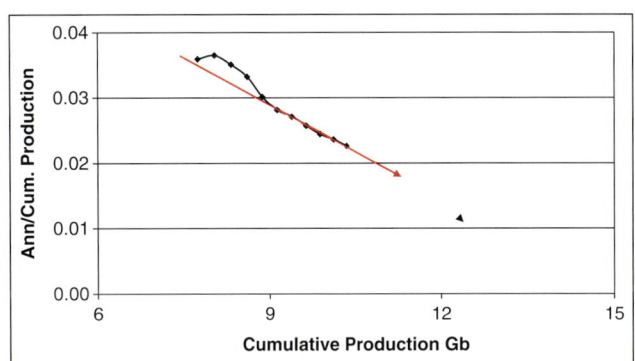

Fig. 48.4 Argentina derivative logistic

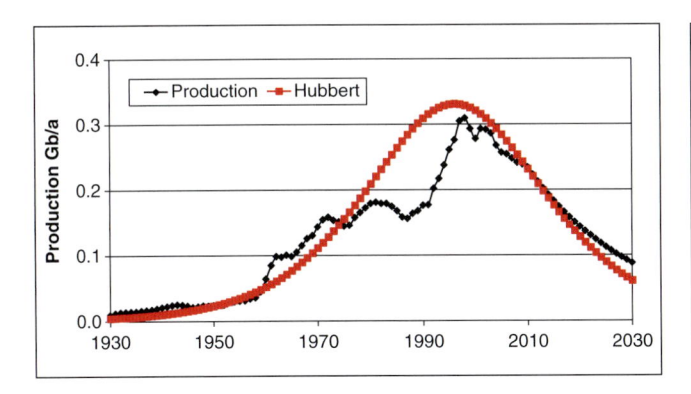

Fig. 48.5 Argentina production: actual and theoretical

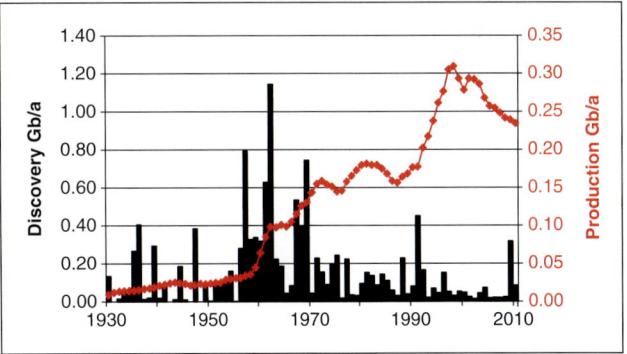

Fig. 48.6 Argentina discovery and production

Bolivia

Table 49.1 Bolivia regional totals (data through 2010)

Production to 2100						Peak Dates			Area	
Amount		Rate					Oil	Gas	'000 km²	
	Gb	Tcf	Date	Mb/a	Gcf/a	Discovery	1999	1999	Onshore	Offshore
PAST	0.56	9.7	2000	11	198	Production	2005	2025	1100	0
FUTURE	0.69	55	2005	19	519	Exploration	1962		Population	
Known	0.58	47	2010	16	526	*Consumption*	Mb/a	Gcf/a	1900	1.8
Yet-to-Find	0.10	8.3	2020	14	800	2010	23	96	2010	10.1
DISCOVERED	1.15	57	2030	11	800		b/a	kcf/a	Growth	6
TOTAL	1.25	65	*Trade*	−7	+430	Per capita	2.2	9	Density	9

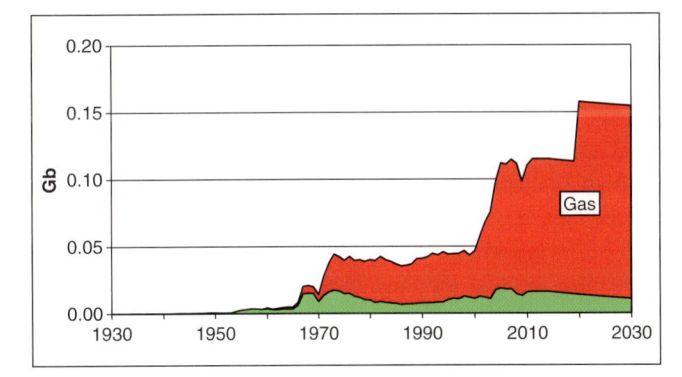

Fig. 49.1 Bolivia oil and gas production 1930 to 2030

Fig. 49.2 Bolivia status of oil depletion

Essential Features

Bolivia is a landlocked country of 1 million km² in the heartland of South America, being bordered by Brasil, Paraguay, Argentina, Chile and Peru. The capital, La Paz, lies at an altitude of about 3,500 m on the Altiplano, which is a barren intermontane valley, flanked by snow-capped Andean peaks, rising to over 6,500 m. It is dominated by Lake Titicaca, an unusual high altitude lake, covering some 8,500 km², on which ply steamers of almost ocean-going size. To the east of the Andes lies the Oriente, an extensive and remote area of foothills, plains and tropical forests, occupying about two-thirds of the country.

Geology and Prime Petroleum Systems

In geological terms, Bolivia is dominated by the great Andean chain. It is made up of the Western Cordillera, capped by recent volcanoes; the Altiplano and Pampean Massifs composed of partly mineralised granitic intrusions; and the Eastern Cordillera comprising folded and faulted Palaeozoic rocks. To the east lies a mildly deformed Sub-Andean zone flanking the Brasilian Shield. The oil prospects are mainly confined to the Sub-Andean Zone, which may be divided into the remote Beni Trough in the north, and an extension of the Neuquen Basin of Argentina in the south.

C.J. Campbell, *Campbell's Atlas of Oil and Gas Depletion*,
DOI 10.1007/978-1-4614-3576-1_49, © Colin J. Campbell and Alexander Wöstmann 2013

Whereas northern South America is blessed with highly prolific Middle Cretaceous hydrocarbon source-rocks, the southern part of the Continent has to rely on much leaner older sources, principally in the Silurian in eastern Bolivia and the appropriately named *Vacas Muertas Formation* (dead cows) of the Jurassic in southern Bolivia and neighbouring Argentina. It is not surprising therefore that the region is gas-prone, with limited amounts of oil.

Exploration and Discovery

Exploration commenced in 1914, with a few boreholes being drilled over the succeeding years, before it picked up in the 1960s, to deliver the peak of 17 boreholes in 1962. Exploration drilling has continued since then at a generally low level although there have been occasional brief bursts of activity.

A number of modest oil and gas discoveries were made in the 1920s, and sporadically thereafter, before rising to two peaks: the first in the 1960s and the second in the late 1990s. Much of the gas is restricted to the exceedingly remote Beni Trough in the northeast of the country, adjoining Brasil, which is an obvious market for it.

The remote location of the country has meant that production, and hence the justification for foreign companies to carry out exploration, was based on the small size of the internal market and the capacity of export pipelines. A total of 395 exploration boreholes have, nevertheless, been drilled, finding some 1.25 Gb of oil and 65 Tcf of gas.

Production and Consumption

Oil production commenced at a low level in 1925 and grew steadily to 1966, when it reached a plateau at around 40 kb/d. Production then fell slightly before recovering to pass an overall peak of 51 kb/d in 2005. It is now expected to continue at about 45 kb/d until around 2014 before starting its terminal decline, being largely controlled by pipeline capacity. Evidently, the remote location of Bolivia has slowed the depletion of its oil and gas, which bodes well for its future. Oil consumption is currently running at about 23 Mb/a, making the country a minor importer.

Gas production commenced in 1960 and grew rapidly to pass 150 Gcf/a in 1978. It then levelled off until 2001, when it began to rise again, reaching another plateau of about 500 Gcf/a in 2005. The resource base allows it to grow in the future, but whether it will do so in practice depends on Government policy and new pipeline capacity. Here, it is forecast to rise with the opening of new export pipelines in 2020 to 0.8 Tcf/a from 2020 onwards until eventual decline sets in around 2040. NGL production from gas plants stands at 10 kb/d with past production amounting to 89 Mb.

The Oil Age in Perspective

Bolivia is a sparsely populated country of about ten million people, of whom some 70% are of pure Quechuan Indian stock. The Altiplano, high in the Andes, was already a centre of population in the seventh century when it formed the seat of the Tiohuanaco Empire that held sway over much of Bolivia and Peru. In 1524, Francisco Pizarro, a Spanish explorer, made his first landing in Peru, returning on a second expedition seven years later to bring the territory into the Spanish Empire with the dual objectives of introducing Christianity and exploiting its gold and silver. The rich Potosi silver mines of Upper Peru (now Bolivia) were found in 1545, and proved a remarkable source of wealth for centuries to come, such wealth being most unevenly distributed. Ironically this flood of gold and silver led to a period of damaging inflation in Spain.

The oppressed Indian population revolted from time to time, with a notable uprising in 1780 being led by a descendant of the last Inca. The fall of Spain to Napoleon in the closing years of the eighteenth century paved the way for Latin American independence, being led primarily by Simon Bolivar, who marched south from Venezuela, and José San Martin, who marched north from Argentina. There were also local uprisings led in Chile by Bernardo O'Higgins, who was of Irish lineage, and in Bolivia by General Sucre. Formal independence was declared in Bolivia in 1825.

The succeeding years were fraught with political difficulties as was the case in most of Latin America. The *guano* deposits (bird excrement) of what was coastal Bolivia were desperately needed as a source of nutrient to support agriculture in overpopulated Europe, leading the Chile Government, which was backed by foreign mining companies, to take the territory from Bolivia in the Nitrate Wars of 1879–1884.

Military government alternated with democratic administrations over the following decades in a highly stratified society. The mineral wealth dominated the country's economy, especially following the discovery of substantial tin deposits. Rubber plantations in the eastern territories also enjoyed a temporary boom before being sold to Brasil in 1903. A disputed boundary with Paraguay in the remote southeast part of the country, which was thought to have oil prospects, led to the Chaco Wars of the 1930s, in which 100,000 men died. Bolivia later nationalised the holdings of Standard Oil (Exxon), the most prominent of the foreign oil companies operating in the country, and established a State company, YFPB, which took a strong position.

The Second World War saw the emergence of a new political confrontation between Fascist and Communist persuasions, reflecting, in a sense, the political divisions of Europe. In 1952, came the National Revolution, in which the foreign mining companies were nationalised, and universal suffrage, together with land allocations, granted to the Indian peasants. A reaction followed in 1964 with a return to military government under General Barrientos, who survived a subsequent attempted coup, organised by Che Guevara, the well-known Argentine Marxist revolutionary with ties to Fidel Castro, and now a T-shirt icon. An even more oppressive regime followed, suppressing the labour movements and placing the mines under military control.

Still another swing followed in 2005 with the election of Evo Morales, again espousing the rights of the downtrodden Indian population. His government moved to legalise the production of coca on which the peasant farmers have relied for many years, and later expropriated the rights of the foreign oil companies. He evidently joins Ugo Chavez of Venezuela in a bid to free Latin America from the chains of global markets, in order to dedicate its resources to its own people.

Bolivia seems to be well placed to face the Second Half of the Oil Age, having a modest population, and relatively under-depleted oil and, especially, gas resources. It must also have plenty of potential for solar and hydro-electric power, and no longer faces any particular conflict with its neighbouring countries. Internal social problems between landowner and peasant may nevertheless erupt in conflict, as they have many times in the past, although the landowners may see a decline in their power, and indeed wealth, under populist governments.

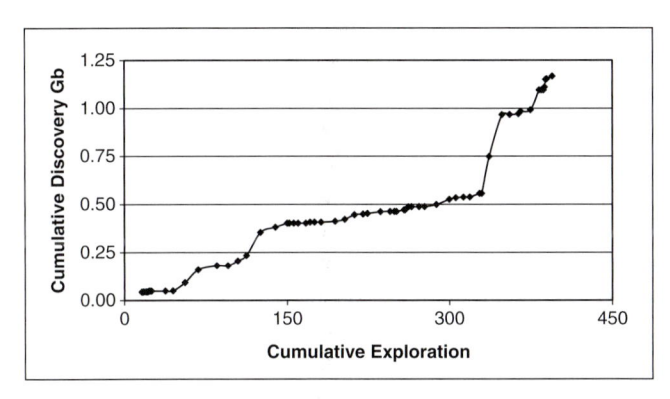

Fig. 49.3 Bolivia discovery trend

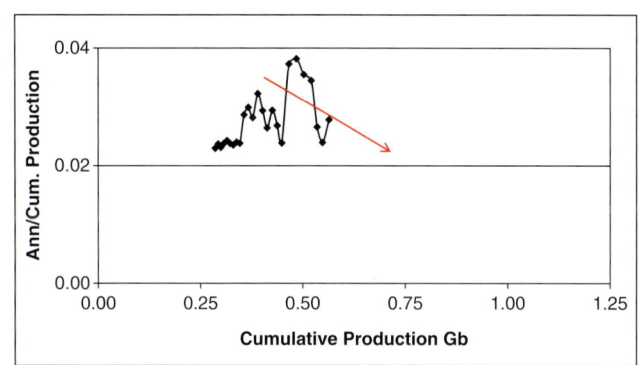

Fig. 49.4 Bolivia derivative logistic

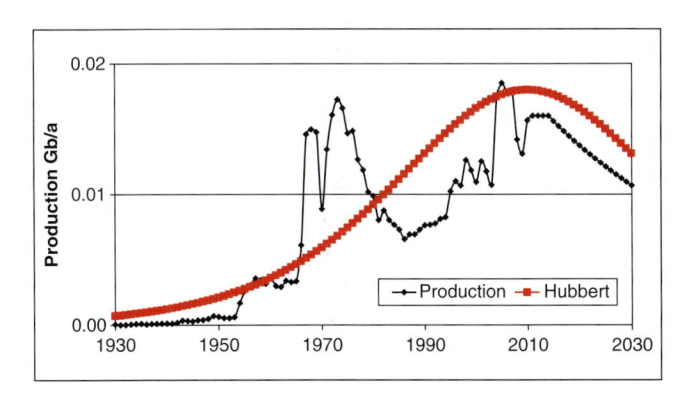

Fig. 49.5 Bolivia production: actual and theoretical

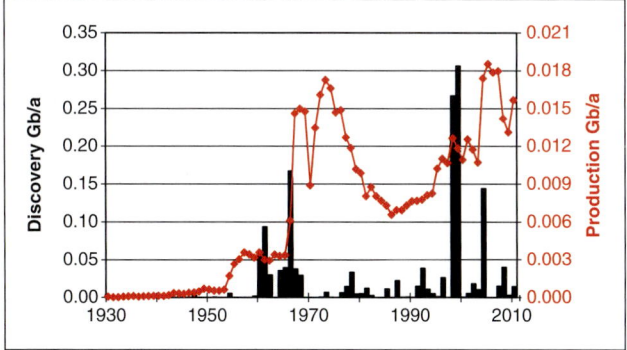

Fig. 49.6 Bolivia discovery and production

Brasil

Table 50.1 Brasil regional totals (data through 2010)

	Production to 2100					Peak Dates			Area	
	Amount		Rate				Oil	Gas	'000 km²	
	Gb	Tcf	Date	Mb/a	Gcf/a	Discovery	1975	2008	Onshore	Offshore
PAST	6.1	11	2000	181	469	Production	1990	2011	8580	1200
FUTURE	2.9	12	2005	169	500	Exploration	1982		Population	
Known	2.6	11	2010	59	500	Consumption	Mb/a	Gcf/a	1900	18
Yet-to-Find	0.3	1.2	2020	48	500	2010	969	890	2010	197
DISCOVERED	8.7	22	2030	39	258		b/a	kcf/a	Growth	10.2
TOTAL	9	23	Trade	−910	−390	Per capita	4.9	5	Density	23

Note: The above table refers to Regular Conventional Oil and Gas only (i.e. excluding the Deepwater)

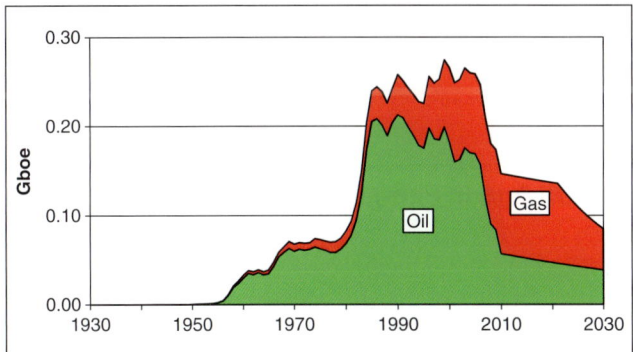

Fig. 50.1 Brasil oil and gas production 1930 to 2030

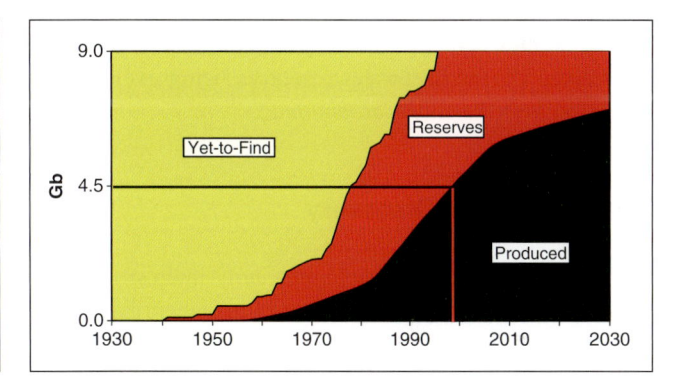

Fig. 50.2 Brasil status of oil depletion

Essential Features

Brasil is the largest country in South America, covering some 8.5 million kms² and supporting a population of about 197 million. Mountain ranges of moderate relief form the northern boundary with Venezuela, giving way to the vast rain forests of the Amazon basin. To the south, lie extensive dissected tablelands of forest and grass. Most of the population is concentrated along the Atlantic littoral, where the largest city, Sao Paulo, with about 20 million inhabitants, and Rio de Janeiro, the former capital, are located. Brasilia is a purpose-built modern capital at an altitude of 1,000 m in the south of the country. It became the seat of government in 1960, and now houses a population of about two million.

The great Amazon River flows eastwards through extensive tropical rain forests. It, in fact, partly reversed its direction of flow following the uplift of the Andes at the end of the Tertiary Period, as previously some of the interior of the continent drained westwards into the Pacific through the Gulf of Guayaquil in Ecuador. Accordingly, the current delta in the Atlantic is comparatively small.

Geology and Prime Petroleum Systems

In geological terms, most of Brasil is made up of ancient crystalline rocks of the Guayana and Brasilian Shields. They are separated by a great left-lateral suture that cut Africa and South America, when they formed the ancient continent of

Pangea. It provided a line of weakness followed by the Amazon River.

Three main petroleum systems may be identified. One comprises the Amazon rift system where Silurian source-rocks have charged a number of modest oilfields; another minor one comprises the margin of the Sub-Andean basin, although lying outside the main belt of Cretaceous generation; and a third, much richer, system follows the eastern seaboard, where rifts formed with the initial opening of the South Atlantic in the early Cretaceous. Oil was generated in stagnant lakes that filled the rifts, and later migrated into rather poor quality reservoirs that were deposited along the rift margins. These systems yielded a large number of small- to moderate-sized fields in what is now a thoroughly explored province, offering limited scope for new discovery. Some interest also attaches to the Palaeozoic rocks of the Parana Basin which lie beneath an extensive cover of Permian volcanics.

Another vastly more important system comprises a series of deepwater basins, including the Campos Basin. Some remarkable recent deepwater discoveries, including the Tupi and Jupiter finds, have been made at extreme depths beneath a thick layer of salt. They are considered separately in Chapter 12: the deepwater resources being excluded by definition from *Regular Conventional*.

Exploration and Discovery

Onshore Brasil did not attract much early interest from the international oil companies in view of its generally adverse geology. The first recorded exploration borehole was sunk in 1939, to be followed by a few more over the next decade. Some minor finds were, however, made offering encouragement for the government to form the State enterprise, Petrobras, to properly evaluate the possibilities. It met with modest success, primarily in the Amazon region, before turning to the more promising offshore, where the first small discovery was made in 1968. Gradually, the deeper water possibilities came into view, and greatly to its credit, Petrobras pioneered the necessary technology, making the first giant discovery in 1985 with the Albacora Field. The Campos Basin became one of the world's great deepwater provinces, described in Chapter 12.

The total discovery of *Regular Conventional oil and gas* amounts to almost 9 Gb of oil and 22 Tcf of gas, leaving little remaining scope for new discovery, given that more than 3,000 exploration boreholes have been drilled.

It is difficult to determine how much of the reported gas is attributable to the deepwater, but it is here estimated that some 22 Tcf have been found in the rest of the country.

It is also worth mentioning that Brasil has substantial oil shale deposits in the Permian Irati Formation, which are estimated to be capable of yielding as much as 50 Gb of oil, although exploitation is not as yet commercially viable.

Production and Consumption

Onshore production commenced in 1940, and grew at a modest pace to reach a peak at almost 250 kb/d by 2004. Adding the conventional offshore (excluding by definition the deepwater), gives an overall peak in 1990 at 583 kb/d. It had fallen to 162 kb/d by 2010, and is expected to continue to decline at about 2% a year.

Gas production commenced in 1949 and rose steadily to 500 Gcf/a by 2002, at which level it is expected to plateau until 2020, when 70% will have been depleted, before commencing its terminal decline at about 7% a year.

Brasil's consumption of oil was rising gently throughout the 1970s but has since doubled to reach 969 Mb/a by 2010. Soaring imports were causing a heavy burden on the balance of trade, but the entry of deepwater production has now almost put the country into balance. Gas consumption has also risen to 890 Gcf/a, exceeding production, making the country a minor importer of Bolivian gas.

The Oil Age in Perspective

Vicente Pinzon, a Spanish explorer, landed near Recife in 1500 to find a vast country sparsely populated by Arawak and Carrib Indians, a few of whose descendants are still to be found in the Amazon headwaters. Although discovered by a Spaniard, the territory lay within what had been declared by the Pope to be a Portuguese sphere of influence under the Treaty of Tordesillas of 1493. A few months later, the Portuguese Government dispatched an expedition under the Italian navigator, Amerigo Vespucci, to confirm its rights, giving his Christian name to the Americas. A programme of Portuguese colonisation followed over the next centuries but the country was also subject to Spanish, French, British and Dutch attentions, which were accompanied by partial settlement. Missionaries extended practical sovereignty into the interior, occasioning disputes with bordering Spanish colonies. Slaves from Africa were imported in large numbers to work the plantations.

When Portugal was occupied by Napoleon's troops in 1807, the Portuguese Government under the Regent, Prince John, moved to Brasil, which became the seat of government for the homeland. On returning to Lisbon in 1816 after the Napoleonic wars, he left behind his son, Dom Pedro, who declared independence for his adopted country in 1822, appointing himself as its Emperor in a grandiose gesture. But political turbulence and boundary disputes with Argentina followed until order was established under his son and successor, Pedro II, in 1840. A period of economic progress ensued with the construction of railways and the abolition of slavery. A new constitution, modelled on that of the United States, established the country as a Federal Republic in 1891, but did not bring to an end the political turbulence, which continued through much of the twentieth century. A dictatorial regime under President Vargas ran the country before the Second World War and again afterwards. Even so, immigration from Europe, and particularly Germany, increased radically, being accompanied by economic growth. A period of military government was followed in 1985 by a return to civilian rule. Although economic progress was made, being partly related to the country's position as the world's largest coffee producer, weak financial management led to periodic epochs of rampant inflation and excessive foreign debt.

Brasil's consumption of oil was rising throughout the 1970s to pass 1 Mb/d by the end of the decade. Soaring imports were causing a heavy burden on the balance of trade. With this incentive, Petrobras turned its eyes oceanward, and with necessity being the mother of invention, decided to try to find out what the deepwater potential might be. To its enormous credit, it pioneered the necessary technology, demonstrating that the so-called developed world has no particular claim to technological prowess. Even so, Brasil opened its doors to foreign oil companies in the late 1990s, breaking the monopoly of Petrobras, which will nevertheless likely retain a dominant position in the prime areas.

In fact, it makes little sense to grant foreign companies the right to export, when the country will shortly need all it can get for itself.

An election in 2002 returned a populist, left-leaning government under Luis Lula da Silva. This is consistent with the general political atmosphere in Latin America as it responds to globalism, which has concentrated wealth still further into the hands of the so-called oligarchs. He was succeeded in 2011 by Dilma Rousseff, as the first female President of the country, who is the daughter of a Bulgarian immigrant. In earlier years she had Communist leanings but has moved to pragmatic capitalism, and significantly had previously been State Secretary for Energy.

Looking ahead, Brasil has other energy options with which to face the future, apart from the entry of deepwater oil and gas. It has solar energy and hydroelectric power, although it has not yet proved adequate to meet electricity demand. It produces substantial amounts of ethylene from sugar cane, but that is a mixed blessing insofar as the expansion of industrialised farming is destroying the virgin forests of the Amazon, which carries adverse environmental impacts, especially in relation to climate change. It has vast deposits of Permian oil shale in the south of the country, as already mentioned, which will likely be exploited in the future at a low rate. It also supports a nuclear industry, having itself resources of uranium.

Although well placed in some respects, the country does face serious tensions as the large predominantly urban population comes to adjust to the Second Half of the Oil Age which will call for a return to the land for greater self-sufficiency. Certainly, it makes no sense to export energy in any form today in the face of its own growing future needs.

Fig. 50.3 Brasil discovery trend

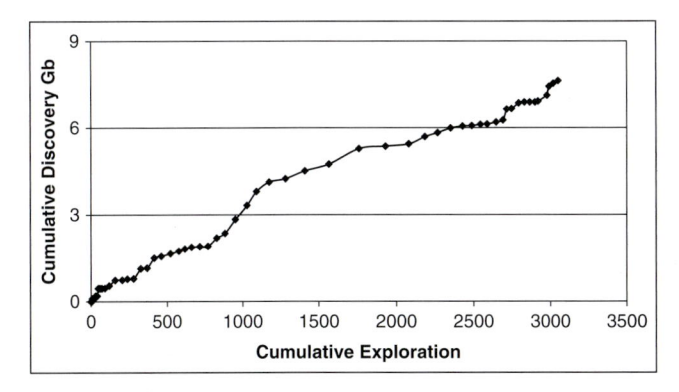

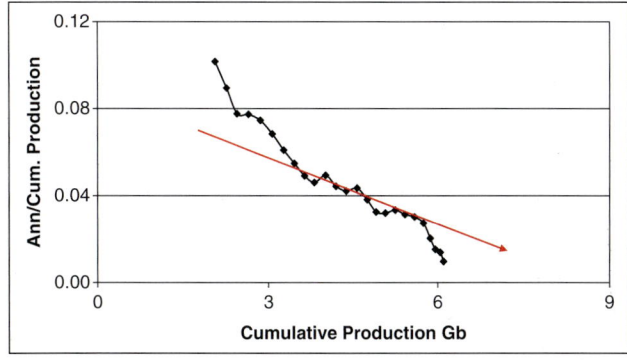

Fig. 50.4 Brasil derivative logistic

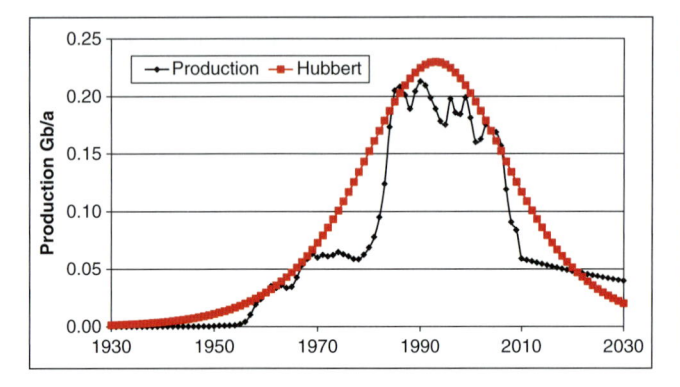

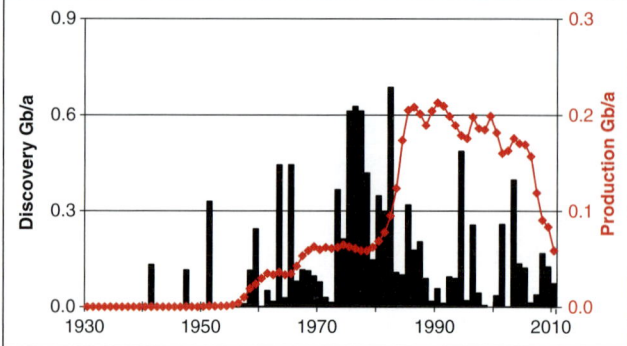

Fig. 50.5 Brasil production: actual and theoretical

Fig. 50.6 Brasil discovery and production

Chile

Table 51.1 Chile regional totals (data through 2010)

Production to 2010						Peak Dates			Area '000 km²	
Amount			Rate				Oil	Gas		
	Gb	Tcf	Date	Mb/a	Gcf/a	Discovery	1960	1960	Onshore	Offshore
PAST	0.43	4.7	2000	2	101	Production	1982	1992	760	225
FUTURE	0.07	5.3	2005	2	82	Exploration	1972		Population	
Known	0.07	5.0	2010	1	63	Consumption	Mb/a	Gcf/a	1900	3
Yet-to-Find	0.00	0.3	2020	1	50	2010	110	188	2010	17
DISCOVERED	0.50	9.7	2030	1	50		b/a	kcf/a	Growth	5.3
TOTAL	0.50	10	Trade	−110	−125	Per capita	6.4	11	Density	23

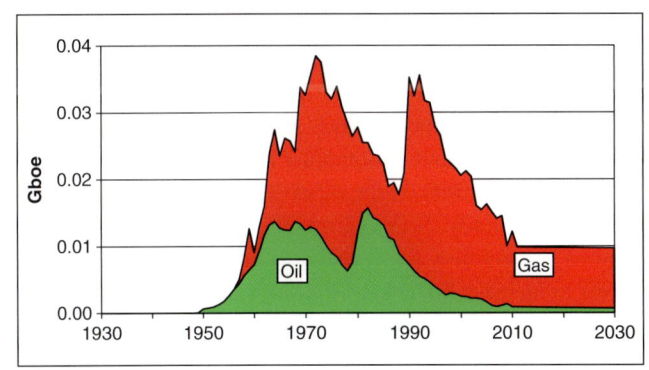

Fig. 51.1 Chile oil and gas production 1930 to 2030

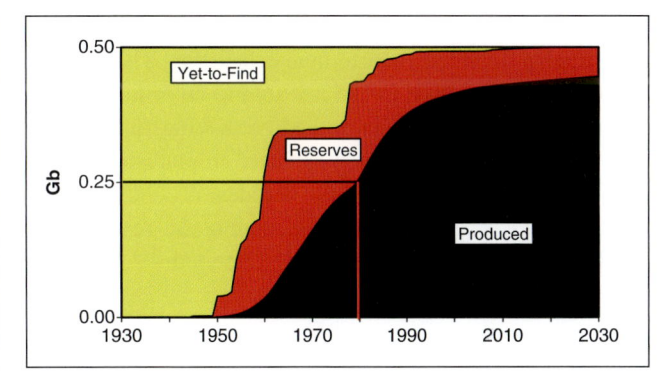

Fig. 51.2 Chile status of oil depletion

Essential Features

Chile is a long, narrow, mountainous country flanking the southern limits of the great Andean mountain chain of South America. Wide barren uplands, forming an extension of the Altiplano of Bolivia, give way southwards to mountain ranges, with the highest peak rising to almost 7,000 m. In the northwest of the country is the coastal Atacamas Desert, but most of the country enjoys a temperate, even Mediterranean climate. To the south is a highly indented terrain of fjords, lakes and islands. About one-tenth of the country is arable land, some providing rich pasture for cattle, with the rest being desert, wild mountain or forested terrain. It is drained by a number of relatively small rivers flowing westwards into the Pacific.

Geology and Prime Petroleum Systems

Most of the country is made of up of Andean ranges, composed largely of granitic and metamorphosed rocks, which are locally mineralised. As a result, Chile is an important mining country having the world's largest deposits of copper, together with iron, gold, silver and lithium. In addition, an extension of the Austral petroleum system of Argentina

C.J. Campbell, *Campbell's Atlas of Oil and Gas Depletion*,
DOI 10.1007/978-1-4614-3576-1_51, © Colin J. Campbell and Alexander Wöstmann 2013

crosses into the southern extremity of Chile, providing a small amount of oil and gas.

Exploration and Discovery

The discovery of the Austral Basin of Argentina naturally led to exploration in the adjoining parts of Chile, where the first exploration borehole was sunk in 1938 leading to a small oil discovery in 1945. Drilling increased over the years at a modest rate to reach a peak in 1972 when as many as 32 boreholes were drilled, but has since declined to almost zero. Oil discovery too was at a modest level, peaking in 1960, with later small finds being made in 1998 and 2008. Gas was discovered in parallel, and included a somewhat surprising find of two giant fields totalling 4.6 Tcf in 1960, almost half of the estimated total discovery of 9.7 Tcf.

Production and Consumption

Oil production commenced in 1949, and rose to a peak of 43 kb/d in 1982 before falling to its present low level, which may drag on for some years to final exhaustion. The production of gas commenced in 1957. It rose to an initial peak in 1973, before in turn declining and recovering to a second lower peak in 1996. It is now expected to plateau at about 50 Gcf/a for the next two or three decades.

Oil and gas consumption is running at respectively 110 Mb/a and 188 Gcf/a, most being imported.

The Oil Age in Perspective

Some 500,000 people, belonging to various native Amer-Indian tribes, occupied Chile at the time of the Spanish Conquest in the sixteenth century. The initial Spanish expedition of 1536–1537 failed to find gold and silver, comparable with that of the Inca Empire of Peru, and left disappointed. The country remained something of a backwater in the Spanish Empire, although a small settlement was established later.

Chile participated in the general movement towards Latin American independence, following the fall of Spain to Napoleon. At first, Spain resisted these moves, causing Chilean patriots, including Bernardo O'Higgins, who was of Irish descent, to flee to Argentina where he rallied support, returning at the head of a victorious army in 1817, which paved the way for the declaration of independence.

The early years of independence were not easy, with growing conflicts between the army, the landowners and the people at large. Nevertheless, the economy began to grow, especially with the development of nitrates from *guano* (accumulated bird excrement) found in coastal regions. The bird-life itself is encouraged by the Humboldt Current, which flows north along the coast, bringing nutrients to feed the abundant fish stocks on which the birds live. This development gave rise to the Nitrate War of 1879–1883, in which Chile, backed by European commercial interests, expanded northwards at the expense of Peru and Bolivia to take the coastal areas rich in nitrates.

US mining companies moved in during the early years of the twentieth century, finding and developing the huge copper and other mineral deposits of the country. These operations overtook the nitrate business, which went into decline when Norway developed a process, based on its hydroelectric power, to produce synthetic nutrients by electrolysis.

A degree of industrialisation followed, being accompanied by the growth of various socialist movements to protect the interests of miners and the working class generally. This was especially stimulated by the effects of the First Great Depression in 1930, when copper prices collapsed causing economic and financial hardship. The Second World War brought new prosperity, as the demand for copper rose. Chile was at first neutral but was persuaded to declare war on Germany in 1942 after the United States entered the war, despite the large number of German immigrants.

The post-war epoch saw growing political conflict between Socialist and Communist factions on the one hand and the ruling powers on the other. There was a rise in immigration, especially from Germany. Salvador Allende, an avowed Communist, was elected to power in 1970. Various moves to increase State intervention, including the nationalisation of mining companies and banks, followed, but ended with his assassination, three years later, in a coup d'état undertaken by the army with (it is widely suspected) the backing of the CIA. That brought General Pinochet to power. He ruled in an authoritarian manner suppressing the opposition, some 250,000 of whom were imprisoned. Torture and killings became commonplace. He supported Britain in its war with Argentina over the Falkland (Malvinas) Islands in 1982, before falling from power in 1990. He survived several subsequent trials, legal processes and accusations of human rights offences and embezzlement until he died of a heart attack in December 2006 at the age of 91. Socialist government followed, with the 2006 elections returning Michelle Bachelet, the first elected lady-President in Latin America. Her father had been put to death by Pinochet, while she herself was subjected to imprisonment, torture and exile. She was in turn succeeded in 2010 by Sebastian Piera of the National Renewal Party in a shift to the right of politics.

Chile has experienced an explosion of population over the past century, which presumably was supported largely by mining and other industry. Probably as a consequence, it has faced a particularly difficult political situation as the workers sought to break the power of the elite. This does not sound like a strong basis upon which to face the Second Half of the Oil Age, when industry declines. The country is already a net importer of oil and gas, the latter from Bolivia, which may become a source of political tensions, but it does have considerable hydroelectric potential. It nevertheless enjoys a temperate climate and its agriculture and fisheries could support a reduced population reasonable well.

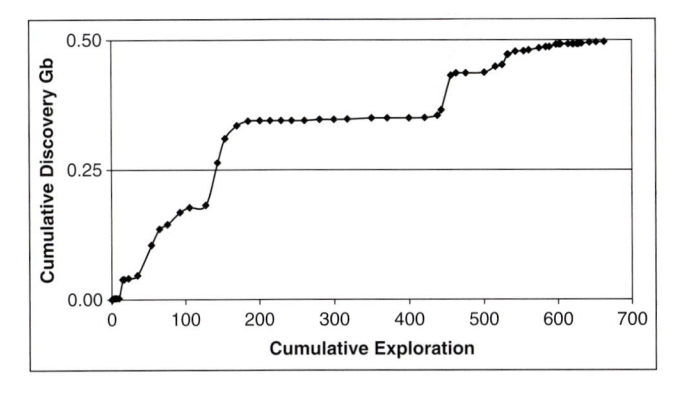

Fig. 51.3 Chile discovery trend

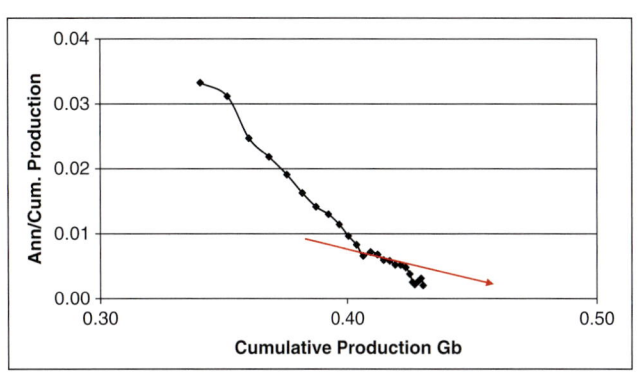

Fig. 51.4 Chile derivative logistic

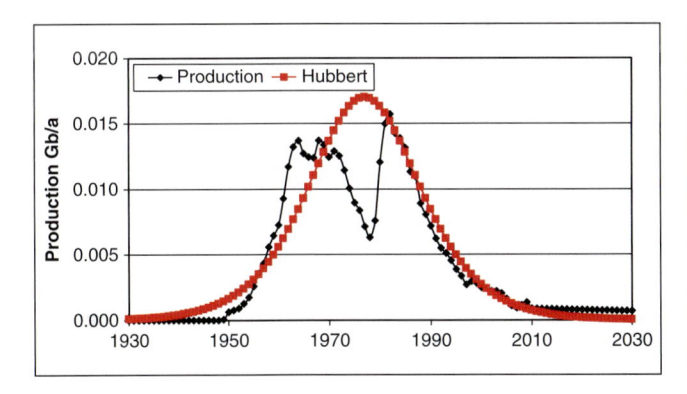

Fig. 51.5 Chile production: actual and theoretical

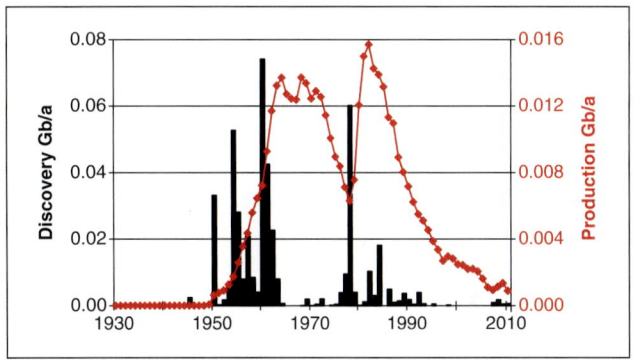

Fig. 51.6 Chile discovery and production

Colombia

Table 52.1 Colombia regional totals (data through 2010)

	Production to 2100					Peak Dates			Area	
Amount			Rate				Oil	Gas	'000 km²	
	Gb	Tcf	Date	Mb/a	Gcf/a	Discovery	1988	1973	Onshore	Offshore
PAST	7.5	15	2000	252	513	Production	1999	2012	1,143	150
FUTURE	5.6	15	2005	192	532	Exploration	1988		*Population*	
Known	4.7	14	2010	287	1,124	*Consumption*	Mb/a	Gcf/a	1900	5
Yet-to-Find	0.8	1.5	2020	167	650	2010	108	321	2010	47
DISCOVERED	12.2	28	2030	104	191		b/a	kcf/a	Growth	9.2
TOTAL	13.0	30	*Trade*	+179	+803	Per capita	2.3	7	Density	41

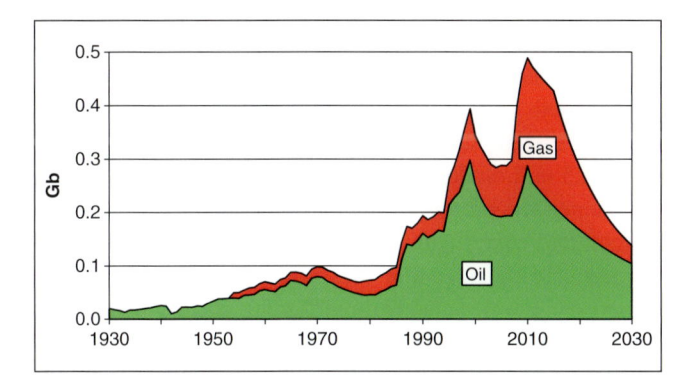

Fig. 52.1 Colombia oil and gas production 1930 to 2030

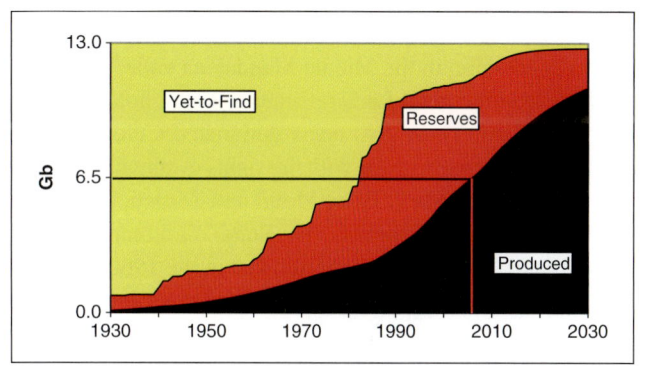

Fig. 52.2 Colombia status of oil depletion

Essential Features

Colombia lies at the northwest corner of South America, next to the Isthmus of Panama. It is cut by three ranges of the Andes, which are separated from each other by the Magdalena and Cauca Valleys. In the southeast are the plains and grasslands of the Llanos which pass into the forests of the Upper Amazon and Orinoco valleys. In the north are coastal lowlands, passing into the arid terrain of the Guajira Peninsula, bordering Venezuela. The country is washed by the Caribbean to the north and the Pacific to the west, and drained by the Magdalena River, flowing northwards into the Caribbean, as well as by the Orinoco headwaters in the deep interior.

Geology and Prime Petroleum Systems

The Central Cordillera forms the core of the Andes in Colombia, being made up mainly of metamorphic and granitic rocks. The Western Cordillera consists of low-grade metamorphic Cretaceous rocks uplifted from the subducted Pacific margin, while the Eastern Cordillera is made up of Cretaceous geosynclinal sediments, some 10,000 m thick, overlying older, highly deformed, strata. Tertiary, mainly non-marine, sediments fill the intermountain valleys. Major transcurrent faulting has affected the Andean ranges. The Santa Marta Fault, with a sinistral movement of over 100 km, has offset an extension of the Central Cordilleras to form the Santa Marta Massif bordering the Caribbean. The Perija

Range on the border with Venezuela also reflects uplift associated with the fault.

Middle Cretaceous anoxic shales form prolific source-rocks, which have charged mainly Tertiary reservoirs in three prime petroleum systems, represented by the Middle Magdalena Valley; the Sub-Andean basin between the Andes and the Guayana Shield and the southwestern corner of the Maracaibo Basin of Venezuela that extends into Colombia territory.

Exploration and Discovery

Exploration commenced in the early part of the last century with the first recorded boreholes being sunk in 1907. It grew at a moderate pace until the 1950s when it increased markedly to reach a peak in 1988 when as many at 78 boreholes were drilled. It then declined to a low of ten in 2002 before surging to 72 in 2008 and then declining to 59 in 2010.

Colombia has had a long oil history, starting in 1905 when General Virgilio Barco secured rights to the Colombian extension of the Maracaibo Basin, and Roberto de Mares took a concession in the Middle Magdalena valley, which in 1918 yielded the giant La Cira-Infantas Field, holding some 830 Mb of oil. These two areas dominated Colombian oil production for many years, with the rights eventually passing to subsidiaries of Shell, Esso, Mobil and Texaco, before the establishment of Ecopetrol, the State oil company. The remote Sub-Andean province followed in the 1960s delivering a series of giant fields. A large number of independent companies have moved in recent years to replace the major companies whose interest seems to be waning.

Associated gas was discovered in parallel with the oilfields, with two major gas fields delivering a peak in 1973, which has been overtaken recently.

Production and Consumption

Oil production, based on the combined results of the two main discovery cycles, reached a peak of 816 kb/d in 1999, close to the midpoint of depletion. It then declined to 531 kb/d in 2006 before surging to 786 kb/d in 2010 as fallow fields in the Llanos were finally linked to a pipeline. It is now expected to decline at about 4.9% a year. Consumption stands at about 108 Mb/a, equivalent to about 35% of production, meaning that the country is presently an important exporter, largely to the United States. It can remain a net exporter for a number of years, if it so decides, but at a declining rate.

The country has modest gas reserves amounting to about 14 Tcf, being produced at 1,124 Gcf/a. It may plateau at about 1,200 Gcf/a until 2015, by which point it will be 70% depleted and commence its terminal decline at about 12% a year.

The Oil Age in Perspective

One of the chieftains of the ancient Chibcha civilisation had the habit of covering his body in gold-dust before bathing in Lake Guatevita, near Bogotá, in which emeralds and other precious stones were thrown to placate the gods. *El Dorado*, as he was known, stimulated the interest of the Spanish Conquistadores in the wake of Columbus, who had reached the northern coast on his last voyage in 1502.

In a remarkably short span of 50 years, the Spaniards had established themselves throughout the country, building towns and monasteries high in the Andes. By 1739, Bogotá had been established as the Vice-Royalty of Nueva Grenada, holding dominion over what is now Venezuela, Ecuador and much of Central America, south of Mexico. But in 1819, Simon Bolivar, the great *Liberator* of South America, who was born in Caracas, defeated the Spanish royalists, bringing independence to the region, which later fragmented into separate republics. Cornelius O'Leary, a mercenary Irish soldier from Cork, wrote the national anthem for the new republic. The last territorial adjustment came in 1903, when the United States engineered the secession of Panama, after Colombia had refused consent for the construction of the Panama Canal. It tried to make amends in 1914 by paying an indemnity of 25 million dollars.

Independence brought the eternal conflict between federalism and centralism, exacerbated by physical mountain barriers and the fact that the disparate regions had been settled by immigrants from different parts of Spain. It sowed the seeds of the violence, often degenerating into banditry which had been endemic for two centuries before the narcotics trade gave it a new stimulus. Large tracts of the country have been at times under the control of war lords, some importing arms by air to support private armies, while surprisingly also sponsoring certain social programmes in their regions. The coca leaf has been grown since pre-Conquest days, without posing any particular local problem, but that changed when cocaine became a commodity on the global market.

Colombia's population has increased by a factor of nine over the past Century to reach 47 million. It is of mixed European, Amer-Indian and African origins, but the people live without any particular racial discord. A few indigenous Amer-Indian communities survive in remote areas but probably face extinction or integration. The violent political situation has driven people to the cities, especially Bogotá, the capital, where more than ten million now live, in many cases in desperate conditions. Their largely untreated effluent flows over the once beautiful Tequendama Falls, releasing a bacteriological fog. But despite every adversity, the Colombians retain their vitality, courage and good humour.

52 Colombia

A left-leaning revolutionary movement, known as FARC, controls remote sections of the country. In early 2008, Government forces attacked a base just across the border in Ecuador, killing its leader and prompting a strong reaction from the Governments of Ecuador and Venezuela, which briefly mobilised a military response, raising international tensions in the region. The FARC has had a serious impact on oil production, delaying production from the remote Llanos Basin and giving the country a somewhat anomalous production profile.

Colombia may be among the first countries to enter post-globalism, having already seen the emergence of small sustainable communities and local markets. In its case, the transition is due not so much to the decline of essential fuel supply from oil depletion, but to the impact of the international narcotics trade which gives it a particularly violent character.

In energy terms, Colombia is in fact well blessed, with many years of, albeit declining, oil supply, which could be conserved for national needs, and a good renewable energy potential from hydroelectric and solar power. It is also endowed with substantial coal deposits. With a change of government, it may well align itself with Venezuela and Bolivia, which are moving to free Latin America from foreign commercial and financial exploitation. One useful step might be to legalise the production of cocaine to free the country from the influence of the drug trade which has so damaged the national life, leaving the foreign importers to police their own addicts at home.

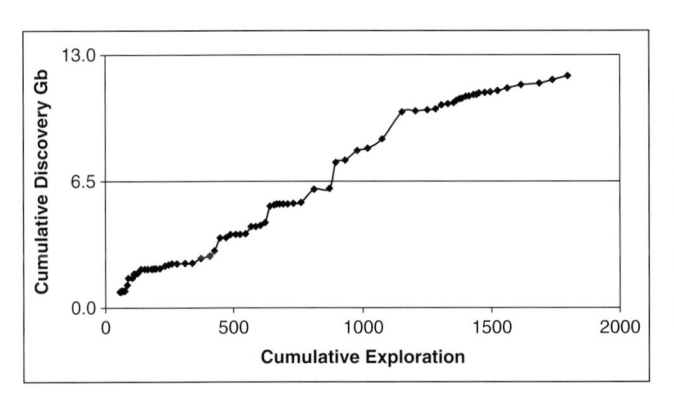

Fig. 52.3 Colombia discovery trend

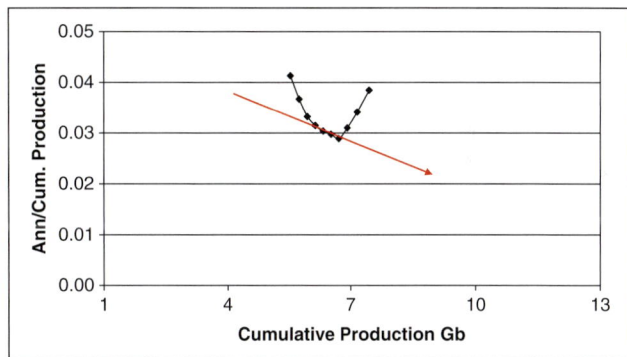

Fig. 52.4 Colombia derivative logistic

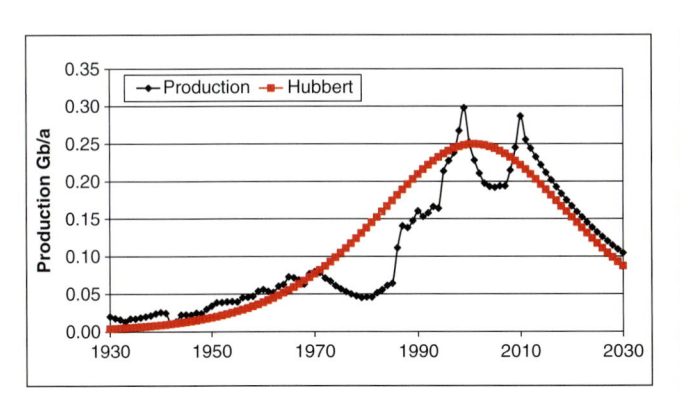

Fig. 52.5 Colombia production: actual and theoretical

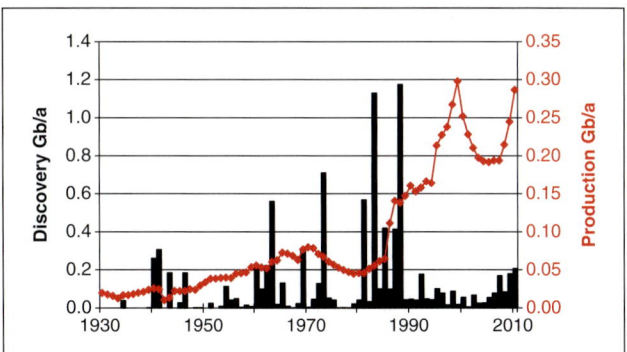

Fig. 52.6 Colombia discovery and production

Ecuador

Table 53.1 Ecuador regional totals (data through 2010)

Production to 2100						Peak Dates			Area	
Amount			Rate				Oil	Gas	'000 km²	
	Gb	Tcf	Date	Mb/a	Gcf/a	Discovery	1969	1969	Onshore	Offshore
PAST	4.7	1.0	2000	144	40	Production	2006	2013	280	50
FUTURE	3.3	1.0	2005	194	44	Exploration	1972		Population	
Known	3.0	0.8	2010	177	50	Consumption	Mb/a	Gcf/a	1900	1.2
Yet-to-Find	0.3	0.1	2020	105	39	2010	73	12	2010	15
DISCOVERED	7.7	1.9	2030	62	18		b/a	kcf/a	Growth	10.8
TOTAL	8.0	2.0	Trade	+104	+38	Per capita	5	1	Density	54

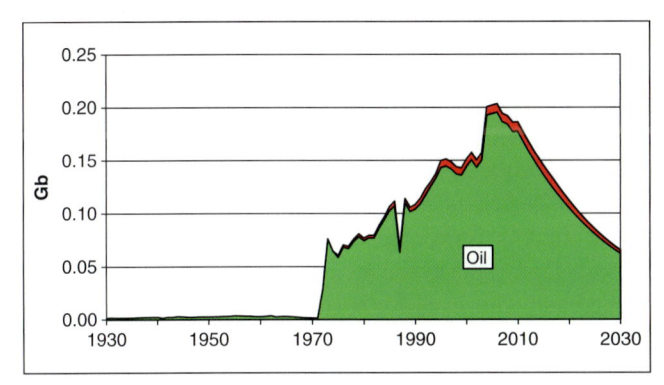

Fig. 53.1 Ecuador oil and gas production 1930 to 2030

Fig. 53.2 Ecuador status of oil depletion

Essential Features

Ecuador is the second smallest country in Latin America, covering some 280,000 km² in the northwest of the continent, between Colombia and Peru. It comprises three very different terrains. In the west, lie partly barren coastal lowlands, supporting extensive banana plantations and the busy port of Guayaquil, at the head of a gulf with the same name. Next follow the impressive Andean ranges, capped by active volcanoes rising to over 6,300 m, with the capital Quito, lying in a verdant intermontane valley. In the east, is the so-called Oriente, a vast tract of tropical rain forest, drained by the headwaters of the Amazon. Also of mention are the Galapagos Islands in the Pacific, whose exceptional fauna was made famous by the pioneering studies of Charles Darwin in 1836.

The country supports a population of some 15 million, most living in the vicinity of Guayaquil on the coastal lowlands and in Quito, the capital. The vast Amazon headwaters are sparsely populated, partly by some surviving Indian tribal groups. In racial terms, about one quarter of the population is of European extraction: one third is native Indian, some still speaking the Quechua tongue, and the balance is a mixture, including a small percentage with African ancestry.

Geology and Prime Petroleum Systems

The Andes of Ecuador, which is made up mainly of metamorphic and granitic rocks, capped by impressive volcanic peaks, are flanked by sedimentary basins. To the west, lies a Tertiary sequence with a high volcanic content and an absence of hydrocarbon source-rock, save at the southern end around the Peninsula of Ancon. That lies on the flanks of the proto-delta of the Amazon River, which flowed into the Gulf of Guayaquil, prior to the uplift of the Andes in late Tertiary times. Here, lower Tertiary source-rocks have fed overlying reservoirs in a modest oilfield. By contrast, the Oriente province to the east is blessed with rich Middle Cretaceous (Turonian) source-rocks. They have charged both over- and under-lying Cretaceous sandstones, as well as, locally, overlying Tertiary reservoirs, with oil and gas. This has given rise to a string of major oilfields, located primarily just in front of the foothills. Towards the south, the main source-rock trend swings westward to be consumed in the Andes, which explains the disappointing results of exploration in southern Ecuador and eastern Peru. Heavy degraded oil has also been found at shallow depth on the eastern margin of the basin.

Exploration and Discovery

The discovery of oil in the coastal region of Peru in 1869 prompted early investigation of the adjoining territory of Ecuador by Anglo-Ecuadorian Oilfields, which later become a subsidiary of Burmah Oil, before being merged into BP. The first borehole was drilled in 1911, leading to the discovery of the Ancon Field, seven years later.

The geological potential of the vast Sub-Andean basin to the east of the mountains was recognised from the earliest surveys. But its remote location was a serious deterrent to exploration, although Shell did mount a heroic campaign in the late 1930s and 1940s, drilling a number of anticlines in the foothills without success. The discovery by Texaco in 1962 of the Orito Field in neighbouring Colombia opened a new chapter, leading to the discovery, by the same company, of a series of giant fields across the border in Ecuador. These fields were located in gentle structures identified by seismic surveys in front of the foothills. Oil discovery, with associated gas, reached a peak in 1969.

Production and Consumption

Oil production from the Ancon Field on the coast commenced in 1918, and grew to a peak of 10 kb/d in 1962 before declining. The construction of a trans-Andean pipeline from the Amazon headwaters to the Pacific allowed production from the Oriente to start in 1973 at 200–300 kb/d, rising with new capacity to around 500 kb/d in 2004. Production started to decline from natural depletion in 2008 and is expected to continue to do so at about 5% a year. A deposit of heavy oil has also been found on the eastern margin of the Oriente Basin, but production is delayed both by its remote location and environmental concerns. The country consumes about 73 Mb/a meaning that it can remain an important exporter for a few more years. Limited amounts of associated gas can also be produced for a long time to come, being dedicated to electricity generation.

The Oil Age in Perspective

The early history of the country, when it was occupied by various Amer-Indian tribes, is little known, but in the fifteenth century it was invaded by the Inca Empire from the south under Atahualpa, giving it some national status. In 1534, it fell to Benalcazar, a lieutenant of Pizarro, the Spanish Conquistador. Quito, the capital, was founded, retaining to this day some fine colonial architecture. The Indians of the Andes accepted Spanish rule, but conflict continued in the rest of the country.

Moves to independence followed during the early nineteenth century, leading to the country's liberation by Simon Bolivar's forces at the Battle of Pichincha in 1822, which was fought at an altitude of 3,300 m on the outskirts of Quito. At first, it formed part of a union with Colombia and Venezuela, but seceded in 1830.

The subsequent history has been fairly typical of Latin America. Wealth has been concentrated within a few families, and weak democracies have alternated with military governments. In this connection, mention may be made of the Velasco Ibarra phenomenon. He was a man of aristocratic background who lived from 1893 to 1979. He was elected President in 1933 on a somewhat socialist platform, having travelled the country on mule-back, being the first politician to try to win the support of remote rural communities. His idealism was not however matched by a practical ability to govern. His rule accordingly tended to deteriorate to a point at which the military was reluctantly forced to intervene, sending him into exile. Calls for a return to democracy would then lead to new elections and the return of Ibarra to power, with the process being repeated five times over the next 40 years.

In terms of foreign policy, Ecuador has fought a losing battle against stronger neighbours for control of the sparsely populated and ill-defined Amazon headwaters. The last adjustment was made in 1942, when it was forced to cede extensive territory to Peru under US pressure.

The country has faced various financial crises in recent years. When attempts to peg its currency (the *sucre*) to the dollar failed in 2000, the country formally adopted the dollar as its currency, which actually conveys benefit to the US economy.

It is also being adversely affected by incursions from various Colombian dissident groups, some connected with the global trade in narcotics, as described in the previous section on Colombia.

53 Ecuador

A State company, CEPE, was formed in 1972, later to be reconstituted as PetroEcuador, which acquired the rights of Texaco in 1990 under a form of nationalisation. Ecuador joined OPEC for a few years, but then resigned before rejoining, as it comes to realise the growing importance of its oil exports.

An election in 2006 returned Rafael Correa, who describes himself as a Christian leftist, having good relations with Venezuela's Ugo Chavez. He has been in conflict with the international oil companies operating in the country, with threats of nationalisation. He seems likely to seek a more independent role for the country, endeavouring to resolve its financial difficulties and to free it from the machinations of the World Bank and International Monetary Fund. He was re-elected in 2009, but in 2010 faced a major strike by the Police Force complaining about a loss of privileges. The strike was eventually put down by the army after road blocks and serious disturbances in which eight people lost their lives and 247 were injured.

Although the population has exploded over the past century, Ecuador remains fairly well placed to face the post-peak world. It could move to restrict oil exports to conserve its resources for it own use, relying on the banana crop for foreign exchange.

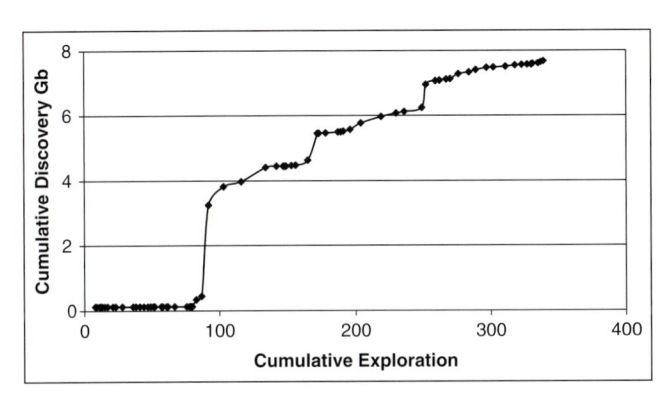

Fig. 53.3 Ecuador discovery trend

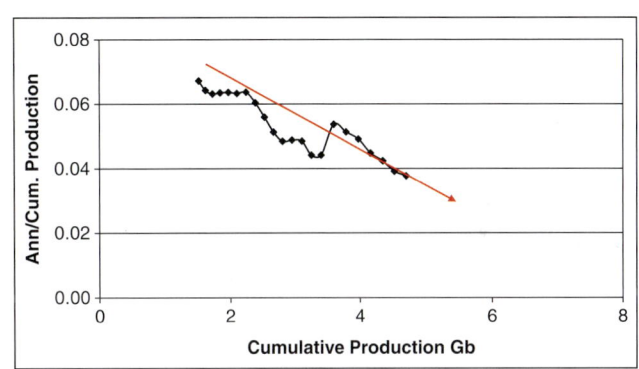

Fig. 53.4 Ecuador derivative logistic

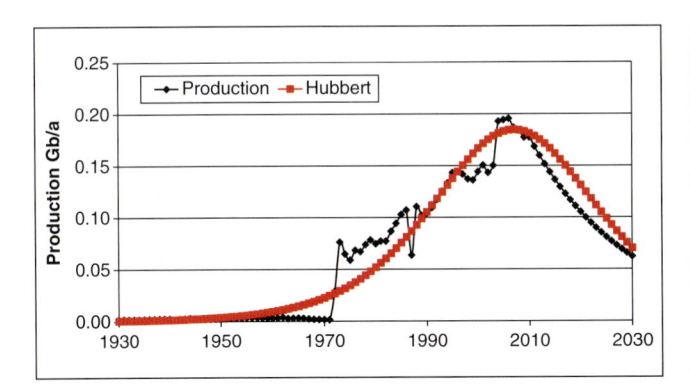

Fig. 53.5 Ecuador production: actual and theoretical

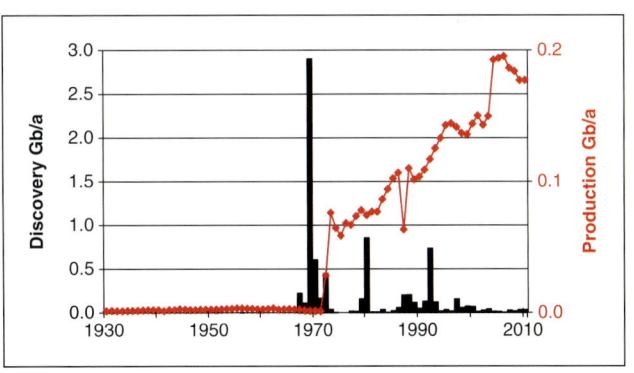

Fig. 53.6 Ecuador discovery and production

Mexico

Table 54.1 Mexico regional totals (data through 2010)

Production to 2100					Peak Dates				Area	
Amount		Rate				Oil	Gas	'000 km²	
	Gb	Tcf	Date	Mb/a	Gcf/a	Discovery	1977	1977	Onshore	Offshore
PAST	39	58	2000	1,099	1,511	Production	2004	2008	1,970	820
FUTURE	13	42	2005	1,217	1,583	Exploration	2003		Population	
Known	12	38	2010	940	1,722	Consumption	Mb/a	Gcf/a	1900	14
Yet-to-Find	1.3	4.2	2020	469	1,443	2010	757	2,135	2010	115
DISCOVERED	51	96	2030	234	836		b/a	kcf/a	Growth	7.6
TOTAL	52	100	Trade	+184	−413	Per capita	6.6	19	Density	58

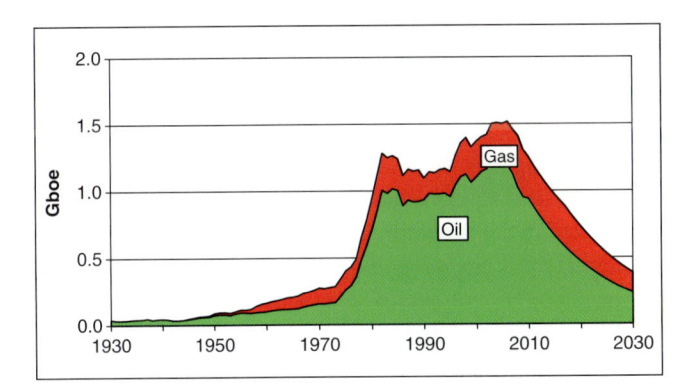

Fig. 54.1 Mexico oil and gas production 1930 to 2030

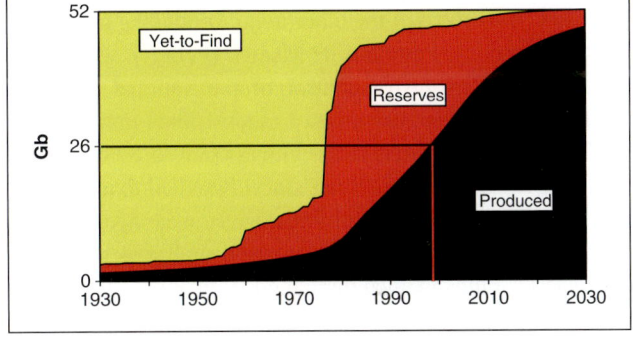

Fig. 54.2 Mexico status of oil depletion

Essential Features

Mexico covers an area of almost 2 million km², being bordered by the United States to the north and Guatemala and Belize to the south. A central plateau, between 1,000 and 2,000 m above sea-level, is flanked by two branches of the Sierra Madre mountain chain, which is capped by volcanic peaks rising to almost 6,000 m above sea-level. Baja California forms a long peninsula on the Pacific margin in the northwest of the country, while the Yucatan Peninsula in the southeast flanks the Gulf of Mexico, being made up, in part, of inhospitable limestone karst country. Much of Mexico is arid, except for the coastal areas and the highlands of the south. The population has risen sevenfold since 1900 to 115 million, of whom about 21 million live in Mexico City, one of the World's largest and most polluted capitals. An additional seven million Mexicans live in the United States, some illegally.

Geology and Petroleum Systems

The southern extremity of the Rocky Mountain chain of North America forms the western margin of Mexico, being made up mainly of highly deformed and partly metamorphosed Palaeozoic rocks, covered by extensive lava sheets. It gives way eastward to strongly folded Mesozoic sediments, including thick carbonates, before passing into a Tertiary basin flanking the Gulf of Mexico. The Yucatan Peninsula in the southeast forms a Palaeozoic Massif capped by flat-lying Miocene carbonates. An important dextral shear zone, associated with volcanic activity, cuts across the country at the latitude of Mexico City, continuing offshore along the northern margin of the Yucatan. Mid-Cretaceous source rocks have charged associated carbonates with oil and gas.

Exploration and Discovery

Oil exploration took off during the early years of the twentieth century being led by Mexican Eagle, a British company, controlled by Lord Pearson, who was involved in constructing railways in the country. Several American companies were also involved. One highly prolific field after another was found in what became known as the Golden Lane, near Tampico, where fractured Cretaceous carbonate reservoirs gave exceptionally high flow rates. As a result, Mexican oil production accounted for as much as one-quarter of the world's total during the early years of the last century. But nationalism and disputes over the rightful share of the national patrimony led the country to expropriate the foreign oil interests in 1938, so setting an example which was later to be followed by most of the other major producing countries. A State company, Petroleos de Mexico (Pemex), was established to take exclusive control of the national industry. Exploration was maintained at a modest level until around 2000 when it increased rapidly, with as many as 87 boreholes being drilled in 2003, but it has since dwindled to about 20.

The offshore extensions of the province were opened up in the 1960s and 1970s, contributing to the overall peak of discovery in 1977. Finds include the famous Cantarell Field, which was found on the strength of reports of seepage by fishermen. It was a prolific producer, holding almost 20 Gb of reserves, whose productive life was extended by a massive nitrogen injection programme, but now heads steeply into decline.

The reporting of oil reserve has fluctuated widely, and is highly suspect. In 1980, reserves were stated to be 44 Gb, rising to 56 Gb in 1989 before declining to 48 Gb in 1998, and then they collapsing to 27 Gb in 2001 and 10.4 Gb in 2010. It is possible that the higher earlier numbers were used as collateral when US banks tried to rescue the *peso,* and may have been related to the treatment of the large but sub-commercial Chicontopec Field. In addition, moves for and against the privatisation of Pemex, and the re-entry of US companies under the NAFTA provisions, may be partly responsible.

Many of the oilfields have associated gas, of which some 96 Tcf have been discovered.

Mexico probably has some deepwater potential in the Gulf of Mexico, here tentatively estimated at 5 Gb, which, if confirmed, could deliver a peak around 2015. It will be addressed in Chapter 12.

Production and Consumption

Oil production commenced in 1901 and grew to an early peak of 530 kb/d in 1921, which was not overtaken until 1975 with the opening of the offshore. It then rose to an overall peak of almost 3.4 Mb/d in 2004, but has since fallen to 2.6 Mb/d, being now set to decline at about of 6.7 % a year. Mexico has been a major oil exporter, principally to the United States, but with consumption standing at 757 Mb/a, that cannot continue much longer.

Gas production stands at almost 1.7 Tcf/a, and may continue at this level until 2017, by which time 70% will have been depleted, before terminal decline at about 5% a year sets in. Consumption is running at 2.1 Tcf a year.

The Oil Age in Perspective

Mexico has been inhabited for 20,000 years, attaining a high level of civilisation towards the end of the first millennium when the Aztecs rose to prominence before falling to the Spanish conquistadores under Cortez in the 1520s. Along with Colombia and Peru, Mexico became one of the administrative centres of Spain's Latin American Empire, its territory extending over what is now the southern United States.

Various inconclusive moves toward independence finally culminated in the declaration of a Republic in 1824. But US immigrants had begun to settle in the northern territories in large numbers, prompting a conflict, which led to the US-Mexican War of 1846–1848. Mexico lost, being forced to give up Texas, California, New Mexico, Arizona, Nevada, Utah and much of Colorado.

As in other Latin American countries, independence brought instability as different factions vied for power. One such faction secured the support of Napoleon III of France, who in 1862 landed an army, establishing Maximilian, an Austrian grandee, as Mexico's new Emperor. He saw himself as a benign despot, but did not quite receive the welcome he expected. The French forces withdrew five years later under pressure from the United States, and the unfortunate Maximilian was executed by a rival leader, General Juarez.

Porfirio Diaz came to power in 1876 following another rebellion. He led an efficient, if autocratic, government that survived until 1911, and is remembered for his famous dictum: *Poor Mexico—so far from God yet so close to the United States.* Growing demands for social reform and a fairer distribution of wealth led to the formation of the Partido Nacional Revolutionario (PNR), which under various leaders has dominated Mexican politics for many of the ensuing years. It can probably be described as a national socialist party, springing from the same pressures as manifested themselves in pre-war Europe. In 2000, it lost the Presidency to Vicente Fox of the National Action Party (PAN). His successor, Felipe Calderon, of the same party, who is a devout Catholic, was in turn almost defeated in 2006 by Sr. Obrador, the former Mayor of Mexico City, representing a left-leaning party with a base in the southern province of Chiapas where inhabitants of Mayan origin seek to recover their lands.

Earlier, in 1992, Mexico entered into a free-trade Treaty (NAFTA) with the United States and Canada, which was supposed to stimulate trade, but in reality led to the greater foreign control of Mexican industry and the outflow of capital through the hidden influence of a foreign trading currency.

Emigration to the United States, both legal and illegal, continues on a massive scale, with remittances to the home country being a major source of income. The country also suffers heavily from the narcotics trade, bringing violence to many of its cities.

The future situation looks grave indeed. The country's oil supply, which previously provided substantial revenues to the Government, is falling steeply, such that the country will cease to have a surplus for export within five years. Even more critical yet would be the impact of a major recession in the United States which might prompt its large Mexican population to head for home exacerbating the tensions and levels of unemployment. Mexico may well prove to be a battleground between the new nationalism of South America, as represented by Ugo Chavez of Venezuela, and advocates of globalization in the north.

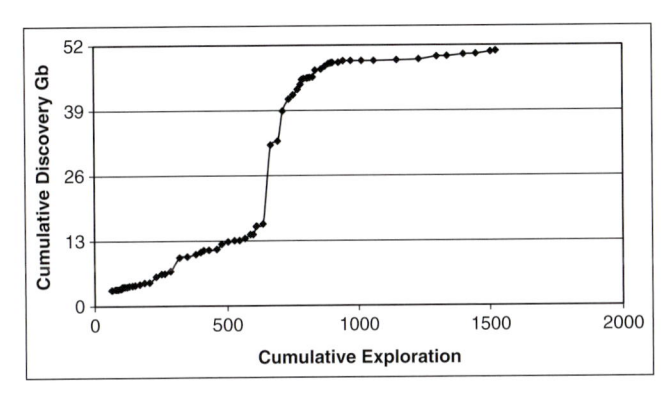

Fig. 54.3 Mexico discovery trend

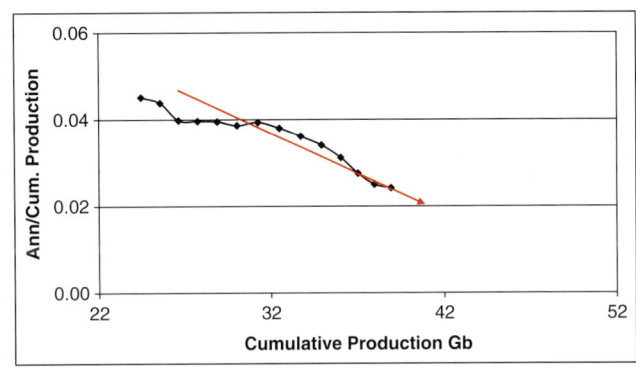

Fig. 54.4 Mexico derivative logistic

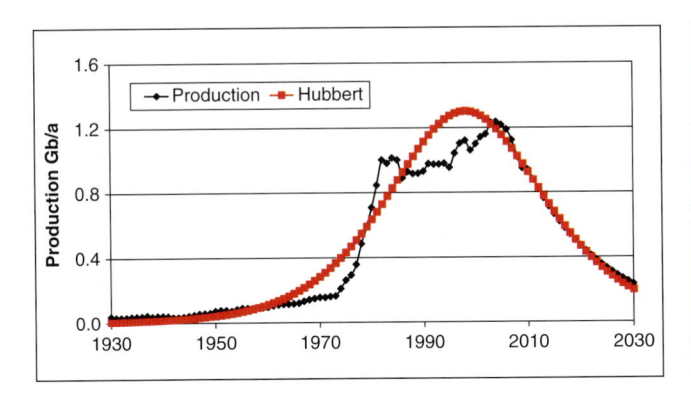

Fig. 54.5 Mexico production: actual and theoretical

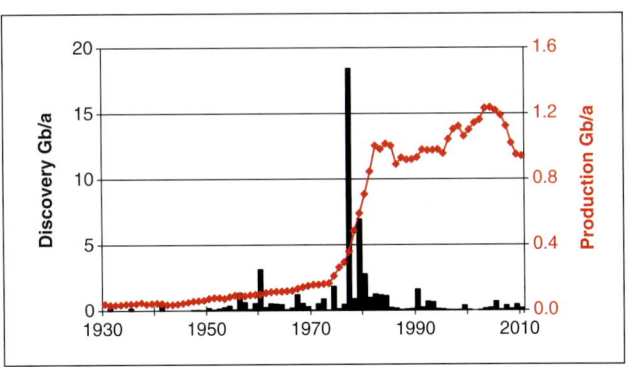

Fig. 54.6 Mexico discovery and production

Peru

Table 55.1 Peru regional totals (data through 2010)

	Production to 2100					Peak Dates			Area	
	Amount		Rate				Oil	Gas	'000 km²	
	Gb	Tcf	Date	Mb/a	Gcf/a	Discovery	1869	1986	Onshore	Offshore
PAST	2.6	4.9	2000	35	29	Production	1982	2025	1,290	100
FUTURE	1.4	20	2005	28	196	Exploration	1975		*Population*	
Known	1.1	16	2010	27	393	*Consumption*	Mb/a	Gcf/a	1900	5
Yet-to-Find	0.3	4.0	2020	22	400	2010	69	191	2010	29
DISCOVERED	3.7	21	2030	18	400		b/a	kcf/a	Growth	5.6
TOTAL	4.0	25	Trade	−42	+202	Per capita	2.3	6.6	Density	23

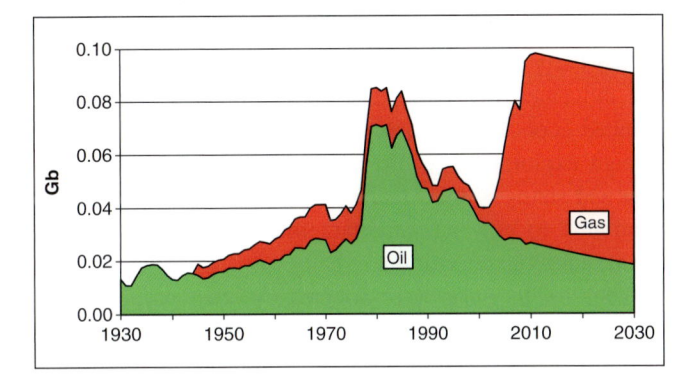

Fig. 55.1 Peru oil and gas production 1930 to 2030

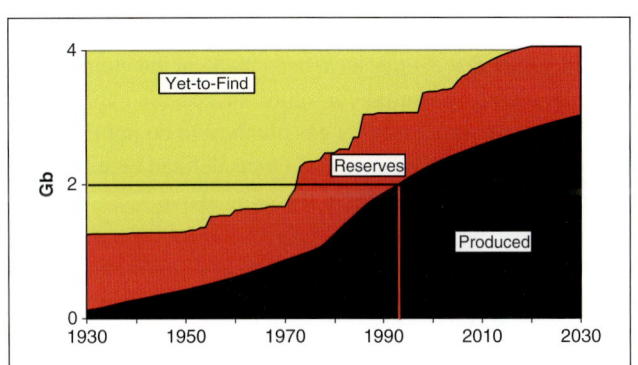

Fig. 55.2 Peru status of oil depletion

Essential Features

Peru straddles the great Andean mountain range, whose snow-capped peaks rise to over 5,000 m. It is flanked to the west by a coastal littoral of semi-desert, caused by the cooling waters of the Humboldt Current; and to the east by a huge area of tropical forest in the Amazon headwaters. The population has exploded over the past century by a factor of 5.6, with Lima, the capital, now supporting, if that is the word, some eight million inhabitants.

Geology and Prime Petroleum Systems

The Andes make an elbow-bend, known as the Huancabamba Deflection, in the north of the country, which is due to a major sinistral shear-zone separating the Guyana Shield to the north from the Brasilian Shield to the south. The geological evolution of the two segments differed from each other, affecting their respective petroleum systems. There are in fact three distinct systems to consider. The first comprises the northwest littoral, where oil was generated in the proto-delta of a predecessor of the Amazon River that flowed westwards into the Pacific prior to the uplift of the Andes in the late Tertiary. The second lies in the Andean foothills to the east and the adjoining present Amazon headwaters, but it is much leaner than the sub-Andean zone of Ecuador and Colombia to the north, because it lays outside the belt of prime Middle Cretaceous generation. A third, gas-prone, system in the southeast part of the country, including the Madre de Dios Basin, depends on Palaeozoic sources within the frontal ranges of the Andes themselves. Recent reports of a similar development offshore await confirmation, here being doubted on the basis of regional trends.

Exploration and Discovery

Tar seepages on the northwest coast have been known for centuries, and prompted early exploration efforts in the nineteenth century, backed by British capital through the Lobitos Company (later to be acquired by Burmah Oil which was in turn merged into BP). The first borehole was sunk in 1863, being duly rewarded six years later by the world's first giant field, La Brea-Parinas, with 1.14 Gb of oil, to which subsequent extensions of like amount were added during the early years of the twentieth century. Some offshore extensions were brought in after 1955.

Attention turned to the Amazon headwaters in the 1950s, yielding a small discovery in 1957, known as the Maquia Field, followed by other modest finds in the 1970s.

A third petroleum system in the southeast Andean ranges delivered a series of major gas-condensate discoveries, starting with Aquaytia in 1962, followed by Camisea, in 1984, with as much as 13.6 Tcf of gas and 725 Mb of Natural Gas Liquid. Other discoveries have been made in the vicinity, and there is clearly potential for more successful exploration. The liquids are processed by gas plants, and do not therefore qualify as *Regular Conventional Oil* as defined herein, being excluded from the statistics in the table above.

Production and Consumption

Oil production commenced in 1879 and rose slowly to pass 10 kb/d by 1921 and 50 kb/d by 1955. It reached an overall peak in 1982 of 195 kb/d, before declining to its present level of 73 kb/d. It is set to continue to decline at about 2% a year, which is a modest decline rate, reflecting remote onshore tail-end operations in difficult terrain.

Recorded gas production commenced in 1945, rising to an early peak of 81 Gcf/a in 1985. It then declined to 29 Gcf/a in 2000 before increasing to 393 Gcf/a in 2010 following expansion of the Camisea pipeline.

The Oil Age in Perspective

Archaeological research shows that human settlement began more than 13,000 years ago. Later came the great Inca Empire, which in the sixteenth century held dominion over a vast territory extending from Ecuador in the north to Argentina in the south. It however fell to the Spaniards under Pizzaro, who landed with no more than 180 men in 1531. The development of the Potosi silver mines, a few years later, made Peru the jewel of the Spanish Empire, and Lima became a wealthy and sophisticated seat of imperial government.

Moves to independence in Latin America developed during the nineteenth century, stimulated in part by the fall of Spain in the Napoleonic Wars and the declaration of independence in the United States, but Peru remained relatively loyal with its aristocratic pretensions and large expatriate Spanish community. Even so, the country eventually fell to General San Martin of Argentina, who had marched north, eyeing the silver mines, to declare independence in 1821. He in turn was replaced by Simon Bolivar, who came south three years later after liberating Venezuela, Colombia and Ecuador.

In 1879, Peru lost a resource war with Chile over control of nitrate deposits in the Atacamas Desert. Europe, in the days before synthetic fertilisers, depended on these critical supplies to feed its growing population. In 1889, the London-based, *Peruvian Corporation* exploited a financial crisis in the country to secure control of the railways and the right to mine three million tons of *guano*. British interests also took a stake in Peru's early oilfields.

Haya de la Torre, a Peruvian exile in Mexico, founded the APRA Party in 1924 as a Latin American parallel to the fascist movements of Europe. It stood for an anti-capitalist, planned economy to improve the lot of the poor, and end foreign exploitation, while aiming to re-establish the heritage and status of the American Indian. The movement has continued to play an important role in Peruvian politics, albeit mainly in opposition.

Some of its ideas were adopted by other parties in the eternal quest to resolve the great social disparities between an increasingly urban Indian poor on one side, and the land-owners, capitalists and foreign investment on the other. Military governments have intervened twice in recent years, in 1948–1956, and again in 1968–1980, as democratic government failed, having sponsored open market policies that often led to burgeoning foreign debt and inflation. A neo-Marxist guerrilla movement, the *Sandero Luminoso* (Shining Path), was a later manifestation of the same conflict but has now fallen from prominence, no doubt to be replaced in due course by other such elements. In 1985, APRA succeeded in winning the Presidency under Alan Garcia, who suffered at the hands of the international financial community, but returned to power in the 2006 election.

In the meantime, a somewhat curious development occurred in 1990 when Alberto Fujimori, a Peruvian of Japanese ancestry, came to power imposing an authoritarian government that lasted for ten years and delivered economic growth and stability. He was later accused of corruption and human rights offences, including forced sterilisation, before going into exile in Chile, where he is subject to extradition proceedings. His daughter contested an election in 2011 but was defeated by Olianta Humala, who had been an officer in the Army and founded the Peruvian Nationalist Party, which seeks to better represent the ethnic Amer-Indian population.

55 Peru

Peru has had a long oil history, being credited with the World's oldest giant oilfield, La Brea—Parinas on the northern littoral, which is attributed to 1869, although Pizarro had caulked his ships with tar from seepages in the area as early as 1528. The original title to the property, which included the mineral rights, was granted by no less than Simon Bolivar himself. Various British companies were involved in the early years, but an Esso (Exxon) subsidiary took a dominant position. Its absolute rights became increasingly anomalous after the Second World War in an environment of increasing socialism, and were eventually expropriated in an action leading to the creation of a State oil company, Petroperu. A second lease of life for this mature province came with the development of offshore extensions by Belco in 1959, but no prolific finds were made.

The development of the giant Camisea Fields is of great importance to the country. It was discovered by Shell, but the local market was not perceived profitable enough to justify a costly pipeline, and the rights lapsed before development could be put in hand. The Government however has now secured the finance to build the processing plant and pipeline to the coast which is now in operation.

In summary, Peru is a mature oil country, well past peak, but has discovered a late-stage gas-condensate play, which will surely expand further. The country faces the challenge of attracting foreign companies to conduct high risk exploration, without losing access to any surprise finds that might be made. A chequer-board licensing policy might succeed in allowing Petroperu to watch from the sidelines and take offsetting acreage where justified.

The country has become a net oil importer on a trend that is set to rise, although gas-liquids from Camisea will become an increasingly important source of fuel. Looking ahead, Peru could be regarded as not well placed to face the Second Half of the Oil Age, primarily due to population pressures, which have driven Indians from the barren highlands to try to make a living in the cities, especially Lima. It does nevertheless have the huge expanses of the upper Amazon which have not so far attracted significant settlement.

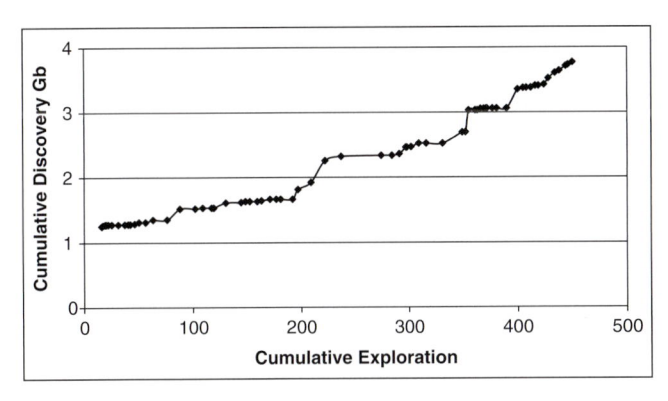

Fig. 55.3 Peru discovery trend

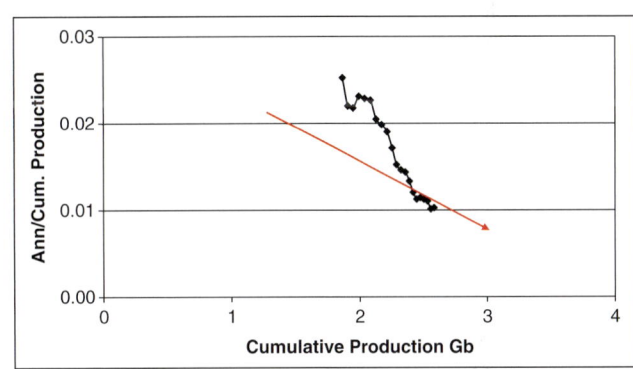

Fig. 55.4 Peru derivative logistic

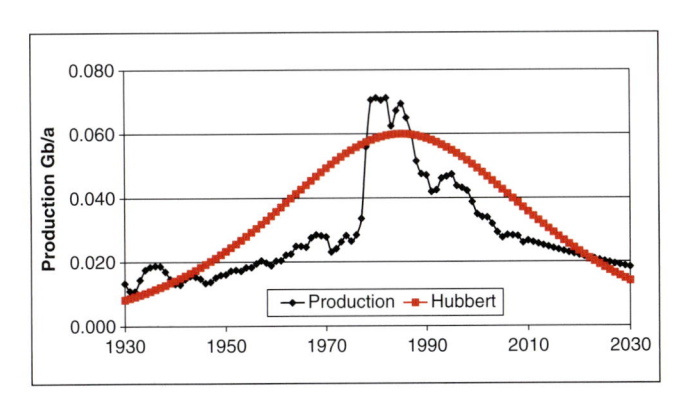

Fig. 55.5 Peru production: actual and theoretical

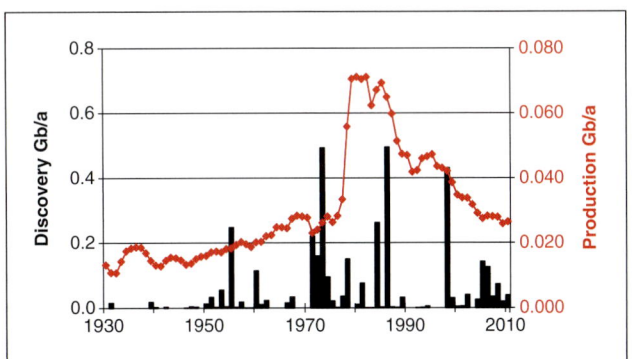

Fig. 55.6 Peru discovery and production

Trinidad

Table 56.1 Trinidad regional totals (data through 2010)

Production to 2100						Peak Dates			Area	
Amount			Rate				Oil	Gas	'000 km²	
	Gb	Tcf	Date	Mb/a	Gcf/a	Discovery	1959	1968	Onshore	Offshore
PAST	3.5	22	2000	45	577	Production	1981	2010	5	12.5
FUTURE	1.0	28	2005	53	1175	Exploration	1972		*Population*	
Known	0.9	24	2010	36	1576	*Consumption*	Mb/a	Gcf/a	1900	0.3
Yet-to-Find	0.1	4.2	2020	25	1358	2010	15	780	2010	1.3
DISCOVERED	4.45	46	2030	18	502		b/a	kcf/a	Growth	4.0
TOTAL	4.5	50	*Trade*	+21	+796	Per capita	11.5	600	Density	254

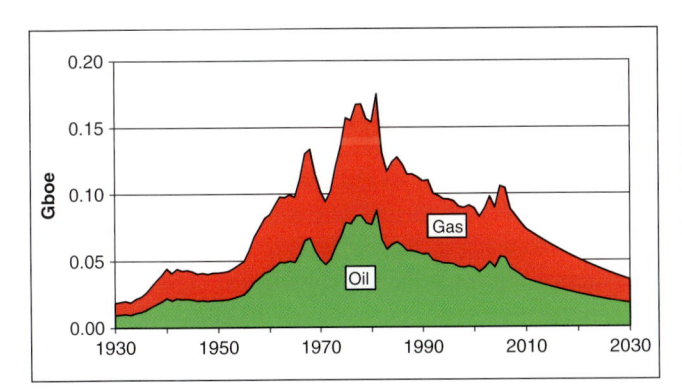

Fig. 56.1 Trinidad oil and gas production 1930 to 2030

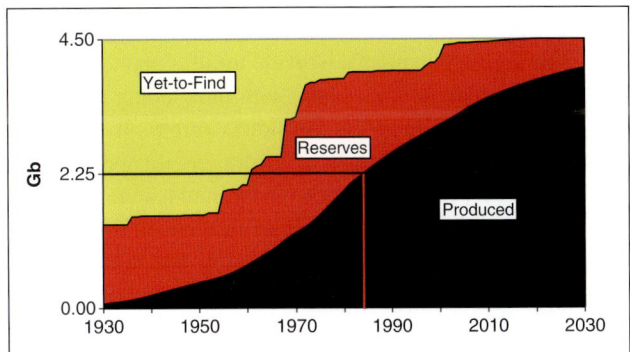

Fig. 54.2 Trinidad status of oil depletion

Essential Features

Trinidad is one of the two islands making up the Republic of Trinidad and Tobago, off the east coast of Venezuela. It covers an area of 5,000 km² and is cut by the Northern and Central Ranges, rising to respectively 900 and 300 m, while in the south are a series of low hills, termed the Southern Ranges. The intervening lowlands are partly swampy, giving the Caroni and Ortoire Swamps. The natural vegetation is tropical rain forest, but most of the lowlands are under cultivation, with sugarcane being a substantial cash crop. The island supports a population of one million people, which has increased by a factor of four over the last century, making it a crowded place with a density of 252/km².

The bulk of the population is of African and Indian origin, being the descendants of slaves and indentured labour brought in to work the sugar estates after the abolition of slavery. The remaining 20% are of mixed Spanish, French, Portuguese, English, Chinese and Amer-Indian descent.

Geology and Prime Petroleum Systems

In geological terms, Trinidad forms an extension of the oil-rich East Venezuelan Basin, being in turn flanked by a continental shelf that extends southwards along the margin of South America. The Northern Range is a direct extension of the Andean Coast Range of Venezuela, being composed of low-grade Cretaceous and Jurassic metamorphic rocks.

To the north of it lie structures associated with the Antillean Island arc. A major transcurrent fault, known as the El Pilar Fault, marks the southern boundary of the Northern Range, before extending offshore into the Atlantic. Another wrench fault, the Los Bajos Fault, cuts obliquely across the southern part of the island, being partly responsible for oil accumulations.

The main oil-bearing region lies to the south of the Central Range including offshore extensions both eastward into the Atlantic and westward into the Gulf of Paria, which separates Trinidad from Venezuela. The principal source-rock in the Caribbean region was a deposit of organic-rich clay laid down under conditions of global warming about 90 million years ago in the mid-Cretaceous. It crops out in parts of the Central Range of Trinidad, and is probably an important source of oil in the basin to the south. In addition are other sources for both oil and gas in the thick sequence of Tertiary sediments that overlie it, especially in the Atlantic offshore, which constitutes the palaeo-delta of the Orinoco River. The depth of the source-rock helps determine whether the fields yield oil or gas.

The geology of the southern basin is exceedingly complex. The Tertiary sequence is made up of a great thickness of deformed, monotonous clays, with intervening and overlying sands, some being of turbiditic origin. It was also affected by slumping as rafts of sediment slipped into the subsiding basins. Another somewhat unusual feature are mud-volcanoes, forming mounds of mud brought to the surface by gas seepages that occasionally catch fire. The famous Pitch Lake is a huge natural seepage of degraded oil.

Exploration and Discovery

Trinidad has had a long oil history, starting in 1866 when two shallow boreholes were drilled, followed by efforts to distil tar from the Pitch Lake. The first truly commercial well was drilled in 1907 which was followed by the discovery of the Forest Reserve Field in 1914, with some 320 Mb. That prompted a more intensive search by several British companies, which was rewarded with a number of small to modest finds. Shell, BP and Trinidad Leaseholds, itself being sold to Texaco in 1956, became the dominant onshore operators. The latter ran a major refinery at Pointe-a-Pierre, which, for a period, refined Middle East, in addition to local, oil for export to the United States and European markets. The refinery was in fact important for the United Kingdom during Second World War, supplying much of the high-octane aviation fuel for the Air Force.

A second cycle of exploration opened offshore in the 1950s leading to the discovery in 1956 of the giant Soldado Field with 600 Mb in the shallow, calm waters of the Gulf of Paria, off the delta of the Orinoco River. It was in turn followed by a third cycle in 1961 when Amoco (now BP) secured rights to the offshore Columbus Basin in the Atlantic, bringing in a series of major oil and gas finds, starting in 1967. A somewhat surprise fourth cycle may be opening with the discovery of about 2.4 Tcf of gas-condensate in an entirely new province to the northwest of the island near the Venezuelan median line. It remains to be seen if any deepwater potential will be identified after the failure of the six deepwater boreholes drilled so far. A State company, Petrotrin, has taken an active role in exploration and refining in recent years, and British Gas and BHP are relative latecomers with a strong position in gas.

A total of 363 exploration boreholes have been drilled making it a very mature area.

Production and Consumption

Onshore oil production commenced in 1909 and grew steadily to an early peak in 1968 at 145 kb/d, since when it has declined to about 20 kb/d. The production of offshore resources commenced in 1955 to peak in 1977 before declining. Overall production peaked in 1981 at 240 kb/d, since when it has declined to 98 kb/d and is set to continue to decline at about 3.5% a year.

Gas production also started early in parallel with oil, but surged upwards with the opening of the offshore, especially after 2000 to reach its present level of 1.5 Tcf/a, which will likely now stabilize at the present pipeline capacity. Oil and gas consumption stand at respectively 15 Mb/a and 780 Gcf/a making the country an exporter, especially of gas, with as much as 796 Gcf/a being sold, especially to the United States. Trinidad operates important NGL plants producing as much as 43 kb/d. Plans for a gas pipeline to Miami, supplying the other West Indian islands along the way, have been under consideration. Present levels of gas production can probably be held until around 2020 before terminal decline sets in.

The Oil Age in Perspective

Christopher Columbus made landfall on Trinidad on his Third Voyage in 1498, giving it its name after three prominent hills in the southeast corner of the island, which was at the time inhabited by a small number of Arawak Indians. It was later visited, in 1559, by Sir Walter Raleigh who came upon the Pitch Lake oil seepage, from which he caulked his ships. It became a Spanish territory for 300 years until the French took it in 1781, to be in turn defeated by a British naval expedition in 1797.

It remained a British colony until 1956 when a degree of autonomy was imposed prior to full independence in 1962. This brought an element of racial conflict, culminating in a failed coup by Muslim Fundamentalists in 1990. Renewed security fears were expressed in December 2002 stimulated by the Afghan and Iraq wars. The two main political parties are the People's National Movement—representing primarily those of African origins—and the United National Congress, representing those of Indian origins. Soaring oil and gas prices have brought great wealth to the country, meaning that the decline of oil in the years ahead will be severely felt. Even so, it appears fairly well placed to face at least the initial years of the Second Half of the Oil Age, having ample remaining oil and gas for domestic use, but in due course will have to revert to sustainable agriculture facing population pressures which are likely to stimulate ethnic tensions.

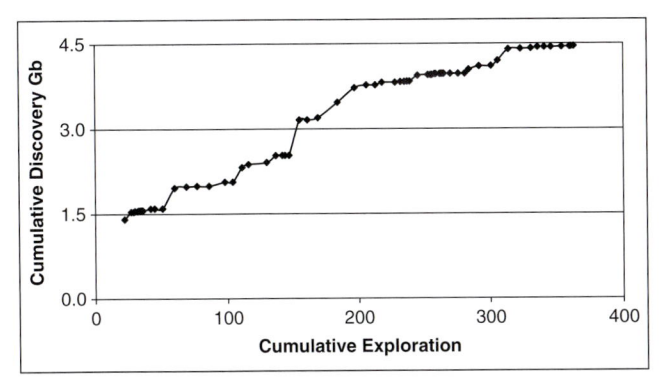

Fig. 56.3 Trinidad discovery trend

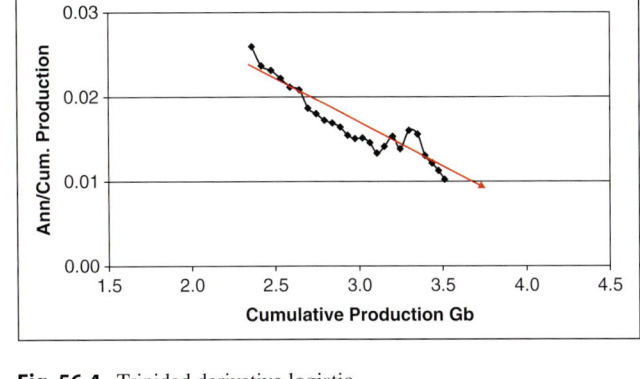

Fig. 56.4 Trinidad derivative logistic

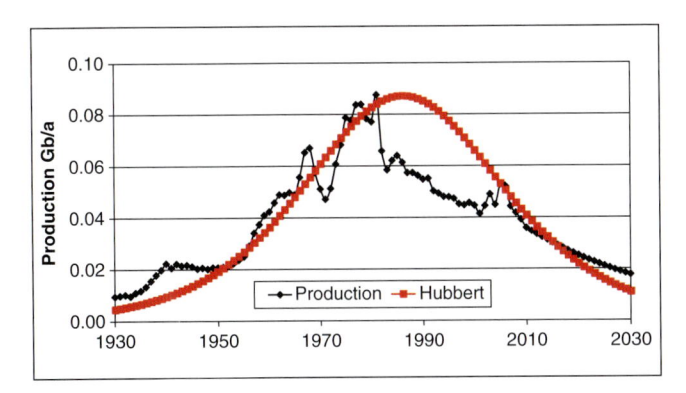

Fig. 56.5 Trinidad production: actual and theoretical

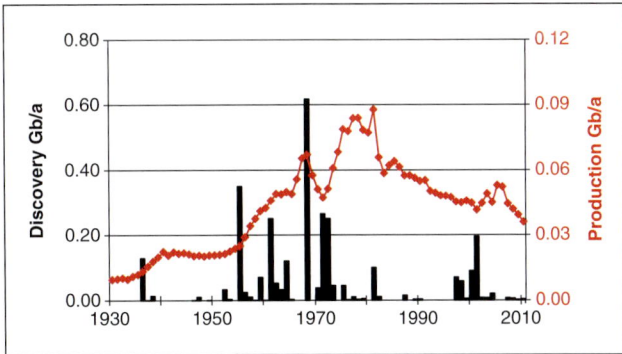

Fig. 56.6 Trinidad discovery and production

Venezuela

Table 57.1 Venezuela regional totals (data through 2010)

Production to 2100					Peak Dates			Area	
Amount		Rate				Oil	Gas	'000 km²	
	Gb	Tcf	Date	Mb/a	Gcf/a	Discovery	1914	1941	Onshore	Offshore
PAST	50	65	2000	901	2,137	Production	1970	2030	916	100
FUTURE	25	185	2005	553	2,041	Exploration	1981		Population	
Known	21	166	2010	304	2,510	*Consumption*	Mb/a	Gcf/a	1900	3
Yet-to-Find	3.7	18	2020	269	2,500	2010	272	748	2010	29.3
DISCOVERED	71	232	2030	238	2,500		b/a	kcf/a	Growth	8.9
TOTAL	75	250	*Trade*	+31	+1,762	Per capita	9.3	26	Density	32

Note: excludes Extra-Heavy oil (mainly from the Orinoco tar-belt) and Heavy Oil (<17.5° API)

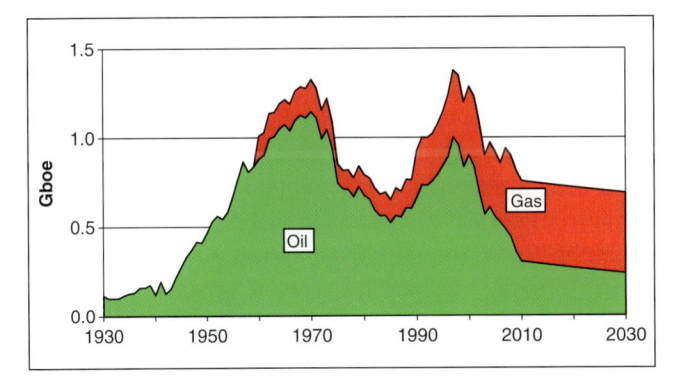

Fig. 57.1 Venezuela oil and gas production 1930 to 2030

Fig. 57.2 Venezuela status of oil depletion

Essential Features

Venezuela is a country with a diverse terrain. In the south lie the tropical rain forests of the high Roraima hinterland and the Orinoco River basin, which pass westwards into the grasslands and plains of the Llanos. Two Andean ranges, capped by Pico Bolivar, at an altitude of 5,007 m, follow to the north, before giving way to the badlands of Falcon and the deserts of Paraguana, complete with sand dunes and cactus. To the West, lies Lake Maracaibo, a large inland shallow sea, while to the East is the Orinoco delta and the Gulf of Paria, which separates Venezuela from Trinidad. The country has rich natural resources, with substantial iron-ore deposits in the interior, in addition to its ample oil endowment. It also owns a number of small Caribbean islands, including Sta. Margarita. It supports a population of 29 million, which is not an excessive number for such a rich county.

Geology and Prime Petroleum Systems

A continuation of the Central Range of the Andes of Colombia is offset by major transcurrent faults to extend eastward along the coast of Venezuela, where it is known as the Coast Range, being made up of low-grade Mesozoic metamorphic rocks. The Eastern Range of Colombia divides into two branches; one, known as the Perija Range, runs northward along the frontier while the other continues northeastwards across Venezuela to abut the Coast Range. These ranges flank the Maracaibo Basin, filled by Cretaceous and Tertiary sediments. Another large sedimentary basin, the East Venezuelan Basin, separates the Andes from the Guyana Shield to the south. The Oca and El Pilar faults are part of a major dextral transcurrent fault system cutting across the mouth of Lake Maracaibo to form the southern boundary of the Coast Range.

Both the Maracaibo and East Venezuelan Basins contain the rich Middle Cretaceous source-rocks, comprising the *La Luna Formation*, which has charged Cretaceous limestone reservoirs in Lake Maracaibo and overlying Tertiary sandstones in both basins. This source was so prolific that large amounts of oil migrated to shallow depths along the southern margin of the East Venezuelan basin where it was weathered and degraded by bacteriological action to give the well-known Orinoco tar-belt, holding perhaps 270 Gb of eventually recoverable Extra-Heavy oil, described in Chapter 12.

A third, much smaller petroleum system with gas potential, has been discovered recently in the Caribbean off the northeast coast.

Exploration and Discovery

The Pitch Lake of Trinidad attracted early interest to the oil potential of this part of the world, leading oil explorers to look across the limpid waters of the Gulf of Paria to wonder what Venezuela might offer, as it too had a pitch lake. The first well was in fact drilled in 1878 to the south of Lake Maracaibo, but it was not until 1907 that local interests secured concessions, which eventually passed into the hands of the major international oil companies. They began exploration in earnest in the years preceding the First World War. Shell was one of the pioneers, being introduced to the country by no less than the legendary Armenian oilman, Calouste Gulbenkian, founder of the Iraq Petroleum Company, who probably understood how to deal with General Gomez, the then dictator. These pioneering efforts were rewarded when a well on the shores of Lake Maracaibo blew out with a flow rate of over 100,000 b/d, having penetrated a highly fractured Cretaceous limestone reservoir. Standard of Indiana (now BP) had substantial holding before selling out to Exxon in the 1930s in exchange for a block of its stock. Gulf Oil (now Chevron) was the third principal operator. Venezuela was for many years the jewel in Shell's crown, which by 1932 had made it Britain's largest supplier.

The industry went from strength to strength both between and immediately after the two world wars, with production rising from 300 kb/d in 1930 to over three million in the late 1960s. A total of 2,300 exploration boreholes have been sunk, finding 71 Gb of oil and 232 Tcf of gas. These discoveries include an impressive list of giant fields, dominated by the Bolivar Coastal fields of Lake Maracaibo, found in 1914, which together hold some 31 Gb of oil.

It is a mature area, discovery having peaked in 1914. Exploration drilling peaked in 1981 when 122 boreholes were sunk, but has fallen steeply since, with no more than one or two a year being drilled in the recent past, due both to political constraints and the dwindling number of remaining prospects.

Production and Consumption

Oil production commenced in 1917, and grew rapidly to peak in 1970 at 3.1 Mb/d, before declining, in part due to OPEC quotas. Overall production stood at 2.2 Mb/d in 2010, of which, it is estimated, 832 kb/d are *Regular Conventional*, the balance being heavier than 17.5° API, coming mainly from the Orinoco tar-belt. It is expected to continue to decline at about 1.2% a year.

Recorded gas production commenced in the 1950s, rising to its current level of 2.5 Tcf a year. Most of it is associated gas coming from the oilfields and is expected to continue at this level for many years.

Oil consumption stands at 272 Mb/a making the country a major exporter. Gas consumption is at 748 Gcf/a, leaving modest exports in the form of NLG.

The Oil Age in Perspective

Little is known about the early history of the country before it was sighted by Christopher Columbus in 1498 on his third voyage to the New World. Spanish settlement began in 1520, when Caracas, the capital, was founded in an Andean valley, being administered until 1819 jointly by the Spanish Vice-Royalty of Peru and the Audencia of Santo Domingo. It was the birthplace of Simon Bolivar, known as the Liberator of South America. After several years of struggle, he brought independence to Venezuela in 1829, only to die in the following year, a disillusioned man, with his notion of a united Latin America having been destroyed by factional disputes. It could be said that Ugo Chavez, the present President, is doing his best to fulfil Bolivar's dream.

The subsequent history was characterised by revolution, counter-revolution and dictatorship, interspersed by brief periods of not very successful democratic government. Venezuela has been something of a backwater in terms of European settlement and immigration, despite its great mineral wealth. Even so, the population, which is of mixed European and African extraction, has increased almost ninefold over the past Century to reach its present level of 29 million, mainly living in the Andean and coastal regions.

The expropriation of BP's Iranian interests in 1951 did not pass unnoticed in Caracas, where the government was already in dispute with the foreign companies over the split of oil revenues. It led Perez Alonso, an idealistic oil minister, to open discussions with the major Middle East producers, to try to form a world equivalent of the Texas Railroad Commission, which in earlier years had regulated the US over-production to support price. He eventually succeeded with the formation of the Organisation of Petroleum Exporting Countries (OPEC) in 1960. The government started passing laws imposing stiffer terms on the existing concessions, which paved the way for a full nationalisation in 1976. That was accompanied by the creation of a national company, Petroleos de Venezuela (PdVSA). By now, exploration was at a mature stage, so the main challenge was to develop the extensive heavy oil deposits that had long been known, and to work in the corridors of power at OPEC to obtain the best price.

The present President, Ugo Chavez, is an ex-paratrooper who won landslide elections in 1998, 2000 and 2006, but follows a long tradition of somewhat authoritarian rulers. Venezuela, like many Latin American countries, has been run by a wealthy elite of the so-called oligarchs, many of whom, no doubt, shift their money overseas, leaving the poor with a minor share of the country's great oil wealth. President Chavez has tried to change this relationship with a decidedly anti-globalist policy, having made well-publicised visits to Cuba. In 2007, his government withdrew from the International Monetary Fund and World Bank, the principal agents of globalism. He is successfully using his oil wealth to forge a new alliance of Latin American countries with a view to breaking free from what has been described as dollar imperialism. Plans to start trading oil in euros have been announced but failed to be implemented. He was almost ousted from power in 2002 in a coup, which was welcomed, if not orchestrated, by the United States, but he outwitted the conspirators. No doubt, further attempts to remove him will be made, despite his popular mandate. The United States imports about 10% of its oil from Venezuela, explaining its interest in the politics of the country.

Gasoline is sold to the domestic market at extremely low prices, which nevertheless yield a satisfactory return to the State Company. The government would no doubt face popular outcry if it should need to cut internal demand to maintain profitable exports, but those days are still far off as world prices soar. Meanwhile, new wealth flows into the country, strengthening its currency and influence, giving it every motive to conserve its petroleum resources, despite external pressure. The country is making substantial arms purchases from Russia to strengthen its defences, but also faces rising inflation, despite its wealth. Indeed there are undercurrents of tension as food and other prices soar, despite Chavez's popular mandate. The country has lately nationalised the assets of the major oil companies in the tar-belt.

Venezuela is very well placed to face the second half of the Age of Oil with its modest population, its fertile lands and its substantial remaining oil and mineral wealth. It is likely to come under increasing pressure from the United States, which depends heavily on it for oil imports, but is well placed to resist such pressure, provided it does not reach military proportions. It is at the same time welding support elsewhere in Latin America for what could evolve into a hemispheric conflict, if Mexico shifts its allegiance south.

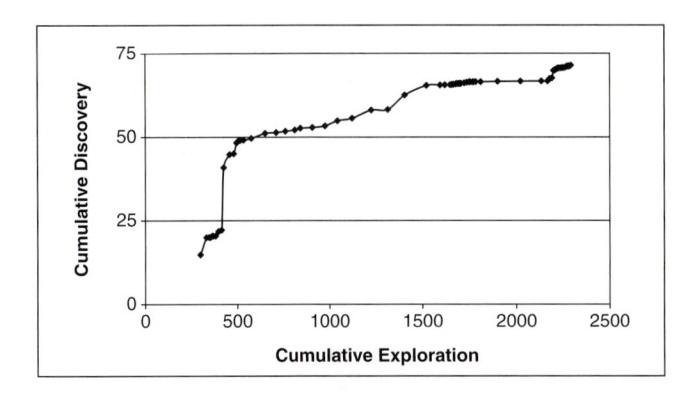

Fig. 57.3 Venezuela discovery trend

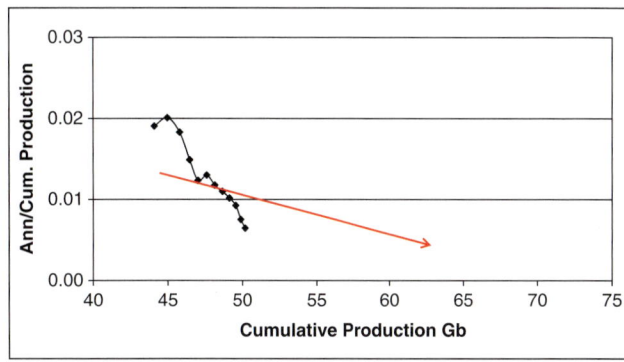

Fig. 57.4 Venezuela derivative logistic

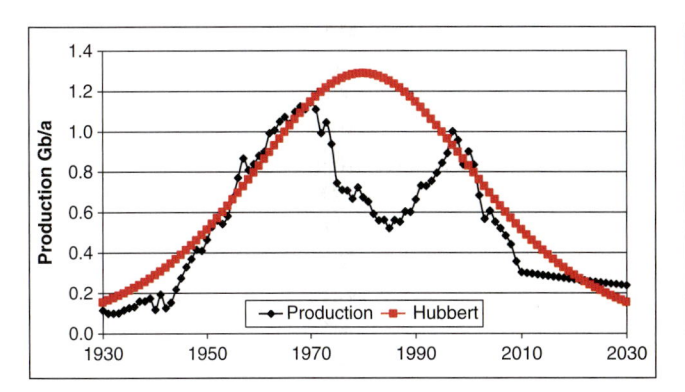

Fig. 57.5 Venezuela production: actual and theoretical

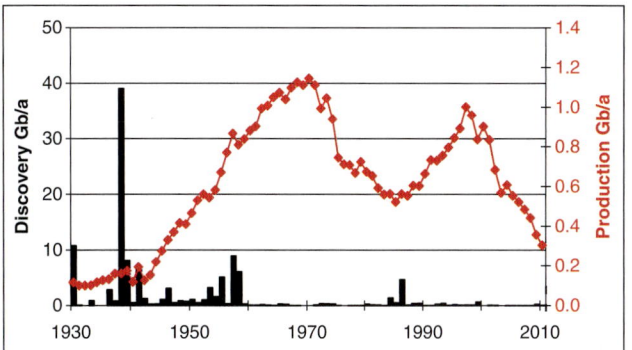

Fig. 57.6 Venezuela discovery and production

Latin America Region 58

Table 58.1 Latin America regional totals (data through 2010)

Production						Peak Dates			Area	
Amount			Rate				Oil	Gas	'000 km^2	
	Gb	Tcf	Date	Mb/a	Gcf/a	Discovery	1914	1941	Onshore	Offshore
PAST	125	234	2000	2948	7159	Production	1998	2015	18840	3770
FUTURE	57	391	2005	2682	8493	Exploration	1982		Population Growth	
Known	50	343	2010	2080	10127	Consumption			1900	64
Yet-to-Find	7.6	48	2020	1265	8770		Mb/a	Gcf/a	2010	501
DISCOVERED	175	577	2030	814	6049	2010	3079	6999	Growth	7.5
TOTAL	182	625	Trade	−969	+3159	Per capita	5.2	11.7	Density	27
Note: Data refer to main producing countries only, **except for consumption and trade data** which also includes other countries.										

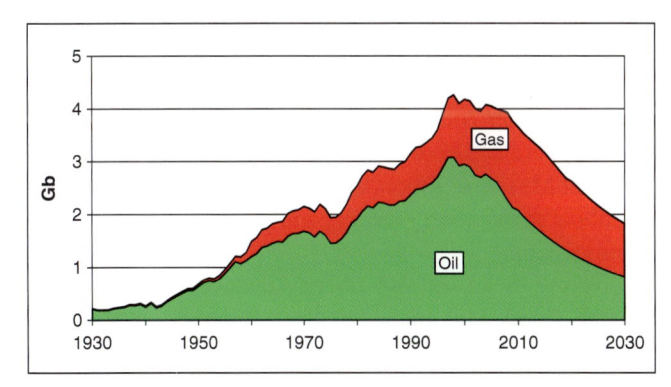

Fig. 58.1 Latin America oil and gas production 1930 to 2030

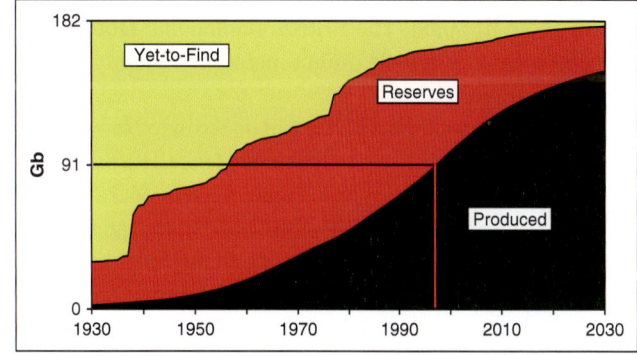

Fig. 58.2 Latin America status of oil depletion

Latin America, which for this purpose is defined as South America, Mexico and the Caribbean islands, includes ten significant oil producing countries. The region is dominated by Venezuela and Mexico, having the largest reserves of *Regular Conventional Oil*. Two of the countries also have substantial *Non-Conventional* deposits in the form of the deepwater oil and gas of Brasil and the heavy oils of Venezuela.

The other countries, making up the region are listed in the table. Those indicated (*) have minor amounts of oil and/or gas which are included with other minor producers in the world, as the amounts are too small to model meaningfully. They are however consumers of oil and gas (see Chap. 11).

The region has an indicated total endowment of 182 Gb of *Regular Conventional* oil, amounting to almost 10% of the World's total. It boasts the world's first giant oilfield found in Peru in 1869, and it also saw major finds in Venezuela and in

Table 58.2 Other countries of Latin America

Antigua	Dominican Rep.	Haiti	Paraguay
Bahamas	El Salvador	Honduras	Puerto Rico
*Barbados	French Guiana	Jamaica	St.Kitts
*Belize	Grenada	Martinique	St.Lucia
Costa Rica	Guadeloupe	Neth. Antilles	St Vincent
*Cuba	*Guatemala	Nicaragua	Suriname
Dominica	Guyana	Panama	Uruguay
*Minor oil and/or gas producers			

Mexico in the early years of the last Century, giving an overall peak of discovery in 1914.

It also has an endowment of about 625 Tcf of gas, principally in Venezuela and Mexico.

The current overall depletion rate of *Regular Conventional Oil* is a modest 3.5% a year reflecting the fact that production is mainly onshore from ageing fields. It is set to decline at about this rate, falling from 5.7 Mb/d in 2010 to about 2.2 Mb/d by 2030.

The population for the entire region, including the additional countries listed in the table, is 596 million, with a population density of 28/km².

Total oil consumption stands at 3,079 Mb/a, meaning that the Region has become a net importer of 969 Mb/a. It has endowment of about 600 Tcf of gas, of which 391 Tcf remain. Production stands at 10 Tcf/a year, having peaked in 2008. It will now likely decline at about 2.5% a year to 6 Tcf/a by 2030. Reported consumption stands at 6.9 Tcf/a, meaning that about 3.2 Tcf/a is exported mainly in the form of LNG and NGL.

The early history of the region was dominated by the great Maya, Aztec and Inca civilisations, well remembered by their impressive archaeological remains, such as Machu Pichu in Peru. Spanish and Portuguese conquest followed in the sixteenth century, primarily in the quest for gold and silver with which to finance European wars. The Pope established spheres of interest for the two countries from which Portugal had rights to Brasil, with the rest of the region going to Spain. The British, French and Dutch were content with a relatively minor presence, mainly in the Caribbean region.

A turning point came with the fall of Spain to Napoleon in 1809 which prompted wars of independence led by Simon Bolivar from Venezuela and General San Martin from Argentina which liberated the Spanish empire, apart from Mexico. Brasil and Mexico later acquired their independence without insurrection as such. Internal wars within the region have been limited, although Bolivia and Peru lost territory to Chile in the so-called nitrate wars of 1879–1883, reflecting Europe's need for imported agricultural nutrients. Mexico also lost its northern territories to America in the war of 1846–1848

Independence brought internal conflicts of many different types but mainly between the oppressed peasants, and in some cases miners, and their landlords. In general, the political evolution has seen alternating epochs of authoritarian rule and ineffectual democracy.

The region has been a victim of globalism in recent years, as foreign interests moved in to exploit near-slave labour, exporting product and profit, leaving the countries burdened with heavy foreign debt. For example, Ecuador found itself having to dedicate its substantial oil revenues in their entirety to servicing foreign debt. The elite of the countries are themselves the agents for foreign commercial and financial engagements, a situation which naturally gives rise to internal political tensions. An extreme variant affects Colombia, where drug barons supply an international market in cocaine, leading to much civil tension and conflict.

Table 58.3

	Standard Deviation Public Reserve Data		Surprise Factor	
	Oil	Gas	Oil	Gas
Argentina	0.05	0.92	2	4
Bolivia	0	1.21	3	10
Brasil	2.13	0.09	2	6
Chile	0.07	1.28	1	6
Colombia	0.28	0.29	2	5
Ecuador	0.89	0.02	4	6
Mexico	0.69	2.58	7	8
Peru	0.35	0.5	2	8
Trinidad	0.08	0.91	1	3
Venezuela	87.55	17.25	7	7

Scale 1-10 of increasing uncertainty

A remarkable feature was the success of Fidel Castro in running a Communist State in Cuba since coming to power in 1959. This country faced its own Peak Oil crisis when imports were suspended on the fall of the Soviets, but it rose to the occasion finding ways to cut demand and build new sustainable life styles, setting an example. Ugo Chavez in Venezuela now seems to be inspired to build a new independence for Latin America, building alliances and special relations with other countries in the region, especially Bolivia, Ecuador, Peru and to some degree Brasil. The scene seems to be set for Mexico to follow this direction, as tensions with its neighbour to the north grow in parallel with the decline in its oil production.

Looking ahead, the region seems relatively well placed to face the Second Half of the Oil Age with the possible exception of Mexico and other Central American republics, which may see a flood of returning emigrants if deep economic depression grips the United States. Clearly, there will have to be a return to more sustainable living, which the region could support albeit with a radical fall in urban populations.

Confidence and Reliability Ranking

The table lists the standard deviation of the range of published data on reserves. A relative confidence ranking in the validity of the underlying assessment is also given: the lower the Surprise Factor the less the chance for revision.

Argentina

Argentina has been thoroughly explored over a long period of time, being at an advanced stage of oil depletion. The data quality is good, leaving few doubts about the oil assessment. The gas situation is somewhat less sure.

Bolivia

Bolivia is a remote land-locked country with a limited oil potential which is probably now well established, but its gas endowment is under-developed and most uncertain.

Brasil

Brasil's *Regular Conventional Oil and Gas* endowment is modest and thoroughly explored, leaving few doubts about the assessment. It has in addition large amounts of deepwater oil, considered in Chap. 12.

Chile

Chile has virtually exhausted its modest oil endowment, making the assessment easy. A reported large gas find in 2003, if confirmed, may speak of more gas than would have been expected

Colombia

Colombia has been thoroughly explored, and its geology is well understood, leaving little room for surprise so far as oil is concerned. Production from remote fields east of the Andes has been delayed partly from terrorist activities distorting the normal profile but now comes online. Gas is another question, with perhaps unrecognised possibilities offshore and in the thrust belts bordering the Eastern Andes. It may also move to develop heavy oil deposits bordering the Guyana Shield.

Ecuador

Ecuador's reported oil reserves unexpectedly doubled in 2003, although no new significant finds were made, which casts doubt on the validity of the data. The change may reflect recognition of the long-known Tiputini Field with its heavy oil in the remote interior, whose reserves may not previously have been counted. The country has been thoroughly explored. It may, however, have more gas potential than recognised offshore in the Gulf of Guayaquil, or perhaps at depth in the Andean foothills.

Mexico

Mexico's reported oil reserves have been subject to much variation, due no doubt to political pressures of various kinds, including the role of the State Company, Pemex. The assessment is accordingly less than sure although the country's underlying potential is well understood. It has considerable deepwater potential, considered in Chap. 12.

Peru

The huge expanses of the Oriente Basin in the Amazon headwaters are relatively unexplored, but generally lack prime source-rocks, although some possibilities remain in isolated minor troughs. The coastal region is well known having been thoroughly explored. The oil assessment is therefore considered to be reasonably reliable. The development of gas, derived from Palaeozoic sources in the Camisea region of the Andes, and in an alleged recent offshore discovery, is at a relatively early stage, leaving much uncertainty.

Trinidad

Trinidad has been very thoroughly explored both onshore and offshore, so that its oil and gas potential is well understood, save perhaps for the recent gas finds in the Caribbean to the northwest. The assessment is therefore considered reliable

Venezuela

Venezuela has been thoroughly explored being at a mature stage of development. The primary difficulty facing the assessment is to draw the boundary with the large Heavy and Extra-Heavy oil deposits, which are excluded from *Regular Conventional* by definition at a density cutoff of 17.5°API. The difficulty affects both reserves and past production. Venezuela is also a member of OPEC, subject to those pressures in respect to reserve reporting. The oil assessment is considered reliable in terms of orders of magnitude although serious uncertainties remain. The gas assessment is less reliable.

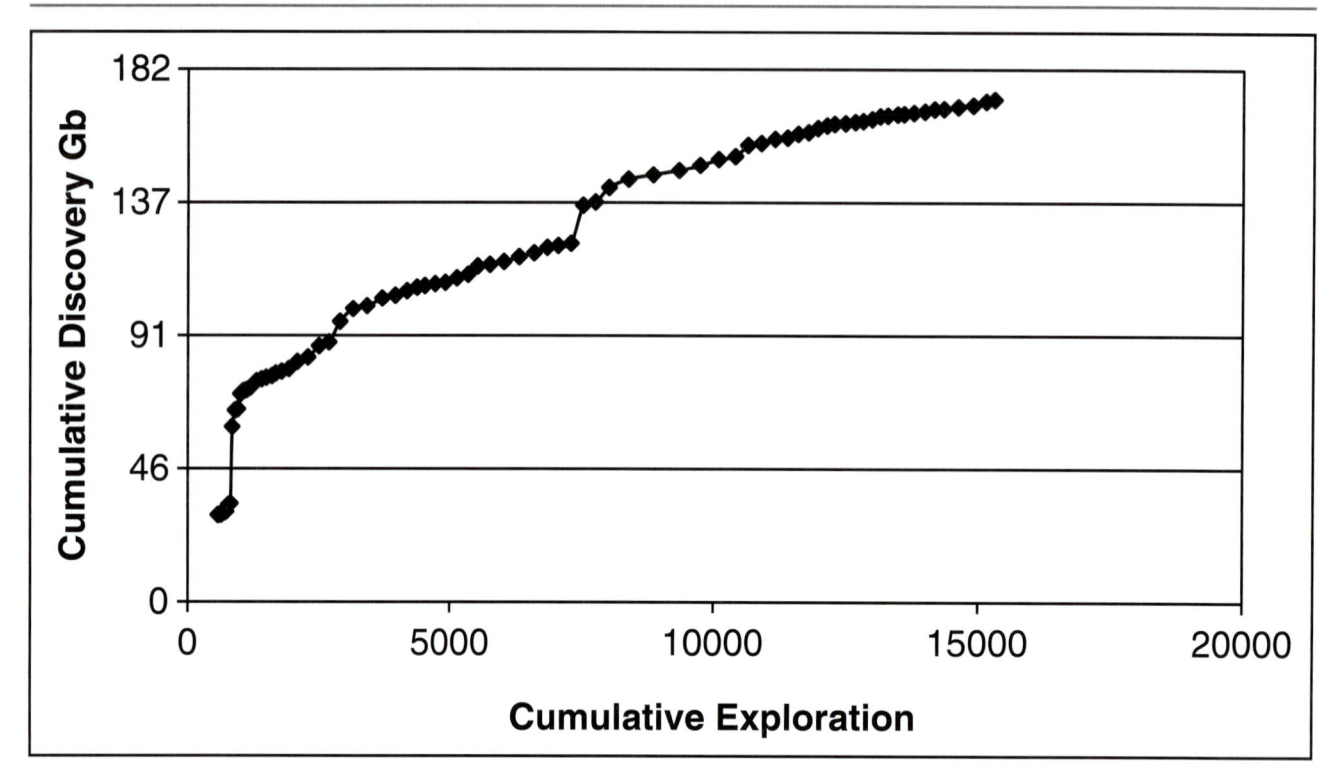

Fig. 58.3 Latin America discovery trend

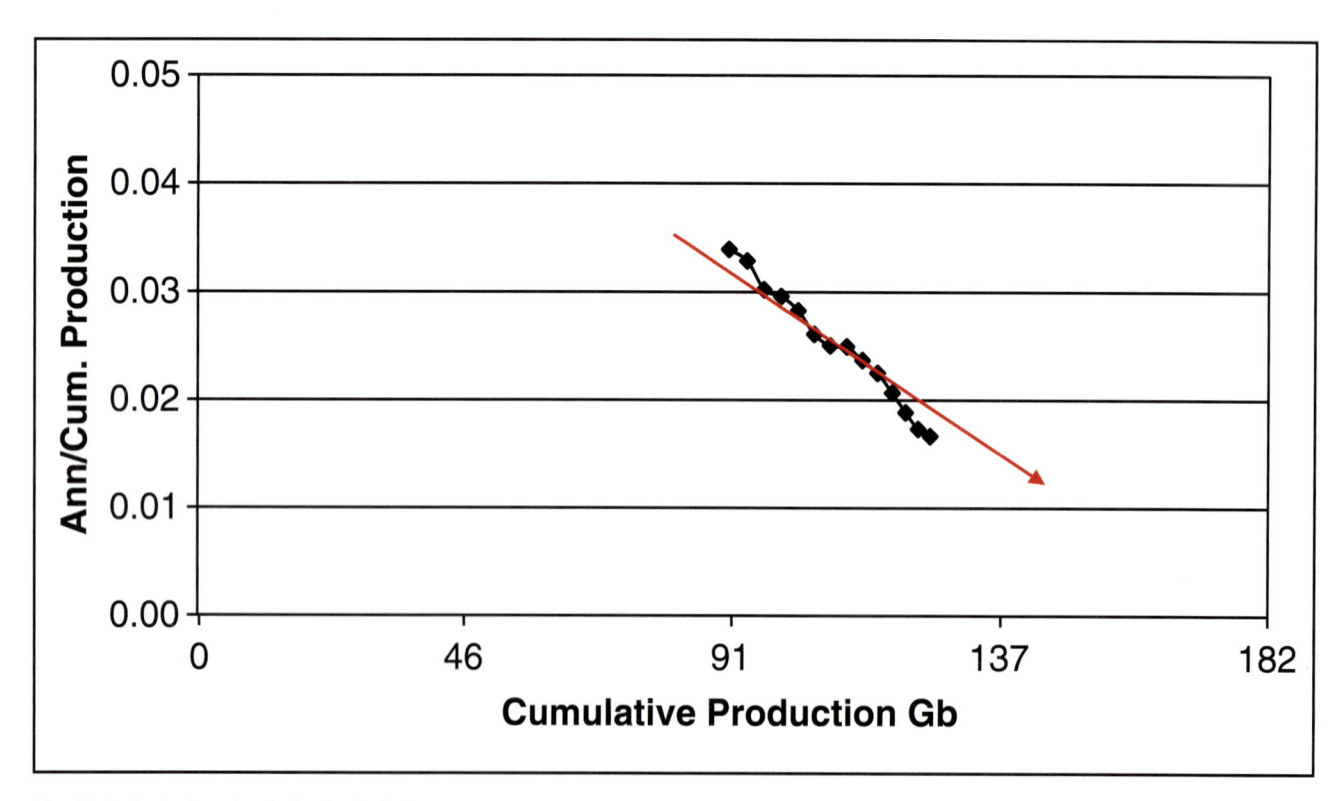

Fig. 58.4 Latin America derivative logistic

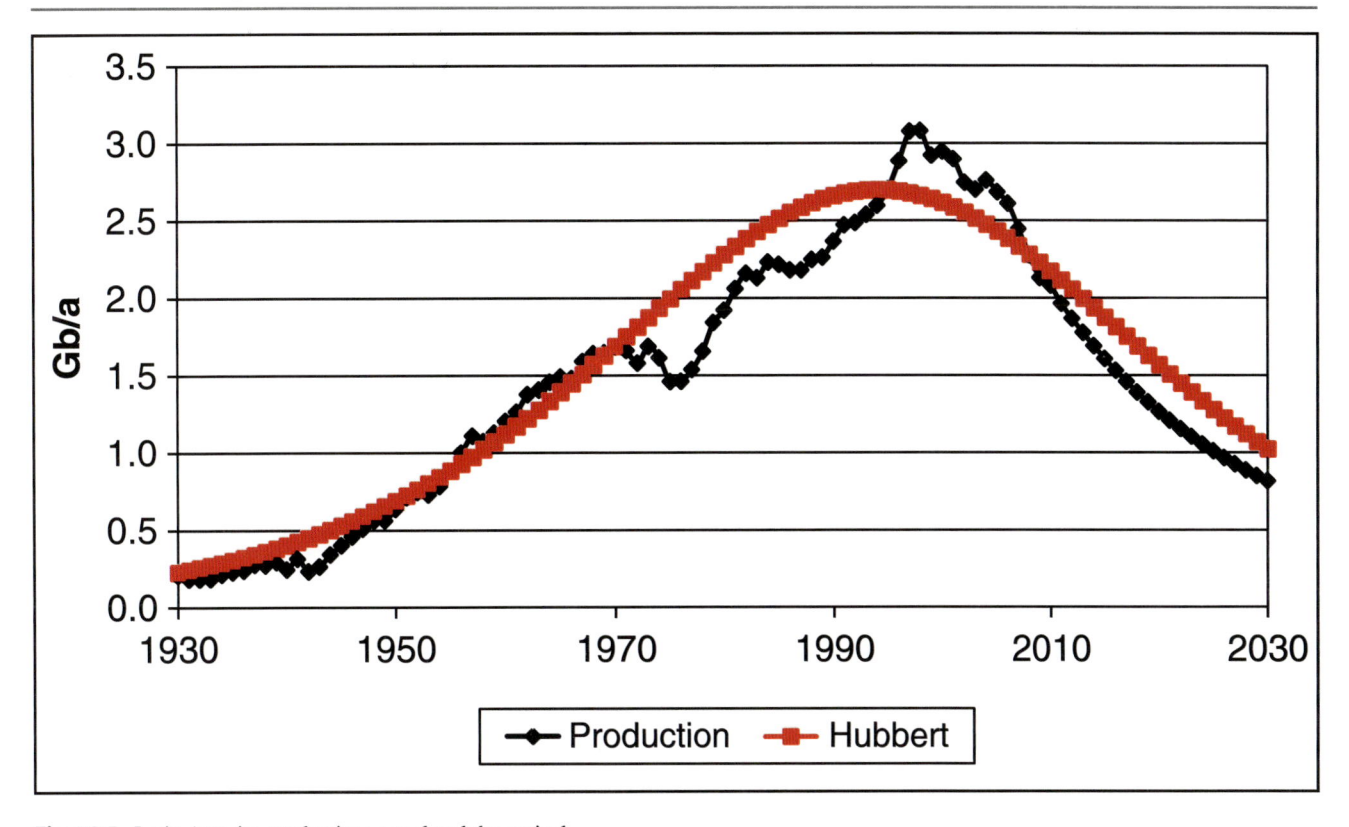

Fig. 58.5 Latin America production: actual and theoretical

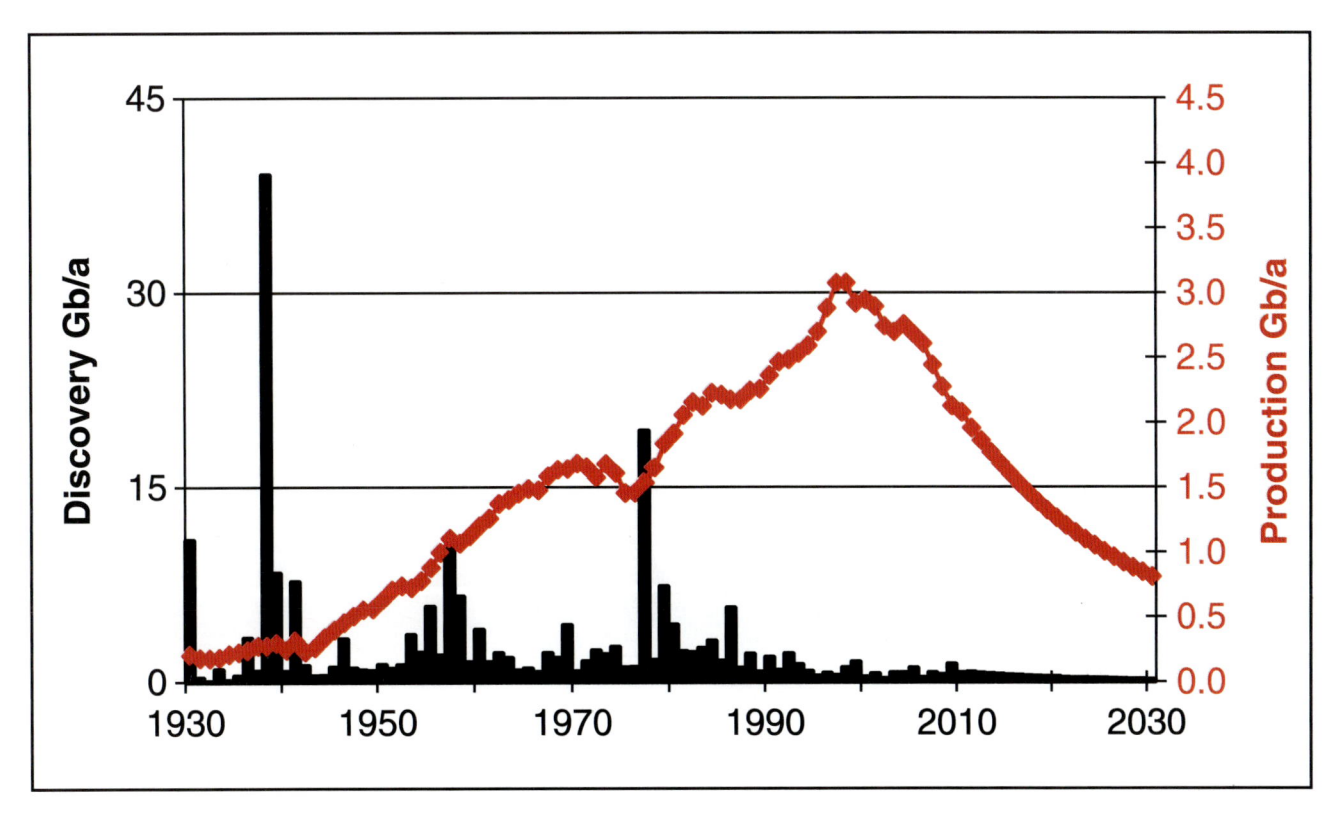

Fig. 58.6 Latin America discovery and production

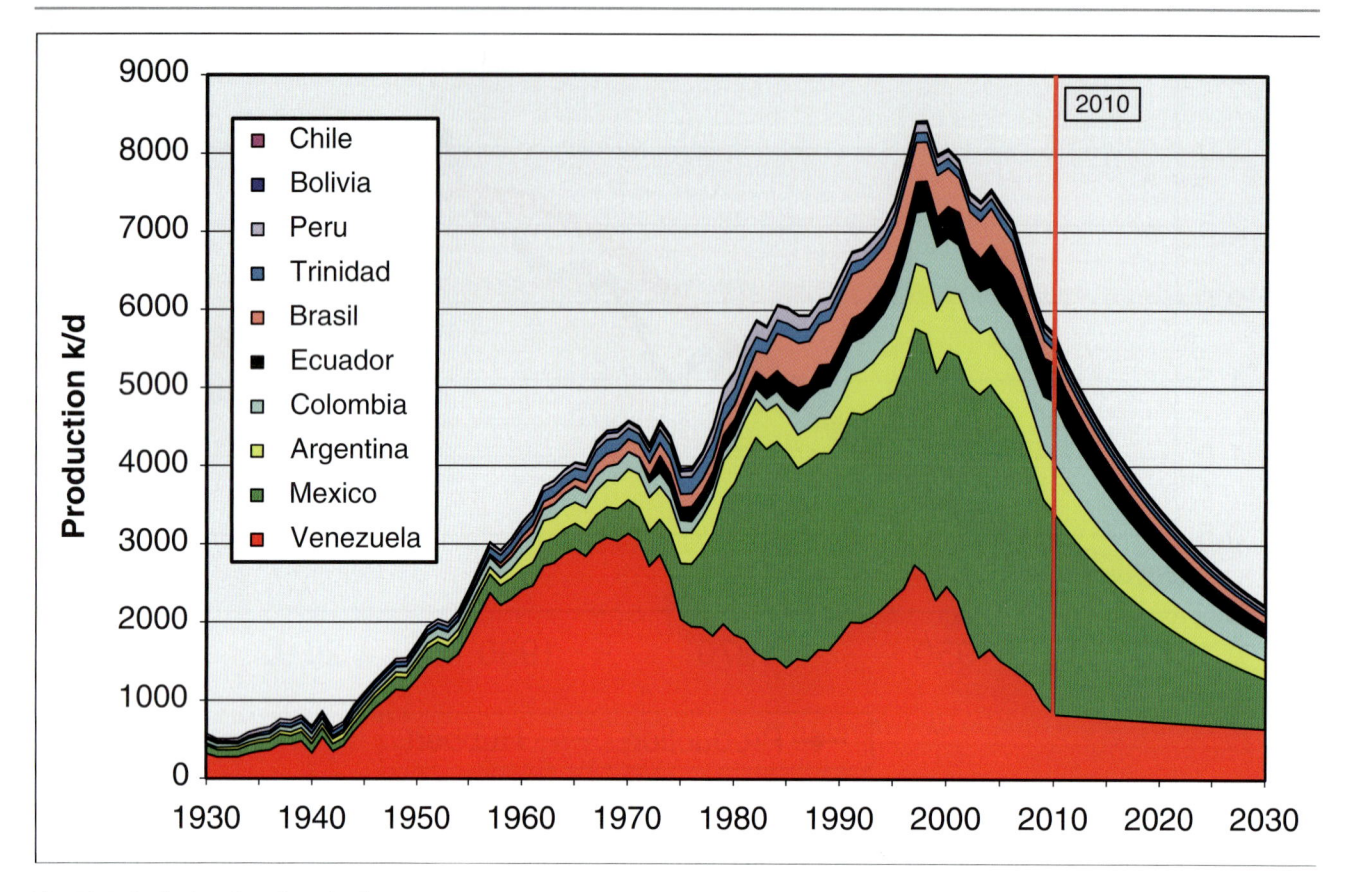

Fig. 58.7 Latin America oil production

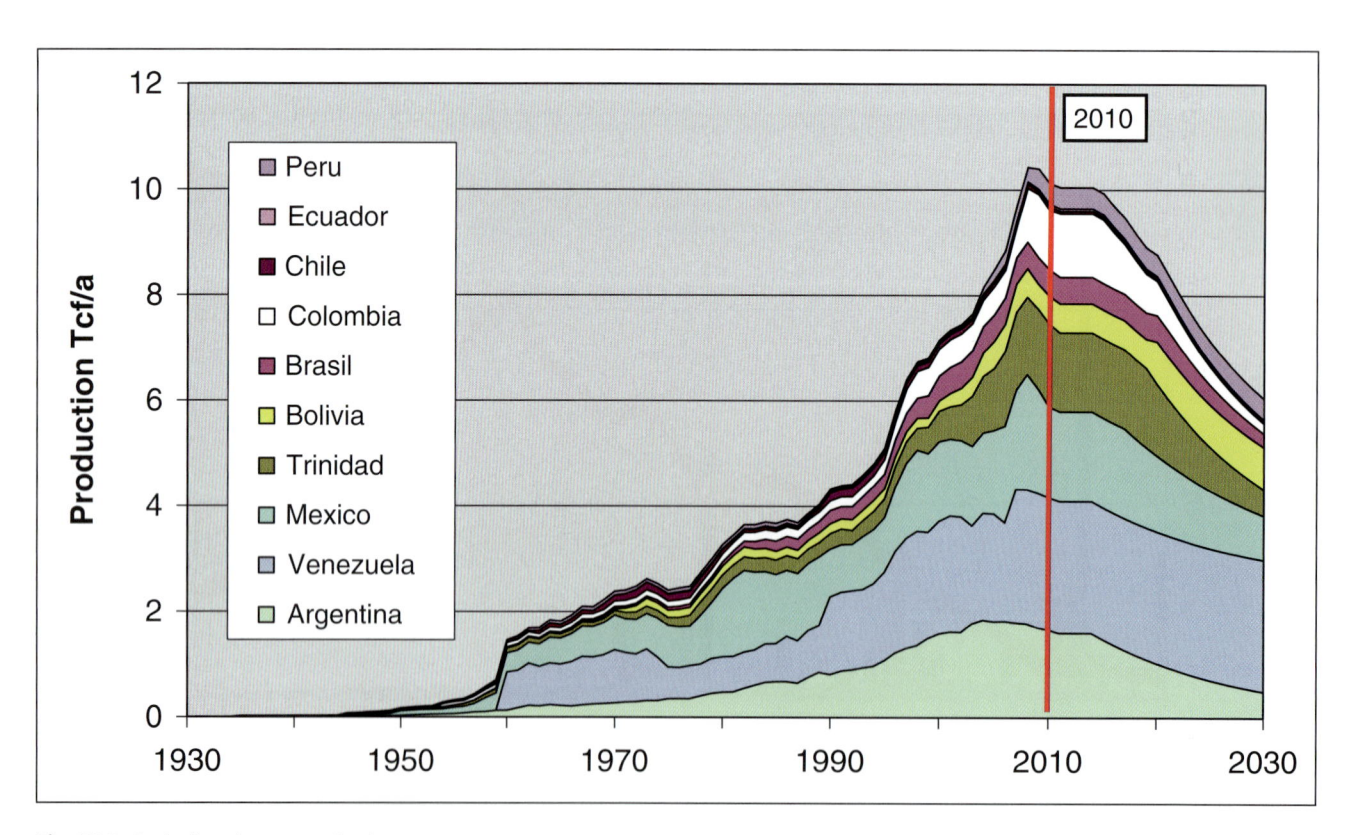

Fig. 58.8 Latin America gas production

58 Latin America Region

Table 58.4

L.America																			
		\multicolumn{8}{c	}{KNOWN FIELDS}				\multicolumn{3}{c	}{Revised 05/10/2011}				
		Present		Past			Reported Reserves		Future	Total	NEW	ALL	TOTAL	DEPLETION		PEAK		
		Kb/d	Gb/a		5yr	Disc	Average	Deduction	%	Found	FIELDS	FUTURES				Mid-			
Country	Region	2010	2010	Gb	Trend	2010		Static	Other	Disc.	Gb	Gb	Gb	Gb	Rate	%	Point	Expl	Disc	Prod
Venezuela	E	832	0.30	50.33	−8%	0.04	181.05	0.66	−60.00	95%	20.97	71.30	3.70	24.67	75.0	1.2%	67%	1991	1981	1914	1970
Mexico	E	2576	0.94	24.67	−4%	0.16	11.00	0.00	0.00	97%	11.76	50.69	1.31	13.07	52.0	6.7%	75%	1999	2003	1977	2004
Argentina	E	641	0.23	10.35	−2%	0.08	2.53	0.00	0.00	95%	3.95	14.30	0.70	4.65	15.0	4.8%	69%	1998	1985	1962	1998
Colombia	E	786	0.29	7.45	10%	0.21	1.59	0.53	0.00	94%	4.72	12.17	0.83	5.55	13.0	4.9%	57%	2006	1988	1988	1999
Ecuador	E	486	0.18	4.70	−2%	0.03	6.24	0.00	−1.00	96%	2.97	7.67	0.33	3.30	8.0	5.1%	59%	2006	1972	1969	2006
Brasil	E	162	0.06	6.10	0%	0.07	14.01	0.00	−27.38	97%	2.61	8.71	0.29	2.90	9.0	2.0%	68%	1998	1982	1975	1990
Trinidad	E	98	0.04	3.51	−6%	0.00	0.73	0.16	0.00	99%	0.94	4.45	0.05	0.99	4.5	3.5%	78%	1984	1972	1959	1981
Peru	E	73	0.03	2.58	−1%	0.04	0.87	0.00	0.00	93%	1.13	3.72	0.28	1.42	4.0	1.8%	65%	1993	1975	1869	1982
Bolivia	E	43	0.02	0.56	−2%	0.00	0.46	0.01	0.00	92%	0.58	1.15	0.10	0.69	1.3	2.2%	45%	2015	1962	1999	2005
Chile	E	2	0.00	0.43	−4%	0.00	0.11	0.03	0.00	99%	0.07	0.50	0.00	0.07	0.5	1.3%	86%	1980	1972	1960	1982
L. AMERICA	E	5699	2.08	124.95	4%	0.64	240.59	1.39	−88.38	96%	49.70	174.65	7.60	57.30	182.3	3.5%	69%	1997	1982	1914	1998

Header (Oil): PRODUCTION TO 2100 — Regular Conventional Oil — 2010

Table 58.5

L. AMERICA																				
		\multicolumn{8}{c	}{KNOWN FIELDS}				\multicolumn{3}{c	}{Revised 15/11/11}					
		Present		Past		Discovery	Reserves		FUTURE	TOTAL	FUTURE	ALL	TOTAL	DEPLETION		PEAKS			
		Tcf/a	FIP	Gboe/a		5yr	Tcf	%	Reported	Deduct	KNOWN	FOUND	FINDS	FUTURE		Current		Mid-			
										Non-											
Country	Region	\multicolumn{3}{c	}{2010}	Tcf	Trend	2010	Disc.	Average	Con	Tcf	Tcf	Tcf	Tcf	Tcf	Rate	%	Point	Expl	Disc	Prod
Venezuela	E	2.51	1.31	0.45	65.25	0.07	0.2	93%	175.56	0	166.28	231.52	18.48	184.75	250	1.34%	26%	2030	1981	1941	2030
Mexico	E	1.72	0.02	0.31	57.78	−0.01	0.27	96%	14.46	0	38	95.78	4.22	42.22	100	3.92%	58%	2006	2003	1977	2008
Argentina	E	1.66	0.06	0.3	42.64	−0.02	0.66	92%	14.05	0	21.89	64.53	5.47	27.36	70	5.73%	61%	2006	1985	1977	2004
Colombia	E	1.12	0.7	0.2	14.8	0.23	0.69	95%	4.1	0	13.68	28.48	1.52	15.2	30	6.89%	49%	2006	1988	1973	2012
Ecuador	E	0.05	0.02	0.01	1.04	0.02	0.03	93%	0.15	0	0.82	1.86	0.14	0.96	2	4.91%	52%	2009	1972	1969	2013
Brasil	E	0.5	0.03	0.09	10.93	0	0.33	95%	12.85	12.6	10.86	21.79	1.21	12.07	23	3.98%	48%	2011	1982	2008	2011
Trinidad	E	1.58	0.08	0.28	22.16	−0.05	0.28	92%	15.51	0	23.67	45.82	4.18	27.84	50	5.36%	44%	2012	1971	1968	2010
Peru	E	0.39	0.24	0.07	4.92	0.11	0.52	84%	11.92	0	16.07	20.98	4.02	20.08	25	1.92%	20%	2030	1975	1986	2025
Bolivia	E	0.53	0.08	0.09	9.75	0	1.34	87%	26.23	0	46.96	56.71	8.29	55.25	65	0.94%	15%	2030+	1962	1999	2025
Chille	E	0.06	0	0.01	4.71	−0.04	0	97%	2.37	0	5.03	9.74	0.26	5.29	10	1.17%	47%	2015	1972	1960	1992
L. AMERICA	E	10.13	9.29	1.58	233.98	0.03	4.32	92%	277.21	12.6	343.23	577.22	47.78	391.02	625	2.52%	37%	2018	1982	1941	2008

Header (Gas): PRODUCTION TO 2100 — Regular Conventional Gas — 2010

Table 58.6

Regular Conventional Oil Production						
Mb/d	2000	2005	2010	2020	2030
Venezuela	2.47	1.51	0.83	0.74	0.65
Mexico	3.01	3.33	2.58	1.29	0.64
Argentina	0.76	0.7	0.64	0.39	0.24
Colombia	0.69	0.53	0.79	0.46	0.29
Ecuador	0.39	0.53	0.49	0.29	0.17
Brasil	0.5	0.46	0.16	0.14	0.11
Trinidad	1.43	0.14	0.1	0.07	0.05
Peru	1.33	0.08	0.07	0.06	0.05
Bolivia	1.21	0.05	0.04	0.04	0.03
Chile	0.98	0	0	0	0
L. AMERICA	0.83	7.35	5.7	3.46	2.23

Table 58.7

Gas Production						
Tcf/a	2000	2005	2010	2020	2030
Venezuela	2.14	2.04	2.51	2.5	2.5
Mexico	1.51	1.58	1.72	1.44	0.84
Argentina	1.58	1.82	1.66	1.03	0.49
Colombia	0.51	0.53	1.12	0.65	0.19
Ecuador	0.04	0.04	0.05	0.04	0.02
Brasil	0.47	0.5	0.5	0.5	0.26
Trinidad	0.58	1.17	1.58	1.36	0.5
Peru	0.03	0.2	0.39	0.4	0.4
Bolivia	0.2	0.52	0.53	0.8	0.8
Chile	0.1	0.08	0.06	0.05	0.05
L.AMERICA	7.16	8.49	10.13	8.77	6.05

58 Latin America Region

Table 58.8

Latin America Region	PRODUCTION AND CONSUMPTION							POPULATION & AREA		
	OIL			GAS				Population	Area	Density
	Production	Consumption	Trade	Production	Consumption	Trade		Growth	Km³	
		p/capita	(+)		p/capita	(+)		Factor		
Major	Mb/a	Mba	b/a	Mb/a	Gcf/a		Kcf/a	Gcf/a	M	M
Venezuela	304	272	9.29	31	2510	748	25.5	1762	29	8.9	0.92	32
Mexico	940	757	6.59	184	1722	2135	18.6	–413	115	7.6	1.97	58
Argentina	234	226	5.57	8	1663	1529	37.8	134	41	7.8	2.79	15
Colombia	287	108	2.3	179	1124	321	6.8	803	47	9.2	1.14	41
Ecuador	177	73	4.89	104	50	12	0.8	38	15	11	0.28	54
Brasil	59	969	4.92	–910	500	890	4.5	–390	197	10	8.58	23
Trinidad	36	15	11.51	21	1576	780	600	796	1	4.5	0.01	254
Peru	27	69	2.38	–42	393	191	6.6	202	29	5.6	1.29	22
Bolivia	16	23	2.24	–7	526	96	9	430	10	5	1.1	9
Chile	1	110	6.48	–109	63	188	11.1	–1.25	17	5.3	0.76	22
Sub-total	2080	2621	5.23	–541	10127	6898	13.8	3229	501	7.5	18.84	27
Minor												
Barbados	0.27	3.29	10.95	–3.02	1.03	1.03	3.43	0	0.3	1.4	0	698
Belize	1.56	2.56	8.52	–0.99	0.00	0	0	0	0.3	1.2	0.02	14
Cuba	17.47	64.24	5.74	–46.77	49.44	40.61	3.63	9	11.2	6.7	0.11	102
Guatemala	4.94	25.92	1.76	–20.98	0.00	0	0	0	14.7	7.1	0.11	136
Nicaragua	0	10.94	1.85	–10.94	0.00	0	0	0	5.9	11.8	0.13	45
Paraguay	0	11.32	1.72	–11.32	0.00	0	0	0	6.6	10.0	0.41	16
El Salvador	0	17.16	2.77	–17.16	0.00	0	0	0	6.2	6.0	0.02	295
Suriname	5.46	5.48	10.96	–0.02	0.00	0	0	0	0.5	3.0	0.18	3
Uruguay	0	18.98	2	–18.98	0.00	2.82	0.83	–3	3.4	4.0	0.18	19
Other	0	297.4	6.47	–297.40	0.00	56.1	1.22	–56	46	0.0	0.5	92
Sub-Total	29.7	457.28	4.81	–427.58	31.00	100.56	1.06	–70	95.1		1.65	58
L. America	**2109.7**	**3078.7**	**5.2**	**–969.0**	**10158.0**	**6998.6**	**11.74**	**3159**	**596.1**	**7.5**	**20.5**	**29.1**

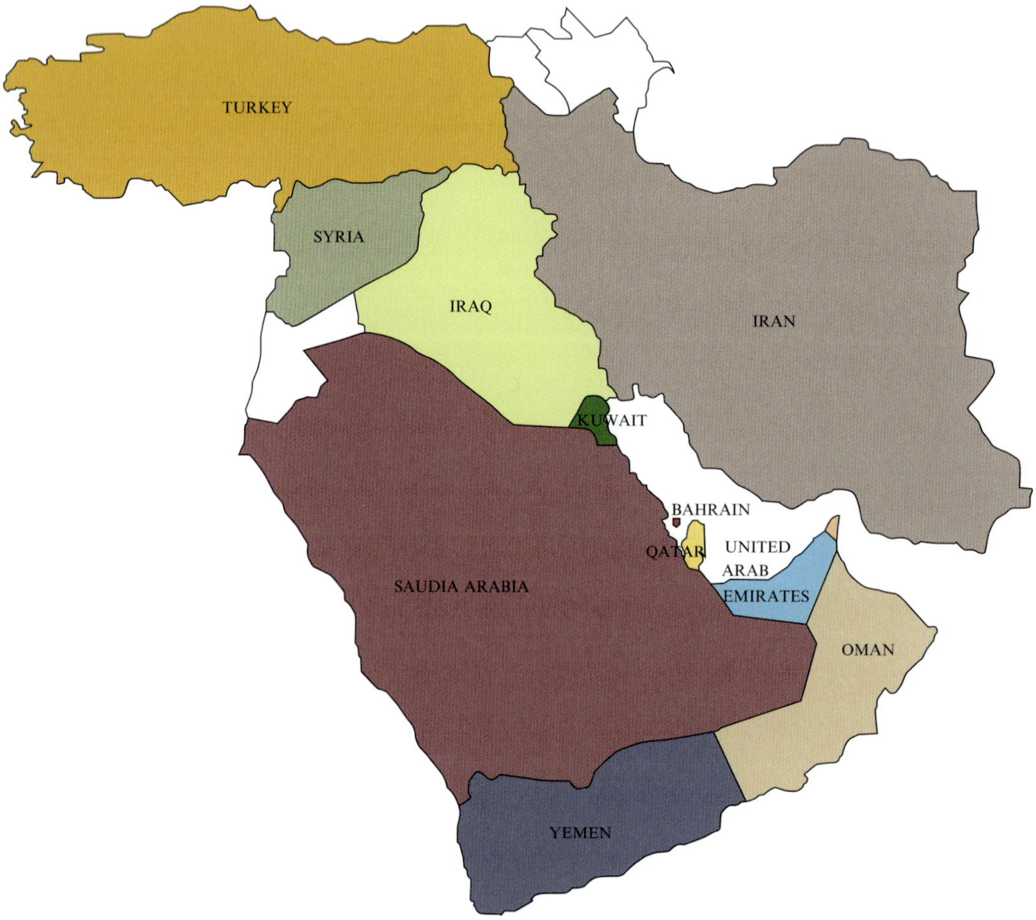
TURKEY
SYRIA
IRAQ
IRAN
KUWAIT
BAHRAIN
QATAR
UNITED
ARAB
EMIRATES
SAUDIA ARABIA
OMAN
YEMEN

Part VII

Middle East

Bahrain

Table 59.1 Bahrain regional totals (data through 2010)

	Production to 2100					Peak Dates			Area	
	Amount		Rate				Oil	Gas	'000 km²	
	Gb	Tcf	Date	Mb/a	Gcf/a	Discovery	1932	1932	Onshore	Offshore
PAST	1.12	16	2000	14	412	Production	1970	2015	1	6
FUTURE	0.23	4.8	2005	13	470	Exploration	1983		*Population*	
Known	0.22	4.6	2010	13	551	*Consumption*	Mb/a	Gcf/a	1900	0.1
Yet-to-Find	0.01	0.2	2020	7	185	2010	17	433	2010	1.3
DISCOVERED	1.34	20.8	2030	4	70		b/a	kcf/a	Factor	13
TOTAL	1.35	21	*Trade*	−4	+118	Per capita	13	333	Density	1,880

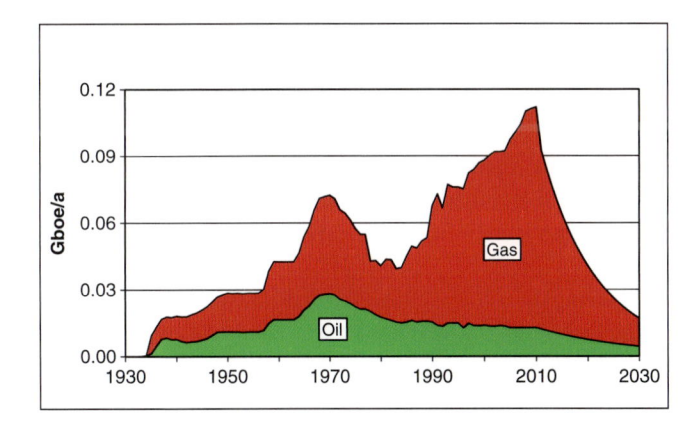

Fig. 59.1 Bahrain oil and gas production 1930 to 2030

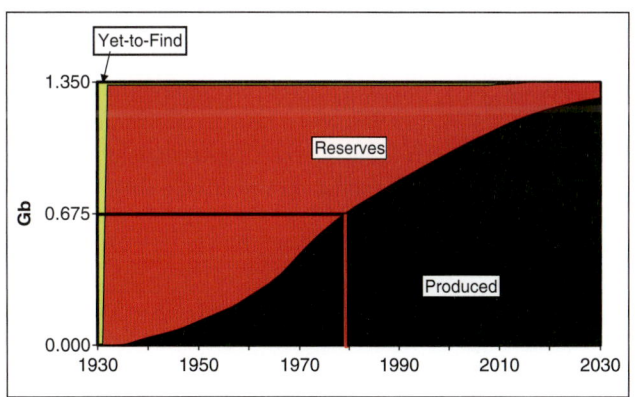

Fig. 59.2 Bahrain status of oil depletion

Essential Features

Bahrain is the largest of a group of small flat-lying islands off Saudi Arabia, to which it is now linked by a causeway. It claims sovereignty to the adjoining waters having now settled earlier disputes with Qatar, Iran and Saudi Arabia. It became a British protectorate in 1861, being used as a base from which to control piracy, before gaining independence as an emirate in 1971. In 2002, it became a constitutional monarchy under Hamad ib Isa al-Khalifah. It is primarily an Islamic country, of whom about half belong to the Shi'ia sect, but it has cosmopolitan society, now enjoying rapid economic development. It has provided the United States with a military base since the 1990s.

Geology and Prime Petroleum Systems

Bahrain lies on the western margin of the Persian Gulf within the prime area of oil generation. As discussed more fully in connection with Saudi Arabia, there were several source-rock intervals, dominated by the late Jurassic and mid Cretaceous intervals. Multiple evaporates in the sequence provided effective seals. Like Qatar to the east it overlies a broad structural uplift bringing the deeper Palaeozoic system of gas-condensate into range.

Exploration and Discovery

Caltex (Chevron-Texaco) took an oil concession to Bahrain in the 1920s, finding a major field at Awali in 1932, with some 1.4 Gb of oil, which proved to be the gateway to Saudi Arabia. Approximately 1.34 Gb of oil and 21 Tcf of gas have been found.

Exploration is at a low level with no more than eight exploration boreholes, out of a total of almost 25, being sunk over the past decade. It suggests that this small island has been exhaustively explored, although there might be some remaining potential offshore and for deep gas-condensate.

Production and Consumption

Oil production commenced in 1938 and rose to a peak of 77 kb/d in 1970. Different databases report widely different numbers, but 2010 production is here taken to be 35 kb/d, being set to decline at 5.2% a year into the future, 83% of the endowment having been depleted.

Gas production commenced in 1935 and reached an initial peak of 246 Gcf/a in 1970, before sinking to a low of 127 Gcf/a in 1978, and then recovering to an overall peak of 551 Gcf/a in 2010. It is now set to decline at about 9% a year. Oil and gas consumption stand at respectively about 17 Mb/a and 433 Gcf/a.

The Oil Age in Perspective

Bahrain held a strategic position in ancient trading routes between the Middle East and the Far East, including India, and itself enjoyed a cultural flowering under the so-called Dilmun Epoch, even being regarded by some as the site of the Biblical Garden of Eden. It was one of the first countries in the region to embrace Islam, being visited by the Prophet in AD 629.

A rich pearling industry brought wealth to the islands, which attracted the Portuguese who occupied the territory from 1521 to 1602. They were in turn replaced by a Persian ruler, Shah Abbas I, who incorporated it into the Safavid Empire. In 1783, the Al Khalifa clan from neighbouring Qatar invaded the islands and, with some interruptions, have remained in power ever since. Britain signed a treaty in 1820 making Bahrain a form of Protectorate, aiming particularly to suppress piracy in the Persian Gulf. An epoch of stability ensued, lasting until after the Second World War, when British influence waned. The discovery of oil in 1932 brought new wealth to the country accompanied a surge of industrialisation, and the development of a financial centre rivalling that of Beirut. There was also massive immigration from especially the Philippines, Pakistan, Egypt and Iran. In parallel with this came a certain political change seeing the rise of socialist and communist pressures combined with growing Arab nationalism, reflected in the evolving political Middle East scene.

The Iranian Revolution of 1979 had a large impact on Bahrain leading to a failed coup in 1981 and conflict over Islamic fundamentalism, accompanied by riots and bomb attacks. These pressures were further stimulated by the Gulf Wars. Serious revolts broke out again in the so-called Arab Spring of 2011 fuelled in part by the deep-seated Sunni–Shi'a conflict, and were suppressed partly with the help of Saudi troops.

Looking ahead, it is clear that Bahrain's oil production is in decline. Initially, its impact has been offset by soaring prices bringing new wealth to the country. In the longer term, however, the country is vulnerable to worldwide financial tensions, and will likely face a difficult future as its barren landscape cannot support its population, which has grown by a factor of as much as 13 over the past century.

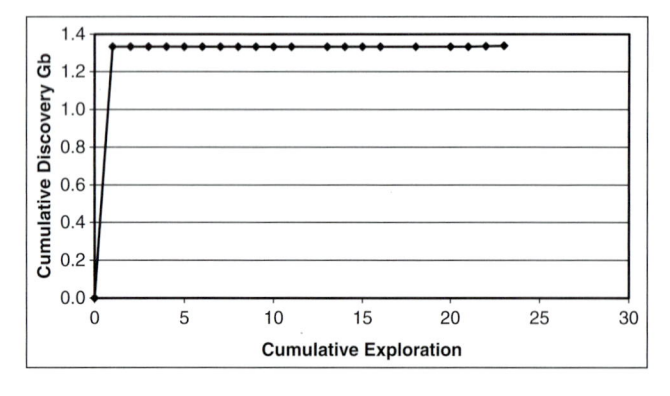

Fig. 59.3 Bahrain discovery trend

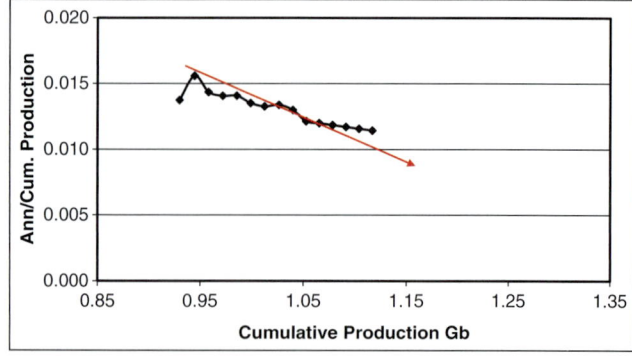

Fig. 59.4 Bahrain derivative logistic

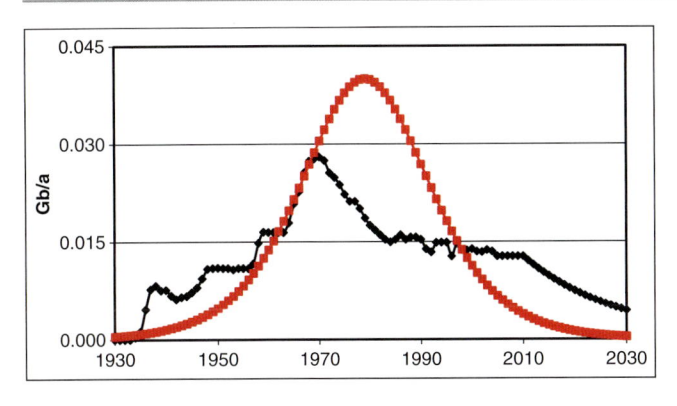

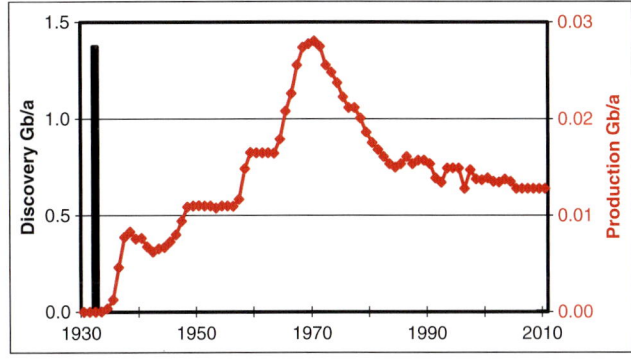

Fig. 59.5 Bahrain production: actual and theoretical

Fig. 59.6 Bahrain discovery and production

Iran

Table 54.1 Iran regional totals (data through 2010)

Production to 2100						Peak Dates		Area	
Amount			Rate				Oil	Gas	'000 km²
	Gb	Tcf	Date	Mb/a	Gcf/a	Discovery	1964	1964	Onshore	Offshore
PAST	66	119	2000	1,349	3,871	Production	1974	2030	1,640	15
FUTURE	84	1081	2005	1,511	5,386	Exploration	1967		Population	
Known	72	865	2010	1,489	7,774	Consumption	Mb/a	Gcf/a	1,900	10
Yet-to-Find	13	216	2020	1,386	7,500	2010	673	5106	2010	78
DISCOVERED	137	984	2030	1,138	7,500		b/a	kcf/a	Growth	7.8
TOTAL	150	1200	Trade	+816	+2,668	Per capita	8.6	66	Density	48

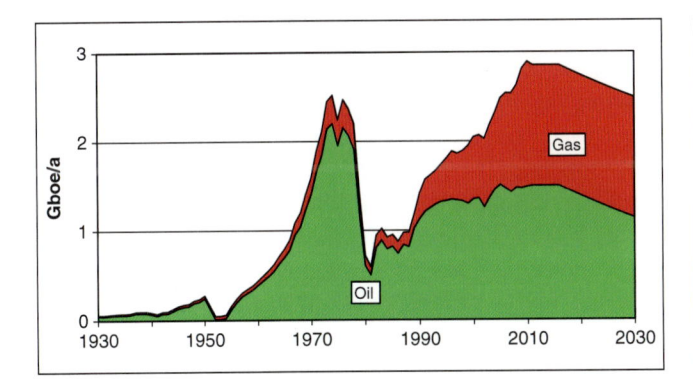

Fig. 60.1 Iran oil and gas production 1930 to 2030

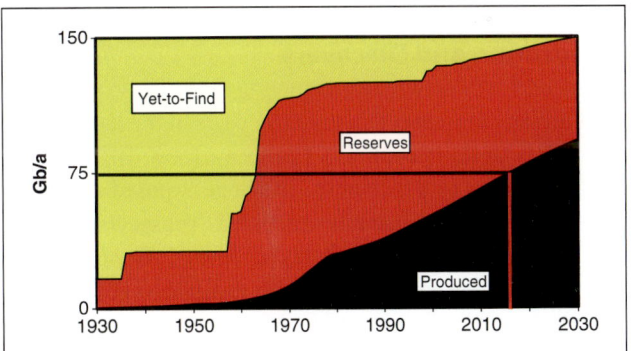

Fig. 60.2 Iran status of oil depletion

Essential Features

Iran is a mountainous country of high relief, covering almost 1.65 million km², which separates the Caspian Sea from the Persian Gulf. The deeply eroded Zagros Mountains run through the country parallel with the Persian Gulf, with peaks rising to over 4,000 m, while the Elburz Mountains form a narrow range in the north rising to over 5,600 m above the Caspian from which they are separated by a narrow, fertile and partly forested coastal belt. To the east, lies an extensive arid plateau, with salt flats, at an altitude of about 900 m, which is in turn flanked farther to the east by ranges of hills along the frontier with Afghanistan and Pakistan.

The climate ranges widely from freezing winter conditions with snow falls in the northern mountains to hot summers over most of the country. Much of it is arid, with the highest rainfall being found in the north on the shores of the Caspian and in some of the high mountain valleys.

The country supports a population of almost 78 million, made up of several distinct tribal and religious groups, mainly of Aryan origins. Some eight million of the inhabitants live in Tehran, the capital, located at the northern end of the interior plateau at the foot of the Elburz Mountains.

Geology and Prime Petroleum Systems

The Zagros Mountains mark the collusion of the Arabian and Asian tectonic plates. From a petroleum standpoint, interest is concentrated on the foothills and the adjoining waters of the Persian Gulf, where a sequence of sedimentary rocks as much as 15,000 m thick has been preserved.

The foothills are characterised by huge anticlinal structures that formed as rock masses slipped off the rising mountains on a glide-plane provided by a layer of Miocene salt. The prime source-rocks are deeply buried Jurassic clays, which in southern Iran have been depressed into the gas-generating window, but there are also leaner, yet still prolific, source-rocks in the Upper Cretaceous and Eocene sequences. The main reservoirs are the fractured limestones of the Miocene, Asmari Formation. Pliocene evaporites give effective seals to the reservoirs, which as a result have exceptionally long oil columns, commonly with substantial gas caps, possibly being partly recharged from depth. They are not easy reservoirs to manage, being prone to high-pressure gas invasion.

The Qatar Arch is an important oblique uplift that cuts across the Persian Gulf, extending from Arabia into Iranian waters. It has brought a deeper petroleum system into range, made up of Silurian source-rocks which have charged the overlying Permian sands with substantial amounts of gas-condensate.

Exploration and Discovery

Exploration commenced in the early years of the last century when the Anglo-Persian Oil Company (now BP) secured exclusive rights to the country. The huge exposed anticlines of the Zagros Mountains provided prospects readily identifiable by the field geologists of the day, whose pioneering work was rewarded when a well at a place called appropriately, Masjid-e-Sulaiman (the Mosque of Solomon), blew out on May 26th 1908, finding not only a giant oilfield with 1.3 Gb of oil but opening the world's largest oil province.

Exploration continued, delivering peaks of discovery in 1928, 1936, 1958, 1964 and 1992, which were dominated by a small number of giant fields, together giving an overall peak in 1964. A total of some 430 exploration boreholes have now been drilled, peaking in 1974, when some 22 were drilled. In fact, this is a relatively modest effort compared with that of other countries, but it is evidently a concentrated habitat in geological terms, with the bulk of its oil lying in comparatively few huge structures.

It is very difficult to assess the endowment due to the extremely unreliable nature of the data, but total discovery is here estimated to be 137 Gb, of which 72 Gb remain, with scope for the addition of an estimated 13 Gb to come from future discoveries. It will be remembered in connection with this assessment, that Iran was reporting reserves of 49 Gb in 1987, consistent with a long prior trend from when the country's oil industry was controlled by international companies. It then announced an increase to 93 Gb to protect its OPEC quota. Total production to 1987 amounted to 35 Gb, which suggests that the increase represented total found (49 + 35 = 84 Gb) with a slightly higher recovery.

The country also has a substantial gas potential, especially in the southern Persian Gulf. It is claimed that some 984 Tcf have been discovered, of which comparatively little has been produced, but the data are very unreliable. Statoil, the Norwegian oil company, reported a new find of as much as 17.5 Tcf in mid 2011, but confirmation is awaited. Despite the substantial endowment, Iran imports gas from Turkmenistan to supply the capital Tehran.

Production and Consumption

Oil production commenced in 1913, and grew to an early peak of 6 Mb/d in 1974 before falling to a low of 1.3 Mb/d in 1981 as a result of OPEC quota and other political factors. It has risen since to 4 Mb/d in 2010. It is difficult to forecast future production, but the estimates adopted here suggest that some 44% of the country's oil has been depleted, which means that there is theoretical scope for production to increase to the midpoint. But, on balance, bearing in mind the political situation, it is here assumed that production will remain flat at about its present level to around 2015 before commencing its terminal decline. The Depletion Rate will have by then risen to 1.9% a year, which is still a comparatively low one. If the country should come under military attack, production would no doubt decline steeply, ironically leaving more in the ground for the future.

Gas production has risen, roughly in parallel with oil, to its present level of 7.7 Tcf a year. In technical terms, it is possible to increase production to draw down the gas cap of existing oilfields towards the end of the lives, and by tapping new deeper sources, especially in the south-eastern Persian Gulf, but on balance it is thought that political constraints will hold it close to its present level.

The Oil Age in Perspective

When *Homo sapiens* stepped out of Africa, some 4 million years ago, one of his first stopping places was the Middle East, the site of the biblical Garden of Eden. There is a theory that the Planet was struck by comets 7,000 and 4,000 years ago which gave rise to tidal waves wiping out the inhabitants of the lowlands, depositing salt flats in America and populating the landlocked Caspian with marine life. If so, the inhabitants of the Zagros Mountains may have been one of the groups to survive: indeed it has been suggested that their "primitive" astronomy was sufficient to anticipate the second comet, leading the people to head for the hills with their cattle, possibly the origin of the legend of the Ark.

At all events, Iran (or Persia as it was previously known) is clearly an ancient country with a long and complex history. The first Persian Empire was established in the fifth and sixth centuries BC by the Medes, from whom the current Kurds are descended, and held dominion over the Middle East, Egypt and parts of Greece. Later came the classical wars in which Greece under Alexander the Great conquered the land, followed in turn by Romans.

But then came a national resurgence under the Sasanid Empire, known for its fire-worship, based on natural gas seepages and burning hydrocarbon source-rocks. The Sasanids were in turn defeated by the Arabs in 636 AD who brought the Muslim faith to the country, prompting an epoch of renewed cultural achievement.

Then in 1218 a devastating invasion by Mongols, led by Genghis Khan, literally decimated the population by famine and massacre. That in turn gave way to the Safavid Empire lasting from 1501 to 1920, whose dominion extended over most of what is now Iraq and eastern Turkey. It was a powerful empire espousing the *Shi'ia* brand of Muslim faith, but it too eventually gave way to factional disputes. Its end was hastened by serious famines that struck the country in 1870–1871 and again in 1917–1919, when millions died.

The growth of the oil industry in the late nineteenth century prompted various entrepreneurs to turn their eyes towards Iran. They included an Englishman by the name of William Knox D'Arcy, who sent an agent to try to negotiate oil rights in 1901, facing competition from the Russians. This in turn prompted moves by Britain to bring Iran into its sphere of influence and frustrate Russian ambitions, which encouraged various elements within Iran to move towards reform, and the so-called modernisation. That led to a *coup d'etat* in which an army officer, Reza Khan (later to be known as the Reza Shah Pahlavi), came to power, initiating the construction of roads and railways, as well as a national education system. Trying to balance the rival pressures of Britain and Russia in the inter-war years, he welcomed German friendship and influence, but that in turn prompted an invasion by Britain and the Soviet Union in 1941 during the Second World War, forcing the Shah to abdicate in favour of his son, Mohammad Reza Pahlavi.

New pressures followed the war with the emergence of a popular nationalist politician, Mohammed Mossadegh. He was the son of a Bakhtiari finance minister and a princess, and had been educated in Switzerland. He sought to establish democracy in the country and nationalise the long standing exclusive oil concession with the Anglo-Iranian Oil Company, the predecessor of BP. The Prime Minister, who had opposed the move on technical grounds, was assassinated in March 1951, and a few days later the Iranian Parliament moved to nationalise the company's rights, appointing Mossadegh as the new Prime Minister.

Although the post-war Socialist government of Britain had nationalised many industries at home, it resisted a comparable move by Iran, withdrawing BP's technical staff and embargoing exports. This had serious adverse economic consequences for the country.

Britain then sought US aid in deposing Mossadegh with the help of its Secret Services under the so-called Operation Ajax, calling on the Shah to exercise his prerogative to dismiss the Prime Minister, having fomented various internal disputes. Mossadegh was successfully deposed and spent the remainder of his life under house-arrest. His successor, Fazlollah Zahedi, soon came to terms with the foreign interests forming the so-called Consortium whereby American oil companies, Shell and CFP of France took 60% of what had previously been BP's exclusive position, eventually being persuaded to pay the company an indemnity for its loss.

The Shah, being fully backed by the United States and British Governments, became increasingly autocratic in the ensuing years, suppressing political opposition in his country with the help of a secret police force. He was however denounced by a religious leader, the Ayatollah Khomeini, who was subsequently exiled. It prompted growing popular opposition to the Shah, which erupted in the form of the Iranian Revolution of 1978, when mass popular demonstrations caused the Shah to flee the land in the following year. The Ayatollah returned from exile to declare an Islamic Republic. Relations with the United States deteriorated when a group of students seized its embassy, taking its staff as hostages, in order to secure the return of the Shah for trial. The hostages were later freed after an abortive US military mission.

Perceiving the level of US opposition to the Iranian Government, Saddam Hussein, the leader of neighbouring Iraq, saw an opportunity to press earlier territorial claims to give his country a greater access to the Persian Gulf, and decided to launch a military attack in 1982. This evolved into the Iran–Iraq War that lasted eight long years, with massive loss of life on both sides. Iraq was actively supported in the war by the United States, Britain and other countries, which furnished military and financial aid.

A new government came to power after the death of Khomeini in 1989 and sought to improve relations with the West, but with little success. In 2003, the United States and British forces invaded Iraq, and President Bush declared Iran to belong to the same *Axis of Evil*, in part because of its sympathy for the dispossessed Palestinians. Tensions remained high over the ensuing years with economic sanctions against the country, and the rising threat of direct military action.

Whatever the pretexts and claimed justifications in respect of Iran's nuclear facilities, there can be little doubt that Iran's control of the Gulf of Hormuz through which Middle East oil exports pass is a strategic factor in the dispute.

Mahmoud Ahmadinejad came to power in 2005 maintaining a fairly hard and independent line, being re-elected in 2009 although facing serious demonstrations and unrest. Rostam Ghasemi, a senior member of the Revolutionary Guards, was appointed oil minister in 2011, and automatically assumed the presidency of OPEC, where he may have an opportunity to strengthen Iran's hand in international relations connected to oil supply.

Looking ahead, Iran, which still has substantial oil and gas resources, now supplemented by nuclear power, appears to be relatively well placed to face the Second Half of the Age of Oil, assuming that it is not subject to the US/Israeli attack. Such an act of aggression might achieve an initial military success, but would undoubtedly alienate the mass of the people numbering in excess of 70 million. The country has had a very long history and may well emerge relatively well from the First Half of the Oil Age that triggered so much foreign intervention.

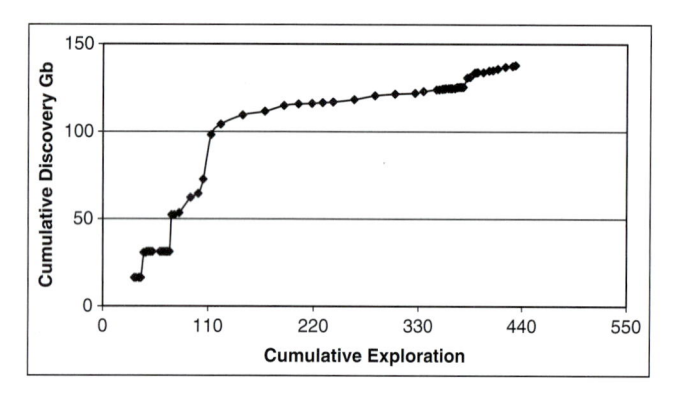

Fig. 60.3 Iran discovery trend

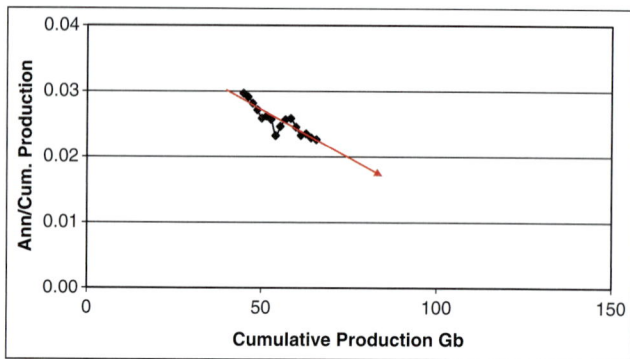

Fig. 60.4 Iran derivative logistic

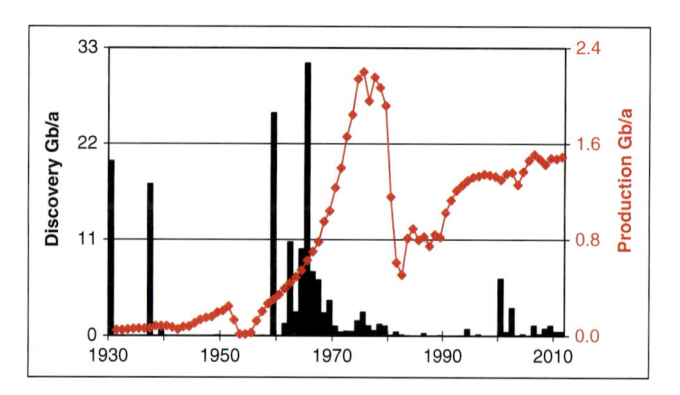

Fig. 60.5 Iran production: actual and theoretical

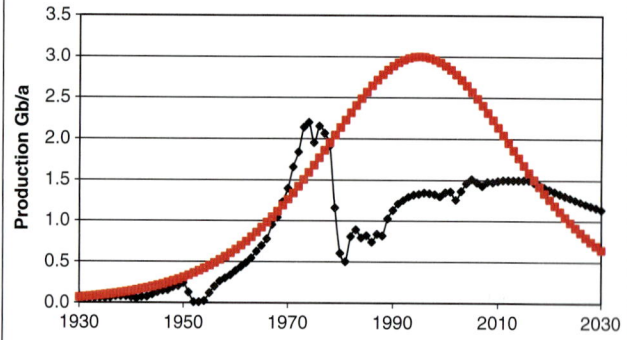

Fig. 60.6 Iran discovery and production

Iraq

Table 61.1 Iraq regional totals (data through 2010)

Production to 2100					Peak Dates			Area	
Amount			Rate			Oil	Gas	'000 km²	
	Gb	Tcf	Date	Mb/a	Gcf/a	Discovery	1928	1953	Onshore	Offshore
PAST	34	14	2000	938	154	Production	2025	2030	440	0.5
FUTURE	81	136	2005	685	401	Exploration	1978		*Population*	
Known	69	95	2010	876	596	*Consumption*	Mb/a	Gcf/a	1900	2.0
Yet-to-Find	12	41	2020	1,426	971	2010	253	46	2010	33
DISCOVERED	103	109	2030	1,290	1,567		b/a	kcf/a	Growth	16.4
TOTAL	115	150	*Trade*	+622	+550	Per capita	7.7	1.4	Density	74

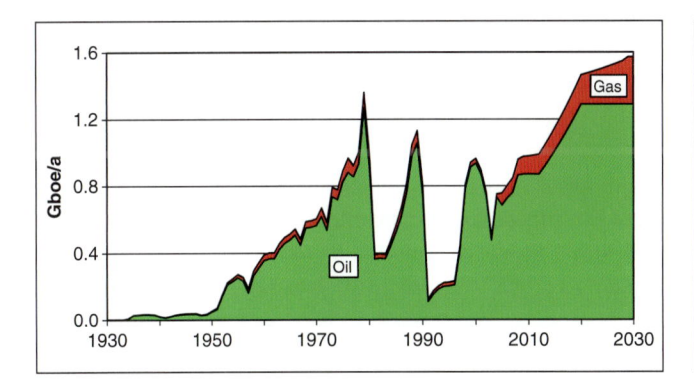

Fig. 61.1 Iraq oil and gas production 1930 to 2030

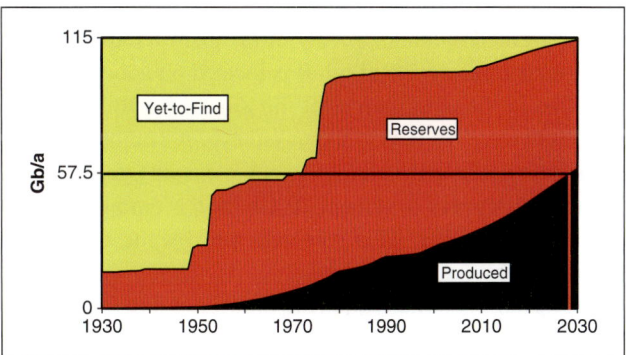

Fig. 61.2 Iraq status of oil depletion

Essential Features

Iraq is a landlocked country of 440,000 km² in the centre of the Middle East. It includes relatively fertile regions along the valleys of the Euphrates and Tigris Rivers, which drain into the Persian Gulf. But they are bordered by an arid plateau to the north and by extensive deserts to the south and west. It is home to almost 33 million people, of whom about 60% belong to the *Shi'ia* sect of Islam. An important Kurdish minority live in the north of the country

Geology and Prime Petroleum Systems

Iraq covers the axis and western flank of the Persian Gulf Basin as well as a section of the foothills of the Zagros Mountains to the east. Precambrian rocks of the Arabian Shield are overlain by a thick sequence of gently deformed sedimentary rocks offering two prime petroleum systems. In the west, Silurian source-rocks have charged the overlying Permian sandstones with hydrocarbons, which may include oil in western regions, but with deeper burial to the east are likely to offer no more than gas-condensate. The western platform of Iraq possessing this system has received only a modest exploration effort, but the results to date are probably sufficient to conclude that its potential is limited.

The main system lies in axial part of the basin where Upper Jurassic and mid-Cretaceous prime source-rocks have charged intervening sandstone and limestone reservoirs with substantial amounts of oil. Other leaner source-sequences are also present. Some of this oil has re-migrated upwards into Miocene fractured carbonate reservoirs in the Zagros Foothills, and some may have migrated long distances westward into the bordering platform. Effective evaporate seals contribute significantly to the entrapment of oil and gas.

In structural terms, the basin is cut by transverse faults, related to the earlier Hercynian structural configuration, giving highly prospective uplifts separated by less promising structural depressions. In general, the basin rises and thins to the north where it swings into Turkey.

Oil migration was primarily vertical giving rise to multiple productive reservoirs containing oil of varying density, reflecting the particular conditions of generation and migration. Multiple seals form an important part of the trapping mechanism.

Exploration and Discovery

The first recorded exploration boreholes were drilled in 1903 and 1905 making small discoveries, which encouraged the formation of the Turkish Petroleum Company, later reconstituted as the Iraq Petroleum Company (IPC) following the defeat of Turkey in the First World War. Exploration was soon rewarded by the discovery of the giant Kirkuk Field in 1927 with some 16 Gb of oil. It is located on a huge surface anticline in the outer ranges of the Zagros Foothills, whose prospects were obvious to the naked eye. A number of other finds were made in the vicinity over the ensuing years, bringing total discovery to almost 20 Gb by the outbreak of the Second World War. This was achieved by the drilling of some 50 exploration boreholes.

Operations resumed following the War as attention turned to the southern end of the country in front of the Zagros foldbelt where seismic surveys were needed to identify and delineate the prospects. It was rewarded in 1949 with a string of giant finds, containing about 10 Gb, followed by further important finds over the ensuing decades. The larger fields of Iraq are listed in the table, illustrating the normal pattern of the larger being found first.

Exploration has been substantially reduced over the past decade because of the political difficulties with no more than 14 exploration boreholes, out of a total of 167, being drilled since 1990.

Table 61.2 IPC-shareholding

Name	Date	Size Gb
Rumailia	1953	22
Kirkuk	1927	16
E. Baghdad	1976	11
W. Qurna	1973	6
Majnoon	1977	6
Zubair	1949	4
Nahr Umr	1949	4
Jambur	1954	3
Bai Hassan	1953	2
Subba	1977	2
Khabbaz	1976	2

Almost 110 Tcf of gas have been discovered in the course of the oil exploration, most of which remains to be produced, save in the case of Kirkuk, where the gas cap was depleted to provide for local electricity generation during the embargo, which has adversely affected the recovery of the underlying oil.

Despite the political constraints, it is evident that the larger fields of Iraq have already been found, with the scope for future discovery being here assessed at about 12 Gb. Much media attention has been directed at the little drilled western platform, but the results are likely to be disappointing as it lacks the merits of the prime trend to the east.

Production and Consumption

Oil production commenced in 1928 but did not reach significant levels until the mid-1930s, not passing 100 kb/d until 1945. It then rose steeply to an overall peak of 3.5 Mb/d in 1979, after which it became volatile in response to various political constraints. It currently stands at about 2.4 Mb/d and is here expected to rise to a plateau of about 3.9 Mb/d around 2020. The Depletion Rate is very low at about 1%, suggesting that in technical terms there is scope to increase, but on balance it is doubted that the political circumstances will so permit.

Consumption stands at 253 Mb/a, giving a per capita consumption of 7.7 barrels a year. The country can accordingly remain a substantial exporter for many years to come. Gas was flared in earlier years with recorded production commencing in 1950. It remained at a modest level thereafter, and is expected to increase in the future if political conditions so permit. Plans for a new gas pipeline from the Middle East to Europe are under consideration. If constructed, it might tap Iraq's large gas resources.

The Oil Age in Perspective

Some of the world's greatest ancient civilisations developed in this area. Indeed, the Garden of Eden, where Adam and Eve disported themselves, is supposed to have been located here. Cyrus the Great of Persia conquered the place in 539 BC, before it fell to Alexander the Great in 331 BC. Greek and renewed Persian dominion followed until it was overrun by Muslim Arabs in the seventh century. It was later subject to Mongol invasions, and the attentions of Persian and Turkish rulers, before the Ottomans established a firm grip in the seventeenth century, operating eventually through three local administrations (*vilayets*) having a fair degree of delegated authority. Various nomadic Arab tribes were never fully integrated, and the Kurds in the north, being descendants of the ancient Medes, have long sought their independence.

The area began to attract the conflicting commercial and political attentions of Britain and Germany during the latter part of the nineteenth century. Britain, as a trading sea-power, was interested in the coastal areas, including what is now Kuwait, and also established a shipping company on the Tigris to serve the interior. Germany, being a land-power, proposed building a railway from Berlin to Baghdad, recognising its importance in a military context. The Middle East itself seems to have been of limited interest to Britain, apart from having a strategic importance as a bastion against Russian expansion threatening communications with India, the jewel in Britain's imperial crown.

Oil had been known in the area since antiquity, being used as a form of mortar in the construction of Babylon. New interest developed in the early years of the twentieth century, when engineers came across oil seepages in the course of surveying the concession granted by the Ottoman Sultan for the proposed German railway. The Sultan called in a young Armenian oilman, by the name of Calouste Gulbenkian, to investigate, launching him on what became his life's work to develop Iraq's oil. To this end, he established the Turkish Petroleum Company in 1912. It was owned by the Deutsche Bank (25%), which controlled the previous railway concession that conveyed the mineral rights, Shell (25%) and the Turkish National Bank (50%). The latter had been set up by British financial interests, with Gulbenkian holding 30%. The British Government then intervened to secure a holding for what is now BP, reducing Gulbenkian's share to 5%.

The rights to the concession were confirmed on 28th June 1914, a few days before the outbreak of the First World War, in which Turkey sided with Germany, with whom it already had close links. The importance of oil became evident during the war, and France and Britain, followed by the United States, began to discuss the eventual carve up of the Middle East while hostilities were still in progress. It was already perceived to hold much of the world's endowment.

Negotiations began in earnest in the peace treaties that followed the war, eventually giving Britain mandated administrative control of the territory. It was declared a Kingdom, with the crown being placed on the head of Prince Feisal, the son of the Grand Sharif of Mecca. He had been Britain's premier ally in the war, and had been promised an Arab Kingdom in return for his contribution. In fact, Feisal had first been put on the throne of Syria, but was recycled when that country came into the more republican French sphere of influence.

It was agreed that Iraq's oil, which had become a central issue, would be produced by what became the IPC with the following shareholding:

Shell (Anglo-Dutch)	23.75%
BP (British)	23.75% (previously Anglo-Persian and Anglo-Iranian)
CFP (French)	23.75% (now TotalFinaElf)
Exxon (US)	11.875% (now Exxon-Mobil)
Mobil (US)	11.875% (now Exxon-Mobil)
Gulbenkian (Independent)	5%

The companies also agreed not to compete with each other throughout most of the previous Turkish Empire, including Saudi Arabia: Exxon and Mobil later reneged on the agreement when they joined Aramco in Saudi Arabia in the 1930s.

Exploration soon commenced, and was richly rewarded with the discovery of the Kirkuk Field in 1927, holding about 16 Gb of oil in the northern, Kurdish, part of the country. Production rose gradually to the Second World War, reaching 100 kb/d by 1947. Iraq was not, accordingly, a particularly important exporter to that point.

The post-war epoch was characterised by growing nationalism throughout the region, which was given more encouragement when the United States opposed an Anglo-French military strike to prevent Egypt sequestering the Suez Canal in 1956. Most of the producing countries nationalised the holdings of the foreign oil companies over the ensuing years: Iraq doing so in 1972.

Saddam Hussein was born in 1937, and joined the Ba'athist Party 20 years later. It had been formed in Syria in 1943, adopting socialist principles, being opposed to colonialism, but was run on authoritarian lines. In the following year, the then King, Feisal II, was beheaded in a coup led by a Colonel Kassim, who was backed by Egypt. He in turn fell in another coup that brought the Ba'athists to power in 1968, appointing Saddam Hussein as President in 1979. As described above, the country was a somewhat artificial construction, comprising Kurds—who have long sought independence—in the north, *Shi'ites* with links to Iran in the south and *Sunni's* around Baghdad, the capital. Evidently, it took a strong leader to hold these disparate groups together as a nation. It previously had a substantial, well-integrated Jewish community, Baghdad having been one of the great centres of Judaic culture in the fifth century, but it was driven out by popular outrage on the creation of the State of Israel.

In 1974, heavy fighting broke out between government forces and Kurdish separatists, who were being backed by Iran, but the dispute was settled when Iran withdrew its support in return for resolution of a long-standing boundary dispute related to the key Shatt al-Arab estuary of the Tigris-Euphrates river system, Iraq's main trade route. Tensions with Iran erupted again due to the fall of the Shah in 1979, when unrest among the Iranian Kurds spilled over into Iraq. It soon developed into a full-scale war, which dragged on for almost eight long years with colossal loss of life to both sides. Although nominally neutral, the United States backed Iraq during this conflict, still smarting from an incident in which American citizens were taken hostage in Tehran, following the fall of the Shah. During the late 1980s, the United States supplied Iraq with substantial bank credits and technology to rebuild its military strength. The Soviets too developed close ties, also furnishing credit and weapons.

Meanwhile, President Reagan of the United States and Mrs. Thatcher, Britain's Prime Minister, resolved to try to bring down the Soviet regime, ending the policy of co-existence. The first step was to rearm the Afghans to end the Soviet occupation, and undermine its military credibility. This was achieved with the help of King Fahd of Saudi Arabia, who funded the covert purchase of arms in Egypt for shipment to none other than Osama bin Laden, who was backing the Taliban with CIA support.

The next step was to persuade King Fahd to step up Saudi oil production to undermine the global price of oil. The Soviets relied on oil exports for foreign exchange, which they now needed in greater amounts, to buy equipment with which to counter the new US *Star Wars* initiative. It was a successful strategy, which contributed to the fall of the Soviets, but was achieved at a cost as the low price of oil was bankrupting not only King Fahd but the Texan oil constituents of George Bush Sr.

While all this was going on, Kuwait arbitrarily increased its reported reserves by 50% in 1985 although nothing particular had changed in its oilfields. It did so in order to raise its OPEC production quota, which was based on reserves. It also began pumping from the southern end of the South Rumaila field that straddles the ill-defined border. Iraq complained bitterly both about what amounted to the theft of its oil across the border, and the subsequent loss of oil revenue as prices fell consequent upon Kuwait's failure to observe its contractual OPEC agreement.

Now, the US strategy moved to strengthen the price of oil by dispatching an emissary, Henry Shuyler, to encourage its ally, Saddam Hussein, to intervene in the councils of OPEC to enforce quota agreements sufficiently to achieve that end. It was recognised that words might not be enough to concentrate the minds of the OPEC ministers. Exactly what was proposed is not known, but it seems clear that a border incident to stop Kuwait producing from the southern end of the shared oilfield was contemplated. This interpretation is confirmed by the words of April Glaspie, the US ambassador to Baghdad, who, on the eve of the invasion of Kuwait, made a statement to the effect that boundary disputes between Arab countries were of no concern to the United States. It was clearly an authorised statement, being released simultaneously in Washington under the signature of James Baker, the Secretary of State.

However, Saddam Hussein, possibly misunderstanding a wink and nod, did not stop with a border incident, and mounted a successful full-scale invasion of Kuwait on 2nd August 1990. April Glaspie, on being woken by journalists with the news while on vacation, reportedly responded *Oh My God, they haven't taken the whole place, have they?* which hints of collusion or at least fore-knowledge.

The US policy now changed to condemn its former ally. A series of UN resolutions called for Iraq to withdraw from Kuwait by 15th January 1991, leading to the US aerial bombardment when it failed to comply. Ground forces, led by General Schwarzkopf, crossed the frontier, killing tens of thousands of Iraqis and destroying most of its military capability, before being ordered to halt at the gates of Baghdad when a cease-fire was agreed. The dissident *Shi'ites* in the south and the Kurds in the north saw this as their moment to rise, but were successfully suppressed by remnant government forces. Hundreds of thousands of Kurdish refugees fled into neighbouring Turkey and Iran, where they were not exactly welcome.

The United Nations was then persuaded to impose a trade embargo on Iraq, making it effectively swing oil producer of last recourse, which provided a useful mechanism for stabilising the world price of oil at no cost to anyone else. It was however relaxed from time to time for *humanitarian* reasons when the price of oil rose to uncomfortable levels.

In common with many countries, Iraq had made certain progress in developing modern nuclear, chemical and biological weapons, but by 1998, UN inspectors had reported that virtually all such facilities had been destroyed. Several European, Russian, Chinese and other companies signed agreements to develop oilfields as soon as the embargo was lifted, committing over $1.7 trillion to do so. Such agreements were subsequently nullified to make way for the eventual entry of US companies.

The United States and Britain invaded Iraq in 2003 on the pretext of its alleged weapons which were not in the event found. Saddam Hussein was executed in 2006 for being responsible for the death of 148 *Shi'ites* in the town of Dujail in the 1980s, albeit a small number compared with the million or so innocent people who directly and indirectly lost their lives as a consequence of the invasion.

A relatively stable, albeit weak and corrupt, government has run the country since the war, allowing the withdrawal of the United States and British troops, save in a training capacity. The people suffered greatly from the wars and remain in an impoverished condition such that serious tensions erupt from time to time. For example, car bombs and killing in the town of Kut in August 2011 left 70 dead and 238 injured. International oil companies are beginning to establish themselves in the country despite the risks and many difficulties, but it is unlikely that significant increases in production will be achieved in the years ahead.

It is very difficult to foresee how Iraq will fare during the Second Half of the Age of Oil. On balance it seems likely that it will fragment into *Shi'ia*, *Sunni* and *Kurdish* communities facing much internal conflict. It may fragment even further as different regions try to protect themselves and find a sustainable future on what are, for the most part, barren lands.

61 Iraq

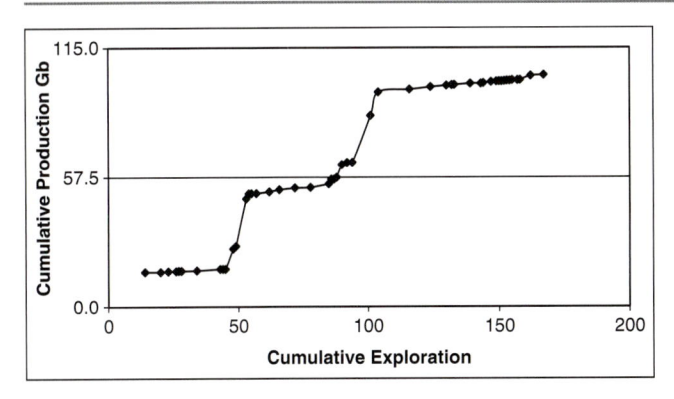

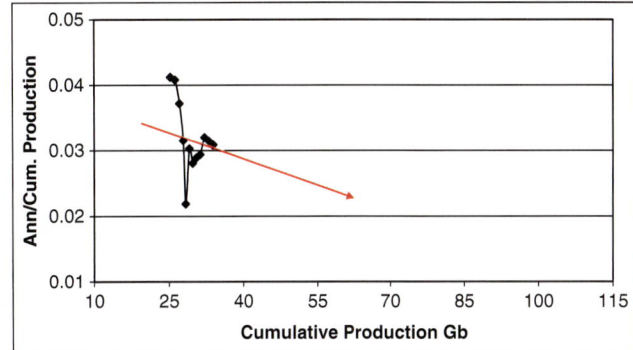

Fig. 61.3 Iraq discovery trend

Fig. 61.4 Iraq derivative logistic

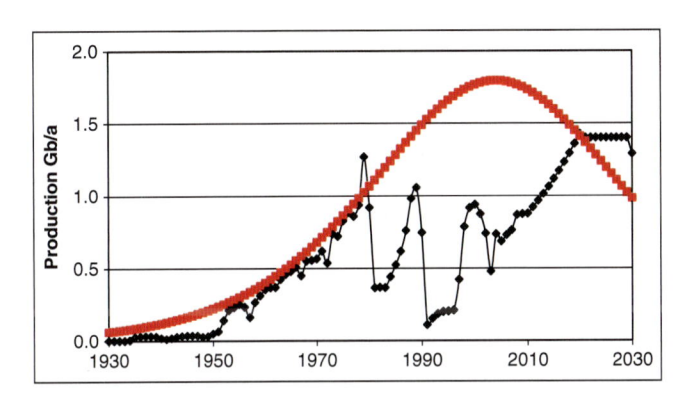

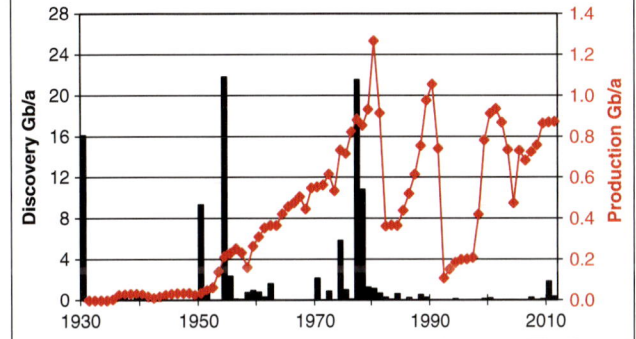

Fig. 61.5 Iraq production: actual and theoretical

Fig. 61.6 Iraq discovery and production

Kuwait

Table 62.1 Kuwait regional totals (data through 2010)

	Production to 2100					Peak Dates			Area	
	Amount		Rate				Oil	Gas	'000 km²	
	Gb	Tcf	Date	Mb/a	Gcf/a	Discovery	1938	1938	Onshore	Offshore
PAST	42	21	2000	644	396	Production	1971	2005	20	15
FUTURE	58	49	2005	817	533	Exploration	1963		Population	
Known	55	46	2010	761	422	*Consumption*	Mb/a	Gcf/a	1900	0.1
Yet-to-Find	2.9	2.4	2020	761	400	2010	129	446	2010	2.8
DISCOVERED	97	68	2030	677	400		b/a	kcf/a	Growth	28
TOTAL	100	70	Trade	+632	−24	Per capita	46	159	Density	157

Note: Excludes Kuwait's share of Neutral Zone

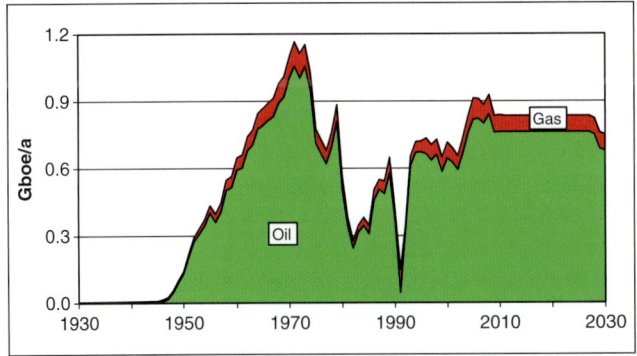

Fig. 62.1 Kuwait oil and gas production 1930 to 2030

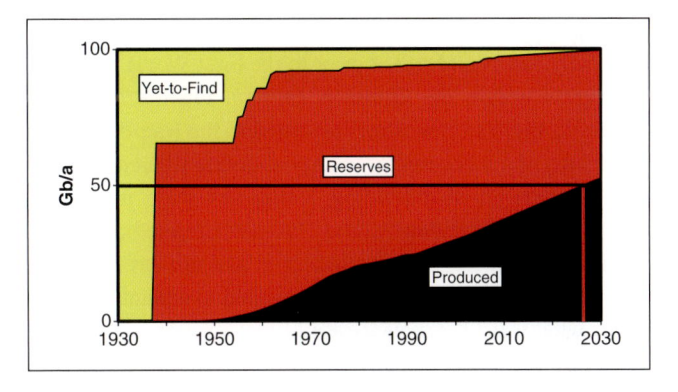

Fig. 62.2 Kuwait status of oil depletion

Essential Features

Kuwait is a small, flat-lying territory of 20,000 km², adjoining the mouth of the Euphrates River at the head of the Persian Gulf. Kuwait City developed as an important trading town on the shores of a bay at the edge of desert which extends inland to the west. There are several islands off the coast of which the largest is Bubiyan. The giant Burgan oilfield was found in 1938 bringing enormous wealth to the town and its inhabitants, especially its ruling family.

Kuwait also lays claim to offshore waters, but the boundaries have been subject to dispute with Iran and Iraq. It also administers half of the so-called Neutral Zone on the Saudi Arabian border, having rights to half of its oil revenue.

The population now numbers at least 2.8 million, making it one of the most densely populated of countries. Almost half the population is made up of non-nationals, of whom 400,000 Indians comprise the largest community. Two-thirds of the Muslim population belong to the *Sunni* sect, with the remainder being *Shi'ia*.

Geology and Prime Petroleum Systems

Kuwait lies on the western side of the axis of the Persian Gulf Basin where the stratigraphic column attains a thickness of about 6,000 m, almost half of which being made up of Mesozoic strata. The prime source-rock is probably the

Middle Cretaceous Magwa Formation, where it thickens and lies at greater depth on the flanks of the productive uplifts. There may, in addition, be deeper Upper Jurassic sources present locally, being within the oil generating window due to a relatively low geothermal gradient.

The principal reservoir is the Cretaceous Burgan Sandstone, which is about 300 m thick, but there are in addition subsidiary reservoirs. In structural terms, there are two major uplifts: the Burgan Uplift, close to Kuwait City, and the Khafji-Nowruz Arch offshore, which forms an extension of the Safaniya structure of Saudi Arabia. A key element responsible for their prolific entrapment was probably their steady growth over geological time due to deep-seated salt movement, which meant that whatever oil was generated in their neighbourhood collected in them.

Exploration and Discovery

Following the discovery of oil in the adjoining Zagros foothills of Iran during the early years of the last century, attention turned to the territories to the west despite a lack of identifiable surface structures. The Turkish Petroleum Company had been formed prior to the First World War having rights within the Ottoman Empire. On the defeat of Turkey in the war, it was reconstituted as the Iraq Petroleum Company, and it became necessary to define its area of interest in which its shareholding companies were prohibited from acting independently. The Ottoman Empire had been a somewhat informal construction so it was not easy to draw firm boundaries. However, it was decided to exclude Kuwait under what was known as the Red Line Agreement.

A New Zealand entrepreneur then arrived on the scene in the 1920s and successfully secured rights first to Bahrain before turning his attention to Kuwait. Gulf Oil was the first to express interest but then faced competition from BP, partly at the instigation of the British Government, which desired to maintain the territory in its sphere of influence. After lengthy negotiations, the two companies agreed to form the Kuwait Oil Company in 1934 securing a 75 year lease to the territory.

Surface outcrops were lacking, so it was necessary to use pioneering seismic surveys to search for prospects. They soon identified the Burgan structure which was successfully tested by a borehole in 1938, finding one of the largest fields in the world with an estimated 60 Gb. The development of this field was enough to occupy its owners for many years to come. Accordingly, exploration drilling did not resume until the 1950s, and even then at a low level, commonly with a single borehole a year, save for 1963 when twelve were drilled. The results were however extremely positive with the discovery of a serious of large fields, including Raudhatain in 1955 with 5 Gb; Sabriya in 1957 with 6 Gb; Minagish in 1959 with 3 Gb; Umm Gudair in 1962 with 5 Gb; and Ratqa in 1977 with 1 Gb, and a number of smaller subsequent finds. Altogether, some 97 Gb have been found in what is by now a very mature area, with limited future potential.

Some 68 Tcf of gas have also been found, and there may indeed be considerable potential for more from deep-seated sources and reservoirs.

It is well to note that the statistics are confused by the treatment of the Kuwaiti share of the Neutral Zone, which differs from dataset to dataset.

Production and Consumption

Oil production commenced in 1946, having been delayed by the Second World War, and grew steadily to reach an early peak of 2.9 Mb/d in 1971, before falling to a low of 666 kb/d in 1982 as a result of the Arab oil embargo and OPEC quota constraints. It built thereafter to a second peak of 1.5 Mb/d in 1989 before collapsing in 1991 as a result of the Gulf War. It has since recovered to 2 Mb/d in 2010, by which time some 37% of the estimated total had been produced. The Depletion Rate is very low at 1.2%, suggesting that in strictly technical terms, production could be increased were massive investment applied to drilling up the smaller fields, especially those in the north of the country. But on balance, it seems more likely that the country will try to hold the present level to preserve its resources for as long as possible.

Gas production has risen in parallel but is now expected to plateau at about 400 Gcf/a.

The country consumes 129 Mb/a of oil giving it a very high per capita level of 46 barrels a year. It consumes 446 Gcf/a of gas, presumably for electricity generation.

Oil Age in Perspective

Kuwait, like the rest of the Middle East, has had a long history being at different times under the sovereignty of the Persian and other empires whose fortunes waxed and waned with shifting power and allegiances. The foundation of the present Sheikhdom may be attributed to 1756, when the territory was invaded by settlers from Arabia, led by the Sabah family that remains in power to this day. It came close to joining the Ottoman Empire in the latter years of the nineteenth century, but that alliance ended under pressure from Britain that made the territory a Protectorate in the First World War. Defining the precise borders with Iraq to the north and Saudi Arabia to the west was of little concern at the time insofar as they were substantially un-inhabited deserts, but the formal boundaries were drawn under British influence in various treaties in the early 1920s, which evidently favoured the claims of Saudi Arabia.

The discovery of oil in 1938, however, changed the position, and prompted Iraq to mount claims based on vague historical associations, and to support an unsuccessful uprising by merchants against the ruling family. These claims resurfaced after the Second World War, especially in 1961 when Kuwait ended its status as a British Protectorate, prompting Iraq to threaten invasion which was countered by British forces. Two years later, Iraq was persuaded to formally recognise the independence of the country.

Oil wealth poured into the country over the succeeding years as did an increasing number of immigrants. The money did not find an easy home and led to the development of a poorly regulated stock market that crashed in 1982, causing much financial grief and certain political tensions.

A brief period of rapprochement with Iraq came in the 1980s when the latter won Kuwait's financial support in the Iraq–Iran War, at a time when Kuwait feared it might be a victim of an invasion by Iran. This was no doubt made easier by the US support for Iraq at the time.

Tensions with Iraq returned in the mid-1980s for two oil-related reasons. First, Kuwait started producing oil from the South Rumaila Field with straddles the ill-defined boundary with Iraq, and in the absence of a unitization agreement may indeed have been able to take some of what was properly Iraq's share of production, or at least to give Iraq reason for such an accusation. The second factor was a decision by Kuwait to increase its reported reserves by 50% from 64 to 90 Gb in 1985, although nothing particular had changed in the oilfields. OPEC production quotas were based partly on reported reserves, so Kuwait's action led to a surge of production, matched two years later by other OPEC countries, which had the effect of reducing price. Iraq reasonably complained that Kuwait's action was in conflict with contractual agreements, and resulted in a loss of Iraqi revenue. It formalised its complaint with a legal claim in the courts.

The United States was supporting Iraq at the time, and evidently gave its leader, Saddam Hussein, reason to believe that it would turn a blind eye in the event that Iraq should take full ownership of the shared oilfield by military action, as confirmed by the remarks of April Glaspie, the US Ambassador, who issued a statement, counter signed by James Baker, the Secretary of State in Washington, to the effect that *boundary disputes between Arab countries were of no concern to the United States*. The subtleties of diplomacy may have been lost on Saddam Hussein who thought that he had *carte blanche* for a full scale invasion which followed on 2nd August 1990.

The United States policy then changed as it mounted the Gulf War against Iraq early in the following year, when the retreating Iraqi forces set fire to the oil wells, which burned unchecked for more than 9 months with the loss of some two billion barrels.

Sheikh al-Sabah soon regained his throne and the country returned to normality as oil production and accompanying oil revenue resumed. But the fear of further conflict with Iraq remained.

The Anglo-American invasion of Iraq followed in 2003, being facilitated to some degree by Kuwait. However, it is obvious that it has created new threats, especially in relation to disputed waterways on the Iraq–Iran border close to Kuwait, being probably accompanied by internal tensions as different elements of Kuwait's society align themselves with the rival factions that arise in Iraq, which may indirectly lead to new opposition to the running al-Sabah family.

An interesting development has been a statement by the Oil Minister that Kuwait's Proved Reserves stand at 24 Gb despite being officially reported at 102 Gb, while its *Proved & Probable* amounts to 51 Gb, which is not too far removed from the estimate of 55 Gb, preferred here. The exact reason for this admission is not known, but it could well have been to counter pressure on the country to raise production in the face of high oil prices. Its oil wealth is indeed colossal, as may be readily computed: if production costs about $10 a barrel and sells at $100, current production yields 68 billion dollars a year, or $22,000 per capita.

It is difficult to know how Kuwait will fare during the Second Half of the Age of Oil. On the one hand, it could continue to enjoy immense wealth over the next few decades if oil prices remain high and it manages to maintain production. On the other hand, it is never easy to be a rich man in a crowd of beggars, and Kuwait may be vulnerable to the evolving tensions of the region. The passions ignited by the Anglo-American invasion of Iraq have spilled over into increased internal conflict, which could overflow into opposition to the ruling family, of which some signs have already been seen. The high price of oil has triggered a world economic recession, and further devaluation of the dollar that could impact Kuwait's substantial foreign investments, with ramifications at home. An epoch of extreme price volatility might be even worse. Kuwait's large immigrant population, lacking any particular roots in the country, could well be another destabilising element. But by the end of the century, when its oil and associated wealth have been substantially depleted, Kuwait may again find a role as an important trading centre at the mouth of the Euphrates River.

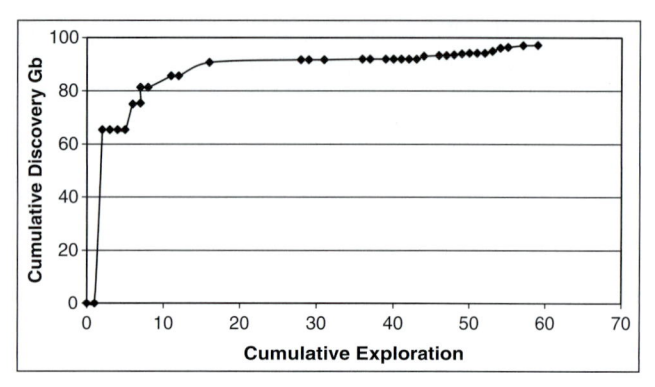

Fig. 62.3 Kuwait discovery trend

Fig. 62.4 Kuwait derivative logistic

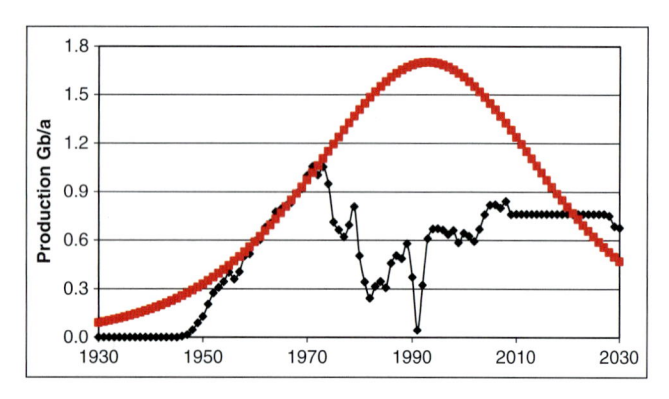

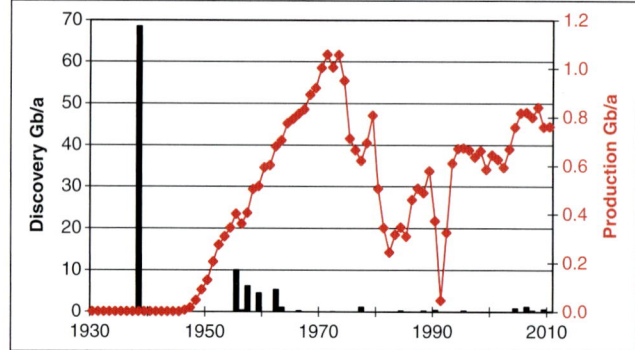

Fig. 62.5 Kuwait production: actual and theoretical

Fig. 62.6 Kuwait discovery and production

Neutral Zone

Table 63.1 Neutral Zone regional totals (data through 2010)

Production to 2100						Peak Dates			Area	
Amount		Rate					Oil	Gas	'000 km²	
	Gb	Tcf	Date	Mb/a	Gcf/a	Discovery	1960	1967	Onshore	Offshore
PAST	8.3	2.7	2000	230	47	Production	2003	2006	5	7.2
FUTURE	4.7	7.3	2005	212	60	Exploration	1962		*Population*	
Known	4.2	6.9	2010	193	42	*Consumption*	Mb/a	Gcf/a	1900	0.05
Yet-to-Find	0.5	0.4	2020	129	50	2010	11	60	2010	0.3
DISCOVERED	12.5	9.6	2030	86	50		b/a	kcf/a	Growth	6
TOTAL	13	10	Trade	+183	−18	Per capita	37	200	Density	61

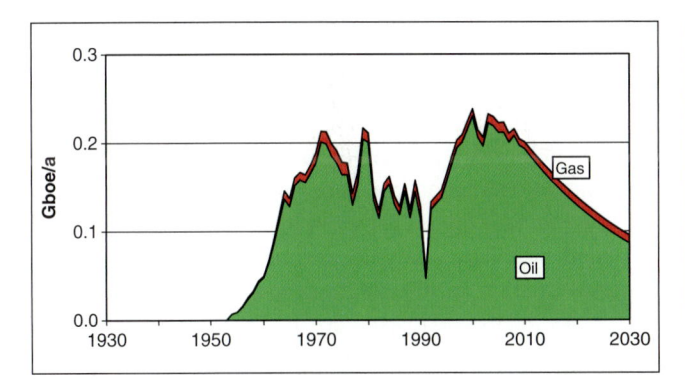

Fig. 63.1 Neutral Zone oil and gas production 1930 to 2030

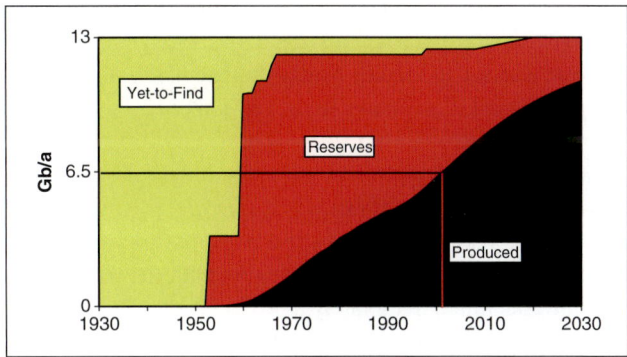

Fig. 63.2 Neutral Zone status of oil depletion

Essential Features

The Neutral Zone is an anomalous territory. When Britain dismembered the Ottoman Empire and sphere of influence at the end of the First World War, it left a sort of no-man's land between Kuwait and Saudi Arabia, to avoid the difficulties of determining a firm boundary between the countries. It was a matter of small import in those days because the desert was occupied by no more than a few wandering Bedouin.

The Saudi–Kuwaiti border was eventually drawn across the territory and petroleum rights awarded to different companies on either side of it, but to avoid tensions, Saudi Arabia and Kuwait agreed to share equally the oil revenues.

It is a sparsely populated region with a few towns along the coast, one of which by the name of Mina al Ahmadi has become an important oil export and trans-shipment port.

Geology and Petroleum Systems

There are two productive trends. The same uplift as that providing the giant Burgan Field of Kuwait continues southwards into the Neutral Zone where it has yielded the Wafra and Fawaris fields, while offshore the Safaniya trend of Saudi Arabia runs northwards into the Neutral Zone, providing the Hout, Lulu and Dorra Fields, the extension of Safaniya itself being known as Khafji.

Exploration and Discovery

Exploration drilling commenced in 1950, and three years later yielded the Wafra Field with 3.3 Gb of oil. Exploration turned offshore in 1960 and was also soon rewarded. Exploration peaked in 1962 when seven boreholes were sunk but has dwindled since. A total of just over 40 exploration boreholes have been drilled.

Peak discovery was in fact represented by the onshore Wafra Field, found in 1953, because the larger Khafra Field is an extension of Safaniya and hence attributable to 1951, when it was found.

A total of 12.5 Gb have been found, and exploration evidently is now at a mature stage, with no more than 470 Mb being expected from future discovery.

Production and Consumption

Oil production commenced in 1954 and rose to an early peak of 560 kb/d in 1979, before declining to a low of 129 kb/d in 1991: the decline being no doubt associated with the invasion of Kuwait. It recovered to 610 kb/d in 2003, before falling to 530 kb/d in 2010. Since some 64% of the estimated total resource has been depleted, it may be assumed that future production will decline at the current depletion rate of 3.9% a year.

Consumption data are not available, being included in those of Kuwait and Saudi Arabia, but are here estimated on a per capita basis, as shown in the table above.

Oil Age in Perspective

As mentioned above and in connection with the sections on Kuwait and Saudi Arabia, Britain faced the challenge of dividing up the spoils of war on the defeat of the Ottoman Empire in the First World War. It was evidently not an easy task, especially in relation to the desert lands having no natural frontiers. It was expedient therefore to establish a sort of no-man's-land between Saudi Arabia, Kuwait and Iraq, which at the time was of no particular significance to anyone, since it was a desert occupied by no more than a few migratory Bedouin tribesmen.

The discovery of oil in the giant Burgan Field of Kuwait prompted definition of the adjoining territories, but this was not achieved until after the Second World War when Kuwait and Saudi Arabia offered concessions on what they perceived to be their share of the territory, to respectively Aminoil and Getty Oil.

The awards themselves in part reflected a policy by the US Government to encourage American independent companies to move into the Middle East to counter the perceived stranglehold by the major companies, who were held to have too dominant a position. Aminoil was a consortium of Phillips, Ashland and Sinclair, and negotiated rights from Kuwait 1947, while Getty successfully negotiated with the Sauds, offering a 50:50 deal, to the dismay of the major companies which were enjoying much more favourable terms. This may have alerted the governments of the Middle East more generally that their concessions could stand better terms, which in turn may have led indirectly both to the formation of OPEC and the eventual nationalisations.

So far as the Neutral Zone itself was concerned, Kuwait and Saudi Arabia formally decided upon the delineation of a common frontier in 1969, dividing the territory, which ceased to have an independent status, save that its oil revenues are shared equally.

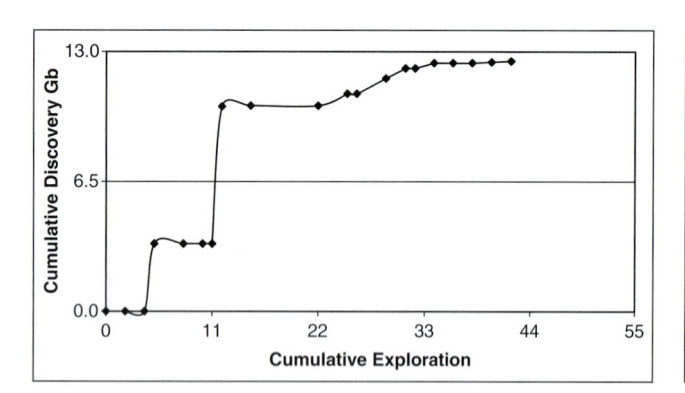

Fig. 63.3 Neutral Zone discovery trend

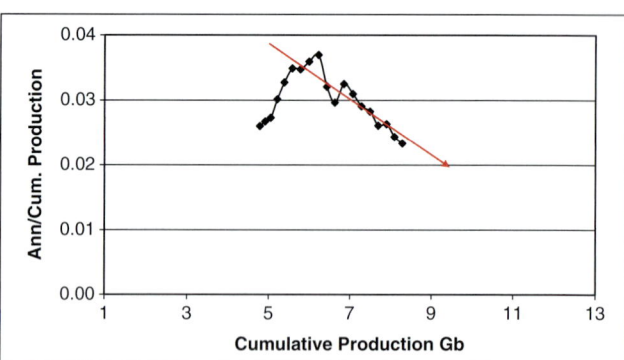

Fig. 63.4 Neutral Zone derivative logistic

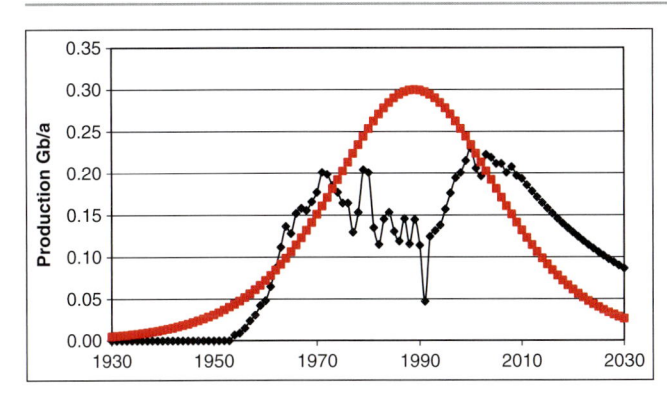

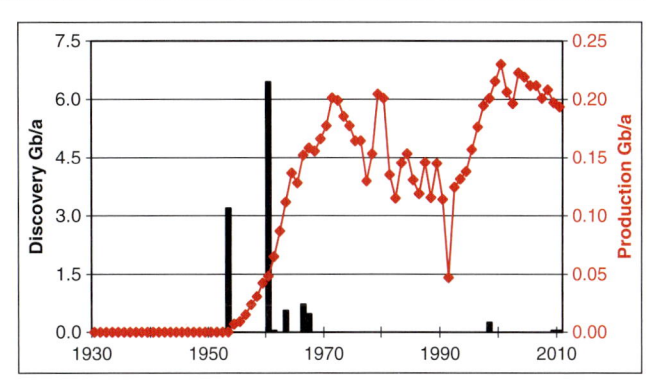

Fig. 63.5 Neutral Zone production: actual and theoretical

Fig. 63.6 Neutral Zone discovery and production

Oman

Table 64.1 Oman regional totals (data through 2010)

	Production to 2100					Peak Dates			Area	
	Amount		Rate				Oil	Gas	'000 km²	
	Gb	Tcf	Date	Mb/a	Gcf/a	Discovery	1962	1973	Onshore	Offshore
PAST	9.3	15	2000	354	484	Production	2000	2021	210	7.5
FUTURE	5.7	45	2005	283	848	Exploration	1991		Population	
Known	5.1	40	2010	316	1176	Consumption	Mb/a	Gcf/a	1900	0.5
Yet-to-Find	0.6	4.5	2020	184	1,300	2010	52	619	2010	3
DISCOVERED	14.4	55.5	2030	107	1,300		b/a	kcf/a	Growth	1.2
TOTAL	15	60	Trade	+264	+557	Per capita	17	206	Density	14

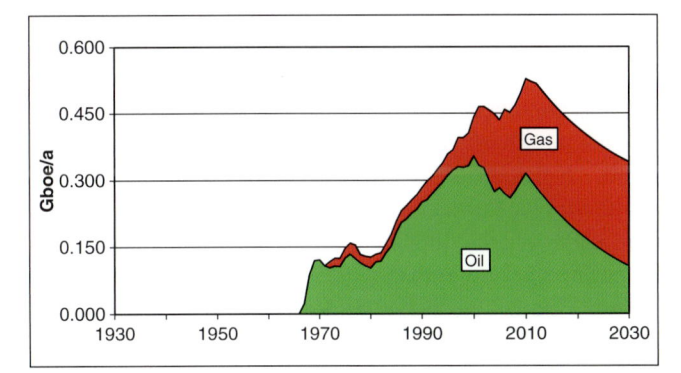

Fig. 64.1 Oman oil and gas production 1930 to 2030

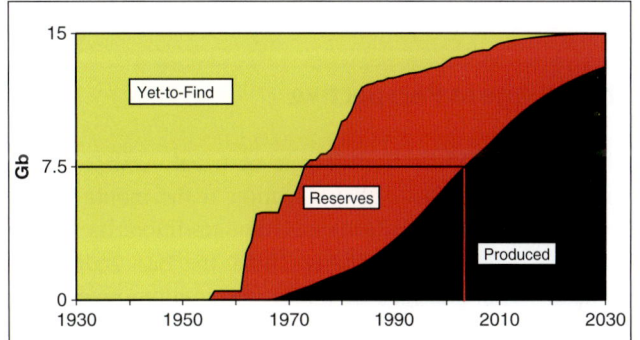

Fig. 64.2 Oman status of oil depletion

Essential Features

Oman lies on the south-eastern limit of the Arabian Peninsula, adjoining the mouth of the Persian Gulf. It covers an area of 309,000 km², made up of a fertile coastal strip along the northeast coast, which is separated from extensive deserts over the rest of the country by a mountain range, which rises to over 3,000 m, The population of three million is concentrated on the northeast coast, which enjoys a hot and humid climate, watered by the SW monsoon. Muscat, the capital, which is an attractive old town of tree-lined streets and parks, supports a population of around 300,000.

Geology and Petroleum Systems

Oman lies at the southern limit of the Persian Gulf oil province, having the same prime Jurassic reservoirs and source-rocks. But, in addition, it has Palaeozoic reservoirs charged by Precambrian source-rocks, being one of the very few occurrences known in the world. Cambrian salt also plays a part in the structural evolution of the area.

Exploration and Discovery

Shell was one of the pioneers in the country, where drilling commenced in 1955. A total of some 730 exploration boreholes have been drilled, peaking in 1991 when 37 were sunk. Discovery peaked in the early 1960s when several giant fields, including Yibal with 1.3 Gb of oil and 3.5 Tcf of gas, were found. It has since dwindled, although small discoveries continue to be made in the central part of the country on a trend that extends northwards into the Emirates. A total of some 14 Gb of oil has been found, of which almost half remain to be produced. As much as 56 Tcf of gas has been discovered, of which 15 Tcf have been produced.

C.J. Campbell, *Campbell's Atlas of Oil and Gas Depletion*,
DOI 10.1007/978-1-4614-3576-1_64, © Colin J. Campbell and Alexander Wöstmann 2013

Shell has used advanced technology, including horizontal wells, to drain the reservoirs, but while holding production levels high, it may have actually diminished the recovery, explaining some reported negative reserve growth. Also, the final decline from such operations tends to be steep.

Production and Consumption

Oil production commenced in 1967 and rose gradually to peak in 2000 at 970 kb/d before declining to 710 kb/d in 2007 and then recovering to 865 kb/d in 2010, at which point some 62% had been depleted. This means that what is left is likely to decline at the current depletion rate of 5% a year, which is relatively high for the Middle East.

Gas production commenced in 1972, being produced at a relatively low level in the 100–200 Gcf/a range until 2000 when it increased sharply to around 1,176 Gcf/a at present.

Oil and gas consumption stand at respectively about 52 Mb/a (142 kb/d) and 619 Gcf/a.

The Oil Age in Perspective

Although the Oman formed part of tribal Arabia during much of its early history, its location at the mouth of the Persian Gulf gave it a strong maritime tradition. Its traders roamed far and wide, reaching China, the East Indies and especially East Africa, where they played a prominent part in the slave trade.

The famous Portuguese explorer, Vasco da Gama, landed in Muscat on his way to India, and the Portuguese later took control in 1507, before being driven out in 1649 by the Imam Sultan bin Saif, who built an empire extending from Pakistan to Zanzibar in Africa. The country was then subject to conflict and strife, being briefly invaded by Persia in 1737, and coming under pressure from Arabia. The sparsely populated interior remained substantially tribal under an Imam, lacking formal frontiers either with Muscat or the other parts of Arabia.

During the nineteenth century, the territories came under British influence, which supported the more secular Sultan of Muscat in his conflicts with the fundamentalist Imam in the interior. The discovery of oil in the interior prompted the British Government to further support the Sultanate, which eventually established sovereignty in 1976 after various internal tensions and conflicts.

The present ruler, Qaboos bin Said al Said, belongs to the ruling dynasty, and is a western-oriented man, running the country as a benign dictator. He has supported British and American political objectives in the area, providing military bases for those powers in the Gulf War and recent tensions. Most of the population belong to the *Ibadi* sect of Islam, with 25% being *Sunnis* and 8% being *Shi'as*. The country was affected by the demonstrations of the so-called Arab Spring in 2011, but to a lesser extent than some of its neighbours.

Looking ahead, the Oman seems to be relatively well placed. It has a useful, albeit declining, supply of oil and substantial barely depleted gas reserves. The fertile coastal strip can probably support the inhabitants of Muscat. The desert interior will probably gradually revert to its tribal status as its oil and gas dwindles. In political terms, it will find itself more aligned with the Sunni's of Arabia than with Iran. Probably, the Oman will remain a useful staging point for any further military operations that the United States may undertake.

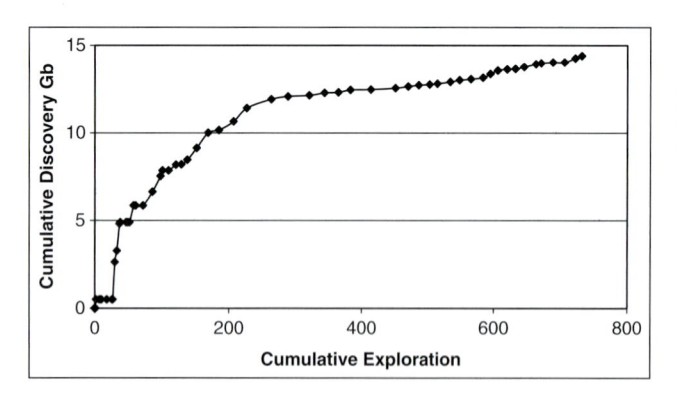

Fig. 64.3 Oman discovery trend

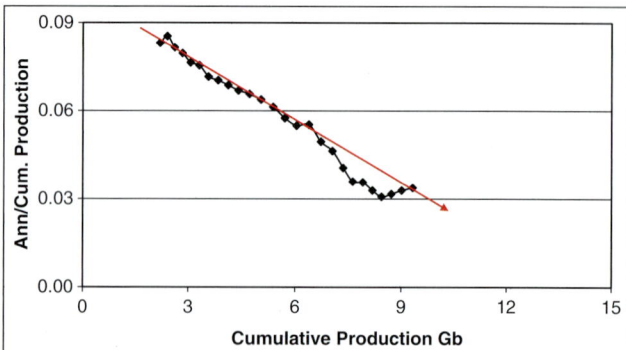

Fig. 64.4 Oman derivative logistic

64 Oman

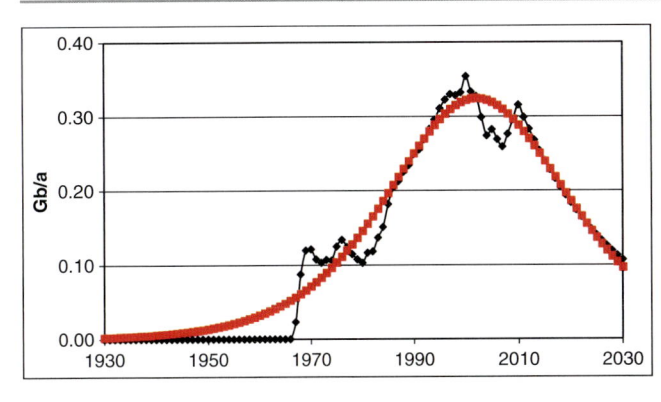

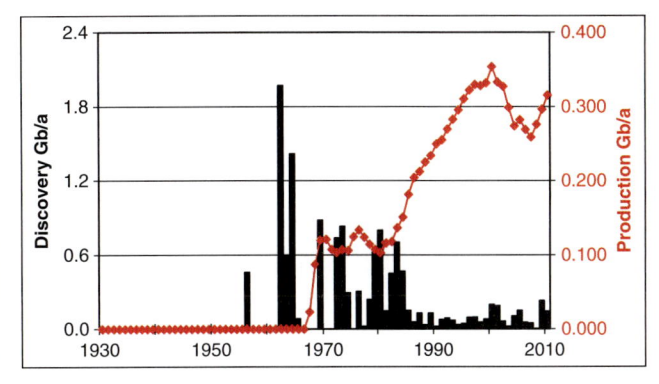

Fig. 64.5 Oman production: actual and theoretical

Fig. 64.6 Oman discovery and production

Qatar

Table 65.1 Qatar regional totals (data through 2010)

Production to 2100						Peak Dates		Area	
Amount			Rate				Oil	Gas	'000 km²	
	Gb	Tcf	Date	Mb/a	Gcf/a	Discovery	1940	1971	Onshore	Offshore
PAST	9.4	37	2000	269	1,312	Production	2030	2030	11	15
FUTURE	26	963	2005	305	2,034	Exploration	1988		Population	
Known	24	915	2010	411	4,611	Consumption	Mb/a	Gcf/a	1900	0.02
Yet-to-Find	1.3	48	2020	410	7,511	2010	61	770	2010	1.7
DISCOVERED	34	952	2030	410	12,235		b/a	kcf/a	Growth	85
TOTAL	35	1000	Trade	+352	+3,841	Per capita	35.6	453	Density	154

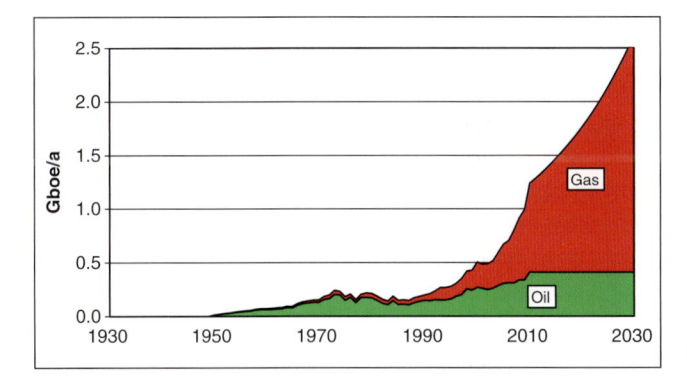

Fig. 65.1 Qatar oil and gas production 1930 to 2030

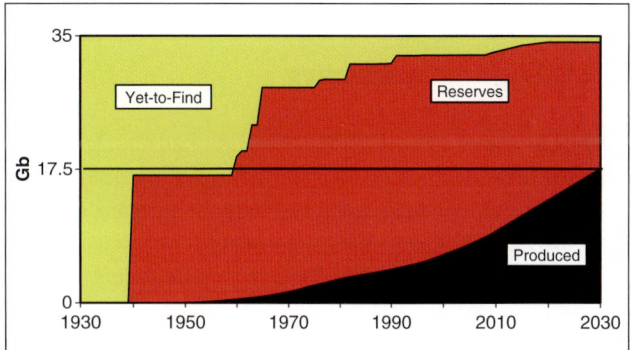

Fig. 65.2 Qatar status of oil depletion

Essential Features

Qatar forms a large promontory extending into the Persian Gulf from Arabia. It consists mainly of low-lying sand dunes, but a low range of limestone hills rises along the western seaboard. It is flanked by a number of islands, some of which have been subject to a, now settled, dispute with Bahrain.

Geology and Prime Petroleum Systems

The promontory of Qatar lies on a major transverse uplift, cutting the Persian Gulf. In fact, it is the most prominent representative of the earlier Hercynian structural grain that has had a strong influence on the evolution of the region, with important implications on the location of oil and gas. The uplift has brought into reach the deep-seated petroleum system that relies on Silurian source-rocks that have been depressed into the gas-generating window. This deep gas has migrated upwards into Mesozoic reservoirs, which have also been charged by leaner oil sources from the flanks of the uplift.

Exploration and Discovery

Exploration commenced in the late 1930s and was rewarded by the discovery of the Dukhan Field in 1940 with some 5.8 Gb of oil and 11 Tcf of gas. This in turn led to the discovery of the North Field, in 1971, the world's largest gas field with 1,000 Tcf of gas and 26 Gb natural gas liquids. (Gasplant Liquids fall outside the *Regular Conventional Oil* as defined herein, and the 26 Gb claimed by Qatar as oil is accordingly excluded from calculations.)

These remarkable finds came from a relatively low level of exploration drilling, totalling some 100 boreholes, with not more than one or two being normally drilled a year.

Production and Consumption

Oil production, consisting largely of condensate, commenced 1949 and has risen steadily to 1.1 Mb/d in 2010, and is expected to plateau at this level far into the future. Gas production commenced in 1946 and has risen steadily to 4,611 Gcf/a, at which point only 4% of the indicated reserves have been produced. It is accordingly expected to rise in the future at 5% a year to reach 10 Tcf a year by 2026, assuming that sufficient liquefaction plant capacity is installed or other outlets found to provide a market.

Oil consumption is reported at 61 Mb/a giving a high per capita usage of 35.6 b/a, but this may conceal the export of product, lost in the database. Gas consumption is reported at 770 Gcf/a, but this too gives an unrealistic per capita usage of 453 kcf/a, and must include liquefaction and export of the product. Generally, the resource base and depletion profile deserves future evaluation, if more reliable information could be obtained.

The Oil Age in Perspective

Qatar, like the rest of the Middle East, has had a long history. It was at first occupied by no more than a few nomads, living in the deserts, but later a natural harbour near an oasis at Qatif developed into an important trading centre, partly when it fell under the control of the Persian Empire during the first millennium. Fishing and pearling were the principal activities at the time.

After various vicissitudes, it fell within the Ottoman sphere of influence during the eighteenth and nineteenth centuries. It faced conflict with the Khalifa Clan of neighbouring Bahrain and sought British help in negotiating a peace which was achieved in 1820. A prominent merchant, by the name of Muhammed bin Thani, later came to power, founding a dynasty that survives to this day. He negotiated the formal status for the territory as a British Protectorate in 1916.

British influence waned after the Second World War, and led to the formation of a federation of Trucial States, the forerunner of what became the United Arab Emirates, but Qatar, now enjoying huge oil wealth, decided to withdraw in 1971 to become a sovereign State in its own right.

In 1995, the crown prince, Emir Hamad bin Khalifa Al Thani staged a successful coup d'etat against his father, and instituted various moves to modernise Qatari society. He also supported the United States by providing a base for the invasion of Iraq in 2003.

The economy had been transformed by the discovery of oil in 1940. A very high standard of living has attracted many immigrants, mainly from India and SE Asia, who now make up some 80% of the population. The absence of corporate income tax has encouraged the establishment of a flourishing international financial centre which provides an annual flow of investment exceeding a trillion dollars to other countries in the region. Furthermore, Qatari interests have invested heavily overseas, now for example owning property, including prominent hotels, in Britain. It also founded the *Al Jazeera* television and satellite network which is a major source of regional news. The country, no doubt as a result of its extreme wealth, seems to have been spared so far the demonstrations of the so-called Arab Spring facing its neighbours in 2011.

Looking ahead, it appears that the Government will, wisely, do its best to conserve its massive gas resources for as long as possible, having recently rejected plans to expand further the Gas Liquefaction plants, which are already some of the largest in the world. Thus, it looks well-placed to face at least the initial stages of the Second Half of the Age of Oil, unless it finds itself embroiled in the military conflicts that may envelop the region. But by the next century, it may again have to rediscover its roots as a fishing and pearling centre. It is likely to face over-population issues and presumably will have to encourage the descendents of the many immigrants to return to their original homes, which will not be easy.

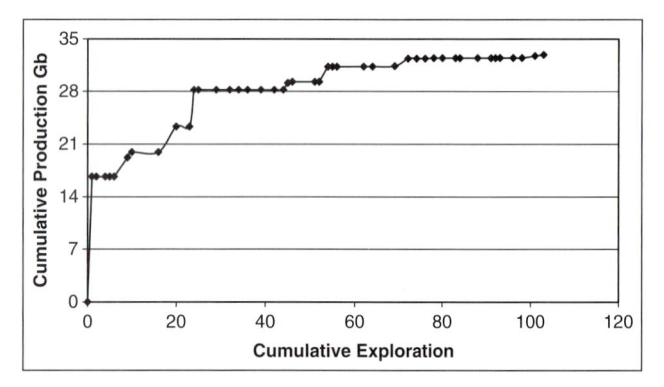

Fig. 65.3 Qatar discovery trend

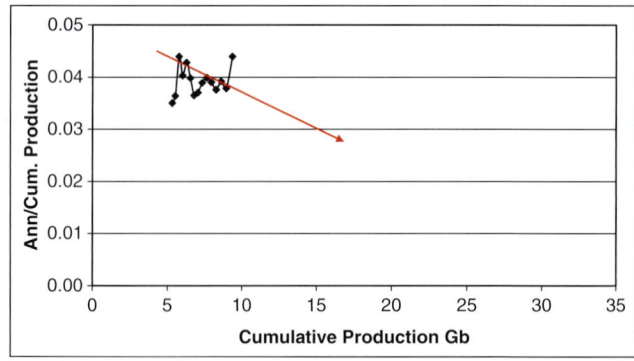

Fig. 65.4 Qatar derivative logistic

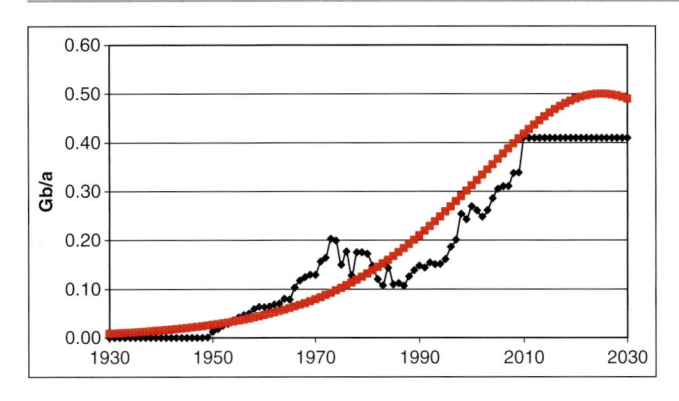

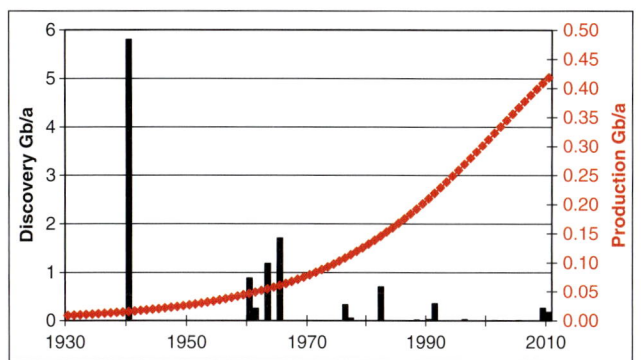

Fig. 65.5 Qatar production: actual and theoretical

Fig. 65.6 Qatar discovery and production

Saudi Arabia

Table 66.1 Saudi Arabia regional totals (data through 2010)

	Production to 2100					Peak Dates			Area	
Amount			Rate				Oil	Gas	'000 km²	
	Gb	Tcf	Date	Mb/a	Gcf/a	Discovery	1948	1948	Onshore	Offshore
PAST	117	84	2000	2,837	1,888	Production	1981	2030	2,160	250
FUTURE	183	316	2005	3,274	2,873	Exploration	1967		*Population*	
Known	174	253	2010	3,055	3,427	*Consumption*	Mb/a	Gcf/a	1900	1
Yet-to-Find	9.1	63	2020	3,000	5,000	2010	965	3,096	2010	27.9
DISCOVERED	291	337	2030	2,451	5,000		b/a	kcf/a	Growth	27.9
TOTAL	300	400	*Trade*	+2,090	+331	Per capita	35	111	Density	13

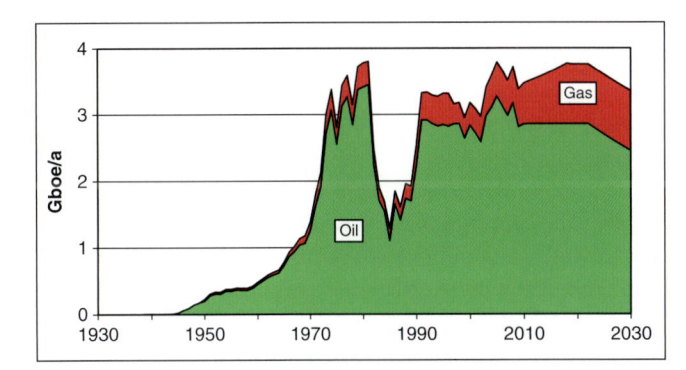

Fig. 66.1 Saudi Arabia oil and gas production 1930 to 2030

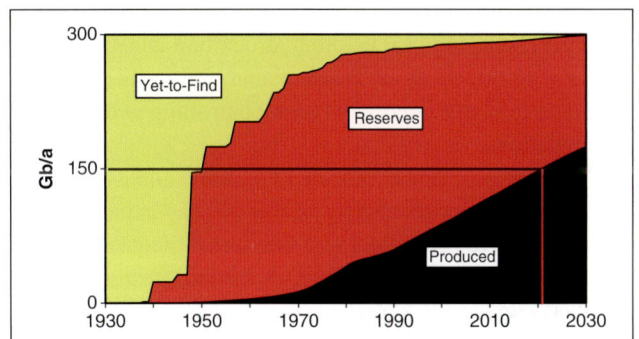

Fig. 66.2 Saudi Arabia status of oil depletion

Essential Features

Saudi Arabia is an arid, barren land of about 2 million km², covering most of the Arabian Peninsula. The western border, flanking the Red Sea, is made up of low mountains rising southwards from about 1,500 m to double that height. The southern end comprises the Asir region, which is relatively fertile thanks to a higher rainfall. To the east, follows an extensive plateau dipping eastwards towards the Persian Gulf, which is however cut by some ranges, including the Tuwayq Mountains, rising to over 1,000 m. The southern part of the country is made up of the so-called Empty Quarter (or *Rub' al-khali*), the world's largest desert covering an area of 650,000 km². It in turn gives way to mountainous country, bordering the Oman. Most of the borders have been subject to dispute in the past.

The population of the country amounts to almost 28 million, of whom about 5.5 million are resident foreigners, many coming from India, Bangladesh and the Philippines. Approximately 100,000 Europeans and Americans live in gated communities, many involved in the oil industry. Most of the indigenous people belong to the various branches of the *Sunni* sect of Islam, including the *Wahhabi* branch, which is a particularly strict calling. Even so, about 10–15% belong to the rival *Shi'ia* sect, with its links to Iran. The practice of Christianity is illegal.

Geology and Prime Petroleum Systems

Precambrian crystalline rocks of the Arabian Shield outcrop in the western part of the country, and are progressively overlain by younger strata to the east, attaining a thickness of more than 10,000 m. There are two prime petroleum systems. The older one relies on Silurian source-rocks, which have charged intermittent Permian sandstones with hydrocarbons. It is responsible for the Hawtah fields in the west of the country where the source-rocks are still shallow enough to be within the oil-generating window, but to the east, deep burial has depressed them below the window to yield gas-condensate, such as is produced from the Qatar Uplift. It probably offers the most scope for additional new discovery.

The main system relies on Jurassic source-rocks and reservoirs, sealed by evaporites, which are responsible for the massive fields of eastern Arabia, including Ghawar, the world's largest. The primary factors responsible for these giant accumulations seem to be as follows. First, is the structural evolution whereby the ancient Hercynian structural grain and deep-seated salt movements led to the gentle growth of the structures over time, such that they were well placed to receive all the oil generated in the vicinity from both prime and secondary sources. Several overlying evaporite layers were in place before hydrocarbon generation and provided effectives seals. It seems that seal in this area has been a particularly important factor controlling hydrocarbon accumulation.

The reservoirs themselves are of variable quality containing both highly porous and permeable zones—in part formed by fossil sponge beds—and less satisfactory sequences relying largely on fracture-porosity. A curious feature of the Ghawar Field is a thick tar deposit at the oil–water contact on the eastern flank of the structure, which has restricted the natural water drive, meaning that production calls for massive water injection to sweep the oil towards the producing wells.

These particular circumstances are partly responsible for the uncertainties regarding estimates of its reserves and future production. The oil-in-place volumes may indeed be very large but the high rates of past production may no longer be attainable in the future.

Exploration and Discovery

Whereas the foothills of Iran exposed promising prospects that were readily identifiable to the pioneering exploration geologists, the barren sands of Arabia concealed the underlying geology so that its potential was not at first recognised. In fact, it wasn't until a discovery was made in Bahrain in 1932 that attention turned to the adjoining mainland, where the first well was drilled three years later. In the absence of outcrops, it was necessary in the early days to rely on shallow boreholes to elucidate the geology, which delivered the first discoveries at Dammam and Abqaiq in 1938 and 1940, respectively. They were followed in 1948 by the discovery of the giant Ghawar Field.

Exploration drilling remained at a relatively low level until the 1960s when it picked up somewhat although, by world standards, it was modest at less then ten boreholes a year. A total of about 180 exploration boreholes have now been drilled, which is a very low number compared, for example, with the United Kingdom where over 3,000 have been drilled, but it has probably been sufficient to identify all the prime prospects, given the concentrated nature of the geological habitat.

There may remain a good potential for finding further deep gas-condensate fields from the Silurian system. It is, however, evident that the country has little incentive to search for minor secondary prospects so long as the prime fields remain in production. Every effort is being dedicated to the application of the most modern technology, including horizontal drilling.

The reporting of reserves and indeed production is unreliable, with some evidence suggesting that the massive increase in reserves reported in 1990 reflected a change in reporting practice whereby the total discovered, as opposed to the remaining reserves, was reported. In addition, there exists the genuine technical uncertainty about the rate of extraction. This problem is somewhat overcome in the classification adopted here, which refers to the *Future Production from Known Fields* to a cut-off in 2100 rather than *Reserves* in absolute terms. Despite the uncertainties, this assessment estimates that 291 Gb have been discovered, of which 117 Gb have already been produced. Future discovery is estimated at 9 Gb.

As much as 337 Tcf of gas have been discovered, of which about 316 Tcf remain. As mentioned above, there may be further potential for finding new deep gas-condensate fields, although the reservoir quality in the Permian *Kuff Formation* is far from ideal.

Production and Consumption

Oil production commenced in 1938 and rose in succeeding years to an initial peak in 1981 at 9.5 Mb/d. It then fell to a low of 3 Mb/d in 1985 due to OPEC quota restrictions before recovering to its present level of 8.4 Mb/d. Some 39% of the assessed resource has now been depleted. The Depletion Rate is very low at below 2%, suggesting either that in strictly technical terms there is scope to increase production, or alternatively that the present assessment is too generous. In any event, the country has little incentive to increase, given the high price of oil, which yields colossal revenues, given that actual production costs are probably still in the $5–15 range. Accordingly, it is here assumed that, in practice, production will plateau at the present level until around 2020 before commencing its terminal decline at 2% a year. This is a middle-of-the-road assessment as the Government claims that it can readily increase to 12 Mb/d, while other analysts conclude that the recent mild decline in production is a prelude to a steeper fall.

The radical increase in population, made possible by the massive rise in oil revenue, has increased consumption radically to 2.6 Mb/d, or 35 b/a per capita, almost matching the level in the United States. This begins to put a certain pressure on exports.

Gas production is reported to have commenced in 1950, much having been previously flared. It has risen roughly in parallel with oil reaching almost 3.4 Tcf/a in 2010. It is expected to continue to rise in the future to a plateau of 5 Tcf a year starting in 2019, when the resource is 30% depleted, given the installation of increased liquefaction facilities.

The Oil Age in Perspective

The deserts of the interior of Arabia were barely habitable in the earlier years, although there is some evidence that the climate may have been somewhat better then. The early civilisations were therefore concentrated on the borderlands of the Red Sea, Indian Ocean and Persian Gulf, outside the frontiers of the current State of Saudi Arabia. A significant trading kingdom developed in Mecca in the western, more fertile part of the country early on, and developed into a religious centre. The religion was based on the early recognition of a monotheistic deity, known in Arabic as Al-allah. The religious dimension was also used as a vehicle for treaties with various tribal groups, in part to facilitate trade for caravans bringing produce from the Yemen to the Red Sea ports.

An important figure by the name of Muhammad was born in Mecca in AD 570, claiming descent from the biblical figures Ismail and Abraham, who feature in the Old Testament. It was not optimal timing as Mecca had been subject to an Abyssinian invasion a few months before his birth, which cost his father his life. He then spent some time as a foster child in the desert, being brought up by relatives after his mother died. He evidently grew up to be a charismatic and good-looking young man, who married a wealthy lady, 15 years older than himself. She bore him several children including Fatimah who married one of his followers, Ali ibn Abu Talib, who became the first Caliph of Islam. They founded the Hashemite dynasty which remained in control of Mecca until the twentieth century. The descendants of this union are recognised by the *Shi'ia* sect of Islam as the true lineage of Muhammad to which they still owe allegiance.

Muhammad was given to taking religious retreats in a nearby cave, where in 610 AD he had a vision of being contacted by the Archangel Gabriel. A blind Christian cousin confirmed the significance of this vision launching him on a career as a divinely inspired missionary for Islam, although the rulers of Mecca, who belonged to the Quraysh clan, were suspicious and resentful. He later had a second vision in which he was transported to Jerusalem on a winged horse by the Archangel to meet God himself. He was then invited to the neighbouring town of Medina where he built a Mosque and attracted a following. This gave rise to much tribal negotiation. Relations with the rulers of Mecca soured further, which in 624 AD led them to attack Medina with an army of 3,000, only to be defeated by much smaller force under Muhammad, further strengthening his local prestige. A second attack from Mecca followed but was also defeated. It is noteworthy that the Jews of Medina sided with the attackers against Muhammad, and were later put to death, prompting a degree of anti-Semitism that survives. His strength and influence spread rapidly, such that emissaries from around the country arrived to express their support. In 632 AD, he returned in triumph to Mecca, setting the precedent for the annual pilgrimages that have continued to the present day, but died shortly afterwards. Whereas Muslims had previously faced Jerusalem in their prayers, he now had them aim in the direction of the temple in Mecca.

Islam subsequently divided into two primary sects: the *Sunni's* and *Shi'ites*. The former are the more pragmatic, accepting the leadership of whoever attains political prominence while giving weight to the records of Prophet's deeds and sayings, known as the *Sunnah*; whereas the latter accept Ali, Mohammed's son-in-law and his descendents as the only true divinely inspired successors to the Prophet. The conflict has a parallel with the notion of the *divine right of kings* that influenced European politics over much of its history, carrying echoes to this day. The division into the two prime sects of Islam has had a far-reaching impact on Middle East history.

The subsequent history of Arabia through the Middle Ages was essentially one of the tribal conflicts of no particular global significance, save for the continued importance of Mecca and Medina as the holy cities of the Moslem world.

Another religious conflict arose in 1703 when Abd al-Wahhabi started preaching a more fundamentalist return to Islam which was later adopted by the Sauds, who were the rulers of Riyadh, the present capital of Saudi Arabia. Their influence grew over time so that by 1804 they were in virtual control of the entire country, facing conflict with the Ottoman Empire of Turkey, which sponsored two invasions by Egyptian troops.

Next on the scene appeared the Rashid tribe, which had been vassals of the Saud's but successfully rose to independence with some support from the Ottomans in the latter part of the nineteenth century, only to be defeated in 1901 when Abdul Aziz Al-Saud (also known as ibn Saud) returned to take Riyadh in a surprise attack, paving the way for the present Saud dynasty. He was helped in part by having secured the support of Britain against the Ottomans.

Meanwhile, the holy cities of Mecca and Medina remained under the control of the Grand Sharif—descended from Muhammad as described above—who nevertheless accepted the supremacy of the Ottoman Caliph. Britain in the First World War was encouraging the Arabs to rise against the Turks, and promised them in return an Arab Empire to cover most of the Middle East outside of Iran. It enlisted the support of the Grand Sharif, who led a revolution in 1916.

But ten years later, ibn Saud captured Mecca, bringing to an end 700 years of Hashemite rule. Britain recognised his claim to the entire region, allowing the Grand Sharif to move into exile in Cyprus, thoughtfully taking the treasury with him, and later offered the thrones of Jordan and Iraq to his sons. The Kingdom of Saudi Arabia was finally recognised by Britain and other powers in 1932, with one of the conditions being an end to any Saudi claims to the oil-rich emirates bordering the Persian Gulf, which were British Protectorates.

Interest in the oil possibilities of the Middle East outside Iran grew in the inter-war years, and the discovery of oil in Bahrain in 1932 whetted the appetite of ibn Saud, whose principal source of revenue at the time was no more than that provided by the pilgrims to Mecca. He was advised by a curious, disaffected British civil servant, Harry St. John Philby, who had the Ford agency in Jidda and was the father of the infamous Cold War double agent, Kim Philby. He successfully persuaded Chevron (then Standard of California) to take up oil rights to the country in 1933 in return for the delivery of 35,000 gold sovereigns. This laid the foundations for the Arabian American Oil Company (ARAMCO) which Texaco, Exxon and Mobil were later invited to join as shareholders. The discovery of the first of a string of giant oilfields five years later more than justified what might have appeared a somewhat speculative investment at the time.

The interruption of the Second World War, during which some of the oilfields were lightly bombed, delayed the development somewhat, with the giant Ghawar Field not coming on stream until 1948. Meanwhile in 1945, a historic meeting between President Roosevelt and King ibd Saud occurred aboard a warship in the Suez Canal. Apparently, the discussion centred around the post-war construction of the Middle East, including the proposal for the creation of the State of Israel, to which the King was much opposed. Exactly what was agreed is not known, but it seems that American support for the Kingdom was offered in return for prior claim to its oil. Mr. Churchill, the British Prime Minister, was reportedly incensed at this incursion into what had been a British sphere of influence.

Ibn Saud died in 1953, to be succeeded by his eldest surviving son, Sa'ud, who agreed in due course for his country to join the Organisation of Petroleum Exporting Countries (OPEC). It was formed in 1960 under Venezuelan impetus, following the earlier example of the US policy, administered by the Texas Railroad Commission, to restrict oil production in order to support the price.

Iran had nationalised its oil industry in 1951, setting an example that the other major producers felt obliged to follow, partly in connection with their OPEC membership. In Saudi Arabia, it was a fairly gentle process as the State increased its stake from 25% in 1973 to full control by 1980. The foreign companies, especially Chevron, continued to provide full technical support, and enjoyed privileged access to production. Texaco, one of the shareholders, even facilitated the direct entry of Saudi Aramco into the US marketing and refining business.

King Sa'ud was deposed in 1964 by Faisal, his more cosmopolitan younger brother, following new challenges by Egypt which sought to create the long contemplated Pan-Arab Empire. Egyptian troops had arrived in neighbouring North Yemen, where they had been strongly opposed by the Sauds. Relations with the United States were also soured over the latter's support for Israel in the 1973 war. This led Saudi Arabia to join with other Arab countries in curbing the export of oil to the United States for a few months, which prompted the First Oil Shock.

King Faisal was assassinated in the following year, to be succeeded by his half-brother Crown Prince Khalid, who was in turn succeeded by Fahd in 1982. Rising oil prices had brought new prosperity to the Kingdom, which led to the arrival of many immigrants, accompanied by radical changes in its social environment, prompted in part by television exposing the people to western influences.

The next challenge came when Iraq invaded Kuwait in 1990, prompting the Saud's to restrengthen their ties with the United States. They had hoped at first to help negotiate a settlement of Iraq's claims against Kuwait, having supported Iraq financially in the preceding Iran–Iraq war, but the efforts failed, and expediency prompted them to facilitate the American engagement.

The next geopolitical development was the Anglo-American invasion of Iraq in 2002, which is widely seen as designed to control Middle East oil supply, whatever the immediate pretexts. It clearly affected Saudi Arabia. On the one hand, many of its people, most of whom are Sunni Muslims, must have great sympathy for the suffering Iraqi people, resenting the fall of its former Sunni Government; on the other hand, the Saud Royal Family has developed close ties with the United States—not least in financial terms as massive royal investments had been placed on Wall Street.

The presence of the US military bases on Saudi territory is resented, and there have indeed been several violent incidents. Some of the culprits were publicly beheaded: their actions being taken as implied opposition to the Monarchy which had permitted the foreign bases to be established. It may be asked if some devout Muslims may have mixed feelings towards the Sauds, who, it will be remembered, ousted the Hashemite rulers of Mecca being the direct descendents of the Prophet Muhammad.

King Fahd died in 2005 at about 80 years of age, and was succeeded by Abdullah, a half-brother, who is of similar age having had 4 wives and 22 children. He is a devout Muslim belonging to the *Salafi* branch of the *Sunni* Sect, which reveres the original teachings of Muhammad, considering later developments to represent a certain debasement.

In summary then, it is clear that Saudi Arabia is an extremely exceptional country run by a virtually feudal theocratic monarchy. But for oil revenue, its barren terrain could support no more than a small fraction of its current population which has increased by a factor of 28 over the past century, to include large numbers of immigrants mainly in servile positions.

Its oil endowment is even more remarkable, as it controls some 13% of the world's supply of *Regular Conventional* oil and owns 22% of what is left in reserves. Despite this remarkable asset however, it faces many challenges. The Anglo-American intervention, which currently focuses on Iran's nuclear aspirations, may escalate with an attack on Iran, which might in turn lead to a general uprising of people throughout the Middle East, affecting all existing governments. On the other hand, the gradual withdrawal of foreign forces from Iraq might prompt a degeneration into further factional conflict which may spread to the other countries in the region. A further difficulty may arise if the dollar-dominated wealth of the Royal family, including national oil revenues, should collapse in the face of economic recession in the United States and subsequent devaluation of the dollar. Such threats might encourage a move to other currencies, which could also have a serious impact on the US position.

It is worth mentioning that oil revenues are little more than profiteering from shortage insofar as actual production costs are still probably in the range of $5–15 a barrel, and form a flood of false liquidity, itself undermining world financial stability. This is another contentious issue, which might prompt external reaction.

Saudi Arabia, despite its massive endowment, is not immune to depletion, such that production is set to fall in the years ahead to perhaps 4 Mb/d in 2050 and 1 Mb/d by the end of the century. According to some analysts, it might fall even faster. But the demand for oil may also fall as the rest of the world is finally forced to come to terms with the depletion of this commodity which has been the driver of the modern economy. It threatens to be a violent transition. The revolutions and tensions of the so-called Arab Spring in 2011 have not so far seriously affected Saudi Arabia but may have a serious influence in the future.

So, on balance, looking ahead to the Second Half of the Age of Oil, it would be reasonable to anticipate growing national, regional and international tensions to affect this archaic kingdom, such that by the end of the century it may have reverted to something not far removed from the tribal conditions it knew in the nineteenth century. The fate of the Royal Family may hang in the balance as they are, by all means, an anachronism in the modern world, however benign they may be.

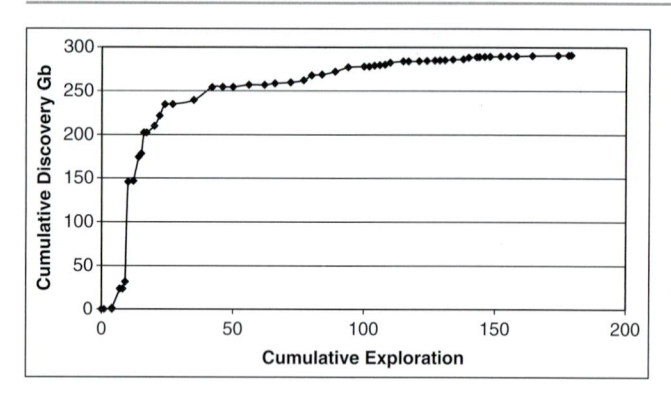

Fig. 66.3 Saudi Arabia discovery trend

Fig. 66.4 Saudi Arabia derivative logistic

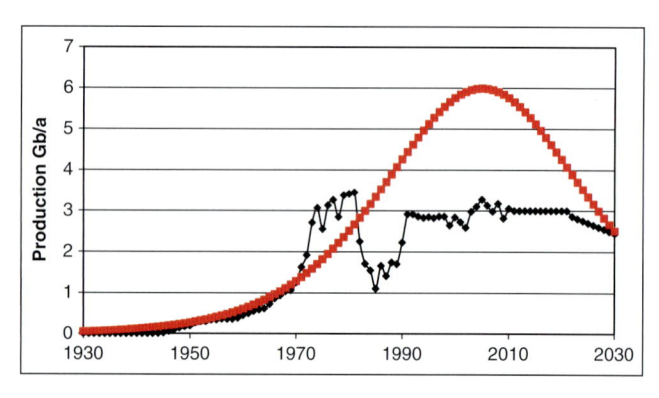

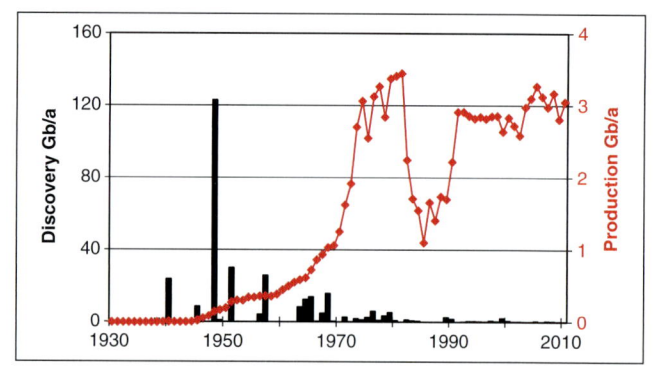

Fig. 66.5 Saudi Arabia production: actual and theoretical

Fig. 66.6 Saudi Arabia discovery and production

Syria

Table 67.1 Syria regional totals (data through 2010)

Production to 2100							Peak Dates			Area	
Amount		Rate					Oil	Gas	'000 km²	
	Gb	Tcf	Date	Mb/a	Gcf/a	Discovery	1966	1987	Onshore	Offshore
PAST	4.9	6.6	2000	191	280	Production	1996	2013	185	2.6
FUTURE	2.3	8.4	2005	158	297	Exploration	1992		*Population*	
Known	1.96	6.7	2010	134	356	*Consumption*	Mb/a	Gcf/a	1900	1.9
Yet-to-Find	0.4	1.7	2020	81	300	2010	107	340	2010	22.5
DISCOVERED	6.9	13	2030	46	195		b/a	kcf/a	Growth	11.8
TOTAL	7.25	15	Trade	+27	+16	Per capita	4.7	15	Density	121

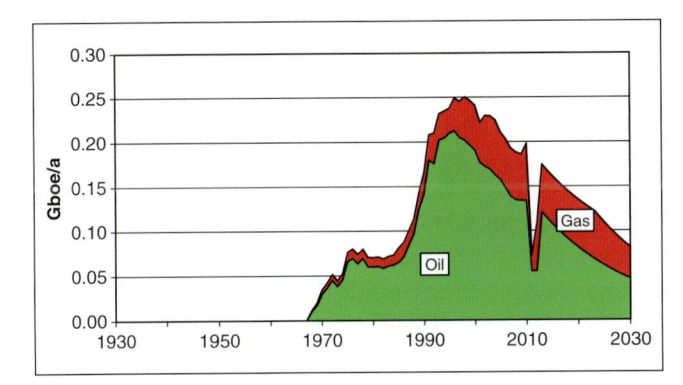

Fig. 67.1 Syria oil and gas production 1930 to 2030

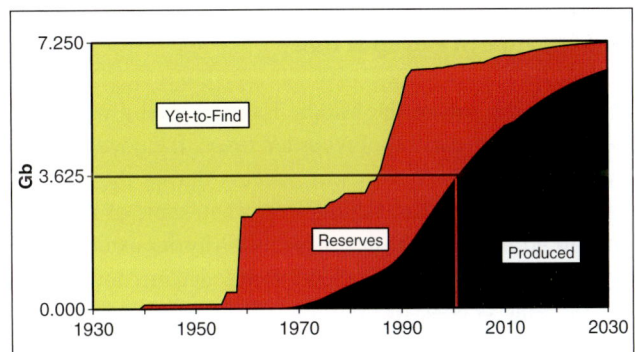

Fig. 67.2 Syria status of oil depletion

Essential Features

Syria, covering an area of 185,000 km², lies on the western margin on the Middle East Region. It is substantially landlocked, being bordered by Turkey, Iraq, Jordan, Israel and the Lebanon, but does have a short Mediterranean coastline. There are two mountain ranges; one, rising to about 2,500 m, extends north eastwards from the Lebanon border in the south, while the other, at a lower elevation, runs northwards, being separated from the Mediterranean by a narrow coastal strip. The country is drained by the Orontes River in the west that flows northwards into Turkey, and by the headwaters of the Euphrates, flowing eastwards into Iraq. The western region enjoys a Mediterranean climate with adequate rainfall and is relatively fertile, contrasting with the barren lands to the east.

The country supports a population of almost 23 million people, about one-quarter of whom live in Damascus, the capital, which is situated at an altitude of almost 700 m.

Geology and Petroleum Systems

Non-prospective Precambrian rocks of the Arabian Shield outcrop in the western part of the country, where they were subjected to several epochs of volcanic activity, but are progressively overlain by younger strata to the east. The Palaeozoic and younger strata of the east fall generally on the margin of the Persian Gulf petroleum province and have yielded a number of modest fields, relying in part on less than ideal Triassic source- and reservoir rocks.

Exploration and Discovery

Exploration commenced in 1939 at a low level, and picked up in the 1980s to reach a peak in 1992 when as many as 28 exploration boreholes were sunk, but has declined markedly since. The early drilling made a number of modest finds prior to the discovery of the giant Souedie Field in 1959 with some

2.4 Gb of moderately heavy oil (27°API). A second phase of discovery followed in the decade after 1985 with annual discovery in the 100–500 Mb range. Minor amounts of associated gas were also found. Some hopes have been expressed for an offshore discovery but the chances are slim.

Production and Consumption

Oil production commenced in 1968 and rose to a peak of 582 kb/d in 1996, since when it has declined to 367 kb/d. Being now 68% depleted, production is set to continue to fall at about 5% a year. Consumption stands at 107 Mb/a making the country a modest exporter for the next few years. Gas production has risen in parallel with oil running at 340 Gcf/a, and is mainly consumed internally. Production of both oil and gas is likely to collapse in 2011 and 2012 due the political strife.

The Oil Age in Perspective

Syria, like the rest of the Middle East, has had a very long history from the earliest of recorded times. It enjoyed a high level of cultural and economic success during the time of both the Greek and Roman empires, when some of its cities came to prominence. At that time, its territories extended far beyond the present frontiers to include Palestine, Jordan and part of, what is now, western Iraq. That chapter came to an end with Arab invasions in 633 AD when it was absorbed into the Muslim Caliphate, but despite various tensions, it continued to prosper and exert a regional influence, thanks in part to trade passing through its Mediterranean ports.

A new chapter opened in 1516 when it was invaded by the Ottomans from Turkey, becoming an important part of that empire over the next 400 years, while still enjoying a fair degree of autonomy. During all those years, the Christian and Jewish minorities were accepted with toleration, taking a strong position in trade. The Ottoman administration progressively lost its drive over time, and in 1831 Syria found itself defeated by an Egyptian army which occupied the country for a decade before being forced to withdraw, partly under European pressure. New conflicts between various religious sects culminated in a civil war between the Druzes and the Maronites, which led to a massacre of Christians by a Muslim mob. This prompted an invasion by France in 1860, when it was decided to separate Lebanon from the rest of the country.

Syria became a military base for the German army during the early years of the First World War, but they were defeated by British and French forces, which established a full occupation by the end of the war. The Sykes-Picot Agreement of 1916 effectively put Syria into the French sphere of influence at the end of hostilities. French colonial rule was generally somewhat autocratic in style, giving scant regard to the growing moves toward independence, which in Syria were compounded by religious tensions between the several Christian sects, the Jews and the Muslims.

With the outbreak of the Second World War and the defeat of France, Syria found itself falling under German influence, which prompted a successful invasion by British and French troops in 1941. It finally won full independence in 1945, after some dispute between the British and French.

The creation of the State of Israel in 1948 enflamed local passions, and economic development distorted the traditional social structure. New political parties came into existence only to be countered by a succession of military and other takeovers. A revival of pan-Arabism arose, leading to a brief union with Egypt from 1958 to 1961, before a Baath'ist government came to power. It was a form of Middle East authoritarian socialism, also embraced by Saddam Hussein in Iraq.

This was followed by growing conflict with Israel, which invaded the Golan Heights of southern Syria in the Six Day War. A new strongman in the form of Hafez al-Assad came to power in 1970 and joined with Egypt to mount a retaliatory war on Israel, the so-called Yom-Kippur War of 1973, taking the opportunity to occupy Lebanon which was in a state of civil war.

Hafez al-Assad died in 2000 after 30 years in power to be succeeded by his son, Bashar. He soon came under pressure when Israel bombed a location near Damascus in October 2003, claiming it to be a terrorist training camp. Syria has also been accused of building a nuclear reactor and came under fire from Israeli aircraft. Significantly, too, the Kurds of Syria have begun to express sympathy with their cousins in northern Iraq and southeast Turkey, opening a new political dimension. Serious political tensions broke out early in 2011 and were vigorously suppressed by government forces, causing death and leading refugees to escape into Turkey. The repression has attracted some negative reaction from other countries including the United States.

It is hard to know how Syria will fare during the Second Half of the Age of Oil, the transition to which threatens to be a time of great regional tension and foreign intervention. But at the end of the day, the benign, western seaboard land will offer a sustainable livelihood for the survivors.

67 Syria

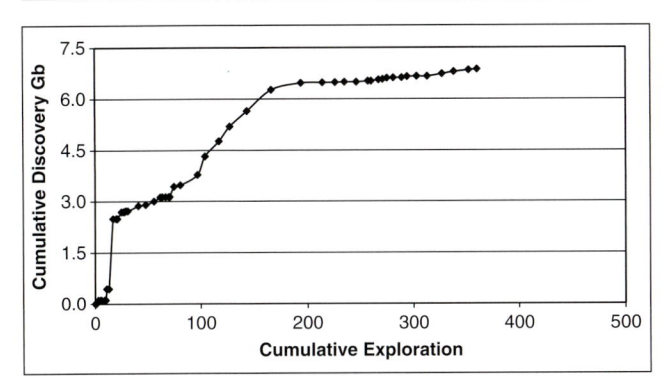

Fig. 67.3 Syria discovery trend

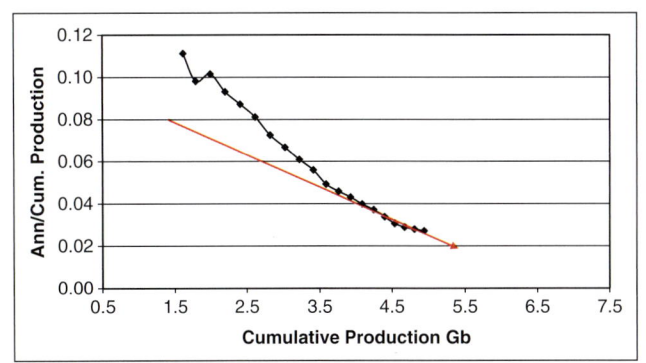

Fig. 67.4 Syria derivative logistic

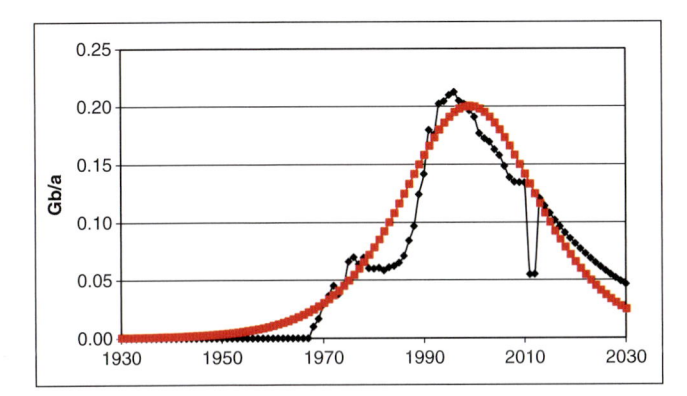

Fig. 67.5 Syria production: actual and theoretical

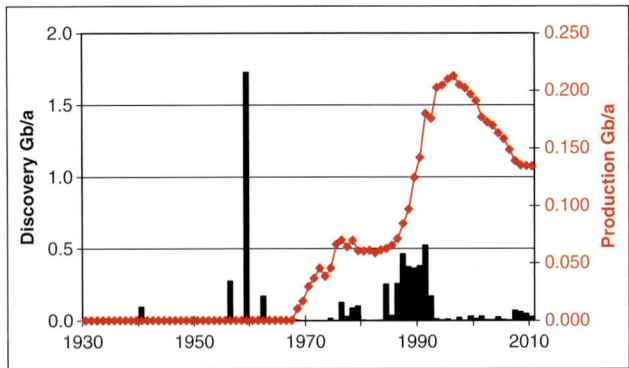

Fig. 67.6 Syria discovery and production

Turkey

Table 68.1 Turkey regional totals (data through 2010)

	Production to 2100					Peak Dates			Area	
	Amount		Rate				Oil	Gas	'000 km²	
	Gb	Tcf	Date	Mb/a	Gcf/a	Discovery	1961	1965	Onshore	Offshore
PAST	0.96	0.45	2000	19	27	Production	1991	2012	783.5	150
FUTURE	0.29	0.55	2005	16	32	Exploration	1975		*Population*	
Known	0.25	0.52	2010	18	23	*Consumption*	Mb/a	Gcf/a	1900	12
Yet-to-Find	0.04	0.03	2020	10	25	2010	236	1346	2010	74
DISCOVERED	1.21	0.97	2030	5	11		b/a	kcf/a	Growth	6.2
TOTAL	1.25	1.0	*Trade*	−218	−1323	Per capita	3.2	18	Density	95

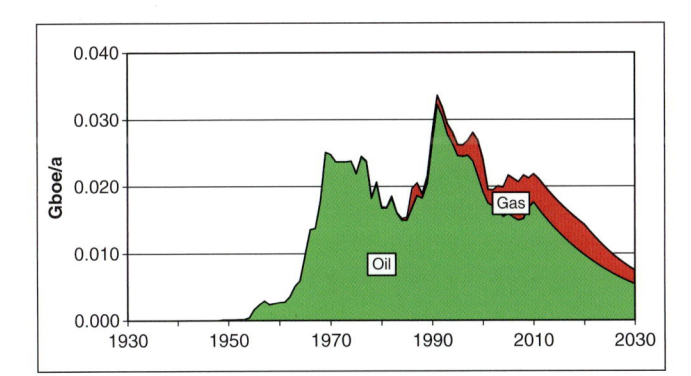

Fig. 68.1 Turkey oil and gas production 1930 to 2030

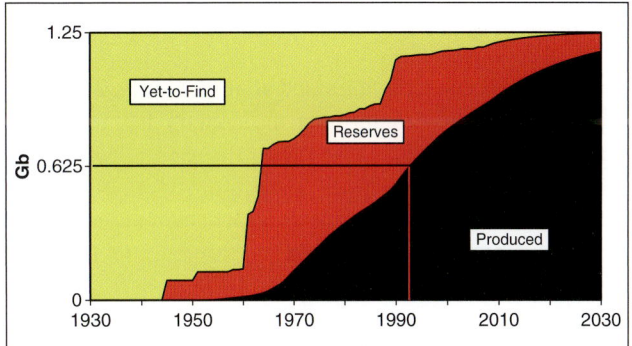

Fig. 68.2 Turkey status of oil depletion

Essential Features

Turkey occupies a territory of 783,562 km² that lies both between the Middle East and Europe, and between the Mediterranean and the Black Sea. The arid Anatolian Plateau, at an altitude of about 1,000 m, is flanked to the north and south by respectively the Pontus and Taurus mountain ranges, the latter rising to about 2,500 m. The highest point in the country is the snow-capped volcanic peak of Mt Ararat, which reaches 5,165 m in the northeast of the country, known for being where the biblical ark supposedly made its landfall. The territory of Thrace in the northwest is separated from the rest of the country by an important waterway, made up of the Sea of Marmara, the Dardanelles and the Bosporus, which connects the Black Sea with the Mediterranean. Relatively narrow and fertile coastal strips adjoin the Mediterranean to the west and the Black Sea to the north. The Mediterranean coast is indented and adjoined by a number of islands.

The country supports a population of 74 million, many of whom live in Istanbul on the shores of the Bosporus and in Ankara, the capital, which is located on the Anatolian Plateau.

Geology and Petroleum Systems

The northern limit of the oil-rich Middle East basin extends into southeast Turkey, but the structures have been generally breached as a result of uplift, such that most of their oil content escaped over time. Nevertheless, a few modest finds have been made in the thrust-belt bordering the mountains. Some hopes have been expressed for finding oil in the adjoining Black Sea, but the evidence to-date suggests that it is no more than a modest gas province, probably lacking effective oil source-rocks.

Exploration and Discovery

Exploration drilling commenced in 1936, and was rewarded by a small find in 1945. Activity built up in the 1960s, when the largest field, Raman Bati with 185 Mb, was found. Discovery reached an early peak around 1961, but then declined to a low of 22 Mb in 2002 before surging to 56 Mb in 2009. Exploration surged too in 1975 when 43 boreholes were sunk. Total discovery to-date amounts to about 1.2 Gb of oil and 1 Tcf of gas.

Production and Consumption

Oil production commenced in 1948 and rose gradually to peak in 1991 at 88 kb/d before declining to 48 kb/d in 2010. Being 77% depleted, production is expected to continue to decline at the current high Depletion Rate of almost 6%. Gas production commenced in 1974 and has increased at a low rate to about 23 Gcf/a in 2010. It is expected to continue at this level until commencing its terminal decline around 2020.

Oil and gas consumption stand at, respectively, 236 Mb/a and 1,346 Gcf/a, meaning that the country has to import some 90% of its needs.

The Oil Age in Perspective

Turkey has always lain at the crossroads between Europe, the Middle East and Russia. It was home for early peoples in Neolithic times, being part of the Hittite Empire which was later brought into the Roman Empire. The Kingdom of Armenia, which occupied much of the country, adopted Christianity in AD 303, becoming the eastern bastion of the Roman Empire. It flourished until 428 AD, when it fell in turn to the Persians and the Arabs.

The country was also subject to settlement by various Turkmen tribes and warriors from Asia who came to prominence, founding the Ottoman Empire. At its prime in the sixteenth and seventeenth centuries, Ottoman dominion extended from the Adriatic to the Caspian and from the gates of Vienna to Egypt and North Africa. Most of the Middle East, apart from Persia, fell within its jurisdiction.

The Ottomans belonged to the *Sunni* sect of Islam, facing the *Shi'ias* of Persia. Christian Armenian communities survived under Ottoman rule, but were treated as somewhat subversive elements subject to periodic repression.

Efforts to modernise the Ottoman Sultanate arose at the end of the nineteenth century with the rise of the so-called Young Turks. This was accompanied by various moves towards independence from the member territories, leading to the formation of the Balkan League of Bulgaria, Serbia, Montenegro and Greece in 1912, which led to war, with varying degrees of backing from the major European powers. A Greek army marched towards Istanbul, while Libya was taken by Italy. The Young Turks orchestrated a successful *coup d'etat* in 1913 and brought the country into the First World War on the side of Germany. This may in part have arisen from Germany's prior interest in building a railway from Berlin to Baghdad to take a trading stake in the Middle East, which was then under Turkish dominion. In fact, the first hints of oil in Iraq were the seepages identified by German engineers who were surveying for the proposed railway.

In the first year of the war, Turkey launched an attack on Russia in the hopes of capturing the oil fields of Baku, but was roundly defeated. The defeat was in part attributed to the Armenians, some of whom had fought with the Russians hoping to regain their ancestral lands, and prompted renewed deportations and massacres. In one sense, the conflict was a civil war between different factions having rival claims to the country, but it smacked of genocide insofar as one side was Christian and the other Muslim.

British forces defeated Turkey in the First World War, having encouraged various Arab elements within the Ottoman Empire to rise in rebellion. Britain then set about dismantling the Empire with the creation of new States in the Middle East which were given somewhat arbitrary frontiers. The division however failed to recognise the claims of Kurds, numbering some 30 million, who are descendents of the ancient Medes and occupy a swath of country from Syria through southeast Turkey and northern Iraq into Iran. The British occupation of what became Iraq did not at first include the predominantly Kurdish northern region of Mosul, where the giant Kirkuk Oilfield is located, but that was added in 1926 following a Kurdish rebellion against the Turkish administration in the preceding year.

We may note in passing that the oil rights of Iraq originally belonged to the Turkish Petroleum Company, formed with capital from Shell and the Deutsche Bank in 1912 by Calouste Gulbenkian, himself an Armenian with links to Baku. It was transformed—as one of the spoils of war—into the Iraq Petroleum Company to be owned by BP, Shell, CFP, Exxon and Mobil. Its founder, Mr. Gulbenkian, managed to hold on to his famous 5%. It had exclusive control of Iraq's oil until 1972 when it was nationalised.

The Russian Army had occupied the eastern part of Turkey, with its predominantly Armenian population, in the early part of the First World War but lost control following the Bolshevik Revolution of 1917. Dispute and conflict continued until the post-war settlement, when the United States recognised the State of Armenia at the Treaty of Sevres in 1920, and drew up its boundaries. That country was then subject to a Soviet invasion, becoming a Soviet Socialist

Republic. The Greeks too had ambitions to take Turkey and launched an invasion in 1921, only to be defeated in the following year.

In 1923, Mustafa Kemel, known as Ataturk, the leader of the Young Turk Movement, came to power adopting a dictatorial style of Government, albeit aimed to modernise the country, disestablishing the long-standing Muslim Caliphate in 1924.

Gradually, the political situation stabilised over the ensuing years as the *de facto* status of Turkey and the countries of its former empire were recognised and accepted, however reluctantly, by those involved, including the major European powers. Ataturk died in 1938, and Turkey came under pressure to join Germany, its former ally, in the early years of the Second World War but it managed to remain neutral until it joined the victorious side just before the end of the war.

The country then came under renewed pressure from Russia, which desired to establish military control of the waterway through Turkey which connects the Black Sea with the Mediterranean, but this was countered by the United States. Turkey became a member of NATO in 1952, being host to some of the US military bases that ringed the Soviet Union in the Cold War.

There was however an ongoing conflict in Cyprus, which had previously been part of the Turkish Empire. It still has a large Turkish community in the northern part of the island which sees itself as Turkish before Cypriot. The tensions prompted a Turkish invasion in 1974, and the issue remains far from settled.

Turkey has enjoyed a degree of economic prosperity over the past few decades, although many of its citizens emigrated to Europe, finding work especially in Germany. It has been eyed by the European Union which desired to bring it into its economic and financial empire, although difficulties remain in relation to the rival claims of Greece and Turkey to the island of Cyprus.

The Anglo-American invasion of Iraq in 2003 opened a new chapter of tension, re-igniting the ancient conflicts between *Sunni*, *Shi'ia*, Kurd and Armenian factions, now supplemented by Zionist pressures. No doubt, the Turks, who are predominantly *Sunni*, have sympathy for the suffering of the people of Iraq, and some may resent the fall of its previous *Sunni* Government. The Kurds however do not share that view, having been oppressed by the previous regime—as well as by Britain in the 1920s when their villages were bombed—and no doubt welcomed the invaders with open arms. Indeed, the establishment of a new Kurdistan might win international support, being consistent with a policy aimed at controlling the Middle East oil supply. The administration of Kurdish Iraq has already secured sufficient independence to grant oil concessions to various companies, including the Hunt Oil Co. of Dallas, Texas. But these moves have ignited the ambitions of the Kurdish separatists in Turkey, who face a military crack-down.

Oil is naturally an important factor in the evolving situation. Turkey, despite having lost control of the oil-rich Middle East outside Iran at the end of the First World War, is still an important transit country. A long-established pipeline from Iraq and the new line from the Caspian both feed the Ceyhan Terminal on the Mediterranean coast. In addition, plans have been made for a new gas pipeline, known as Nabucco and costing $6 billion, to bring Caspian gas to Europe via Turkey. Construction has been delayed in the light of the political situation.

Looking ahead, it would not be surprising if Turkey again began to exert a greater role in Middle Eastern affairs, in a sense rediscovering its imperial past. In 2011, its relations with Israel soured as it gave new support to the Palestinian cause, also strengthening ties with Egypt's new government. But in the meantime, it faces serious challenges from its Kurdish separatists, who may indeed win international support for their ambitions in whatever emerges from the Middle East situation. It becomes increasingly unlikely that Turkey will join the European Union, which may itself fragment in the face of the new economic and financial difficulties. In general, the Turks are a strong people who can be expected to face their future with courage and determination, and on balance are unlikely to accede to the separatist pressures without a fight.

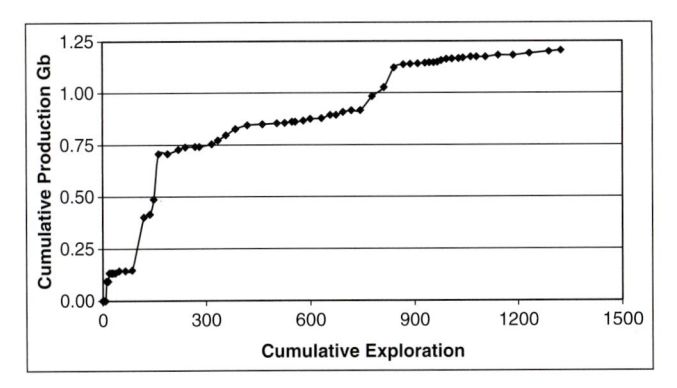

Fig. 68.3 Turkey discovery trend

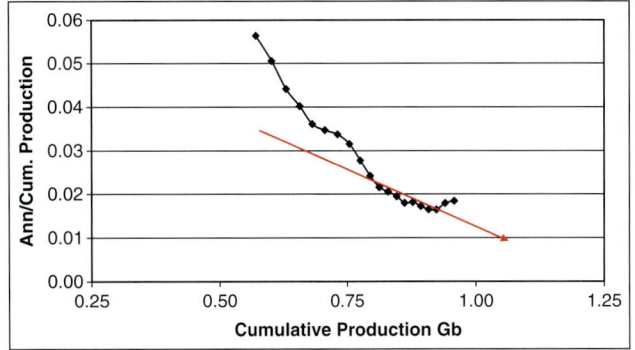

Fig. 68.4 Turkey derivative logistic

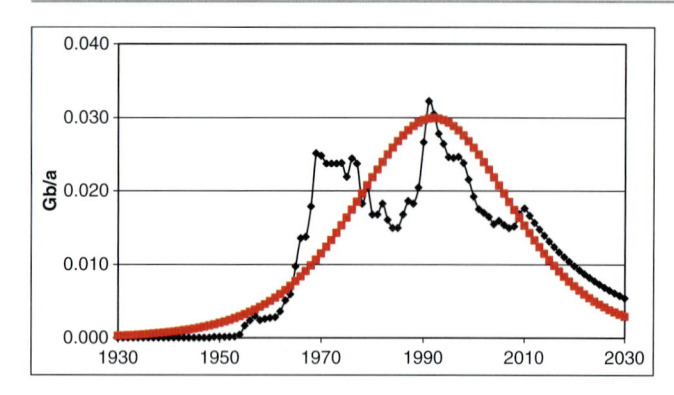

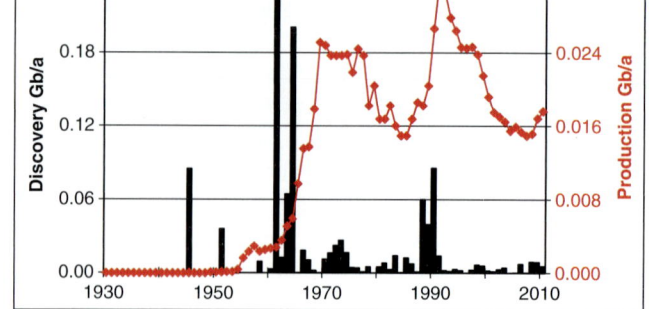

Fig. 68.5 Turkey production: actual and theoretical

Fig. 68.6 Turkey discovery and production

United Arab Emirates

Table 69.1 United Arab Emirates regional totals (data through 2010)

Production to 2100					Peak Dates			Area		
Amount		Rate				Oil	Gas	'000 km²	
	Gb	Tcf	Date	Mb/a	Gcf/a	Discovery	1972	1978	Onshore	Offshore
PAST	30	53	2000	864	1,784	Production	2016	2023	83	10
FUTURE	55	122	2005	925	2,410	Exploration	1952		Population	
Known	50	110	2010	881	2,817	*Consumption*	Mb/a	Gcf/a	1900	0.1
Yet-to-Find	5.5	12	2020	1,000	2,750	2010	199	2,138	2010	7.9
DISCOVERED	79	163	2030	849	2,750		b/a	kcf/a	Growth	79
TOTAL	85	175	*Trade*	+682	+679	Per capita	25	271	Density	95

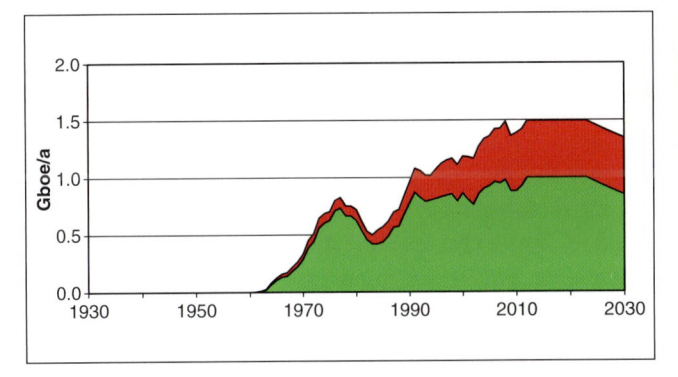

Fig. 69.1 United Arab Emirates oil and gas production 1930 to 2030

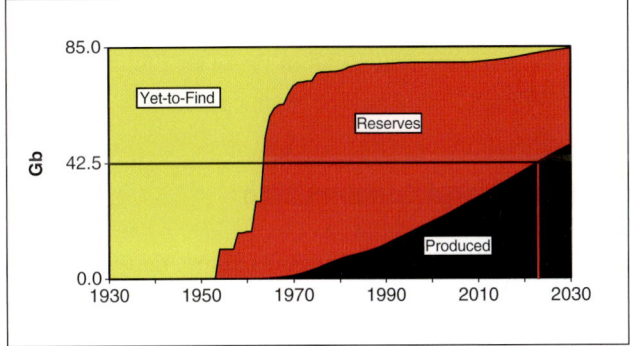

Fig. 69.2 United Arab Emirates status of oil depletion

Essential Features

The United Arab Emirates comprise seven territories at the southern end of the Persian Gulf bordering Saudi Arabia. They are made up of Abu Dhabi, which is by far the largest, Dubai, Sharjah, Ajman, Umm al-Qaywayn, Ra's al-Khaymah and Al-Fujairah. Of these, only Abu Dhabi, Dubai and Sharjah have significant oil resources, with those of Abu Dhabi being by far the largest.

Topographically the area is made up of deserts containing some of the world's largest sand dunes. A mountainous peninsula, being an extension of the Hajar Mountains, forms the border with Oman, and a number of shoals and small islands lie offshore, some of which are claimed by Iran.

The territories support a population of 7.9 million, most of whom live in Dubai and Abu Dhabi, which have become thriving commercial towns. Only about 20% of the population are of indigenous origins, the rest coming especially from India, Bangladesh, Pakistan and Palestine. Most of the population belong to the *Sunni* branch of Islam.

Geology and Petroleum Systems

The Persian Gulf generally flanks the plate boundary responsible for the Zagros Mountains which separates Arabia from the Eurasian continent. It has been a shallow subsiding trough through most of its geological history, but was cut by transverse uplifts, which came into existence during the Hercynian orogeny and influenced subsequent sedimentation. The Emirates are flanked by two such uplifts: the Qatar Arch to

C.J. Campbell, *Campbell's Atlas of Oil and Gas Depletion*,
DOI 10.1007/978-1-4614-3576-1_69, © Colin J. Campbell and Alexander Wöstmann 2013

the north and the Hajar Uplift to the south, both of which are cut by important fault zones. By late Jurassic times, the Emirates lay in a setting comparable to that of eastern Arabia when prolific oil source-rocks were deposited, but during the Cretaceous and Tertiary period somewhat deepwater shelf conditions prevailed. This prime source charged both Jurassic and Cretaceous carbonate reservoirs in structural traps, partly caused by salt movement.

Exploration and Discovery

Exploration commenced in the 1930s, being centred in Abu Dhabi where the Bab Field with 11 Gb was found onshore in 1954. It was followed by a move offshore where a number of giant fields were found, including Umm Shaif Field (1958) with about 6 Gb and Zakum (1964) with over 20 Gb.

A total of about 360 exploration boreholes have been drilled, having passed peaks in 1952 and 1984 when 21 were drilled. It is accordingly a fairly mature area. Lesser amounts of oil and gas were also found in the other Emirates, especially Dubai.

Production and Consumption

Oil production commenced in 1962 and rose fairly rapidly to pass 1 Mb/d in 1971 and reach an early peak of 2 Mb/d in 1977. It then slumped to a low of 1.1 Mb/d in 1983, partly in response to OPEC quota constraints, before recovering to its present level of 2.4 Mb/d. It is expected to rise further to plateau at just over 2.7 Mb/d from 2012 to 2023, the midpoint of depletion, before commencing its terminal decline at about 2.3% a year.

Gas production rose in parallel with oil to pass 1 Tcf/a in 1990, and 2 Tcf/a in 2001. With 30% having been depleted, it is expected to rise to a plateau at about 2.7 Tcf/a from 2010 until around 2035.

Oil consumption is at 199 Mb/a giving an excessively high per capita consumption of 25 barrels a year, with exports standing at 682 Mb/a. Gas consumption also seems high, but the data are probably unreliable.

The Oil Age in Perspective

The barren lands of the Emirates were barely populated and of little significance during the earlier history of the area. The offshore islands and shallow waters had become a base for pirates who were attacking Britain's trade with India, but were suppressed by British naval forces in 1819. By 1853, the various tribal leaders had come to accept British dominion, and indeed protection, forming what became known as the Trucial States. A British military force, known as the Trucial Scouts, maintained law and order over the next two centuries.

The Second World War effectively brought the British Empire to an end, and led to moves towards a new independence, which naturally sparked conflict from the rival powers within the region. A major discovery of oil in Abu Dhabi in 1954 opened a new chapter of prosperity, and was followed in due course by lesser finds in Dubai and the other sheikhdoms. The oil wealth of Abu Dhabi gave it an advantage in welding together the neighbouring territories, whose leaders finally agreed to a union in the early 1970s. Sheikh Zayad bin Sultan al Nahyan became the first President, governing until his death in 2004 when he was succeeded by his son.

The Gulf Wars and the invasion of Iraq have raised tensions somewhat as the bulk of the population are *Sunni* Moslems, and the simmering territorial disputes with Iran over offshore islands may well heat up. But, for the moment, the Emirates are enjoying the benefits of the high price of oil, whose revenues have allowed them to develop major industries, a world airline and an important regional financial centre. Abu Dhabi and Dubai now boast some of the world's highest skyscrapers. The emirates do not seem to have been badly affected by the Arab Spring of 2011, but there have been some disturbances.

It is thought that they can maintain their current level of oil production for another 20 years or so, meaning that they will continue to enjoy ever larger revenues, albeit subject to volatility in the face of global economic recessions. They can accordingly remain a prosperous territory for years to come, although it is never easy to be a wealthy man in a crowd of beggars. Furthermore, there is a potential source of tension from amongst the large immigrant community reflecting circumstances in their home countries. By the end of the century, however, when the oil is gone, little will be left of this anomalous situation.

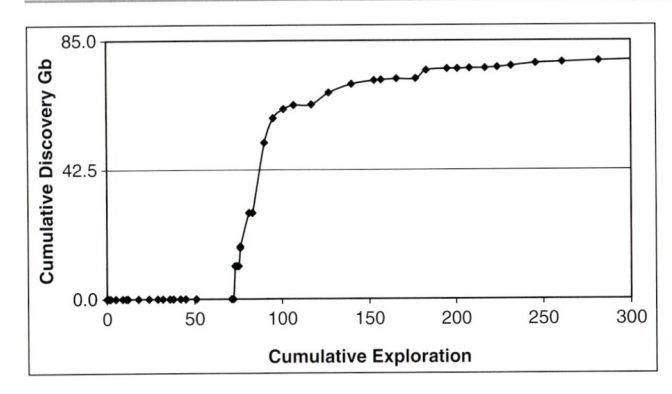

69 United Arab Emirates

Fig. 69.3 United Arab Emirates discovery trend

Fig. 69.4 United Arab Emirates derivative logistic

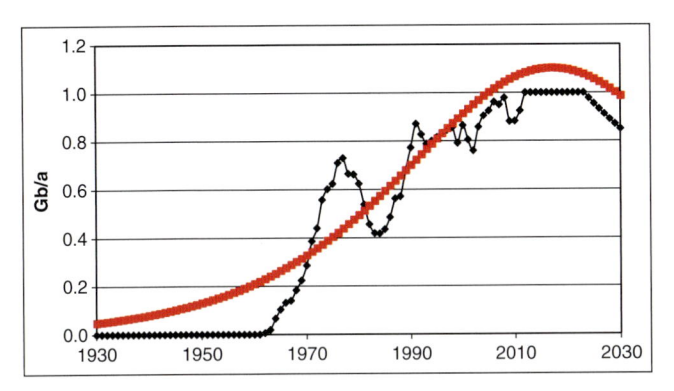

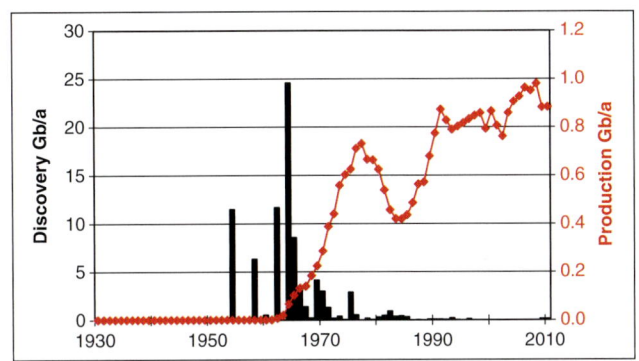

Fig. 69.5 United Arab Emirates production: actual and theoretical

Fig. 69.6 United Arab Emirates discovery and production

Yemen

Table 70.1 Yemen regional totals (data through 2010)

Production to 2100							Peak Dates			Area	
Amount			Rate				Oil	Gas	'000 km²	
	Gb	Tcf	Date	Mb/a	Gcf/a	Discovery	1984	1989	Onshore	Offshore
PAST	2.7	12	2000	160	667	Production	2001	2013	530	33
FUTURE	2.0	13.3	2005	146	727	Exploration	1992		Population	
Known	1.6	12.7	2010	94	1,153	Consumption	Mb/a	Gcf/a	1900	2.0
Yet-to-Find	0.4	0.7	2020	65	691	2010	57	27	2010	23.8
DISCOVERED	4.3	24	2030	42	202		b/a	kcf/a	Growth	11.9
TOTAL	4.75	25	Trade	+37	+1,126	Per capita	2.4	1.13	Density	45

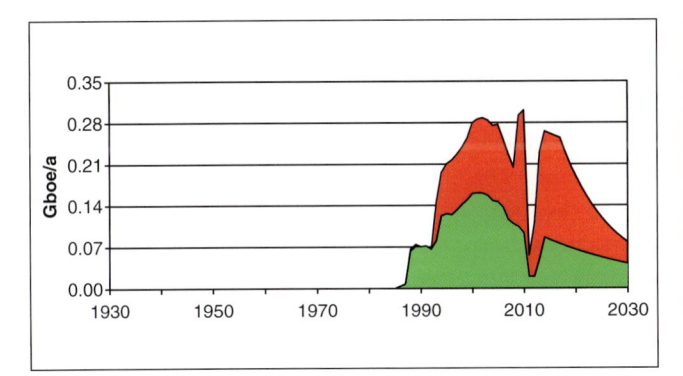

Fig. 70.1 Yemen oil and gas production 1930 to 2030

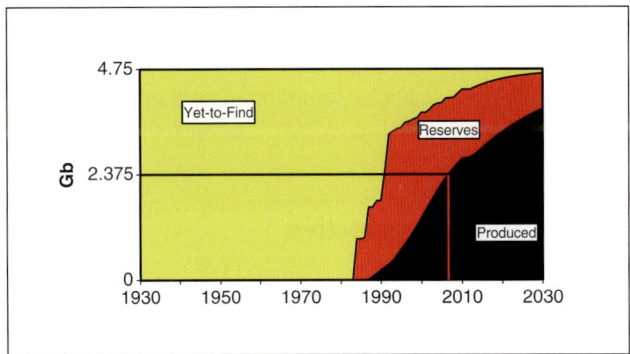

Fig. 70.2 Yemen status of oil depletion

Essential Features

Yemen covers an area of 530,000 km² on the southern margin of the Arabian Peninsula, bordering the Indian Ocean and the Red Sea. A coastal strip gives way to a high mountainous massif rising to almost 4,000 m, which in turn passes into the deserts of the Rub' al-Khali to the northeast. Monsoon rains from the Indian Ocean have given fertile soils along the coastal strip and the flanks of the massif, where most of the population, amounting to 24 million, live. The barren interior is sparsely populated. The capital is Sana'a in the northwest, but Aden on the south coast is the commercial centre, having been a coaling station in earlier years for shipping to the East.

Geology and Petroleum Systems

The western part of the country is formed of ancient Precambrian rocks, which are flanked to the east by a succession of younger sediments, being locally cut by Tertiary to Recent volcanics. The interior basins in the eastern part of the country are mainly rift basins that formed during the late Jurassic and early Cretaceous, as a result of the break-up of the Gondwanaland Continent, but there was also some later deformation during the Tertiary. Prolific Jurassic source-rocks have charged overlying sandstone and carbonate reservoirs with oil and gas in mainly fault-related traps. The three main basins extend in a southwesterly direction and have offshore extensions.

Exploration and Discovery

Some initial seismic surveys were shot before the Second World War, leading to exploration drilling in the 1960s, but the effort was not rewarded until 1984 when the giant Alif Field was found with about 650 Mb of oil and 1.7 Tcf of gas. It prompted further exploration, which peaked in 1992 when as many as 39 exploration wells were drilled, yielding a number of small discoveries. Both exploration and discovery have since dwindled. Some 22 boreholes have been sunk offshore but have failed to deliver positive results.

Production and Consumption

Oil production commenced in 1986 and grew to a peak in 2001 at 441 kb/d, since when it has declined to 257 kb/d. Having passed the midpoint of depletion in 2006, production is likely to continue to fall at about 4% a year, apart from 2011 and 2012 which may see a radical fall as a result of political problems.

Gas production rose in parallel with oil. It is now running at about 1.1 Tcf a year. Consumption of both oil and gas is at a low level, providing some 37 Mb/a of oil for export.

The Oil Age in Perspective

Yemen established an early trade in frankincense and myrrh, being also the home of the Queen of Sheba, who was attracted to King Solomon of Biblical fame. It is also known for having had some of the world's earliest coffee plantations, introducing the drink in the ninth century.

The country fell to the Romans, and gradually lost its importance, especially after 525 AD when an important dam controlling an irrigation system failed. Another low point occurred when King Dhu Nuweas became a convert to Judaism and massacred the Christian population, but was avenged by the Christian leader of Ethiopia who invaded the country, only to be in turn ousted by the Persians.

The country then converted to Islam, and the Prophet sent his son-in-law to act as Governor, building an impressive mosque in Sana'a. It laid the foundations for the later adoption of the *Shi'ia* sect of Islam, which recognises the son-in-law's claim to the Prophet's mantle. Various conflicts followed over the ensuing centuries.

The Middle East became subject to new rivalries in the nineteenth century as Britain wished to bring the region into its sphere of influence, facing opposition from the Ottomans. These pressures prompted Yemeni traders to move eastwards, some to settle in Indonesia. Britain established a base at Aden in 1839 while the Ottomans took the interior of the country. The opening of the Suez Canal in 1869 intensified the tensions, leading to a border commission to set a frontier between North Yemen (Ottoman) and South Yemen (British). North Yemen became independent after the defeat of the Ottomans in the First World War, but did not recognise the frontier with British-controlled South Yemen. There were further boundary disputes with Saudi Arabia to the north, especially after the discovery of oil.

Tribal warfare continued, but was partly held in check by British forces which had the support of local commercial interests in the prosperous Port of Aden. Renewed tensions developed after the Second World War, which eventually led to the re-establishment of two separate States, termed North and South Yemen, in 1967, when Britain withdrew from Aden. South Yemen then moved politically to the left, declaring itself to be the People's Republic of South Yemen, enjoying close relations with the Soviet Union but faced brief wars in 1972 and 1979.

The discovery of oil and a changed policy by President Gorbachev were factors that eventually led to the reunification of the two countries on 1990 under Ali Abdullah Saleh, who had previously been President of North Yemen. The country opposed foreign intervention in what became the Gulf War, which led to the curtailment of foreign financial aid, and alienated some of the other Middle East countries, especially Saudi Arabia, which expelled some 850,000 Yemeni immigrants. About 70% of the population belong to the *Sunni* faith. Yemen faced serious disturbances in 2011 as part of the Arab Spring. The President was injured in one of the attacks and flew to Saudi Arabia for treatment. A change of government becomes a distinct possibility.

Yemen's oil production is already in decline, so that the country is facing the onset of the Second Half of the Age of Oil, which may well see the revival of the ancient tribal conflicts that have afflicted the country for much of its history. At root, they probably reflect the ownership of the available fertile land or grazing rights in the barren interior.

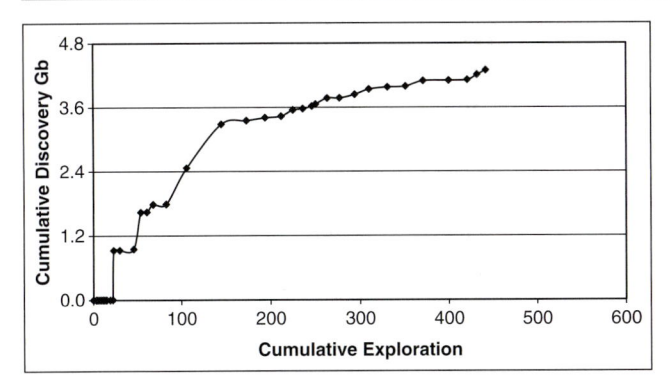

Fig. 70.3 Yemen discovery trend

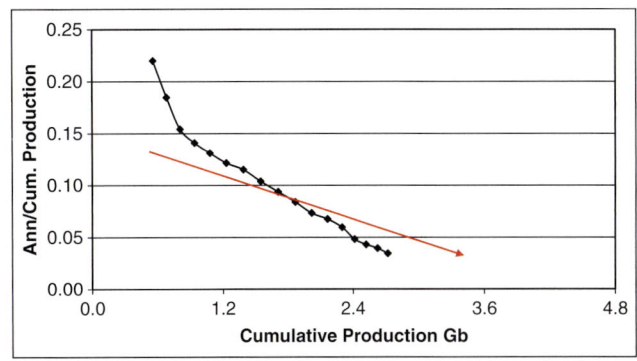

Fig. 70.4 Yemen derivative logistic

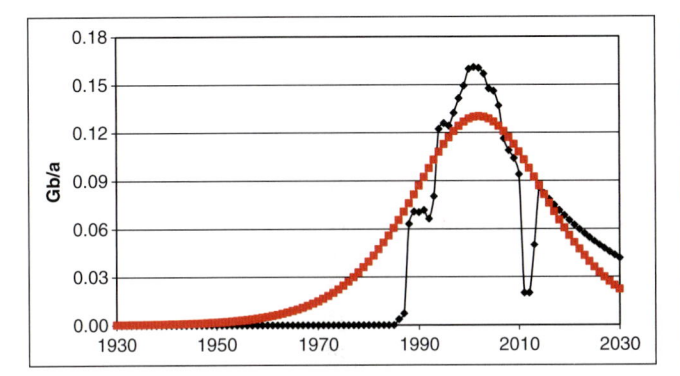

Fig. 70.5 Yemen production: actual and theoretical

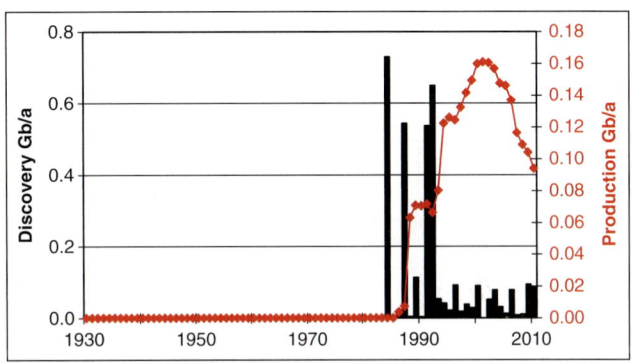

Fig. 70.6 Yemen discovery and production

71 Middle East Region

Table 71.1 Middle East regional totals (data through 2010)

Production						Peak Dates			Area	
Amount			Rate				Oil	Gas	'000 km^2	
	Gb	Tcf	Date	Mb/a	Gcf/a	Discovery	1948	1971	Onshore	Offshore
PAST	320	381	2000	7870	11322	Production	2020	2030	6060	510
FUTURE	507	2746	2005	8344	16070	Exploration	1992		*Population*	
Known	457	2355	2010	8241	22948	*Consumption*	Mb/a	Gcf/a	1900	32
Yet-to-Find	50	390	2020	8461	26684	2010	2760	14427	2010	276
DISCOVERED	777	2737	2030	7106	31280		b/a	kcf/a	Growth	8.6
TOTAL	828	3127	*Trade*	+5481	+8521	Per capita	10	52	Density	45
Note: Data refer to main producing countries only.										

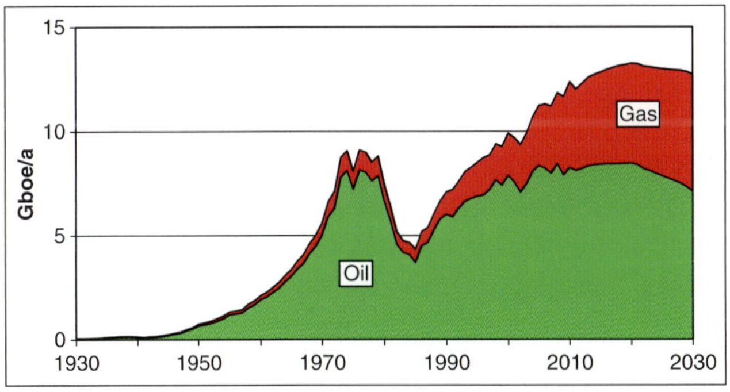

Fig. 71.1 Middle East oil and gas production 1930 to 2030

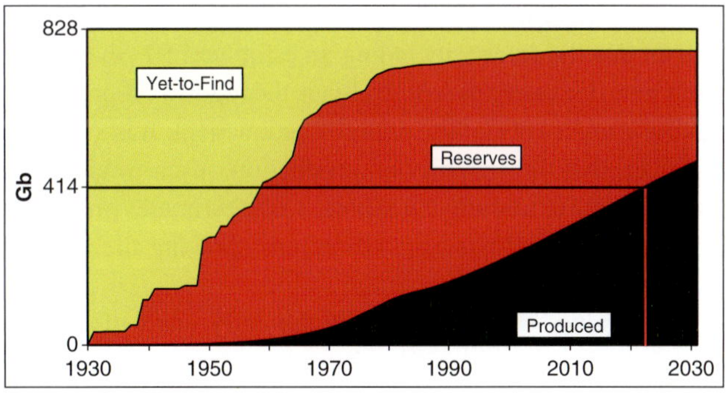

Fig. 71.2 Middle East status of oil depletion

The Region, as herein defined, extends from the Black Sea and the Mediterranean to the Indian Ocean, including the Persian Gulf. It is made up primarily of barren lands including the famous deserts of Arabia, together covering some six million square kilometres and also includes some mountain ranges, including especially the Zagros Mountains of Iran.

In geological terms, the ancient Arabian Shield is flanked by a thick sequence of sedimentary rocks spanning the Phanerozoic time scale. An important plate-boundary between the Eurasian Continent and the Arabian Shield is responsible for the Zagros Mountains and their extensions into Turkey. Of particular relevance are the relics of an earlier transverse structural grain, attributed to the Hercynian Orogeny, which have partly controlled the subsequent generation, migration and entrapment of oil and gas throughout the region. The sedimentary environment was generally one of the restricted basins and high evaporation in a warm climate which has resulted in the deposition of salts and other evaporates providing exceptionally effective seals for oil and gas reservoirs. Carbonate reservoirs are also predominate. A combination of these exceptionally favourable conditions has given the Region a rich endowment which represents 40% of the world's conventional oil and 30% of its gas.

The producing countries are divisible into two groups: the first tier comprises the five principal producers made up of Iran, Iraq, Kuwait, Saudi Arabia and the United Arab Emirates, which dominate world production, having a special depletion profile; and the second tier comprises Bahrain, Oman, Qatar, Syria, Turkey and the Yemen, which are more modest producers, save in terms of gas, with Qatar holding the world's largest field. In addition is the so-called Neutral Zone whose oil revenues are shared equally between Kuwait

C.J. Campbell, *Campbell's Atlas of Oil and Gas Depletion*,
DOI 10.1007/978-1-4614-3576-1_71, © Colin J. Campbell and Alexander Wöstmann 2013

and Saudi Arabia. As discussed further below, most of the countries are somewhat artificial constructions with disputed borders and an uncertain identity.

In addition to the twelve main oil producers, whose oil and gas position is summarised in the table above, are Israel, Jordan, Lebanon and Palestine having negligible petroleum resources, but are naturally consumers. Israel has recently made an important gas discovery in a deepwater area that it claims in the Eastern Mediterranean.

The Region supports a population of almost 276 million people, with an average density of 45/km². However, Bahrain with its small land area and high population is extremely over-crowded, having a density of around 1,880/km². Life expectancy ranges from 55 years in Iraq to 79 in the Emirates and Israel, and fertility rates fall in the range of 2.2 children per woman in Turkey to 6.2 in the Yemen, the latter country also having a high rate of infant mortality at 75 per 1,000. Population growth is relatively high due mainly to immigration into the richer countries of this region.

The total production of *Regular Conventional Oil* to the end of this Century is estimated to be about 828 Gb, of which 320 Gb have been produced through 2010, leaving some 500 Gb for the future, including an estimated 50 Gb coming from new discovery. Saudi Arabia is the highest oil producer, having almost 40% of the region's endowment. It is stressed, however, that the statistics on production, reserves and consumption are extremely unreliable, so the estimates presented here are no more than approximations showing the general situation.

Oil production commenced in the early years of the last century, following the pioneering endeavours of BP in Iran. It expanded greatly after the Second World War with the entry of Saudi Arabia to reach an early peak in 1976 at 22.2 Mb/d, but then declined to a low of 10.1 Mb/d in 1985, due to the OPEC quota and a world surplus driven primarily by the entry of North Sea production. It then rose in the late 1990s to about 20 Mb/d and is expected to remain at about this level for about 20 years before terminal decline sets in at about 2% a year.

It is very difficult to analyse the gas situation, but a preliminary estimate suggests that almost 2,750 Tcf have been found, of which only 381 Tcf have been produced, with some 390 Tcf yet to be found. Current production stands at almost 23 Tcf/a, and is expected to increase in the years ahead, assuming a massive expansion of liquefaction plants and pipeline links to Europe.

Oil consumption has been growing consistently from 773 Mb/a in 1980 to about 2,760 Mb/a today. Per capita consumption stands at a high level of about ten barrels per year. The Region is the world's prime oil exporter, with exports running at about 5,481 Mb/a. Gas consumption in the main countries is reported at 14.4 Tcf/a but it is unsure how much is actually consumed locally and how much is exported in the form of liquids.

The Oil Age in Perspective

The region has had a very long history, being perhaps the site of the biblical Garden of Eden where modern man is said to have been conceived. It supported impressive empires long before the birth of Christ, 2,000 years ago. Of particular significance was the Jordan Valley, whose people adopted a monotheistic religion, from which both Christianity and Islam are descended. The barren lands may have stimulated the people's need for religious inspiration, which has given rise to a much temporal conflict as different factions declared themselves to be closer to divine power.

In broad terms, the history of the region seems to have been dominated by Persia (now Iran) and Turkey, whose empires waxed and waned, as they exerted varying degrees of control over the largely tribal and sparsely populated region. The countries of the Mediterranean coast had the benefit of both foreign trade and more fertile lands which stimulated their early development as demonstrated by the growth of such cities as Jerusalem and Damascus. Muscat in Oman also has had a long history of trade, both with India, the East and the East African slave market. Certainly, prior to the Oil Age, the region represented something of a backwater, with piracy and pearling being the principal activities in the Persian Gulf.

The region came under external pressures during the nineteenth century. Russia was seeking to expand from the north and gain an access to world trade through the Bosphorus Straits which link the Black Sea with the Mediterranean. Meanwhile sea-faring Britain established bases in Aden and the Persian Gulf to protect its trade with India, especially after the opening of the Suez Canal in 1869. Of particular importance in the years preceding the First World War was a German plan to build a railway from Berlin to Baghdad to open up trade and counter the dominance of global trade by the British Empire, supported by its Navy. It secured the rights from the Ottoman Empire and also moved to provide military support for the Turkish army. Britain, not to be outdone, offered to supply Turkey with two battleships to deter Russian expansion. Also on the eve of 1914 war, the British Government took a 51% interest in the Anglo-Iranian Oil Company (now BP) to secure a firm supply of oil for the British Navy, which was converting from coal to oil.

Turkey sided with Germany in the war, being opposed by British forces based in Egypt, which eventually marched north, having encouraged various Arab tribal leaders, including the Grand Sharif of Mecca to revolt against the Ottoman Empire, sowing the seeds for the subsequent political evolution of the region. This was largely accomplished in the peace treaty that followed the war, by which the region came under the British sphere of influence, save for Syria, which was given to the French. Persia and Turkey remained more or less intact, but the rest of the region was broken up into

Kingdoms and Emirates under varying degrees of formal British protection. Most of the frontiers were artificial, not always respecting the local tribal, religious and cultural divisions. For example, the Kurds, who are descendents of the ancient Medes, did not end up with their own territory. These matters were of little material consequence in the barren lands until the discovery of oil brought discord over rival claims to ownership.

The United States, which had joined the victorious allies in 1917 towards the close of the First World War, was not left out of the carve-up of the Middle East oil rights. Exxon, Mobil and other American companies joined BP and Shell to exploit Iraq, while BP and Gulf Oil (now ChevronTexaco) took Kuwait. Saudi Arabia, which later became the jewel in the crown went to Standard Oil of California (also now ChevronTexaco), later to be joined by other American companies. Its prospects were not evident to the pioneering British oil geologists with their experience of the massive surface prospects of the Iranian foothills. BP managed to retain its exclusive position in Iran.

Stability and progress marked the inter-war years under the benign protection of the British Empire, and the new royal families of the oil countries found their coffers filling handsomely with oil revenues. They operated semi-feudal regimes, but faced no particular tensions as the people of the region went about their daily lives.

The region was not materially involved in the Second World War, although its oil exports were an important factor fuelling the war effort of Britain and the United States. Russia again eyed Iran, briefly occupying its eastern regions. The British Empire was extinguished by the Second World War and could no longer preserve the status quo in the region.

The post-war epoch also saw the growth of socialism in Europe as the workers sought a greater share of the proceeds of their endeavours, and there were moves to independence around the world encouraging more democratic forms of government. These pressures also affected the Middle East, although the extreme oil wealth that flowed to the governments helped them continue to operate substantially autocratic forms of government, which were perhaps consistent with the tribal traditions of the region.

The unilateral declaration of the State of Israel in 1948, following the end of the British protectorate of Palestine, opened a new chapter of discord affecting the region, carrying additional religious overtones between Judaism and Islam. Thousands of displaced Palestinians were driven into exile in refugee camps, but perhaps the real conflicts went deeper as the neighbouring countries began to resent the financial and military support supplied to Israel by the United States, which might in some way have appeared threatening. Arab nationalism grew in parallel being led principally by Egypt, which for a time formed a union with Syria, with the idea of building an Arab Empire to match the power of the West. More serious yet were moves to nationalise resources, led by Iran in 1951, when it nationalised its oil rights which BP had enjoyed exclusively for some 50 years. This set the stage for the creation in 1959 of OPEC, albeit under Venezuela's initiative, whereby the world's principal established producers agreed to limit their output under a quota system to support the global price of oil, principally because they were facing new competition from offshore discoveries around the world—including the North Sea. This move led in turn to outright nationalisation in the prime Middle East oil countries: Iraq in 1972; Kuwait in 1975; and Saudi Arabia in 1979.

Iraq, in the heartland of the Ottoman Empire, was perhaps one of the more artificial constructions of the carve-up, and consequently faced internal conflict between the Kurds in the North, the *Shi'ias* with links to Iran, and the *Sunnis*. In 1920, Britain had given the throne of Syria to Faisal, the son of the Grand Sharif of Mecca, because of his father's support in the First World War, but he later fell out with the French administration, and it was arranged that he should take the throne of Iraq instead. He was succeeded in turn by his son and grandson, but the latter was killed in 1958 in a military coup, related to the conflict with Israel and the idea of a new pan-Arabic nation. The resulting government faced internal dispute, one faction proposing to re-exert Iraq's ancient claim to Kuwait.

Meanwhile, the *Ba'ath Party* strived for a blend of Arab nationalism and socialism, in part as a reaction to the formation of the Israeli State. It successfully established its position in the shifting political scene, and eventually came to power, for a time courting friendship with the Soviet Union. Saddam Hussein was a prominent member of the party and became Prime Minister of Iraq in 1979. His ambition was to supplant Egypt as leader of the Arab world, and build a new empire, especially over the oil-rich Persian Gulf. As mentioned above, there were various conflicts with Kurds in the north and a *Shi'ia* community on the Iranian border in the south, which together with the disputed definition of the southern frontier along the Shatt-al-Arab waterway, led to the opening of hostilities with Iran that escalated into full-scale war lasting from 1980 for eight years, being accompanied by appalling loss of life. The US relations with Iran had deteriorated with the fall of the Shah and an incident in 1978 when hostages were taken from the US Embassy to secure his return for trial. The United States accordingly sided with Iraq in the war supplying it with military equipment, financial help and intelligence.

In 1985, when world oil prices were low, Kuwait added 50% to its reported reserves in order to increase its OPEC quota, which is based partly on reserves. It also started pumping oil from the South Rumaila Field straddling the ill-defined frontier with Iraq, which complained bitterly with some justification. It then moved to a military resolution, with some implied support from the US Ambassador, who stated that boundary disputes were of no consequence to Washington. Saddam Hussein evidently took this as a green light for a successful full-scale invasion, to which the United States did react by launching the so-called Gulf War to free Kuwait.

Iraq was then subjected to a partial oil export embargo, sanctioned by the United Nations. It was an expedient mechanism to regulate world oil prices by making it the swing producer of last recourse. This succeeded in taking some of the pressure off the other OPEC countries and leaving the United States in control of the situation. The embargo was relaxed from time to time for *humanitarian reasons* when oil prices rose uncomfortably. It was a very difficult time for the people of Iraq, but Saddam Hussein managed to marshal sufficient support with anti-American rhetoric to remain as leader despite the suffering of the people.

The last chapter opened on 17th March 2003 when President Bush invited the Iraqi Government to step down or face war, which was followed three days later with a successful Anglo-American invasion. Saddam Hussein was executed in December 2006 on the grounds that he had authorised the killing of 148 people in the town of Dujail, which had moved against the Government proposing *Shi'ia* separatism. However regrettable, it was a minor loss of life compared with the millions of deaths caused directly and indirectly by the invasion, which President Bush later justified with the words *our energy supply was at risk*.

The invasion has had a destabilising effect throughout the region by exacerbating the long history of ill-feeling between the *Sunni* and *Shi'ia* communities. There are also moves towards a new Kurdish nation, which would unite the communities separated by the division of the Middle East after the First World War. The Kurds inhabit a contiguous region across Syria, northern Iraq, southeast Turkey and northern Iran.

The long established hostility between the United States and Iran also reaches new levels of intensity being accompanied by threats of military action. Plans have even been considered to reopen a pipeline from Iraq to the Israeli port of Haifa on the Mediterranean.

It is difficult to imagine how this complex situation will evolve. Today, the Middle East Region supplies about 35% of the world's *conventional* oil but that will have risen to almost 50% by 2030, presumably increasing the pressures and tensions. The physical oil supply is important but the financial implications may be even more serious. The Region produces a rounded 23 Mb/d of oil, costing $10–15 a barrel to produce. Selling it at $100 a barrel yields something like 65 trillion dollars a year of artificial liquidity, being nothing more than profiteering from shortage, which is pumped into the world's financial system. It may be an important factor contributing to a financial crash.

Midas-like wealth will continue to flow to the region in the short to medium term, perhaps stimulating the growth of new financial centres, such as already exist in Kuwait, Dubai, Abu Dhabi and Qatar, which will exert an increasing role on regional investment. Such investment may in turn trigger further waves of immigration, especially from the Indian subcontinent and SE Asia. The new wealth may be difficult to absorb, however, leading to further jealousies and conflicts between the people at large and the privileged beneficiaries— including the sundry royal families. These pressures are compounded as the rest of the world faces economic recession, which may evolve into the Second Great Depression. Such an event might be triggered by the realisation that world debt, which is premised on continued economic expansion, loses its collateral if a declining oil supply means economic contraction. A volatile situation has already erupted around the world seeing some extreme examples in the Middle East Region, which is so heavily dependent on oil revenue.

It is also worth remembering that the oil and gas resources of the Middle East—despite their size—will not last for ever, and at current rates of depletion will be approaching exhaustion by the end of the century. The region is for the most part made up of deserts and barren lands with relatively few fertile tracts to provide a sustainable livelihood. Logic therefore suggests that the population will have to fall precipitately to something approaching what it was a century ago—before oil distorted the natural order of things—in order to maintain a workable society. The manner of this contraction does not bear thinking about.

71 Middle East Region

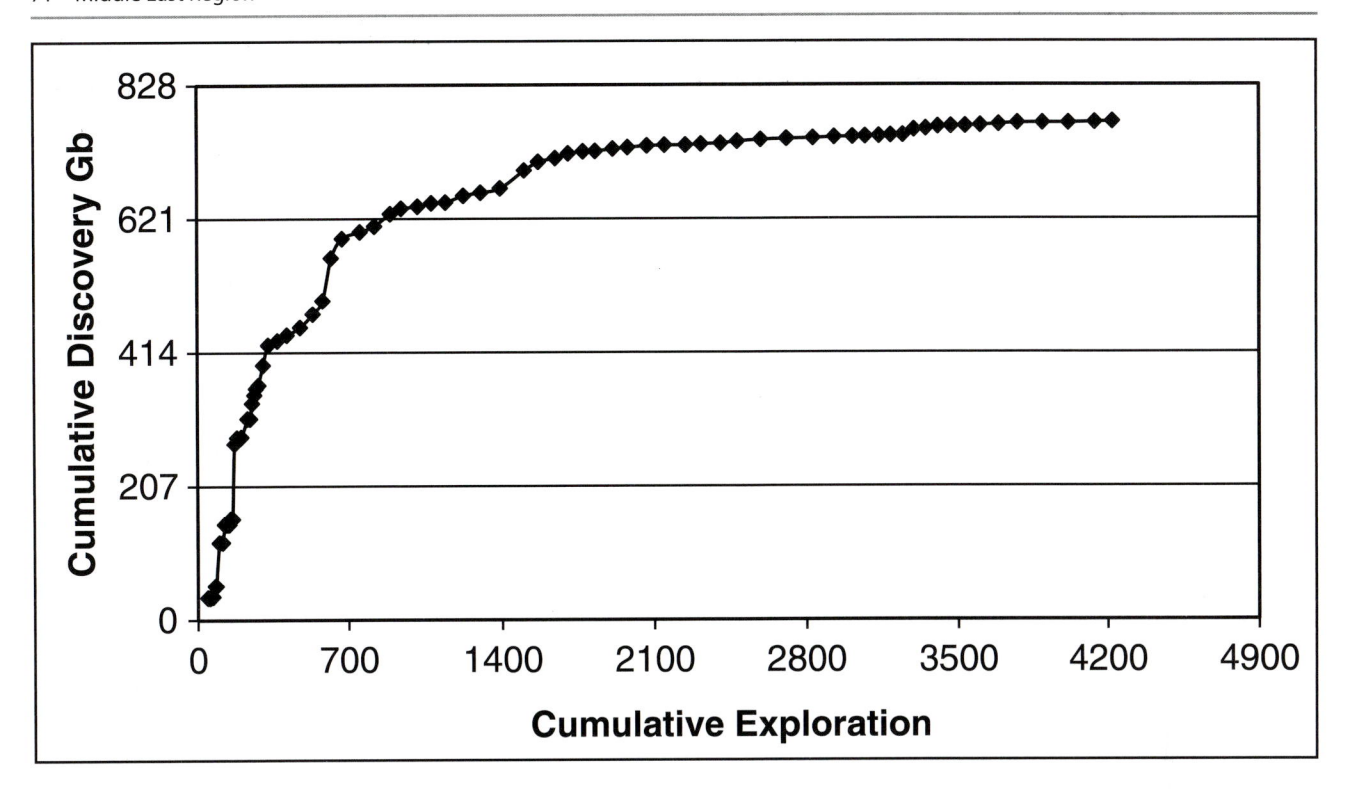

Fig. 71.3 Middle East discovery trend

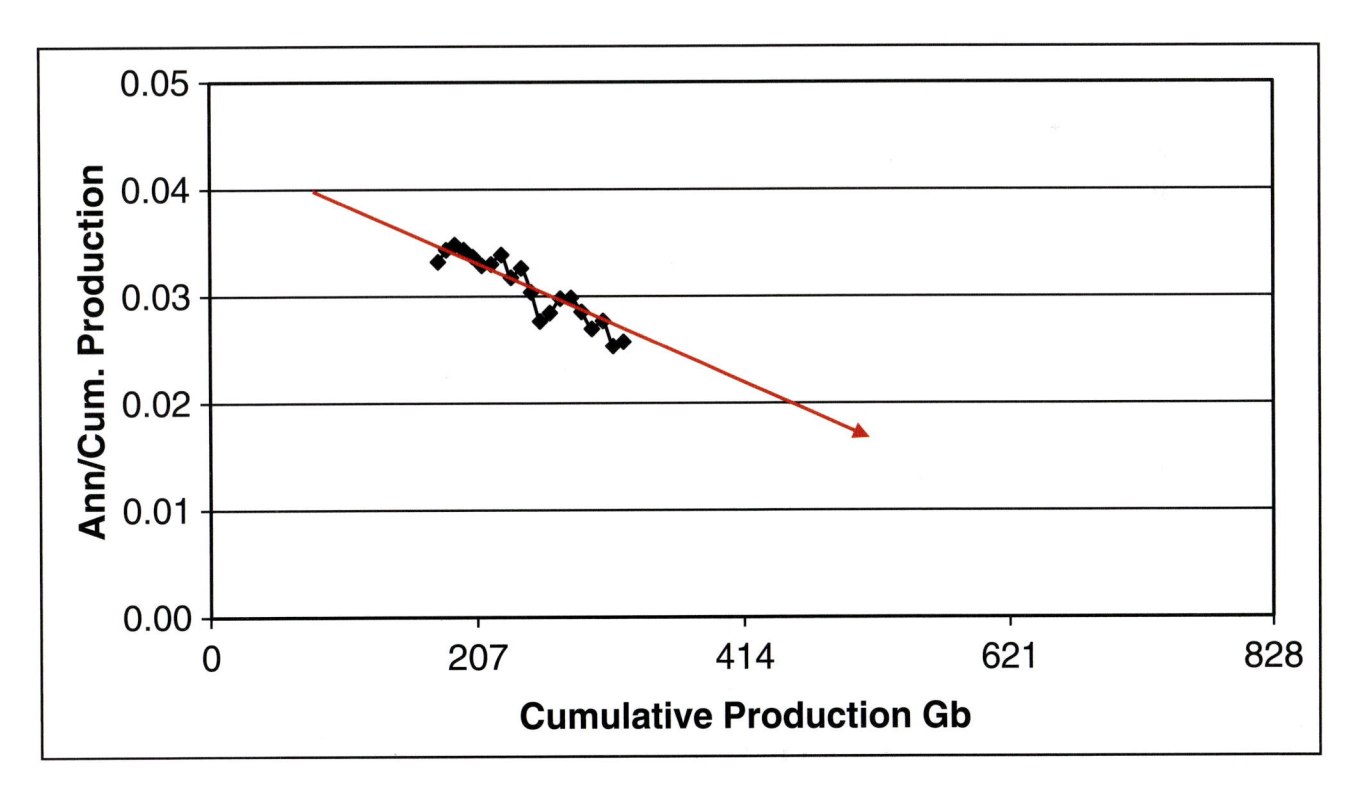

Fig. 71.4 Middle East derivative logistic

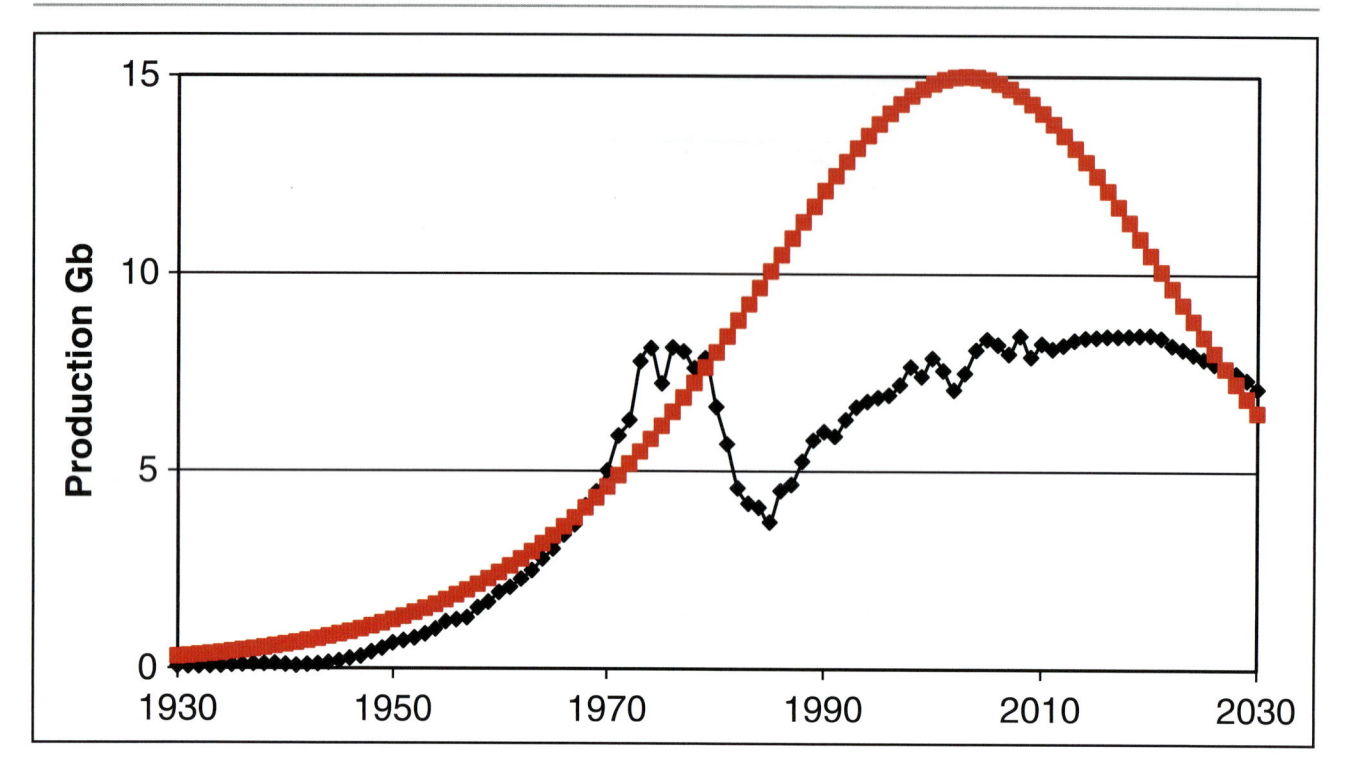

Fig. 71.5 Middle East production: actual and theoretical

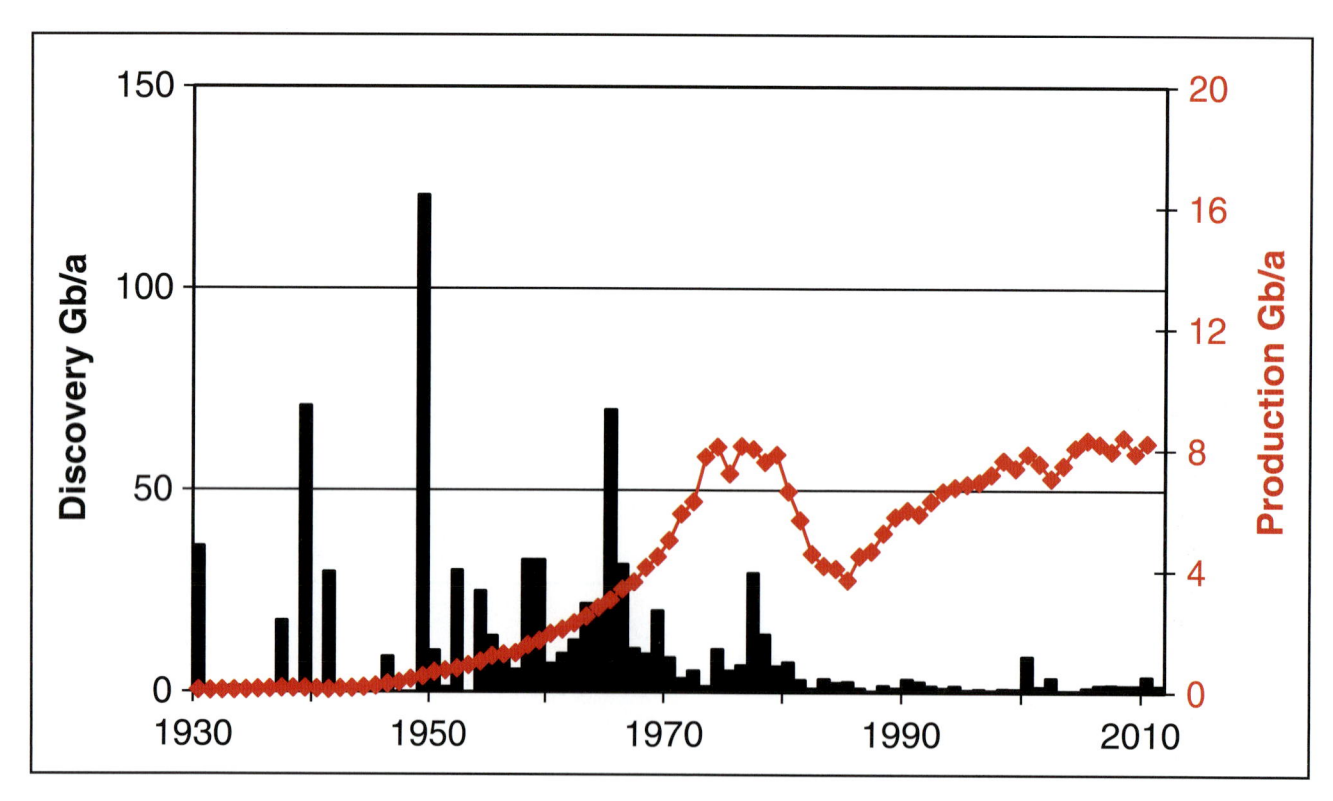

Fig. 71.6 Middle East discovery and production

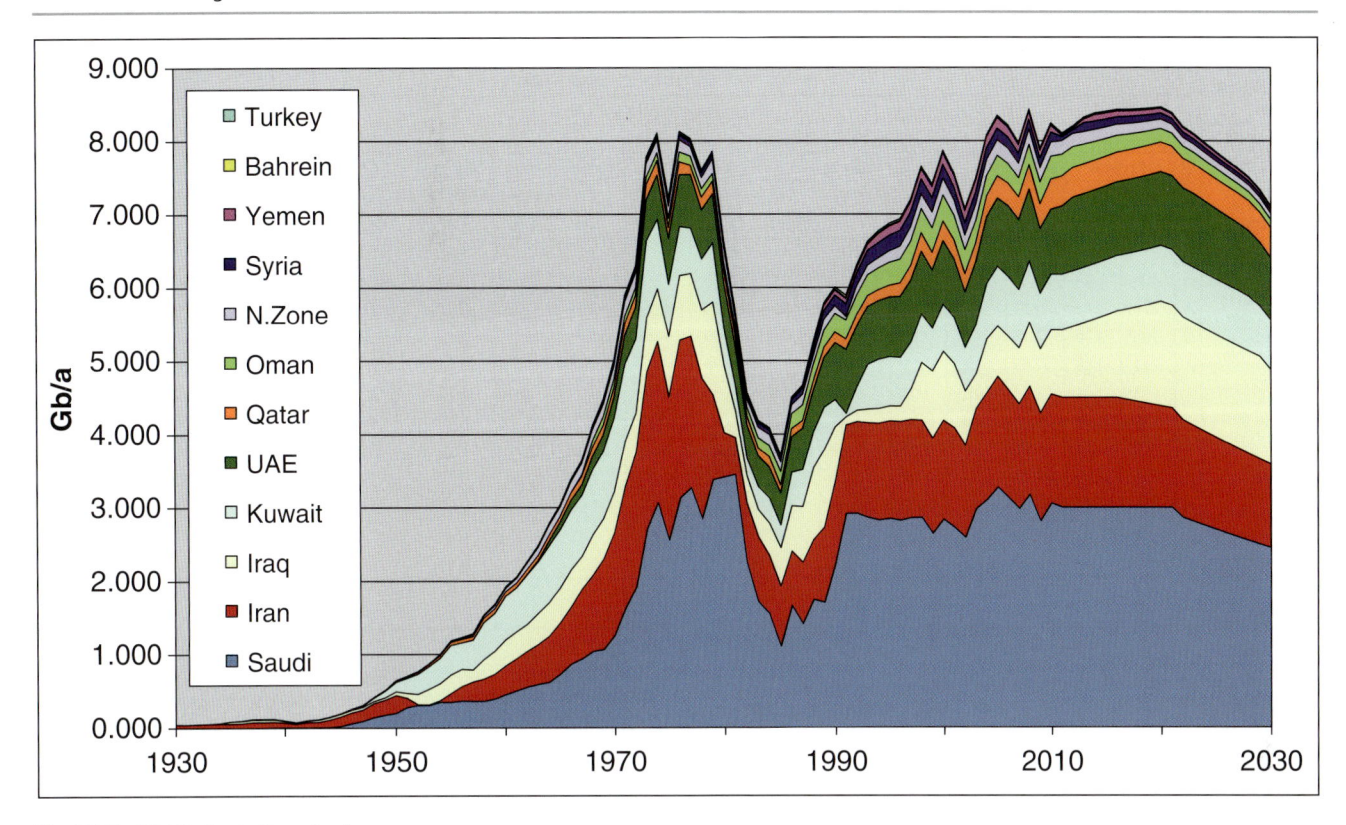

Fig. 71.7 Middle East oil production

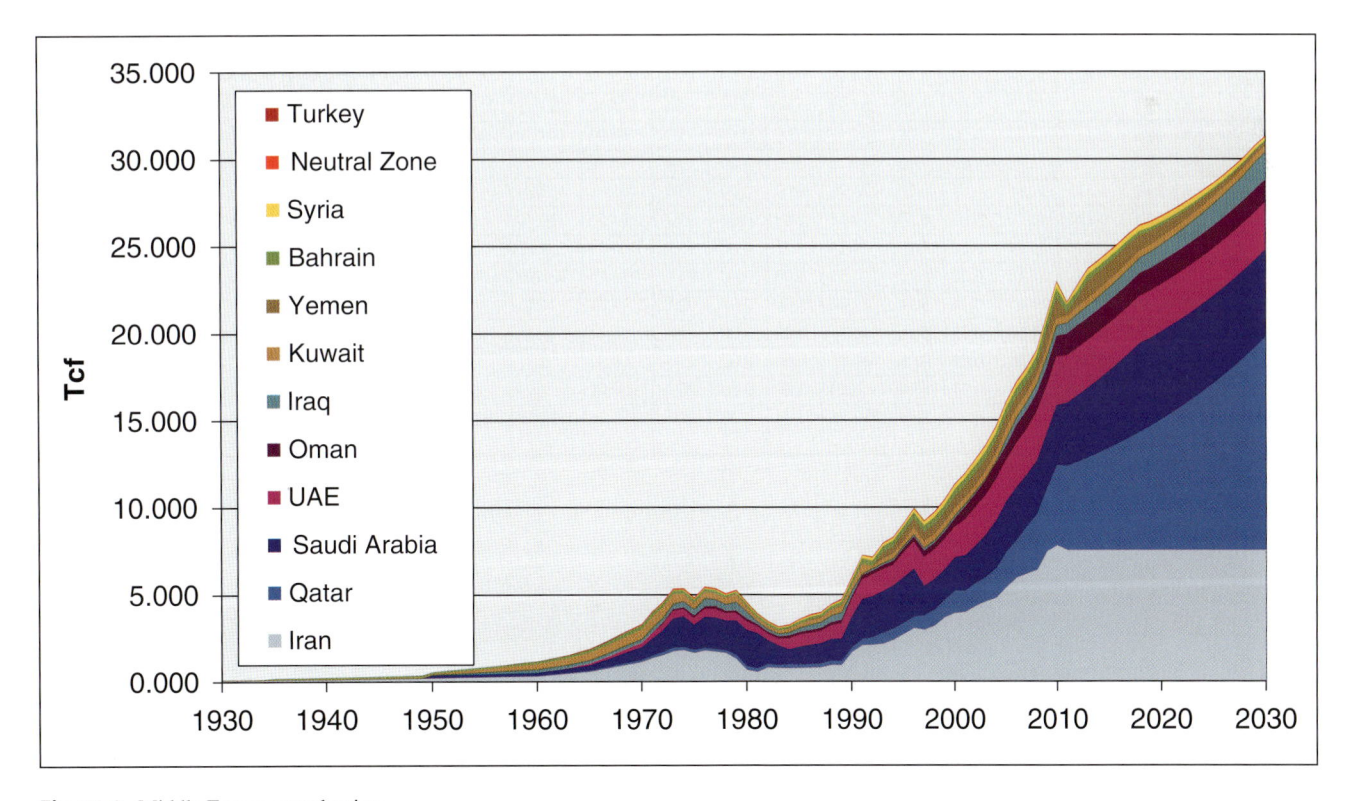

Fig. 71.8 Middle East gas production

Table 71.2

Middle East Region		PRODUCTION TO 2100															
		Regular Conventional Oil														2010	
		KNOWN FIELDS									Revised					23/11/2011	
	Region	Present		Past			Reported Reserves			Future	Total	NEW	ALL	TOTAL	DEPLETION		PEAK
Country		Kb/d 2010	Gb/a 2010	Gb	5yr Trend	Disc 2010	Average	Deduction Static	Other	% Disc.	Gb	Found Gb	FIELDS Gb	FUTURE Gb	Gb	Rate	Mid-Point	Disc Prod
Saudi Arabia	A	8370	3.06	117.31	0%	0.05	262.30	0.00	0.00	97%	173.55	290.87	9.13	182.69	300.0	1.6%	2022	1948 1981
Iran	A	4080	1.49	65.76	0%	0.45	140.32	0.00	0.00	92%	71.60	137.36	12.64	84.24	150.0	1.7%	2010	1964 1974
Iraq	A	2399	0.88	33.84	4%	0.33	122.82	−13.86	0.00	89%	68.99	102.83	12.17	81.16	115.0	1.1%	2030	1928 2025
Kuwait	A	2085	0.76	42.18	−1%	0.12	101.12	0.00	0.00	97%	54.93	97.11	2.89	57.82	100.0	1.2%	2027	1938 1971
UAE	A	2415	0.88	29.69	−2%	0.13	97.44	0.00	0.00	93%	49.78	79.47	5.53	55.31	85.0	1.6%	2023	1972 2016
Bahrain	A	35	0.01	1.12	0%	0.00	0.07	−0.07	0.00	99%	0.22	1.34	0.01	0.23	1.4	5.2%	1979	1932 1970
Oman	A	865	0.32	9.33	3%	0.14	5.52	0.00	0.00	96%	5.10	14.43	0.57	5.67	15.0	5.3%	2004	1962 2000
Qatar	A	1127	0.41	9.36	7%	0.17	24.41	−0.79	0.00	96%	24.36	33.72	1.28	25.64	35.0	1.6%	2030	1940 2030
Yemen	A	257	0.09	2.72	−6%	0.09	2.71	−2.15	0.00	91%	1.62	4.34	0.41	2.03	4.8	4.4%	2004	1984 2001
Syria	A	367	0.13	4.94	−2%	0.03	2.56	−0.47	0.00	95%	1.96	6.90	0.35	2.31	7.3	5.5%	2001	1966 1996
Turkey	A	48	0.02	0.96	3%	0.01	0.27	−0.03	0.00	96%	0.25	1.21	0.04	0.29	1.3	5.2%	1993	1961 1991
N.Zone	A	530	0.19	8.29	−2%	0.05	4.83	−2.88	0.00	96%	4.24	12.53	0.47	4.71	13.0	3.9%	2001	1960 2003
ME REGION	A	22579	8.24	320.49	2%	1.55	35.50	−20.25	0.00	96%	456.61	777.10	50.50	507.11	827.6	2.7%	2015	1948 2020

Table 71.3

Middle East Region		PRODUCTION TO 2100															
		Regular conventional Gas														2010	
		KNOWN FIELDS									Revised					28/12/2011	
	Region	Present		Past			Reported Reserves			Future	Total	NEW	ALL	TOTAL	DEPLETION		PEAK
Country		Tcf/a 2010	Gboe/a 2010	Tcf	5yr Trend	Disc 2010	Average	Deductions Static	Other	% Disc.	Tcf	Found Tcf	FIELDS Tcf	FUTURE Tcf	Tcf	Current Rate	Mid-Point	Disc Prod
Saudi Arabia	A	3.43	0.62	84.03	0.03	0.33	204.48	0.00	0.00	84%	252.78	336.81	63.19	315.97	400.00	0.01	2030+	1948 2030
Iran	A	7.77	1.40	118.55	0.06	3.32	261.95	0.00	0.00	82%	865.16	983.71	216.29	1081.45	1200.00	0.01	2030	1964 2030
Iraq	A	0.60	0.11	14.37	0.08	1.09	104.95	0.00	0.00	73%	94.94	109.31	40.69	135.63	150.00	0.00	2030+	1953 2030
Kwait	A	0.42	0.08	21.48	−0.03	0.10	55.78	0.00	0.00	97%	46.10	67.57	2.43	48.52	70.00	0.01	2030+	1938 2015
UAE	A	2.82	0.51	52.96	0.02	6.10	220.71	0.00	0.00	93%	10983	162.80	12.20	122.04	175.00	0.02	2023	1978 2023
Bahrain	A	0.55	0.10	16.20	0.03	0.05	4.17	2.63	0.00	99%	4.56	20.76	0.24	4.80	21.00	0.10	1998	1932 2015
Oman	A	1.18	0.21	15.23	0.02	1.13	37.24	4.41	0.00	93%	40.29	55.52	4.48	44.77	60.00	0.03	2021	1973 2015
Qatar	A	4.61	0.83	37.10	0.22	6.42	1124.53	0.00	0.00	95%	914.75	951.86	48.14	962.90	1000.00	0.00	2030	1971 2015
Yemen	A	1.15	0.21	11.67	0.16	0.14	21.41	4.00	0.00	97%	12.67	24.33	0.67	13.33	25.00	0.08	2013	1989 2013
Syria	A	0.36	0.06	6.58	0.03	0.13	12.13	0.00	0.00	89%	6.74	13.32	1.68	8.42	15.00	0.04	2013	1987 2015
Turkey	A	0.02	0.00	0.45	−0.05	0.00	12.13	0.00	0.00	97%	0.52	0.97	0.03	0.55	1.00	0.04	2012	1965 2015
N.Zone	A	0.04	0.01	2.72	−0.07	0.04	4.50	0.00	0.00	96%	6.91	9.64	0.36	7.28	10.00	0.01	2030+	1967 2006
M EAST GULF	A	22.95	4.13	381.33	0.07	18.96	2052.16	11.04	0.00	88%	2355.26	2736.59	390.41	2745.67	3127.00	0.01	2030	1971 2030

Table 71.4

Regular Conventional Oil Production					
Mb/d	2000	2005	2010	2020	2030
Saudi Arabia	7.77	8.97	8.37	8.22	6.72
Iran	3.70	4.14	4.08	3.80	3.12
Kuwait	1.76	2.24	2.09	2.08	1.86
UAE	2.37	2.54	2.42	2.74	2.33
Bahrain	0.04	0.04	0.04	0.02	0.01
Oman	0.97	0.77	0.87	0.50	0.29
Qatar	0.74	0.84	1.13	1.12	1.12
Yemen	0.44	0.40	0.26	0.18	0.11
Syria	0.52	0.43	0.37	0.22	0.13
Turkey	0.05	0.04	0.05	0.03	0.01
N.Zone	0.63	0.58	0.53	0.35	0.24
ME Region	21.56	22.86	22.58	23.18	19.47

Table 71.5

Gas Production					
Tcf/a	2000	2005	2010	2020	2030
Saudi Arabia	1.89	2.87	3.43	5.00	5.00
Iran	3.87	5.39	7.77	7.50	7.50
Iraq	0.15	0.40	0.60	0.97	1.57
Kuwait	0.40	0.53	0.42	0.40	0.40
UAE	1.78	2.41	2.82	2.75	2.75
Bahrain	0.41	0.47	0.55	0.19	0.07
Oman	0.48	0.85	1.18	1.30	1.30
Qatar	1.31	2.03	4.61	7.51	12.23
Yemen	0.67	0.73	1.15	0.69	0.20
Syria	0.28	0.30	0.36	0.30	0.19
Turkey	0.03	0.03	0.02	0.03	0.01
N.Zone	0.05	0.06	0.04	0.05	0.05
ME Region	11.32	16.07	22.95	26.68	31.28

Table 71.6

Middle East Region	OIL Production Mb/a	OIL Consumption Mb/a	OIL Consumption p/capita b/a	OIL Trade (+) Mb/a	GAS Production Gcf/a	GAS Consumption kcf/a	GAS Consumption p/capita	GAS Trade (+) Gcf/a	Population Growth Factor	Area Km2 M	Density
Saudi Arabia	3055	965	34.58	2090	3427	3096	111	331	27.9	27.9	2.16	13
Iran	1489	673	8.64	816	7774	5106	66	2668	77.9	7.8	1.64	48
Iraq	876	253	7.75	622	596	46	1	550	32.7	16.4	0.44	74
Kuwait	761	129	46.15	632	422	446	159	−24	2.8	28.0	0.02	140
UAE	881	199	25.18	682	2817	2138	271	679	7.9	79.0	0.08	95
Bahrain	13	17	13.20	-4	551	433	333	118	1.3	13.0	0.00	1880
Oman	316	52	17.28	264	1176	619	206	557	3.0	1.2	0.21	14
Qatar	411	61	35.64	351	4611	770	453	3841	1.7	85.0	0.01	155
Yemen	94	57	2.41	37	1153	27	1	1126	23.8	11.9	0.53	45
Syria	134	107	4.74	27	356	340	15	16	22.5	11.8	0.19	118
Turkey	18	236	3.19	−2.18	23	1346	18	−1323	74.0	6.2	0.78	95
N.Zone	193	11	36.50	183	42	60	200	−18	0.03	6.0	0.00	61
Total	**8241**	**2760**	**10.01**	**5481**	**22948**	**14427**	**52**	**8521**	**276**	**8.6**	**6.07**	**45**

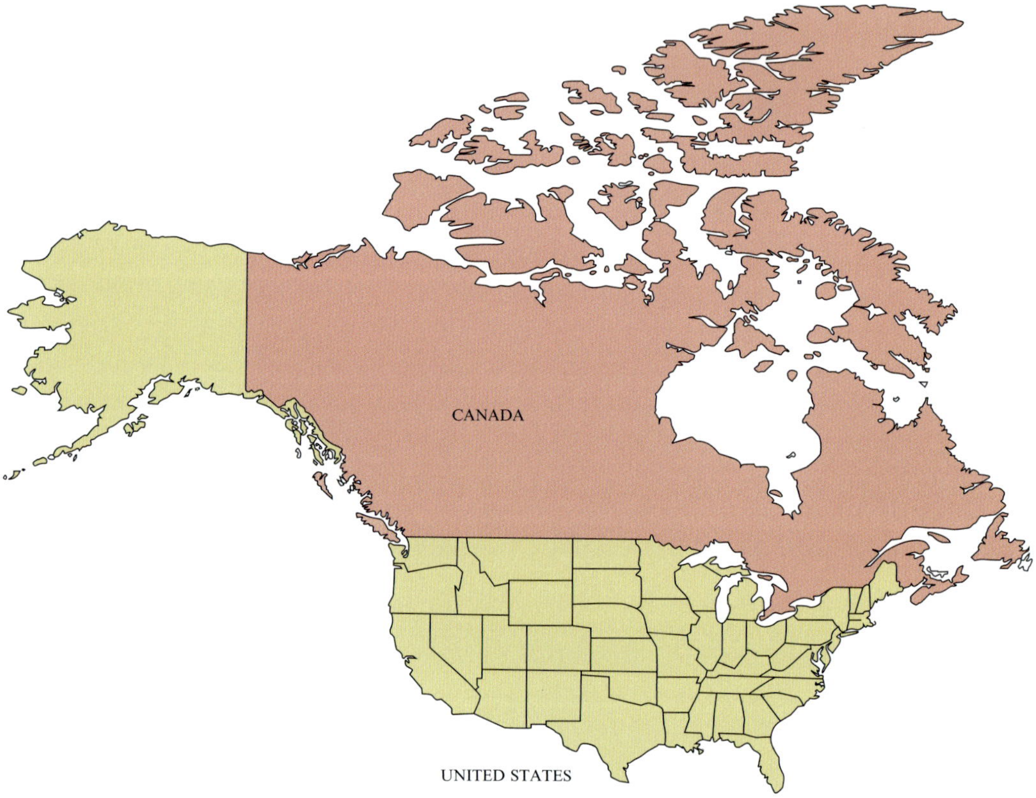
CANADA
UNITED STATES

Part VIII

North America

Canada

Table 72.1 Canada regional totals (data through 2010)

	Production to 2100					Peak Dates			Area	
	Amount		Rate				Oil	Gas	'000 km²	
	Gb	Tcf	Date	Mb/a	Gcf/a	Discovery	1958	1993	Onshore	Offshore
PAST	21	212	2000	333	7,680	Production	1973	2001	10,009	2.530
FUTURE	7.6	38	2005	334	7,734	Exploration	1980		Population	
Known	6.1	34	2010	323	6,695	Consumption	Mb/a	Gcf/a	1900	6
Yet-to-Find	1.1	3.8	2020	178	1,430	2010	806	3,759	2010	35
DISCOVERED	27	246	2030	98	377		b/a	kcf/a	Growth	5.8
TOTAL	28	250	Trade	−484	+2,936	Per capita	23	86	Density	3

Note: Excludes arctic oil and gas, bitumen and extra heavy oil

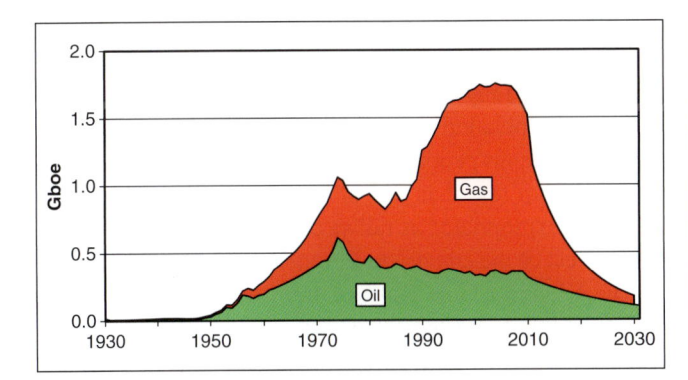

Fig. 72.1 Canada oil and gas production 1930 to 2030

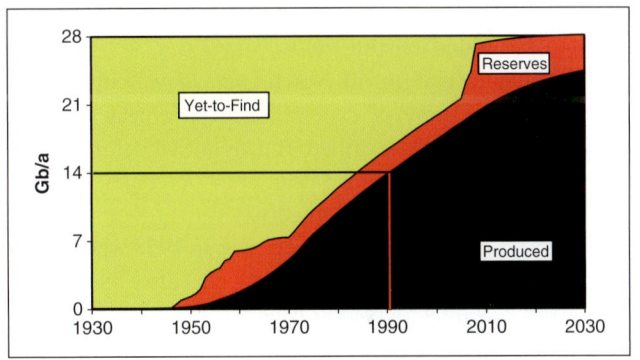

Fig. 72.2 Canada status of oil depletion

Essential Features

Canada is the second largest country in the World covering 10 million km², much of it being north of the Arctic Circle. The central plains and lakes, which overlie the Canadian Shield, are flanked respectively by the Rocky Mountains to the west, the older Appalachian chain to the east, and an archipelago of Arctic Islands to the north. The country has an extensive common border with the United States, much drawn along the somewhat artificial line of the 49° Latitude. It is flanked to the northeast by the autonomous Danish territory of Greenland and to the north by the Arctic Ocean. It is a very sparsely populated country supporting no more than about 35 million.

It is not at all easy to evaluate Canada's oil position in international terms because of its particular commercial environment and reporting practices. First, it applies an exceptionally high cut-off for *Heavy Oil* at 25° API because of the flow constraints imposed by the cold climate. Second, it has a highly fragmented industry, such that the term *wildcat* is generously applied, partly for tax reasons. It has huge deposits of bitumen and extra heavy oil from which synthetic oil is made, and lastly much of the country lies within the Arctic Circle, whose oil and gas are here excluded by definition from what is termed *Regular Conventional Oil*. Much more study is needed to unravel the statistics. This assessment is accordingly no more than an approximation.

Geology and Prime Petroleum Systems

Most of country overlies the non-prospective rocks of the Canadian Shield, but there are three petroleum systems. First is the Western Canadian Sedimentary Basin, lying mainly in Alberta, which relies largely on Palaeozoic source-rocks, but is now at a mature stage of depletion so far as *Regular Conventional Oil* is concerned. Second is the Atlantic Margin off Newfoundland, which has Upper Jurassic source-rocks, and has been developed over the past two decades delivering the giant Hibernian Field with about 700 Mb. It too is approaching maturity. A third province is the Mackenzie Delta and Arctic Islands which are predominantly gas prone. Hopes for new discovery off the Pacific Coast are likely to be dashed as this does not seem to be a very prospective geological setting.

Exploration and Discovery

Oil exploration commenced in the early years of the twentieth century without significant result until 1947 when the Leduc Field was found in a Devonian reef in Alberta, by which time some 250 boreholes had been drilled. Exploration was then stepped up to reach a peak in 1980 when as many as 750 so-called *wildcats* were drilled. In international terms, however, many such exploration boreholes would be treated as out-steps around existing fields.

The Atlantic margin off Newfoundland was opened in 1966. An early surge of exploration gave a peak in 1973 when 27 boreholes were drilled. It then lapsed before recovering in the 1980s to a second peak of 24 boreholes in 1985, since when it has declined markedly.

The overall peak of oil discovery was in 1958 when some 5 Gb of oil were found. Gas discovery rose after the Second World War reaching an overall peak in 1993 when 10.7 Tcf were found.

Canada is now a very mature province with little remaining potential for new discovery of *Regular Conventional Oil and Gas*. The Polar potential is considered in Chapter 12.

Production and Consumption

It is again difficult to draw the boundary, but, as assessed here, the production of *Regular Conventional Oil* commenced in 1868 and grew at a low rate over the succeeding years to pass 170 kb/d in 1951. It then increased rapidly to a peak in 1973 of 1.7 kb/d before declining to 0.9 Mb/d in 2010. It is now set to decline at an indicated Depletion Rate of 5.2% a year. Canada consumes some 806 Mb/a year of which it has to import 62%.

Gas production has been increasing progressively since the early years of the last century to reach a plateau at about 6 Tcf a year. The resource is now 80% depleted, meaning that production is likely to decline at the present Depletion Rate of around 15% a year.

The Oil Age in Perspective

The earliest settlers of North America crossed the Bering Strait from Siberia at the end of the Ice Age, more than 20,000 years ago. Then in the ninth century, came the Vikings, establishing a few small settlements, to be in turn followed in the fifteenth and sixteenth century by British and French fishermen attracted by the rich catches off Newfoundland.

In 1534, a French explorer, Jacques Cartier, made his way up the St. Lawrence River to reach the sites of what are now Quebec and Montreal. Settlers followed developing the fur trade, later supported by the French Government wishing to incorporate Canada into its Empire. The Jesuits also arrived with the intent of converting the indigenous population.

British commercial interests likewise appeared, partly expanding northwards from colonies in America, which led eventually to armed conflict, echoing contemporaneous wars in Europe. The American Revolution and secession brought further conflict and confusion, but British control was progressively established, leading in 1867 to the formal declaration of Canada's Dominion status within the British Commonwealth. French-speaking Quebec remains as a somewhat less than fully committed province of the country.

The extermination of the bison contributed to the decline of the indigenous people who depended on it for food. They now make up less than 2% of the population. Their lands were also expropriated as railways were constructed to open up the west of the country to European settlement.

The Klondike Gold Rush of 1896 drew attention to the mineral potential of the North, which was followed by the discovery of massive deposits of iron, copper, lead, zinc, and other minerals in the ancient rocks of the Canadian Shield. The prairies too were opened up as a rich source of wheat. The resulting economic prosperity encouraged further settlement and immigration such that the population increased sixfold from about five million in 1900 to its present level. Canada supported Britain in both World Wars, with its forces playing heroic roles in several battles.

Canada, perhaps unwisely, has signed up to the North American Free Trade Association, allowing its resources to be drained by its neighbour. Its inhabitants may accordingly soon have to face freezing winters as they keep the hairdryers of Houston going. It is reported that a new impetus to further reduce the barriers under NAFTA-Plus is under active consideration between the governments Canada, the USA, and Mexico, premised on the demands of *homeland security*. Under such pressure, it would be easy to forgive the citizens of Quebec if they were again to seek their independence.

Even so, with a population of only 34 million, Canada seems well placed to face the Second Half of the Age of Oil, provided it can protect its border.

72 Canada

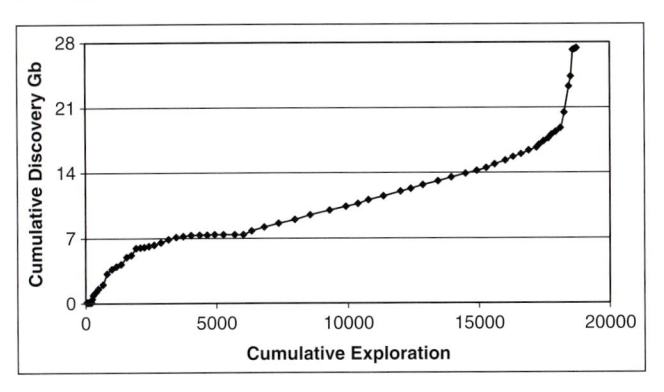

Fig. 72.3 Canada discovery trend

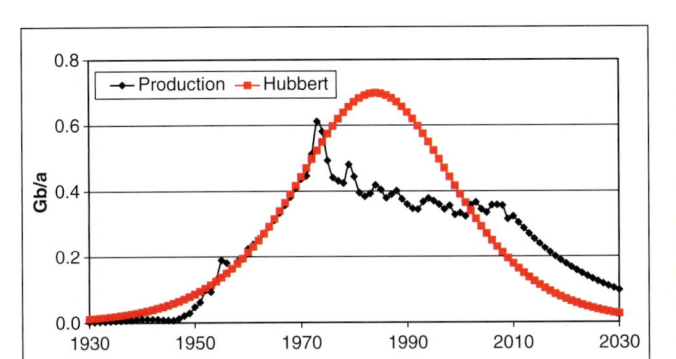

Fig. 72.4 Canada derivative logistic

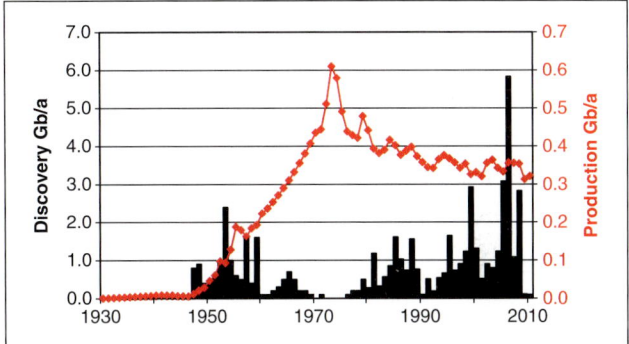

Fig. 72.5 Canada production: actual and theoretical

Fig. 72.6 Canada discovery and production

USA

Table 73.1 USA regional totals (data through 2010)

Production to 2100						Peak dates			Area	
Amount		Rate					Oil	Gas	'000 km²	
	Gb	Tcf	Date	Mb/a	Gcf/a	Discovery	1936	1969	Onshore	Offshore
PAST	179	1181	2000	1,538	23,715	Production	1970	2009	9,670	975
FUTURE	21	219	2005	1,218	22,970	Exploration	1981		Population	
Known	19	197	2010	1,225	23,166	Consumption	Mb/a	Gcf/a	1900	85
Yet-to-Find	2.1	22	2020	694	8,487	2010	7001	24,088	2010	312
DISCOVERED	198	1378	2030	394	3,109		b/a	kcf/a	Growth	3.7
TOTAL	200	1400	Trade	−5,776	−922	Per capita	22.5	77	Density	32

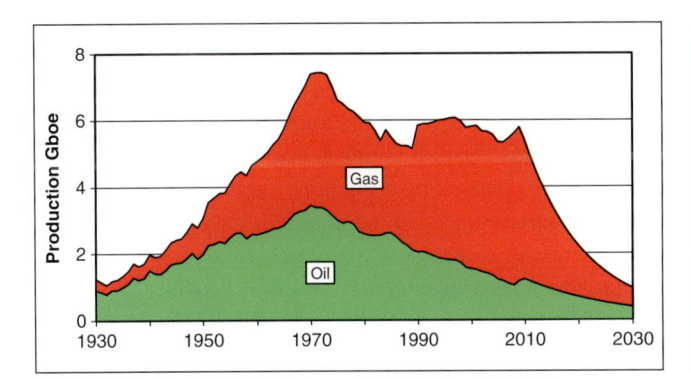

Notes: Excludes heavy oils, shale-oil and gas and deepwater

Fig. 73.1 USA oil and gas production 1930 to 2030

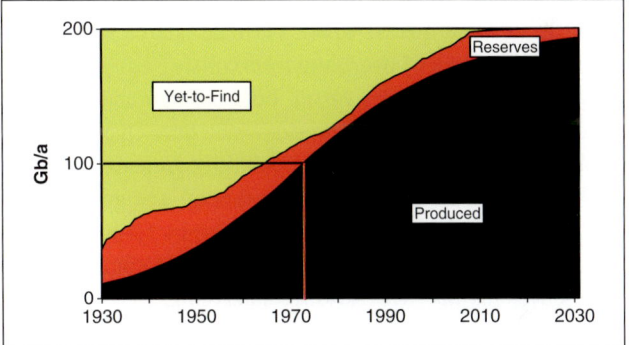

Fig. 73.2 USA status of oil depletion

Essential Features

The United States of America covers an area of 9.6 million km², comprising the southern half of the North American Continent, being bordered to the north by Canada and to the south by Mexico. It also owns Alaska and Hawaii, as well as a number of islands in the Pacific and Caribbean, which are not included in this assessment. The central plains are flanked to the west by the Rocky Mountains, rising to 4,300 m, and to the east by the older Appalachian Range, rising to 2,000 m. The Mississippi River flows southwards to the Gulf of Mexico at New Orleans, and a number of large lakes on the Canadian border drain into the Atlantic through the Hudson River.

The country comprises 50 States that started to come together as a union in the eighteenth century, and currently support a population of 312 million. Most arrived as immigrants over the past century, coming first mainly from Europe and later from Latin America. An estimated 12 million are illegal immigrants. The indigenous population was virtually exterminated, now amounting to less than 1% of the total population, while some 12% of the population are descended from slaves brought from Africa.

Geology and Prime Petroleum Systems

The key elements of this large diverse region are summarised here, excluding the deepwater and Polar regions of Alaska, which do not qualify as *Regular Conventional Oil and Gas* by definition.

Prior to the opening of the Atlantic, some 200 million years ago, North America formed the western part of the Continent of Laurasia. The ancient Appalachian chain and much of the interior is built of Palaeozoic rocks, laid down in

C.J. Campbell, *Campbell's Atlas of Oil and Gas Depletion*,
DOI 10.1007/978-1-4614-3576-1_73, © Colin J. Campbell and Alexander Wöstmann 2013

geosynclinal troughs, being locally intruded by volcanic rocks, and deformed in mountain building movements in late Silurian and Permian times.

The Mesozoic Period opened with the development of continental deserts and massive volcanic activity, responsible, for example, for the 100 m thick Palisades volcanic sill which borders the Hudson River. It was followed by gradual subsidence during the Jurassic and Cretaceous and the deposition of mainly shelf deposits. The western margin was however deformed and intruded in both the late Jurassic Navadan Orogeny and the late Cretaceous Laramide Orogeny, the latter seeing also the emplacement of granite massifs.

Most of the interior was emergent during the succeeding Tertiary period, which saw much volcanic activity which has left thick lava flows and ash beds over vast areas. Marine deposition was confined to the Gulf Coast, bordering the Caribbean, and the Atlantic and Pacific margins.

In terms of Petroleum Systems, the following areas stand out:

The *Rocky Mountain Foredeep* contains several productive basins, of which the most prominent is the Permian Basin of West Texas, where Permian source-rocks have charged a range of overlying reservoirs.

The *Gulf of Mexico Basin*, both onshore and offshore, relies on Mesozoic source-rocks which have charged reservoirs in the overlying Tertiary deltaic sequences. The geothermal gradient is low, meaning that oil was generated at great depth. Slump-faults and salt diapirs form the main trapping mechanisms, in an environment of complex pressure systems that gave rise to secondary migration. Extensive Jurassic salt deposits play an important role in the structural development, and interest now turns to sub-salt prospects, primarily for gas. Mesozoic carbonates form an additional reservoir along the eastern margin of the basin.

The *Mid-Continent* province, running northwards from Texas through Oklahoma, relies on Palaeozoic source-rocks and reservoirs, which have yielded a large number of generally small fields.

The *Pacific Margin of California* has provided a number of fields relying on Tertiary source-rocks and reservoirs in generally complex structures, some of which are related to the major San Andreas transcurrent fault.

Exploration and Discovery

The United States is almost unique in that its mineral rights generally belong to the landowner, based on legal principles inherited from the Spanish Empire of Latin America. As a result, oil operations were highly diversified with individual fields being subdivided amongst many owners. Reserve reporting was subject to strict Stock Exchange rules, based primarily on what individual wells were expected to yield. Tax considerations also influenced what was deemed to be an exploration borehole. For all these reasons, it is unique environment which is difficult to analyse in world terms.

The first discovery was in 1859 in Pennsylvania: an event that is widely regarded as the birth of the oil industry. The other provinces were later opened up, but the overall peak of discovery is attributed to the 1930s, when the East Texas province was brought in. The country has been exhaustively explored and developed.

It is reported that over 400,000 exploration boreholes have been drilled, but in international terms most would probably be considered out-steps to existing fields. The overall peak was in 1981 when over 9,000 were drilled, and the number has now fallen to about 2,000, as the list of remaining viable prospects falls, even though the economic threshold is extremely low.

The country also has several categories of *Non-Conventional* oil but they are not readily identifiable in the database. They can continue to support low levels of production for a long time and have become viable with high oil prices. They include *Heavy Oils*, such as supply the Midway Sunset Field in California; extensive *Oil Shale* deposits, none of which has yet proved to be commercial but may become so in the future despite giving a negative net energy yield; and *Shale-Oil* derived by artificially fracturing inferior reservoirs in source-rock sequences, which have attracted much new interest in recent years, becoming viable with high oil prices. The *Non-conventional deepwater* province in Gulf of Mexico has also delivered some major finds in recent years, being considered in Chap. 12.

The country also had a substantial endowment of natural gas. It was widely flared in earlier years before a market was developed, but is now treated as a prime fuel, especially for electricity generation. If the numbers are to be believed, discovery peaked around 1996, giving a corresponding peak in production in 2009, when 84% had been depleted, which is an unusual pattern.

Production and Consumption

The United States dominated world oil production in earlier years, being also the home to several of the world's major international oil companies. In 1930, it supplied about 65% of the World's production, but its share has since slipped to 21% in 1970 to less than 1% to-day. With its burgeoning domestic demand for oil, the country had become a net importer by 1950. Imports began to rise rapidly after peak production in 1970, such that they have now passed 70%. The irreversible decline of its production means that even if demand were to be held static, the country would be importing 90% of its needs by 2020. It explains why access to foreign oil has long been officially declared a vital national interest, prompting military intervention.

Different databases give different values, but based on that provided by the Energy Information Agency it is assessed here that production of *Regular Conventional* oil commenced in 1859, passing 100 kb/d by 1890. It then rose to an overall peak of 9.4 Mb/d in 1970. It has since declined to 4.2 Mb/d and is set to continue to decline at about 5% a year.

Gas production commenced in the 1930s as a market developed for it. It depletes differently from oil, its production being generally capped below capacity by the pipeline infrastructure. The resulting plateau of production is now coming to an end, giving rise to high prices, prompting a new drilling boom. But the new wells have had to be produced at maximum rate and, as a result, are depleted within a matter of months. Some extra late-stage gas is being obtained by the tapping of the gas caps of oilfields during their dying days.

Gas production increased from 21 Tcf in 1990 to almost 23 Tcf in 2010, some 84% of the total endowment has now been produced, such that the production is likely to fall steeply at about 10% a year. The production of natural gas liquids, now running at about 1.9 Mb/d, will fall in parallel with the gas. There are, in addition, large amounts of non-conventional gas in the form of coal-bed methane, and in the so-called tight reservoirs, contributing about 10% of total supply. Electricity demand is growing, with many gas-fired generators under construction. As a result, the United States will have an increasingly desperate need to tap Arctic gas, possibly draining Canada in the process.

The Oil Age in Perspective

The New World started drifting away from the Old some 200 million years ago. Early Man was able to reach it by crossing what is now the Bering Strait some 20,000 years ago, when the sea-level was lower during the Ice Age. He found a new continent with a very different animal fauna that had evolved in isolation. Little is known about the early inhabitants who are thought to have numbered some ten million when European occupation began in the fifteenth century. The Spaniards established a settlement in Florida in 1565, to be followed by various British settlements along the eastern seaboard. France too took a serious interest, founding Quebec in Canada in 1608, and controlling much of the Mississippi valley.

Many of the colonists went to the New World to escape famine and poverty at home, with in some cases religious persecution being an additional motivation. European wars in the eighteenth century also had their consequences in the New World, with Britain emerging as the dominant power in 1763, when France surrendered its North American territories. The settlers, however, soon moved towards independence, not being enthusiastic for various forms of British taxation, and declared full independence in 1776 after a series of conflicts. A centralised system of government did not come easily as the various settlements, which had evolved into independent States, were reluctant to surrender their autonomy. Constraints on the power of the federal government were established under the Bill of Rights, but have been progressively eroded. The conflicts culminated in a Civil War from 1861 to 1865 between the agrarian South and the industrial North, with slavery being one of the issues. Like most civil wars, it was a vicious affair, costing over 600,000 lives.

A great westward migration of people occurred during the nineteenth century, leading to the virtual extermination of the indigenous tribes. New waves of immigrants flooded in from over-populated Europe, including particularly Scandinavians, Italians, Jews seeking to escape anti-Semitism, and Irish following a devastating famine in 1845–1850.

Texas had been a lightly populated province of Mexico until 1836, when new settlers from the north revolted, declaring it a republic. This prompted a successful war with Mexico, by which the United States acquired Texas, New Mexico, Arizona, California, Nevada, Utah and much of Colorado. Another successful war with Spain followed in 1898, when the United States supported Cuban independence, partly for commercial motives. As a result, it acquired the Philippines, Guam and Puerto Rico, becoming a world power with the imperial aspirations of the day. The sinking of the US warship, *Maine*, in Cuba, which was the pretext for the war, was later found to have been due to an explosion on the ship itself, which some think was orchestrated. The United States later engineered the secession of Panama from Colombia in order to build the Panama Canal to facilitate trade between the east and west coasts.

The territorial limits of the country eventually stabilised into 48 contiguous States, bordered by the residual territories of Mexico to the south and Canada to the north. Two additional territories were added in 1959; Alaska, which had been purchased from the Russians in 1867, became the 49th State; and Hawaii, which had been seized in 1893 over a sugar dispute, became the 50th.

The Industrial Revolution of Europe spread to the United States during the nineteenth century, as its huge natural resources of iron, coal and, later, oil gave it the essential resources for manufacturing. Floods of new immigrants provided cheap labour. Capitalism took off with a vengeance, throwing up dynasties with extreme wealth, including the houses of Astor, Carnegie, Rockefeller, Morgan and Dupont, to name a few. Wall Street emerged as a premier world financial centre, and the dollar began on its path to world financial domination. These excesses were to some extent countered during the early years of the twentieth century when Theodore Roosevelt brought in the so-called Square Deal with various conservation and regulatory measures, breaking up some of the industrial and financial empires.

Banks from the City of London, which had a dominant position in world trade, thanks to the British Empire, succeeded in persuading the US Government to allow them to establish the Federal Reserve Bank in 1913. It discharged the role of a national bank but was privately owned and unaccountable.

The United States entered the First World War in 1917 on the side of Britain, its former colonial master, despite having a substantial number of German immigrants. It may be no coincidence that its entry, which proved decisive, coincided both with the publication of the Balfour Declaration for the Jewish homeland in Israel and the issue of dollar loans to Britain and France.

Having been spared the ravages of war, it emerged as the dominant economic power in the world, and the post-war years saw an industrial boom, stimulating a speculative bubble on the Wall Street. It burst in 1929 bringing on the Great Depression that in large measure lasted until the Second World War gave rise to another boom. The Depression caused great suffering that has left a searing memory, deep in the national psyche

The United States entered the Second World War in 1942 following a Japanese attack upon its naval base at Pearl Harbour in Hawaii, seen by some as a pretext, as the Government did not react to fore-knowledge from radio intercepts. After successful campaigns from North Africa to mainland Europe and the Pacific, it brought the war to an end by devastating two Japanese cities with atomic bombs. The British and French empires were extinguished, leaving the United States to be countered only by the Soviet Union. These two super-powers then glowered at each other for the ensuing 45 years in the Cold War, with active military conflicts being confined to Korea and Vietnam. The country moved to the conquest of outer space, largely for military reasons.

The collapse of the Soviets in 1991 left the United States as a solitary super-power, primed for world economic and financial hegemony. Its vast industrial-military complex faced declining sales unless new wars, or the threat of them, should stimulate demand for military products. The country's financial dominance was a mixed blessing, attracting flows of foreign capital that gave rise to possibly unsustainable levels of foreign debt. A critical event was the abandonment of the Gold Standard in 1975, which removed solid foundations for the currency. A post-Cold War speculative bubble burst in 2000 in a situation reminiscent of the crash of 1929.

On 11th September 2001, two buildings in the World Trade Center in New York were struck by airliners, and the Pentagon in Washington was also hit. The incidents were attributed to Muslim activists. The government thereupon declared a worldwide *War on Terror* toppling the government of Afghanistan after a short bombing campaign, before turning on Iraq. Afghanistan lies on a proposed route for the export of oil from the Caspian region, while Iraq is at the heart of the Middle East, which holds almost half the world's remaining supplies of *Regular Conventional Oil*.

The United States, like Britain, operates a form of democracy dominated by two political parties, which select the candidates for election. There is widespread political patronage by vested interests. The Presidency has relatively strong powers and security of tenure. A bill passed by Congress is sent to the President who may approve or disapprove at his discretion, although the veto can be over-ridden by a two-thirds majority. The election itself is governed by both direct vote and an Electoral College, furnished by the individual States, under complex rules. The Congress consists of a Senate, which represents the individual states, and a House of Representatives, both of which are directly elected.

It is difficult to avoid the conclusion that the United States faces a dire energy crisis that will radically affect its entire way of life, as indeed the Energy Secretary confirmed before attention was diverted by the events of September 11th. This realisation would at least offer a logical explanation for its foreign policy, whatever other factors may also be at work. It furthermore adds weight to the expectation of deepening economic recession. There is a certain logic in expecting the United States, which led the world into the oil age, to also be the first to experience its decline. The dollar indeed lost its AAA ranking in 2011. Serious budgetary constraints may limit foreign military engagements. Most troops have already left Iraq, and are set to withdraw from Afghanistan.

The transition to the Second Half of the Age of Oil threatens to be a time of great social and political tension, especially as the urban and suburban societies, many with large recent immigrant communities, find their survival threatened.

But at the end of the day, the United States is a large and well-endowed country, whose people are rightly known for their courage, pragmatism and initiative, so there are good hopes that they will eventually find a sustainable future that relies increasingly on local rather than national or global solutions.

73 USA

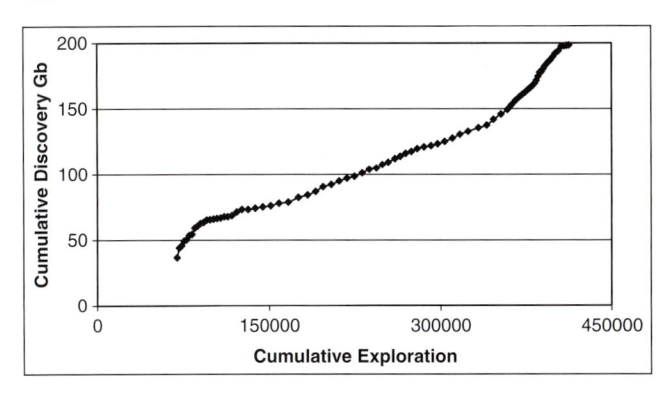

Fig. 73.3 USA discovery trend

Fig. 73.4 USA derivative logistic

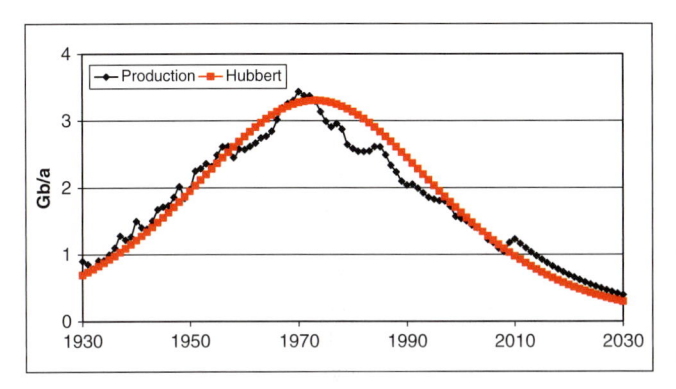

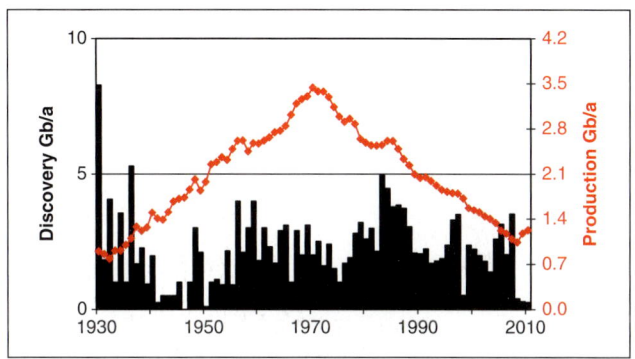

Fig. 73.5 USA production: actual and theoretical

Fig. 73.6 USA discovery and production

North America Region

Table 74.1 North America regional totals (data through 2010)

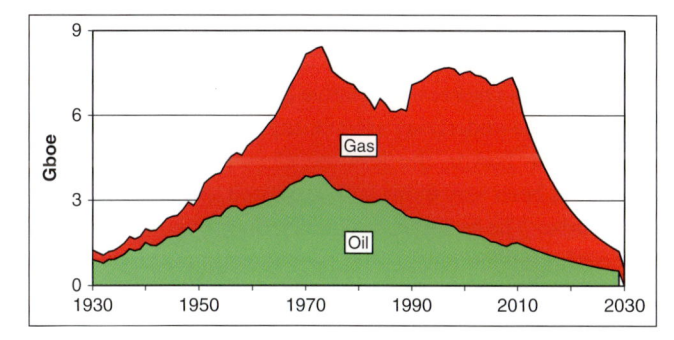

	Production to 2100					Peak Dates			Area	
	Amount		Rate				Oil	Gas	'000 km²	
	Gb	Tcf	Date	Mb/a	Gcf/a	Discovery	1930	1930	Onshore	Offshore
PAST	200	1393	2000	1871	31395	Production	1973	2008	19675	3505
FUTURE	28	257	2005	1553	30704	Exploration	1981		Population	
Known	25	232	2010	1547	29861	Consumption	Mb/a	Gcf/a	1900	91
Yet-to-Find	3.2	26	2020	872	9917	2010	7807	27024	2010	346
DISCOVERED	225	1624	2030	491	3486		b/a	kcf/a	Growth	3.8
TOTAL	228	1650	Trade	−6260	+2837	Per capita	23	78	Density	18

Fig. 74.1 North America oil and gas production 1930 to 2030

Fig. 74.2 North America status of oil depletion

North America is defined for this purpose as the United States and Canada, with Mexico being treated as part of Latin America. There is not much to add to the descriptions of the two countries comprising the Region.

The Region is widely regarded as the birthplace of the oil industry, as described in connection with the section on the United States, although oil wells had in fact been drilled earlier in Romania and on the shores of the Caspian. The Region is dominated by the United States which is home to about 90% of the population, but new importance attaches to Canada both for its large resources of *Non-Conventional Oil* in the tar-sands of Alberta and for its Arctic gas.

Surprisingly, it is perhaps the most difficult region to analyse because of its unique oil environment, with the mineral rights belonging to the landowner which led to an unusual pattern of development. Ownership of the fields, and even the reservoirs within them, was fragmented, giving rise to exceptional development and reporting practises. Oil reserves were treated as financial assets and rightly became subject to strict Stock Exchange rules which effectively recognised the expected future production of current wells (*Proved Producing Reserves*) and the anticipated production from infill wells before they had been drilled (*Proved Undeveloped Reserves*). These factors in turn affected the definition of what precisely constituted the boundaries of an oilfield, and hence what was reported as a discovery. Small extensions or secondary reservoirs tended to be treated as independent fields. Tax treatment also affected the definition. There is accordingly no comprehensive public data base giving discovery dates and exploration drilling, comparable with those available in other regions. For example, it is reported that as many as 430,000 exploration boreholes have been drilled, but in reality most simply tested extensions or secondary pools within what would be considered as single fields in other regions.

The total endowment as here assessed is 228 Gb of producible *Regular Conventional* oil, amounting to 11% of the World's total. The industry grew rapidly and stimulated economic growth, such that it is the most depleted region of the world, having no more than 11% of its original endowment of *Regular Conventional Oil* left. Oil production peaked in 1973 at 10.6 Mb/d, since when it has fallen to 4.2 Mb/d, being set to continue to decline at about 5% a year. It is also the highest consumer with a per capita consumption of almost 23 barrels a year, meaning that it is a substantial importer of 6,260 Mb/a on a rising trend, which carries severe geopolitical implications.

It also has an endowment of about 1,650 Tcf of gas, of which 1,393 Tcf are thought to have been produced.

It has additional substantial resources of *Non-Conventional* oil and gas, which are considered in Chapter 12.

Other Countries

The Region is dominated by continental North America, but there a number of other islands such as Hawaii and Puerto Rico that fall within the US dominion, which can be ignored for this purpose.

The Oil Age in Perspective

It is probably true to say that North America has been more affected by the Oil Age than any other region. It fuelled an expanding industry which in turn led to a financial hegemony as the dollar replaced the pound sterling after the effective collapse of the British Empire following the Second World War.

The United States in particular just about runs on oil, meaning that the world peak of production will be felt more severely here than anywhere else. Indeed, the last decade has seen how its economic and foreign policies, including the use of military force, have been designed to secure access to world oil supply under the principles of globalism, whereby the resources of any country are deemed to belong to the highest bidder.

Its financial hegemony has been seriously affected by the financial collapse of recent years, such that the dollar has lost its ranking as a Triple A rated currency.

It is very difficult to imagine how this oil-based empire will collapse, but as Nature reveals her hand we may anticipate an eventual positive reaction, as people again come to accept the new circumstances. One obvious step would be a new regionalism, as already manifesting itself in other regions, whereby the individual states and cities gain more authority at the expense of a failing central government. Urban life will likely become increasingly difficult, and tensions with the large numbers of recent immigrants, lacking roots, will likely rise.

But eventually by the close of the century a new benign age may have dawned for the survivors, based on the well-known national attributes of enthusiasm, enterprise and realism.

Confidence and Reliability Ranking

The table lists the standard deviation of the range of published data on reserves. A relative confidence ranking in the validity of the underlying assessment is also given: the lower the Surprise Factor, the less the chance for revision. The region has been very thoroughly explored leaving little scope for surprise. The high standard deviation, as regards Canada's reported reserves, reflects the anomalous reporting on non-conventional oil in the tar-sands by the *Oil and Gas Journal*.

Table 74.2

| | Standard deviation Public reserve data || Surprise Factor ||
	Oil	Gas	Oil	Gas
Canada	85.44	15.28	1	2
USA	5.04	24.19	1	2
Range in published data			Scale 1–10	

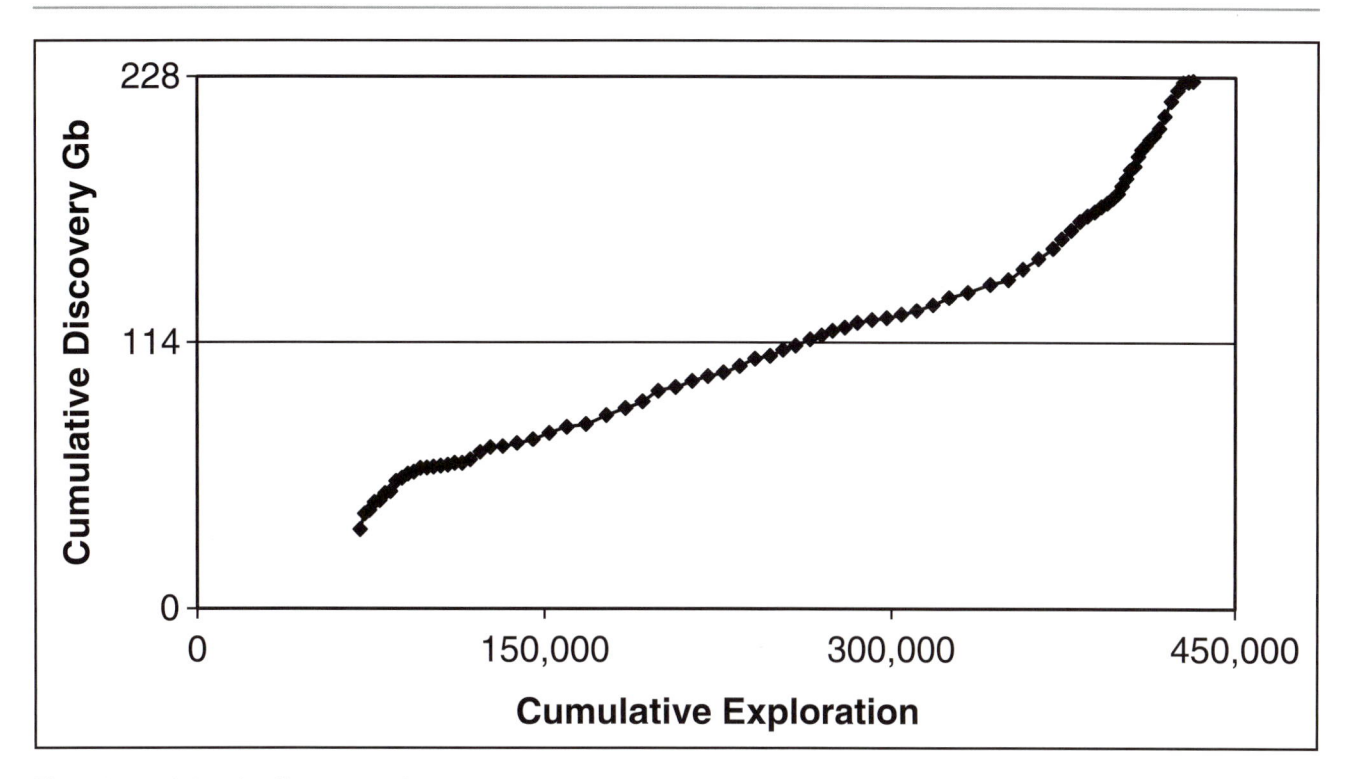

Fig. 74.3 North America discovery trend

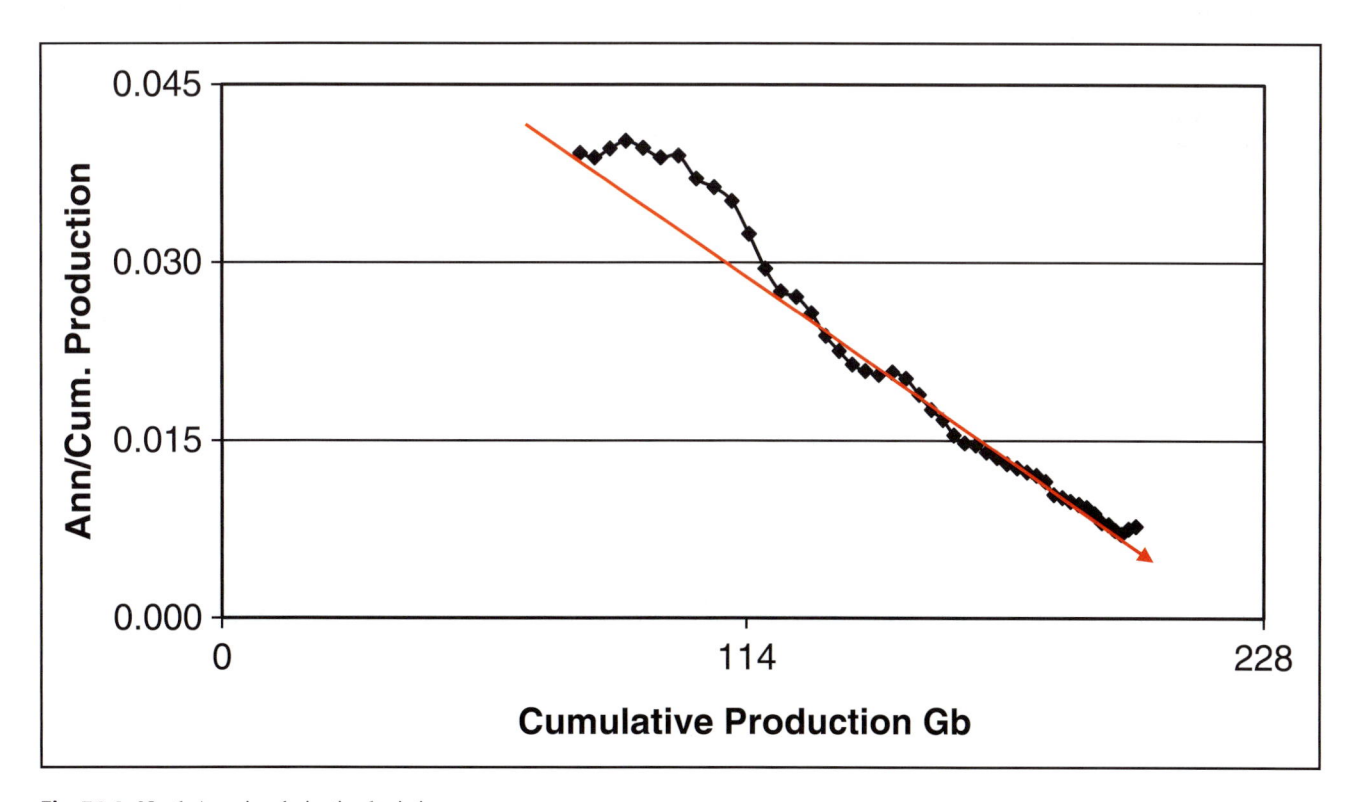

Fig. 74.4 North America derivative logistic

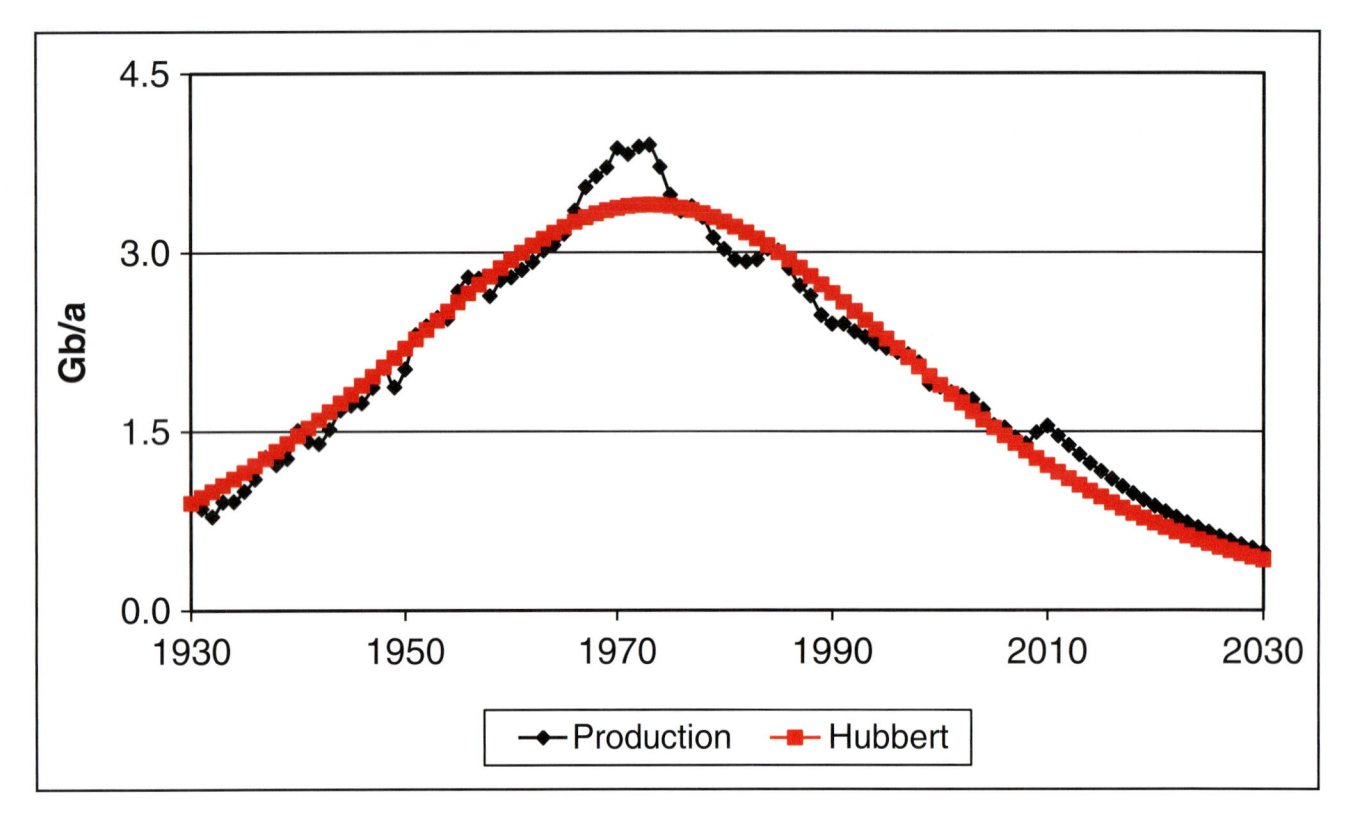

Fig. 74.5 North America production: actual and theoretical

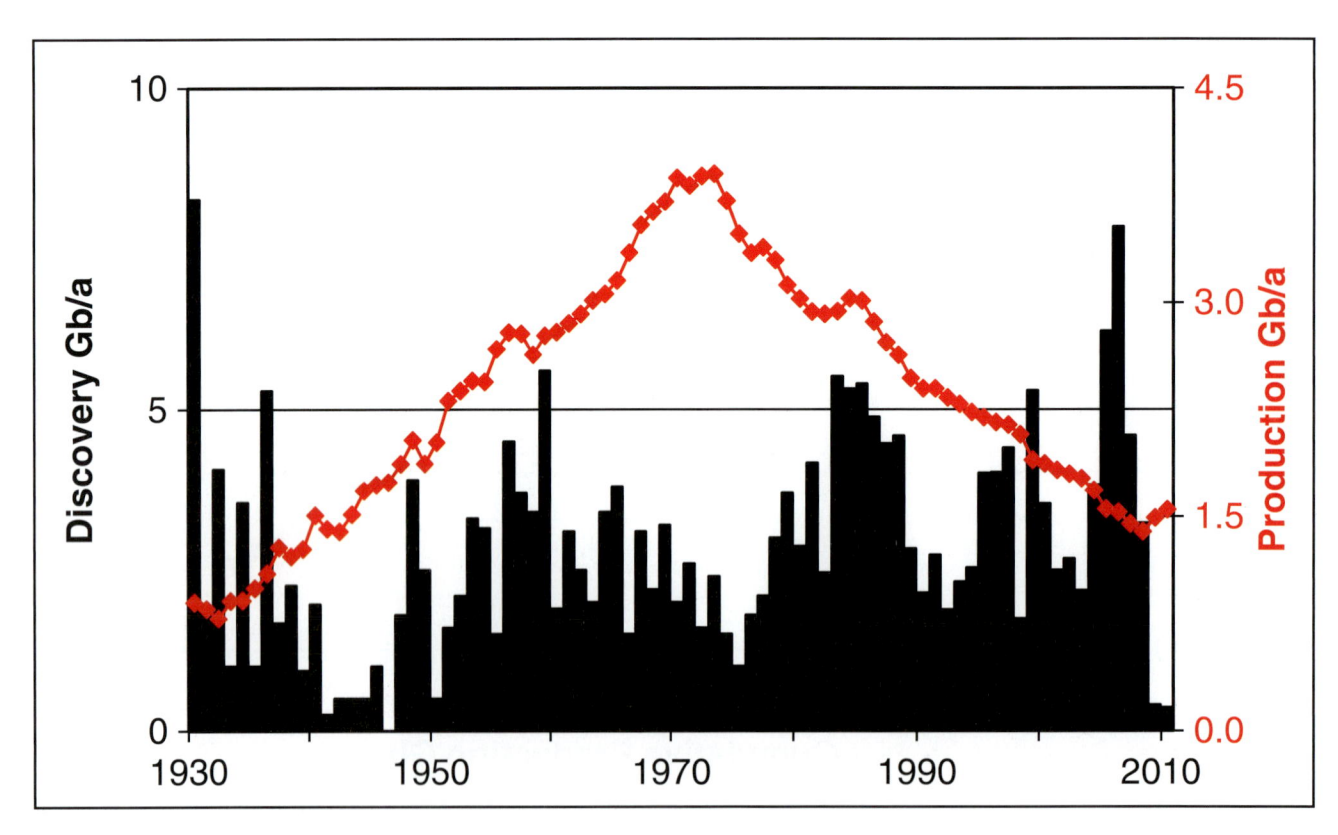

Fig. 74.6 North America discovery and production

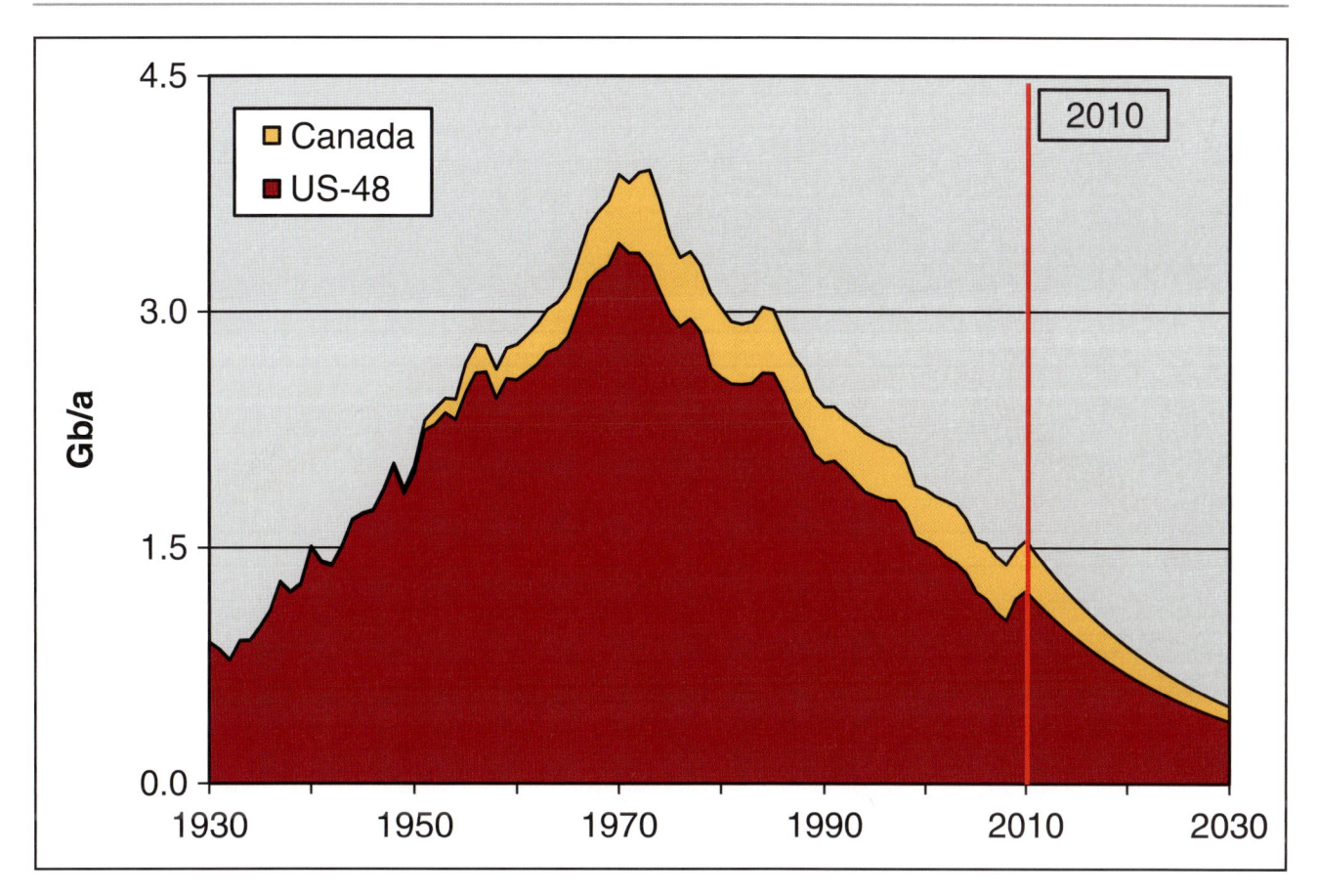

Fig. 74.7 North America oil production

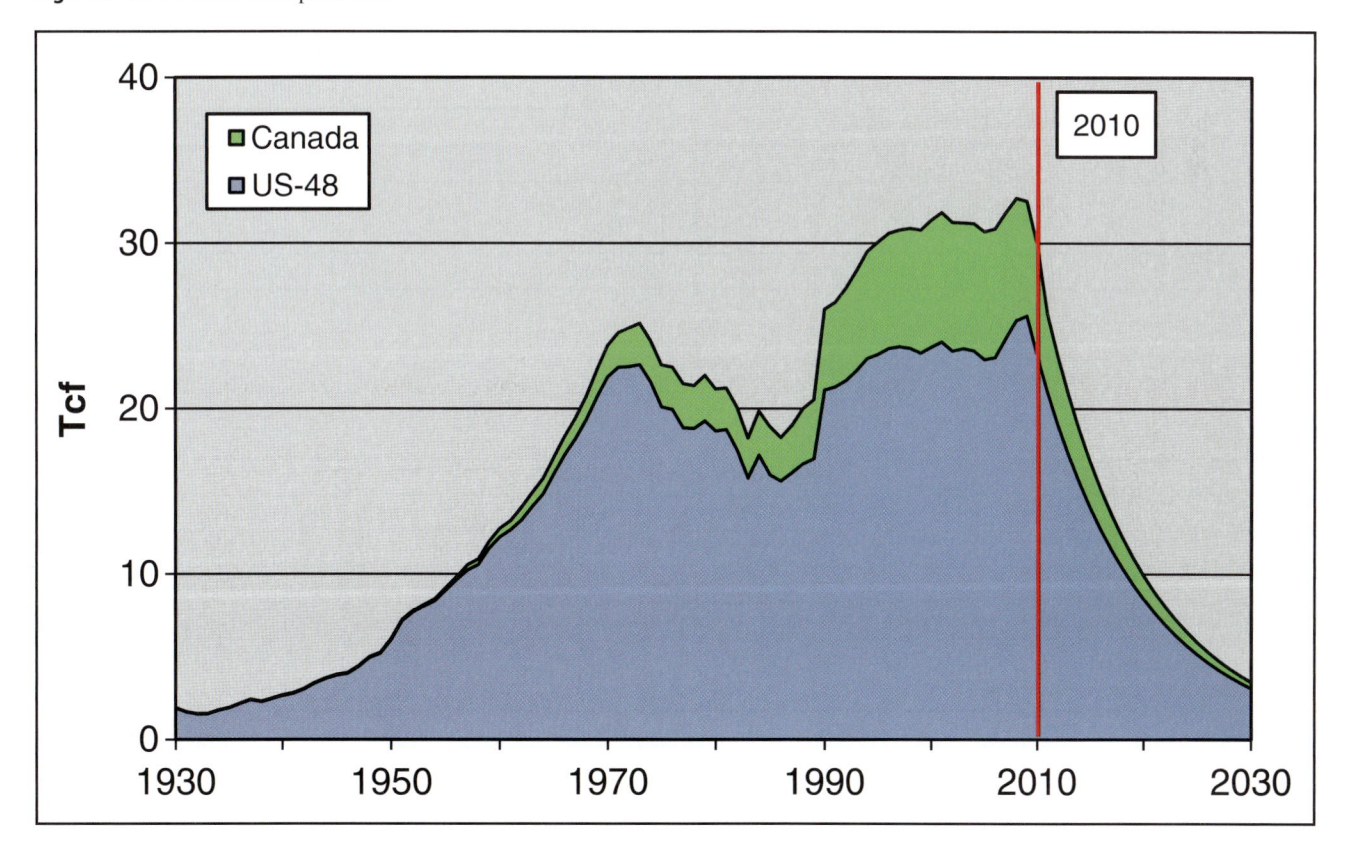

Fig. 74.8 North America gas production

Table 74.3

Country	Region	PRODUCTION TO 2100 — Regular Conventional Oil — KNOWN FIELDS										Revised: 15/10/2011						2010
		Present		Past		Discovery		Reserves		FUTURE KNOWN	Total Found	NEW FIELDS	ALL FUTURE	TOTAL	DEPLETION		PEAK DATES	
		Kb/d 2010	Gb/a 2010	Gb	5yr Trend	Disc 2010	% Disc.	Reported Average	Deduct Non-Con	Gb	Gb	Gb	Gb	Prod	Current Rate	Mid -Point	Disc	Prod
US-48	D	3355	1.22	179.03	1%	0.26	99%	24.07	−11.43	18.87	197.90	2.10	20.97	200.0	5.5%	1972	1936	1970
Canada	D	884	0.32	20.75	−2%	0.05	97%	83.75	−174	6.16	26.91	1.09	7.25	28.0	5.2%	1987	1959	1973
N.AMERCIA	D	4240	1.55	199.78	0%	0.36	99%	107.82	−185	25.03	224.82	3.18	28.22	228.0	5.2%	1973	1930	1973

Table 74.4

Country	Region	PRODUCTION TO 2100 — Regular Conventional Gas — KNOWN FIELDS											Revised: 15/10/2011							2010
		Present			Past		Discovery		Reserves		FUTURE KNOWN	TOTAL FOUND	FUTURE FOUNDS	ALL FUTURE	TOTAL	DEPLETION		PEAKS		
		Tcf/a	FIP	Gb/a 2010	Tcf 5yr	Trend	Tcf 2010	% Disc.	Reported Average	Deduct Non-Con	Tcf	Tcf	Tcf	Tcf	Tcf	Current Rate	% Mid-Point	Expl	Disc	Prod
Canada	D	6.70	0.70	1.21	212	−3%	0.46	98%	54.09	0.00	34.24	246.20	3.80	38.04	250	14.97%	85% 1999	1980	1993	2001
US-48	D	23.17	2.17	4.17	1181	0%	2.74	98%	284.54	0.00	197.37	1378.07	21.93	219.30	1400	9.55%	84% 1987	1956	1996	1979
N.AMERCIA	D	29.86	2.86	5.38	1393	−1%	3.21	98%	338.62	0.00	231.61	1624.27	25.73	257.35	1650	10.18%	84% 1986	1981	1930	2008

Table 74.5

Regular Conventional Oil Production					
Mb/d	2000	2005	2010	2020	2030
US-48	4.21	3.34	3.36	1.90	1.08
Canada	0.91	0.92	0.88	0.49	0.27
N.AMERICA	5.13	4.25	4.24	2.39	1.35

Table 74.6

Gas Production					
Tcf/a	2000	2005	2010	2020	2030
Canada	7.68	7.73	6.70	1.43	0.38
US-48	23.72	22.97	23.17	8.49	3.11
N.AMERICA	31	31	30	10	3

Table 74.7

Country	PRODUCTION AND CONSUMPTION													
	Oil				Gas				POPULATION AND AREA				
	Production	Consumption		Trade	Production	Consumption		Trade	Population	Growth	Area			Density	
			p/capita	(+)			p/capita	(+)		Factor	Mkm2			Population	Oil
	Mb/a	Mba	b/a	Mb/a	Gcf/a		kcf/a	Gcf/a	M		Onshore	Off	Total	per km2	kb/km2
US-48	1225	7001	22.5	−5776	23166	24088	77	−922	311.7	3.7	9.666	0.975	10.641	32	18.8
Canada	323	806	23.0	−484	6695	2936	84	3759	35	5.8	10.009	2.530	12.539	3	2.2
N.AMERICA	1547	7807	22.6	−6260	29861	27024	78	2837	346	3.8	19.675	3.505	23.180	18	9.8

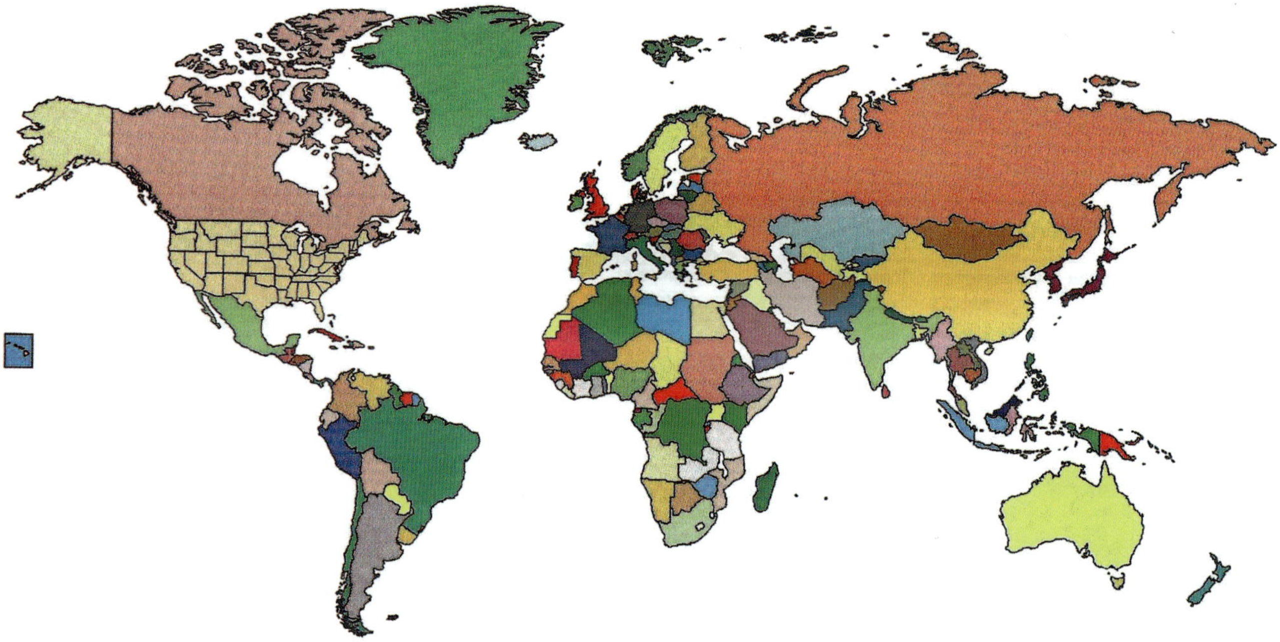

Part IX

Global Analysis and Perspective

The World

Table 75.1 The World regional totals (data through 2010)

Production					Peak Dates			Area	
Amount		Rate				Oil	Gas	'000 km²	
	Gb	Tcf	Date	Mb/a	Tcf/a	Discovery	1964	1971	Onshore	Offshore
PAST	1093	3682	2000	23416	98	Production	2004	2015	105800	-
FUTURE	907	6317	2005	24349	113	Exploration	1981		*Population*	
Known	795	5225	2010	23047	128	*Consumption*	Mb/a	Tcf/a	1900	1500
Yet-to-Find	113	1093	2020	18590	126	2010	31981	99	2010	6903
DISCOVERED	1887	8907	2030	13972	103	Per capita	b/a	kcf/a	Growth	4.6
TOTAL	2000	10000	*Trade*	−8934	+28972		4.6	14	Density	65
Note: All numbers to be generously rounded									

Introduction

The above table sums the regional assessments in the preceding chapters, adding the combined amounts from the minor producing countries. It is stressed that the databases are weak, inconsistent and unreliable, so the estimates should be generously rounded as indicated in the following table.

Table 75.2

Production	Oil (Gb)	Gas (Tcf)
PAST	1090	3600
FUTURE	900	6300
Known	800	5200
Yet-to-Find	110	1100
DISCOVERED	1900	9000
TOTAL	2000	10000

Exploration and Discovery

Oil and gas had been known since antiquity, but deliberate exploration using drilling equipment began in the 1850s, mainly in Pennsylvania, in Romania and on the shores of the Caspian. The next major step came in 1908 when a well in the Zagros foothills of Iran came in; opening what was to prove to be the world's most prolific province around the Persian Gulf. Exploration was then stepped up throughout the world. Great advances in petroleum geology and production technology were made, so that virtually all the world's major onshore petroleum provinces, outside the Arctic, had been identified by the middle of the last century. Attention then turned to the offshore as the advanced technology for such operation was developed, bringing in another domain, which has now also been thoroughly explored. As a result the peak of all discoveries was passed in the mid 1960s.

Now attention turns to the remaining *Non-Conventional* oil resources, comprising the heavy oils and tar deposits, oil in poor reservoirs, now captured by artificial fracturing (*Shale Oil*) and those in deep water and Polar Regions. They are generally more difficult, costly and, above all, slower to produce, being considered in Chapter 12.

Gas was found in parallel with oil. Most oilfields have a gas cap, which tends to be produced after most of the oil has been taken, but there are also fields charged exclusively with gas from gas-bearing source-rocks.

A great deal is now known about the occurrence of oil and gas in Nature, and there have been great advances in the technology of extraction.

Production and Consumption

Oil production commenced in the middle of the nineteenth century and has grown progressively ever since, save for brief downturns following the oil shocks of 1973 and 1979 which were politically driven. Consumption grew in parallel. Although oil data are too unreliable to be sure, it is estimated that the peak of the production of *Conventional Oil* was passed in 2004 or 2005 and that the peak of all liquids followed around 2008. The subsequent decline will initially be at less than 2% a year, growing to about 2.5% by 2030.

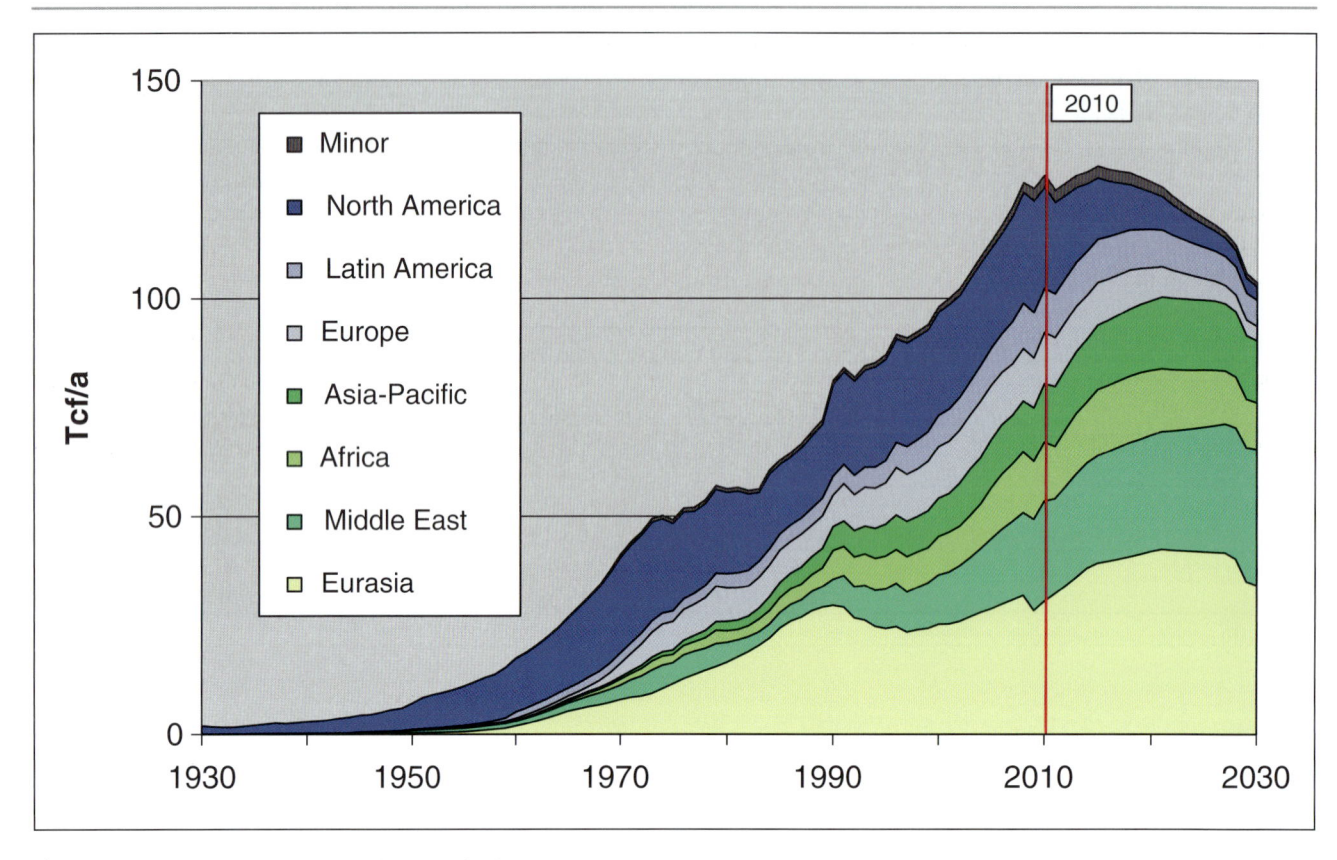

Fig. 75.1 World Regular Conventional Gas Production

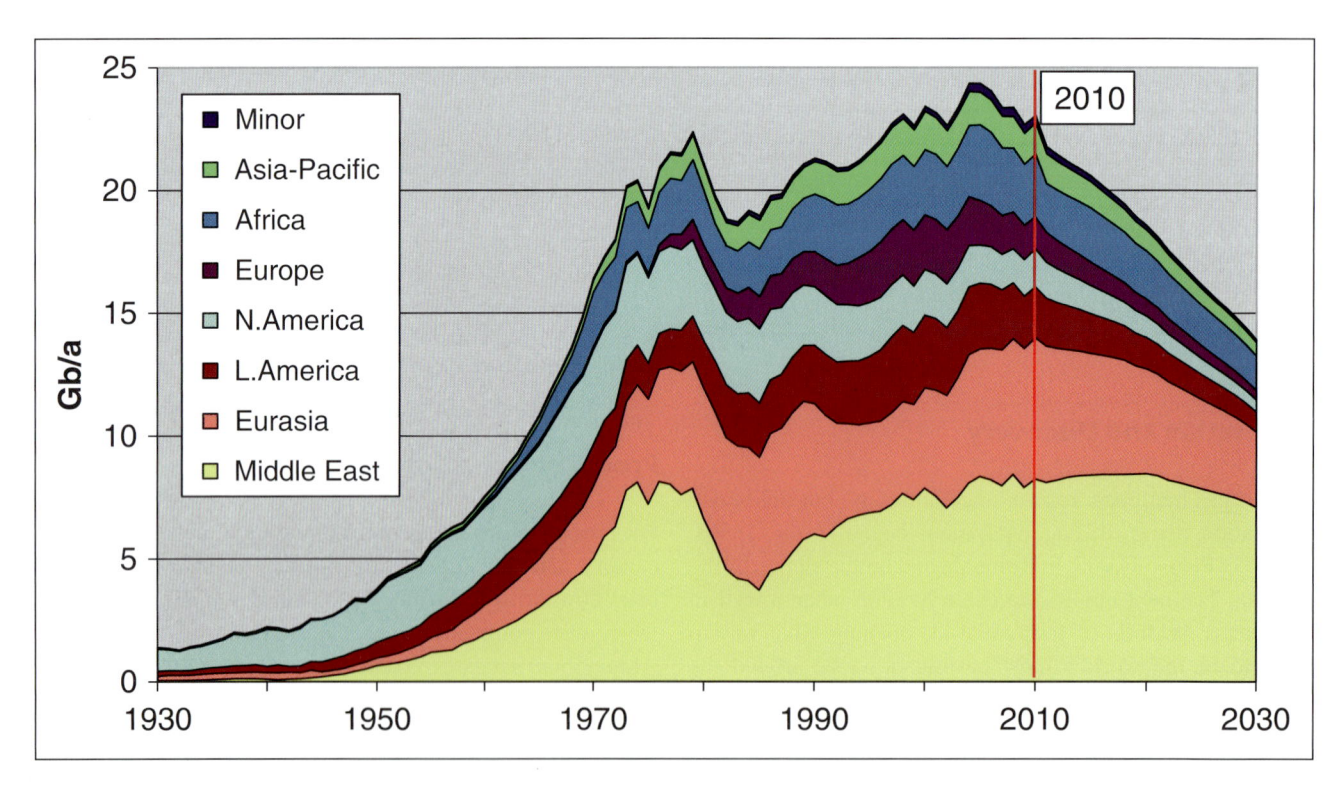

Fig. 75.2 World Regular Conventional Oil Production

The decline is imposed ultimately by the immutable physics of the reservoirs, but there will be many irregularities imposed by above-ground factors.

Gas, being a gas not a liquid, depletes differently from oil. In earlier years, production was restricted in the absence of a market, being in many cases flared or re-injected into the reservoir. Gas markets later developed in industrialised countries, but even so, production was generally constrained by pipeline capacity, which was designed to provide as long a plateau as possible. Much gas still remains in remote locations, now relying on liquefaction to bring it to market.

It is difficult to forecast gas supply, the data being even less reliable than is the case for oil, and the trends less clear, but it is here expected that global gas supply will continue at approximately present levels for another two or three decades before collapsing. It partly depends on the rapid construction of liquefaction facilities in remote areas. Much of the potential gas supply lies in the Middle East Region, which is subject to particular political pressures that may restrict exports. Naturally, if depletion is stepped up by the construction of liquefaction plants or new export pipelines, the decline will come sooner.

It is again stressed that the data are unreliable and the resulting analysis no more than an approximation. That said; the general patterns of depletion can be presented with confidence. The wider significance of the issue is discussed in Chapter 13.

Table 75.3

																				2010	
		PRODUCTION TO 2100																			
		Regular Conventional Oil																			
		KNOWN FIELDS									Revised 20/12/2011										
		Present		Past			Reported Reserves				Future	Total Found	NEW FIELDS	ALL FUTURE	Total	DEPLETION			PEAK		
Country	Region	Kb/d 2010	Gb/a 2010	Gb	5yr Trend	Disc 2010	Average	Deduction Static	Deduction Other	% Disc.	Gb	Gb	Gb	Gb	Gb	Rate	%	Mid-Point	Expl	Disc	Prod
Middle East	F	22579	8.24	320.49	0%	1.55	35.50	−20.25	0.00	96%	457	777	50.5	507.11	828	2.7%	44%	2015	1992	1948	2020
Eurasia	C	15758	5.75	225.20	1%	3.27	123.36	−13.28	−45.00	92%	136	362	30.1	166.55	392	3.3%	57%	2005	1982	1960	1988
N.America	H	4240	1.55	199.78	0%	0.36	107.82	0.00	−185.43	99%	25	225	3.2	28.22	228	5.2	88%	1973	1981	1930	1973
L.America	E	5699	2.08	124.95	4%	0.64	240.59	1.39	−88.38	96%	50	175	7.6	57.30	182	3.5%	69%	1997	1982	1914	1998
Africa	A	7051	2.57	103.81	−3%	1.19	122.00	4.52	−17.37	95%	78	182	10.1	88.19	192	2.8%	54%	2006	1981	1961	2005
Europe	D	3528	1.29	56.08	−4%	0.39	11.87	0.00	0.00	96%	18	74	2.9	21.17	77	5.7%	73%	1999	1986	1974	2000
Asia-Pacific	B	3472	1.27	53.76	−1%	0.47	23.47	3.17	0.00	97%	27	80	2.8	29.49	83	4.1%	65%	2001	1990	1974	2000
Other		817	0.30	8.63	−4%	0.00	0.00	0.00	0.00	83%	4	12	2.5	6.37	15	4.5%	58%	2003			2006
Rounding											0.0	0.0	3.0	3.00	3						
Non Middle East		40564	14.81	772		1259					338	1110	62	400	1173		66%				
World		63142	23.05	1093	−1%	7.88	1294	−24	−336	94%	795	1887	113	907	2000	2.5%	55%	2005	1981	1964	2004

Table 75.4

	Regular Conventional Oil Production				
Mb/b	2000	2005	2010	2020	2030
Middle East	21.56	22.86	22.58	23.18	19.47
Eurasia	11.21	14.23	15.76	11.70	8.39
N.America	5.13	4.25	4.24	2.39	1.35
L.America	8.08	7.35	5.70	3.46	2.23
Africa	7.29	8.45	7.05	5.36	3.84
Europe	6.07	4.95	3.53	1.92	1.06
Asia-Pacific	4.27	3.63	3.47	2.33	1.49
Minor	0.55	0.98	0.82	0.52	0.33
Unforeseen	0.00	0.00	0.00	0.07	0.14
Non M East	42.59	43.85	40.56	27.75	18.81
World	**64.15**	**66.71**	**63.14**	**50.93**	**38.28**

Table 75.5

	OIL EXPORTS & CONSUMPTION						
	Production	Consume		Export	Population	Area	Density
			p/capita	(+)		Km²	
Country	Mb/a	Mba	b/a	Mb/a	M	M	
Middle East	8241	2760	10.01	5481	276	6.1	45
Eurasia	5752	4932	3.02	820	1633	31.5	52
N.America	1547	7807	22.56	−6260	346	19.7	18
L.America	2080	2621	5.23	−541	501	18.8	27
Africa	2574	712	1.64	1861	434	12.6	34
Europe	1288	3351	11.01	−2063	304	1.9	156
Asia-Pacific	1267	2834	1.55	−1567	1825	15.1	121
Minor/Other	298	6935	4.38	−6637	1585		
Non Middle East	14806	29022	−5	−14216	6711	−6	−45
Total	**23047**	**31981**	**4.63**	**−8934**	**6903**	**105.80**	**65**
Other	0	−199	−2.32				
WORLD	**23047**	**31782**	**4.55**	**−8735.1**	**6987**		

Note: Refinery gain deducted from "Other"

Table 75.6

World					REGULAR CONVENTIONAL GAS PRODUCTION											2010	
			KNOWN FIELDS										Revised: 20/11/11				
		Present			Past	Discovery		Reserves		FUTURE KNOWN	TOTAL FOUND	FUTURE FINDS	TOTAL FUTURE		DEPLETION		PEAK
Country	Region	Tcf/a	FIP	Gboe/a	Tcf	5yr Trend	Tcf 2010	% Disc.	Reported Average	Deduct Non-Con	Tcf	Tcf	Tcf	Tcf	TOTAL Tcf	Current Rate	%	Mid-Point	Disc	Prod
Africa	A	13.6	8	2	257	2%	4.3	92%	511	−22	434	690	62	495	752	2.66%	34%	2018	1957	2018
Asia-Pacific	B	13.4	0	2	257	3%	18.5	89%	534	51	482	720	90	573	810	2.28%	29%	2023	1973	2021
Eurasia	C	30.6	28	6	970	0%	35.9	85%	2124	−992	1207	2176	386	1593	2563	1.88%	38%	2020	1966	2020
Europe	D	11.7	12	2	373	0%	1.7	98%	151	0	176	549	12	188	561	5.88%	67%	2004	1959	2004
Latin America	E	10.1	1	2	234	3%	4.3	92%	277	13	343	577	48	391	625	2.52%	37%	2018	1941	2008
Middle East	F	22.9	0	4	381	7%	18.9	88%	2052	0	2355	2737	390	2746	3127	0.83%	12%	2030	1971	2030
North America	H	23.2	27	4	1181	0%	2.7	98%	285	0	197	1378	22	219	1400	9.55%	84%	1987	1996	2009
Other	I	2.8	0.0	0.5	49	15%		80%			30	80	20	51	100	3.48%	49%	1993		
Unforeseen	J	0.0	0.0	0.0							27.9		0.0	62	62		0%			
WORLD	K	**128**	**0**	**23**	**3682**	**2%**	**86.3**	**89%**	**5934**	**0**	**5225**	**8907**	**1093**	**6317**	**10000**	**1.99%**	**37%**	**2023**	**1971**	**2015**

Table 75.7

	Gas production				
Tcf/a	2000	2005	2010	2020	2030
Africa	9	12	14	15	11
Asia-Pacific	9	11	13	16	14
Europe	25	29	31	42	34
Latin America	7	8	10	9	6
Middle East	11	16	23	27	31
North America	24	23	23	8	3
Other	1.2	1.6	2.8	2.3	0.9
Total	98	113	128	127	104
Rounding	0	0	−1	1	−2
WORLD	**98**	**113**	**127**	**128**	**102**

Non-Conventional Oil and Gas

A major cause of confusion in oil statistics is that there is no standard definition of the boundary between the so-called *Conventional* and *Unconventional* oil and gas. It is clearly important to make clear distinctions because the different categories have different distributions, rates of extraction, costs and other characteristics. In this study, it has been found expedient to recognise what is termed *Regular Conventional Oil (>17.5o API)* and *Gas,* defined to exclude the following categories, which are designated as *Non-Conventional*:

1. Oil from coal and organic-rich clays (*kukersite*), commonly termed *oil shale*
2. Oil extracted by the artificial fracturing of low permeability reservoirs (*Shale Oil* or *Tight Oil*)
3. Extra-Heavy Oil (<10° API) and bitumen
4. Heavy Oil (10–17.5°API)
5. Deepwater Oil and Gas (>500 m water depth)
6. Polar Oil and Gas
7. Liquids from gas plants
8. Gas from coal (coalbed methane), tight reservoirs (*shale gas*), deep brines and hydrates. (Note: °API is a measure of density, with water having a density of 10° API)

The resources of *Non-Conventional Oil and Gas* are large, but extraction is normally difficult, costly, environmentally damaging and, above all, slow. The entry of supply from these sources in the future is clearly important, serving to ameliorate the post-peak decline, but it is doubted if they will have much impact on the date or height of the overall peak itself. They are described briefly below.

Coal and Oil Shale

Two German chemists, Franz Fischer and Hans Tropsch, developed early methods of extracting liquids from coal. The technology involves gasifying coal at high temperature and pressure, and applying catalysts, but there are various different procedures. The process was developed in Germany during the Second World War, and later used in South Africa, when it was subject to an oil embargo. It was sufficient to provide about 30% of that country's needs.

The so-called Oil Shales are immature normal source-rocks that have not been heated enough in Nature to give up their hydrocarbons. Strictly speaking they are not shales at all in a geological sense, their scientific name being *kukersite*. They were first exploited in Scotland around 1860, which led to one of the earliest refineries to extract lamp-oil. Another early development was in Estonia where they are still used as a fuel for power stations.

Interest in the development of Oil Shale grew rapidly in the aftermath of the Oil Shock of 1980, especially in the United States, which has large deposits in the Green River Basin of Colorado and neighbouring States. The traditional method of extraction was simply to excavate the material, and then place it in retorts at 350–1,000 °C: the higher the temperature, the lighter the product. One drawback was the large amount of fine-grained toxic ash produced in the process, whose disposal posed environmental problems. Attempts have also been made at in situ retorting with the help of underground combustion, and the injection of hot natural gas. A recent project, developed by Shell in Colorado, involved inserting electric elements, using electricity from a dedicated coal-fired power station, and cooking the deposit for several years, after which it is expected to deliver production to conventional wells. In addition it has been necessary to surround the workings with a refrigerated underground *ice-wall* to prevent the escape of the liquids. It sounds as if it will be subject to an extremely low, if not negative, net energy yield.

There are many other large deposits around the world, especially in Russia, China, Australia, Morocco, Zaire, South Africa, Egypt, Argentina, Chile, Uruguay and Brasil. The resource is enormous, perhaps capable of providing as much as three trillion barrels of oil, but so far none has proved commercially viable, despite in some cases Government subsidy.

Of growing importance are the so-called tight reservoirs, known as *Shale Oil* and *Shale Gas.* In essence they consist of beds of sandstone, siltstone or dolomite with very low porosity and permeability lying within source-rock sequences, commonly at relatively shallow depth. Highly deviated boreholes are drilled into them to run parallel with the formation and thereby be in contact with more of the oil-bearing rock. Liquids, charged with various chemicals, are then injected under high pressure to fracture the rock and cause artificial permeability, which allows the flow of oil and gas from the adjoining source-rocks. The wells are then placed on production and can produce profitable amounts of oil and gas in the current high price environment, although subject to relatively rapid depletion. There are some environmental hazards where the reservoirs are at shallow depth, as the liquids used in the fracturing process may poison the overlying aquifers or cause minor earth tremors. Interest in this new source of oil and gas has expanded rapidly in recent years, especially in the United States, where, the Barnet and Bakken Shales are of particular interest. The environmental hazards have raised objections to developments in several other countries. Obviously there is a wide range of geological circumstances with the more favourable being exploited first.

Bitumen, Extra-Heavy Oil and Heavy Oil

Oil moves upwards from the source-rocks in which it is generated, and can collect in overlying structures bounded by faults and other barriers to migration, but in some cases where the flank of the basin is relatively un-deformed, the oil was able to migrate long distances to approach or even reach the surface. The two largest and best-known examples are in Western Canada and the Orinoco region of Venezuela, but there is also a lesser known and undeveloped major deposit in eastern Siberia. In addition, minor deposits occur in many other producing basins.

There are many subtle differences between these occurrences. *Extra-Heavy Oil* is defined as having a density greater than $10°$ API, whereas *Bitumen* is defined in terms of viscosity, having a value greater than 10,000 mPa. Both commonly occur together being generally referred to as *tar sands*. They grade into ordinary *Heavy Oils*, which are here defined to have a density in the $10–17.5°$ API range.

The Canadian deposits lie in three principal areas of Alberta at Peace River, Cold Lake and Athabasca. The oil itself was generated in the foothills of the Rockies, probably mainly in Mid-Cretaceous source-rocks, although deeper sequences may have also contributed. There were no serious barriers to long range migration in this basin, and much of the oil moved to shallow depths where it was weathered and attacked by bacteria. The lighter fractions were removed in this way, leaving behind a tarry residue that fills the pores of the outcropping sandstones.

There are two principal methods of extraction: surface mining and in situ production. The mining operations, which are long established, involve the excavation of colossal pits up to 75 m deep from which giant dump trucks transport the raw material to processing plants, fuelled by natural gas from fields in the vicinity. New techniques involving the use of catalysts are being tried. A large amount of water is used in the process which is lowering the water-table of the aquifer to the concern of the authorities. The local gas fields used to fuel the plants are also being depleted, prompting the idea of building nuclear power stations to supply their energy needed to process the oil and tar.

The in situ method involves drilling highly deviated wells to run parallel with the productive formation at depth. Two such wells are drilled, one above the other. Steam is then injected into the upper one, which mobilises the heavy oil and tar that seeps down to be produced from the underlying well. New technology, involving pumping down air and igniting the tar underground, are being developed.

The high oil prices of the recent past have stimulated a huge expansion of the operations, with most major oil companies taking an interest. There are consequential staff shortages, such that workers even commute by air from Montreal. Present production amounts to about 2 Mb/d, and is expected to rise to a plateau of about 3.5 Mb/d by 2025, assuming that oil prices remain high.

The situation in Venezuela is rather different, because the deposits lie at a depth of 500–1,200 m, which precludes opencast mining. The conventional procedure is to drill patterns of five wells on a closely spaced grid. Steam is injected into the peripheral wells, driving the mobilised oil to a central producer. But recent developments have also seen the application of long-reach highly deviated wells, with downhole pumps that manage to extract some of the slightly more mobile oil.

Another approach involved mixing the bitumen with water and surfactant to yield a product known as Orimulsion, which can be readily transported and used directly as bunker fuel. There are however environmental concerns as the heavy oils contain sulphur, nickel and vanadium, whose disposal poses difficulties. Production has been declining due in part to government policy.

There are at present four main projects in Venezuela, known as Petro-Zueta, Cerro Negro, Sincor and Hamaca, whose combined production is thought to amount to about 1 Mb/d.

The in-ground resource in both regions is enormous and effectively unquantifiable, but it has been suggested that the Canadian deposits may hold about 2,500 Gb and those of Venezuela about half that amount. About 15% is considered recoverable under present technology but may increase in the future.

Ordinary heavy oils are present in virtually all producing basins, commonly occurring at a relatively shallow depth in marginal areas. Yet again, there is confusion in the absence of a standard definition of the upper limit. Canada prefers $25°$ API, because the cold climate there adversely affects the

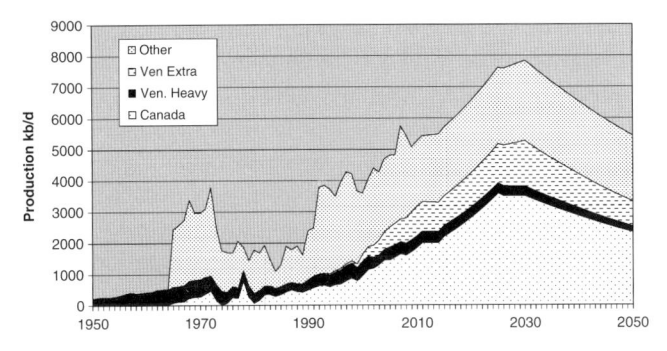

Fig. 76.1 Heavy oils

flow properties of oils below this level. By contrast, Venezuela, with the warmth of the Orinoco, prefers a cutoff at 22° API. A rounded 20° cut-off would be reasonable, but a slightly lower one at 17.5° API is preferred here, because there are many fields with oils slightly heavier than 20° that are in normal production.

Great hopes have been expressed that the development of these heavy oils will more than offset the decline of *Regular Conventional*. It is however perhaps wise to adopt a more sanguine position in view of the mammoth investments, the environmental costs and the low net energy yields, which, as mentioned, have even prompted the idea of exploiting the deposits with the help of nuclear energy. The illustration, built admittedly on most unreliable data, gives the general picture of what can be expected.

Deepwater Oil and Gas

Production in the early days of the oil industry was exclusively onshore, but in due course eyes turned seaward to tap the extensions of fields lying on the coast. At first, that was achieved by deviating wells from the shore, and later, by drilling from steel platforms in shallow waters, as for example in Trinidad, Peru and Lake Maracaibo. Later, a rig was mounted on a barge that could be moved from location to location, which was pioneered in 1949 in the shallow waters of the Gulf of Mexico, off Louisiana.

The breakthrough then came with the idea of building the rig on two submerged pontoons that could rest in relatively tranquil water beneath the wave-base. The first such rig, *Blue Water No. 1*, came into operation in 1962, and began to extend the range of drilling to as much as 200 m of water. The design was subject to continual improvement such that it soon became possible to drill routine wells in the stormy waters of the North Sea. There are two other types of offshore rig worth mentioning: the jack-up which sits on long retractable legs resting on the seabed; and the drillship, having a rig mounted mid-ships on an ordinary vessel, held in place by anchors or thrusters.

Gradually, the continental shelves of the world were explored, and delivered some substantial finds both from the extensions of existing onshore basins and from entirely new provinces. Offshore seismic technology also saw great advances from the early days when a seismic boat let off explosive changes, the echoes of which bounced off formations far below the seabed to be recorded on receptors towed behind the vessel. New sources of energy were developed and computing power brought great sophistication, such that offshore surveys are now cheaper and give better results than onshore ones. Despite these advances, great challenges remain in installing the production facilities, such as the massive steel and concrete platforms of the North Sea.

As the opportunities of the continental shelves were gradually exhausted, attention began to turn to the deepwater domain. The Brasilian State Company, *Petrobras*, pioneered this development, as the country was facing the high cost of imports during the early 1980s. To its enormous credit, it successfully began to find and develop fields in exceptionally deep water. Parallel developments came in the Gulf of Mexico, and later off Angola, Nigeria and other countries on the other side of the South Atlantic.

The technical achievements of installing wellheads on the seabed and developing floating production facilities have been truly impressive. The operations are constrained by the limit of the floating facilities, giving a plateau rather than a peak of production. Only large fields are commercially viable, given the high operating costs. Secondary recovery techniques, such as water-injection, are also constrained in the circumstances. It is even more difficult and expensive to produce deepwater gas. Deepwater operations test technology to the limit and there have been occasional accidents, including the serious Macondo accident in the Gulf of Mexico in 2010, when 11 men lost their lives and widespread pollution had a serious economic and environmental impact on the US coastline.

The move to the deepwater heralded another wave of optimism, as economists, looking at their office atlas, concluded that there were vast oceans about to deliver a limitless new supply of oil, but again the geological constraints began to manifest themselves, as it became evident that very special combination of geological circumstances had to be met. The deepwater finds off South America, Africa and in the Gulf of Mexico rely on oil generated in the rifts that opened as the continents began to move apart, 150 million years ago. At first, the rifts were filled with fresh water to become lakes, resembling those of East Africa today, but then the sea broke in. It was subject to a high level of evaporation under the warm climate of the time, which led to the deposition of a thick layer of salt, which sealed the underlying oil. Later, about 60 million years ago, sands and clays, which had been deposited at the mouths of rivers on the adjoining continents, slumped down the continental slope. In some areas they were

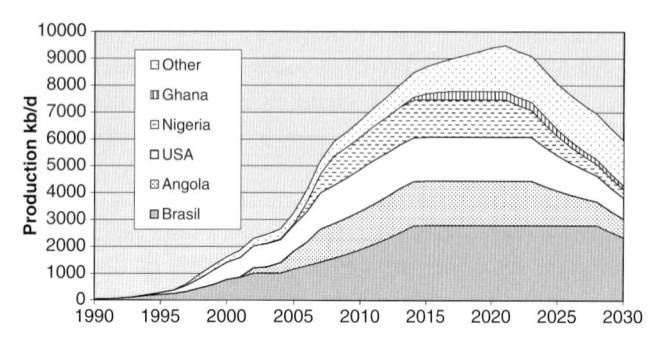

Fig. 76.2 Deepwater oil (>500m water depth)

then taken back into suspension by ocean currents which winnowed out the fine-grained material depositing pods of porous sand on the ocean floor. Still later, structural movements locally ruptured the salt seal to allow the oil to migrate upwards and collect the pods of sandstone, which formed excellent reservoirs for oil. The remarkable combination of circumstances is obvious. Successful attempts are now being made to penetrate the salt seal itself and find what is left beneath it, with some promising results in Brasil, albeit at a depth of about 5,000 m.

In other words, the deepwater finds in the South Atlantic and Gulf of Mexico lie in what is called a divergent plate-tectonic setting, formed as the continents moved apart. Considering the Eastern Hemisphere, we generally find convergent tectonic conditions, lacking the critical early rifts in which the oil was formed. This means that prospects are confined to generally lean and gas-prone deltas, whose fronts have locally extended into deep water. It is the sort of province giving hints of encouragement that fail to materialise.

That said it has to be recognised that the foregoing describes no more than the generality of the position, and until more exploration has been conducted, the possibility of some exceptional new finds, even substantial ones, cannot be excluded.

The known deepwater provinces are already becoming mature, for the larger fields were found first as is almost always the case. In forecasting future production, we again face the confusion of there being no standard definition of the boundary of the deepwater domain, here drawn at an arbitrary water depth of 500 m. Despite the uncertainties, the above graph is thought to give a reasonable approximation of what can be expected. Again, deepwater oil will make a useful contribution to global supply, helping to ameliorate the rate of overall decline, but not having any great long term significance.

Polar Oil and Gas

It will, at the outset, be remembered that oil is derived from algae that proliferated in the warm sunlit waters of tropical and subtropical lakes and constricted seas. It follows that such oil-bearing rocks as now lie in Polar regions must have been transported to high latitudes by plate-tectonic movements, which is clearly a serious overall constraint. Gas was also formed both from vegetal material, as found in the deltas of tropical rivers, and from ordinary oil that has been overheated by deep burial. In short, the Polar Regions were far removed from prime territory for oil and gas generation.

Secondly, it is obvious that operating conditions, both onshore and especially offshore, are extremely harsh, meaning that no one would look there if there were any other options left.

Antarctica can probably be dismissed as oil or gas territory for geological reasons, insofar as the bulk of the relevant plate-tectonic movements, needed to provide a source, were northward. Besides, the region is closed to exploration by international agreement.

The Arctic is rather more promising, led by the remarkable results in Alaska. BP was one of the pioneers, and following its tradition of foothills exploration, learnt originally in the Zagros Mountains of Iran, at first concentrated on the foothills of the Brooks Range. But when that failed to deliver, the company began to look at the platform to the north, where it identified a large uplift at Prudhoe Bay. A lease to the structure was open to competitive bidding, and BP was less enthusiastic than its competitors Arco and Esso, who successfully bid for the crest of the structure, leaving BP with the flank. But by an ironic twist of fate, the crest was found to hold the gas cap, while the flanks in BP's domain held most of the oil. It was a remarkable find, which the company internally estimated might hold between 12.5 and 15 Gb of oil, although, displaying laudable caution, reported only 9 Gb to the Stock Market, in recognition that producing the tail end in this difficult and remote location would be costly and dependent on high oil prices that could not then be forecast. It is now at an advanced stage of depletion and the trend shows that it will not deliver more than originally estimated despite the application of the most advanced technology and higher oil prices.

Significantly, the source-rock is relatively old, being of Triassic age, laid down some 220 million years ago, having had ample time for tectonic movements to transport it northward.

This discovery prompted other smaller finds in the vicinity. Attention also turned to the McKenzie Delta and the Arctic Islands of Canada, but the results were mainly confined to finds of gas.

The Soviets too investigated their Arctic regions. They had already developed substantial oil and gas reserves in the West Siberian Basin, where Upper Jurassic source-rocks had charged overlying Cretaceous sandstones in a series of troughs resembling in fact those responsible for the North Sea. The search was extended northwards into the Timan-Pechora Basin but it was found to become progressively more gas-prone, thanks in part to the deep burial of the source-rocks. Even so, some oilfields were found, relying on Devonian source-rocks, laid down some 380 million years ago, which had been transported northward by plate-tectonic movements.

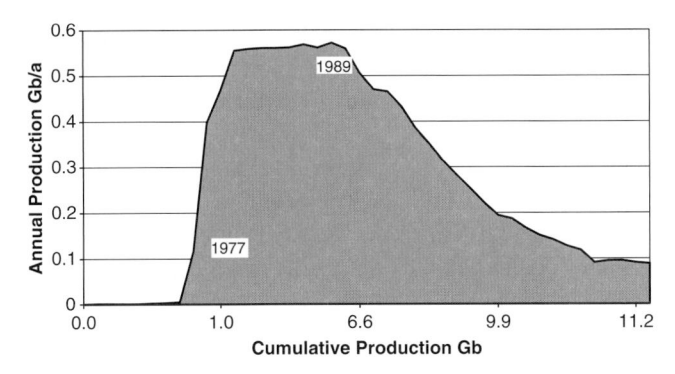

Fig. 76.3 Prudoe bay field, alaska

Norway also looked at the Arctic expanses of the Barents Sea and Spitzbergen (Svalbard), but again the results so far have been disappointing, being mainly of gas. Greenland too is an area of some interest, although exploration to-date has failed to deliver.

A geological assessment of the Arctic regions would, therefore, suggest that they are only locally endowed with effective source-rocks. Another major drawback was its subsequent structural evolution which has been subject to substantial vertical movements of the crust thanks to the fluctuating weight of ice caps that expanded and contracted in response to climate changes in the geological past. As a result, such effective source-rocks as are present, were previously more deeply buried than today, meaning that any oil as they may have had has likely been converted to gas and gas-liquids. Furthermore, the vertical movements had an adverse impact on seal integrity with the result that such accumulations as did form were subject to re-migration and dissipation. Again, it is the sort of domain that gives hints of encouragement but eventually fails to deliver significant results. That said, it has to be admitted that anomalous combinations of circumstance, such as that responsible for the remarkable giant Prudhoe Bay Field in Alaska, can occur.

Great interest is now be expressed in the possibilities of the Arctic Ocean, including offshore Siberia, which may come into range as the Polar ice melts from climate change, prompting some of the adjoining countries to start staking claims.

It is too soon to be sure, but the prognosis is not very favourable, leading to the assessment given here suggesting that it would be realistic to assume for global planning purposes that not more than about 50 Gb will be found in the Polar regions, and that production will peak at about 2.5 Mb/d around 2030.

Gas Liquids

The gas that forms the cap above an oil accumulation normally contains a certain amount of dissolved liquids which condense at surface conditions of temperature and pressure, being termed *Condensate*. Also gas derived from the breakdown of oil on deep burial also contains substantial amount of liquid: indeed there is a category of fields, generally occurring in the deeper parts of a basin that are termed *Gas-Condensate Fields*.

Once the liquid condenses at the surface, it may be fed back into the gas stream to increase its calorific value, or it may be fed into the oil stream, where it simply serves to lighten its gravity slightly. Locally, it may even be marketed directly—and indeed, its composition is close to that of inferior gasoline (petrol) and can be used to power engines.

The treatment of *Condensate* ranges widely from field to field and country to country. In practical terms for this purpose there is no particular reason to distinguish it from conventional oil, although some databases attempt to do so, however difficult the task.

In addition to the liquids that condense naturally, there are specialised plants that can process gas so as to extract butane, pentane and other liquids. The products are known as *Natural Gas Liquids* (or NGL), on which reasonably accurate records are maintained, as for example furnished by the Energy Information Administration (EIA). Such liquids are increasingly important offering a means of exporting product from remote gas fields. It is here estimated that their production will rise in parallel with gas, based on a current yield of 1 Gb of liquid for every 40 Tcf of gas, with perhaps some modest improvement of extraction in the future.

In passing, it is worth stressing that NGL are not to be confused with *Liquefied Natural Gas* (LNG), as is easily done. The latter refers to a process of liquefying gas at low temperatures for transport in specialised insulated tankers from remote locations: the liquid being re-gasified on arrival at its destination.

Non-Conventional Gas

In addition to the categories of gas already mentioned are a number of other sources of gas that truly qualify as *Unconventional*. They include the following subspecies.

Coalbed Methane

Coal is carbonaceous material that commonly contains methane and other gases, including carbon dioxide and nitrogen, held in the molecular lattice of the coal itself. Miners were only too familiar with these occurrences, which could cause deadly explosions, and in earlier years prompted them to take canaries in cages to the coal face to alert them to any dangerous build-up.

The production of coalbed methane is growing rapidly in several of the world's coal basins, especially in the United

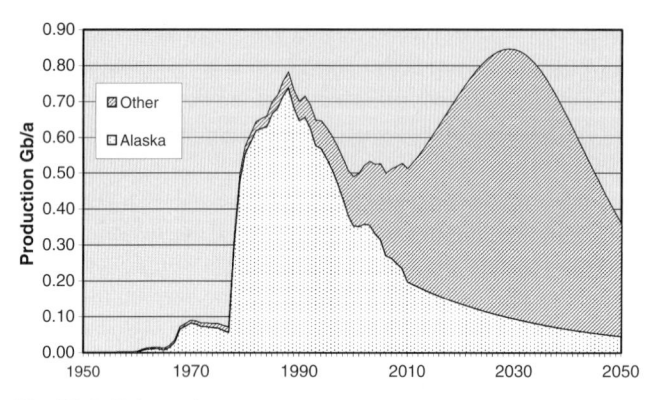

Fig. 76.4 Polar regions

States. There are several methods of extraction: in some cases, cavities are excavated into which the methane can seep naturally from the rocks; and in other cases, the coal-seams are subject to hydraulic fracturing to facilitate the desorption of the gas from the weak molecular bonds that hold it to the coal.

The resources of coalbed methane are considerable worldwide, amounting perhaps to as much as 7000 Tcf, principally in the following countries: China (1,000 Tcf); Russia (4,000 Tcf); Canada (3,000 Tcf); Australia (500 Tcf) and the United States (500 Tcf). Development is most advanced in the United States where it supplies approaching 10% of total gas supply.

The future scope for tapping this source of energy is clearly considerable.

Tight Gas (or Shale Gas)

Certain impermeable carbonaceous rocks may hold substantial amounts of gas, held in the constricted pore space. In some cases, natural fractures allow the gas to seep through the rock, and it can also move along any thin stringers of interbedded permeable sands that are present, forming conduits to conventional wells. Artificial fracturing can also be performed to help extract the gas. The gas content ranges widely from, for example, about 5–10 Gcf per square mile in the Appalachian Basin to as much 35 Gcf in the Fort Worth Basin, where it is currently receiving much attention in the so-called Barnet shale play. It is of course very difficult to assess the resource base, but the world total might amount to about 4,000 Tcf. Again, development is most advanced in the United States where some 20,000 wells are producing from such deposits, albeit at low rates. The resource in the ground is large, but production is subject to low net energy yield, high operating costs and environmental hazards, which probably means that the contribution to global supply will remain insignificant for many years to come. The EIA estimates that production of these non-conventional gases in the United States will rise from about 10 Tcf/a in 2010 to as much as 17 Tcf by 2030. No doubt there will be comparable developments in many other countries. But it is early days to evaluate the real potential. As already mentioned there are some environmental objections as the fracking can cause minor earthquakes and pollute the water supply.

Gas may also occur dissolved in deep brines, due to fact that the solubility of methane increases with pressure. Some such deposits may charge shallower reservoirs. The best-known example is in the depths of the fore-deep of Rockies in the Alberta region of Canada. Again, the resource worldwide is considerable and essentially unquantifiable.

Biogenic Gas

The so-called biogenic gas is formed at shallow depth by the degradation of organic matter in sediments at low temperatures. It can collect in pockets near the surface where it forms a hazard to offshore drilling operations, as the sudden release of gas can reduce the density of the overlying seawater, causing the rig to lose buoyancy. Special seismic surveys are conducted at rig sites to minimise this risk. Biogenic gas is continually recharging certain conventional reservoirs as, for example, in the Po Valley of Italy and in the Gulf Coast Region. This phenomenon is mistakenly taken by some advocates of the abiotic origin of oil and gas as evidence that hydrocarbons emanate from deep in the Earth's crust and will continually replenish the world's supplies.

Hydrates

Much misplaced interest and research has been dedicated to methane hydrates. They comprise a crystallised mixture of water and methane that occurs as an ice-like solid in cold conditions, being found in deepwater and polar areas. They are well known to the oil industry, as they tend to clog offshore pipelines. Interest was in part sparked by reports in the 1970s that the Mossoyakha Field in Siberia was producing from hydrates, but it later transpired that the reports referred to no more than normal hydrate accumulation in cold pipelines. The supposed occurrence of hydrates in the deep oceans was based on an erroneous correlation with a seismic event, known as the *bottom simulating reflector (BSR)*. When this was finally investigated by drilling, it was found that the correlation was flawed, and that the hydrates themselves occurred in no more than isolated granules or thin laminae. Thicker accumulations have however been found locally, where they are related to natural seepages of conventional gas. It is well said that *hydrates are the fuel for the future and likely remain so.*

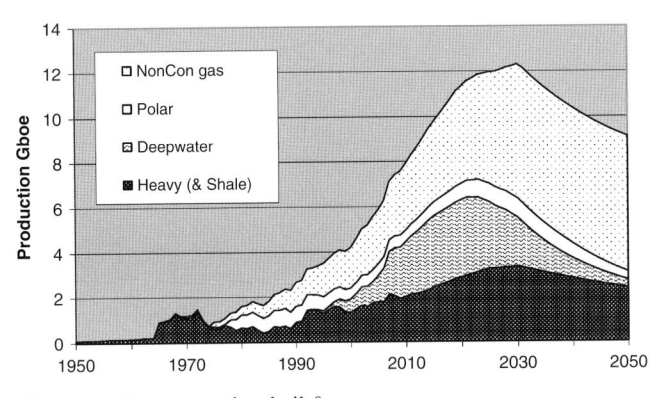

Fig. 76.5 Non-conventional oil & gas

Conclusions

It is evident from the foregoing that the world resources of *Non-Conventional* oil and gas are indeed considerable, but subject to low and costly rates of extraction. Their entry does not delay the date of the peak of overall oil production by much, but they do ameliorate the post-peak decline. Furthermore, they may be considerable local significance when world shortage and soaring prices bite in earnest. It is difficult to forecast a production profile but the above graph depicts a reasonable expectation.

The Oil Age in Perspective

Introduction

The main aim of this book has been to evaluate the world's resources of oil and gas, and their distribution by country. It is a difficult task because of lax reporting standards and ambiguous definitions, yet it is an important issue given the central place of these energy supplies in the modern world. From this standpoint it seemed useful to try to place the Oil Age in an historical context both by country and for the world as a whole. History in turn leads to current affairs, in which oil and gas supply appear to be increasingly important elements. Such an evaluation suggests that there are several sensitive and as yet ill-defined factors influencing the current political situation, but speculation on their nature has been omitted from this edition in the interests of diplomacy.

Placing the oil age in an historical perspective is by all means a big subject, so we can hope to no more than touch on some of the essential elements. This does not aim to be a scientific examination, redolent with indigestible references, but rather a simple overview. It is difficult to come to terms with the Universe and think of all those celestial bodies rotating within their galaxies. We have all seen pictures of the barren surfaces of the Moon and the planets in the solar system. But so far as we know, Planet Earth developed somewhat differently, especially insofar as it had an atmosphere which helped living organisms develop.

Evidently, there are some basic electronic forces that bond matter together into atoms, which in turn attract each other to become molecules. The ingredients of our primordial planet were gradually assembled to form rocks, which cracked under the heat of the sun where uplifted on early mountain ranges. The eroded fragments then slid downwards mainly under the influence of gravity to be deposited as sands, clays and other sediments in valleys, lakes and early seas. At some point, further elemental combinations led to the first forms of what could be described as life in the primeval ooze. It evolved gradually into more complex forms as the oceans filled, and the air above provided an atmosphere, which moderated the climate and shielded the organisms from solar radiation. The circumstances favoured organic evolution, such that, by the opening of the Cambrian Period, some 540 million years ago, the Planet supported a rich fauna, including hard-shelled species, whose remains have been preserved as fossils.

The succeeding years witnessed many fluctuations of climate and oceanic conditions as fragments of the Planet's crust, known as tectonic plates, separated and collided under the influence of deep-seated convection currents. It may indeed be that the phenomenon of plate-tectonics is what distinguishes Planet Earth from most other celestial bodies. Things were far from stable. The atmosphere, even now, is no more than a few kilometres thick, forming but a thin and sensitive protective skin.

But life did evolve. Simple species tended to survive well: for example the Cambrian limpet, or *patella* to give it its scientific name, has remained little changed to this day. More sophisticated forms evolved in places where they found an environmental niche that suited them, only to die out when the niche collapsed from natural causes. Few, if any, reverted to simplicity. There were periods of mass extinction, as for example at the end of the Permian, 250 million years ago, when only the hardiest managed to survive. Probably, the cause was widespread volcanic activity as tectonic plates split apart, which led to climate changes destroying the ecological habitat on which most species depended.

The Arrival of Man and Settled Agriculture

This long process saw the appearance of mammals, including the apes, one enterprising subspecies of which swung down from the trees and learned how to walk on his hind legs. He in turn evolved into the first hominid, which made an appearance only some 6 million years ago, probably in Africa. At first, he lived by hunting and gathering, for which he developed primitive stone tools and weapons. He took to living in and by caves, and also learned how to make fire for warmth and cooking, possibly from the friction caused by rubbing two dry sticks together. Some primitive tribes in remote areas still survive with comparable life-styles.

Climate changes may have stimulated Early Man's travels. Encroaching deserts could have caused some of his members to head north along the Nile Valley in search of better conditions. At a certain point, about 12,000 years ago, an enterprising group found out how to plant seeds and husband animals for meat and milk, leading to settled agriculture. This was a great turning point for the species now known as *Homo sapiens*.

Settled agriculture stimulated a cultural flowering and a new political dimension to life, as individuals now had property to protect and, if need be, fight for, perhaps gaining a longer view of the future for themselves and their offspring. All the other species of the animal kingdom have relied solely on the energy of their muscles, but these early farmers became the first species ever to use external sources when they employed slaves or draught animals to plough their fields and carry their produce to market.

They also became the only species to indulge in organised trade. At first, it was simply barter until someone found a nugget of gold in a river bed. Its shiny surface may have impressed a neighbour sufficiently to offer a sheep in exchange for the nugget, laying the foundation for currency. Settled agriculture also called for storehouses to hold the food between harvests, which in turn required accounting to record how much was received and returned to the farmers. Control of the storehouse brought political power, as it could give preferential supply to the privileged members of the community, or even charge interest. In other words, the storehouse became the foundation of banking.

Not Plain Sailing

But it was not plain sailing for these early people, some of whom had become fairly culturally advanced. Comets may have struck the planet, and no doubt there were tsunamis from earthquakes. Devastating floods swept the lands such that many people perished, save for those who found themselves on high ground, perhaps including those in the mountains of the Middle East.

The survivors may have asked themselves why they had been chosen to be spared, which may have led them to imagine that they were the beneficiaries of some divine power. This sense of divinity helped them design their own social norms for their common good. They built astronomical observatories in the form of stone circles.

While the relatively more advanced civilisations may have been located in the Middle East, early tribes also migrated through Asia, Europe and the rest of Africa, finding new territories in which to settle. Some even managed to cross the Bering Strait and populate the Americas.

Gradually man established himself, and empires developed when a particular group came to pre-eminence. He also learned how to read and write, which allowed him to pass on ideas, especially religious ones defining social norms, to his descendents. As their numbers grew to exceed what their lands could support, they were motivated to occupy new territories, sometimes making wars to do so. In addition to tools, they now needed better weapons. A technological progression brought them from the Stone Age to the Bronze and Iron Ages, as their skills at smelting improved. The Mediterranean was a cradle for such early civilisations and empires, represented notably by the Greek and Roman Empires, but there were also important similar developments in Persia, China and India. The domesticated horse provided transport for trade and armies, while the sailing ship conquered the oceans.

Pre-Modern History

These few observations provide the backdrop for the pre-modern age which dawned, it could be said, about 2,000 years ago, when some 300 million people occupied the Planet. This chapter lasted for about 17 centuries. Primitive tribal conditions remained in much of the world, but there was a flowering of civilisation in Europe, the Middle East, India and China, where cities were built and sophisticated societies constructed. Some had advanced cultural attributes, including scientific and philosophical learning. There were also notable developments in the Americas with the great Aztec, Maya and Inca civilisations as well as the independent civilisations that formed in Africa during this period.

These were, however, years of perpetual conflict as kingdoms waxed and waned, in some cases under religious banners. The notion of an all-powerful deity often prompted temporal leaders to claim preferential divine inspiration, and hence power. Beautiful and impressive cathedrals were often used to crown the kings underlining what was termed the *divine right of kings*. Trade, transport and banking expanded, but the people were still subject to environmental constraints, including those due to climate changes, which may have prompted the wholesale migrations out of Asia that periodically occurred. Minor Ice Ages affected the Planet from time to time, notably between 1350 and 1850, due to changes in solar radiation.

The population no more than doubled during these 17 centuries, as people lived within the sustainable limits of their specific region. But there were domestic tensions as land-ownership became a critical element of power. For energy, people still largely relied on muscle power but water and wind was used to drive mills for grinding wheat and other industry. Firewood and charcoal provided the primary fuels.

The bow and arrow gave way to the sword and canon, following the discovery of gunpowder and new skills in metallurgy. Banking and financial instruments in the form of currency, debt and usury developed in parallel, often having important political links. For the most part, countries were run by kings and nobles, although pressures towards democratic representation grew.

The Steam Engine ushers in the Modern Age

It could be said that the modern age dawned only some 250 years ago, when coal became the premier source of energy, transforming society. Coal had been known in earlier years from outcropping seams, or as lumps washed up on beaches, but now people started digging for it in open pits, which were progressively deepened into regular mines. Minerals were mined in a similar way, but the mines were subject to flooding when they hit the water tables. Draining the mines led to one of the most remarkable technological developments of all time. The simple bucket had given way to the hand pump with its piston, cylinder and shaft, which in turn evolved into the steam pump and the steam engine. It was soon adapted to power transport, providing locomotives for railways, which allowed the rapid expansion of transport and trade. Sail gave way to steam for world trade.

The next step in this development came in the 1860s when Nikolaus Otto, a German engineer, found a way to inject fuel directly into the cylinder of an engine, perfecting the appropriately named *Internal Combustion Engine,* which was much more efficient. At first, it used benzene distilled from coal, before turning to petroleum refined from crude oil, for which it developed an unquenchable thirst. The first automobile took to the road in 1882 and the first tractor ploughed its furrow in 1907, providing food for a rapidly expanding population.

The Oil Industry Is born

Oil from the early wells was refined to provide paraffin (kerosene) as a fuel for lamps, replacing whale oil which was becoming scarce from the depletion of whale stocks. It affected many people's lives by adding an evening to the working day. But the coming of the *Internal Combustion Engine* greatly increased the demand for petroleum which led to the rapid growth of the oil industry. It changed the world in then unimaginable ways.

A coal deposit covers a wide area, being mined where the seams are thick and accessible, meaning that more becomes viable if costs fall or prices rise under normal economic rules: it being effectively a matter of concentration. Oil—or *Conventional Oil* to be more exact—is different, being either present in profitable and often very profitable quantities, or not there at all. Admittedly there are a few so-called *fallow fields* for discoveries that have yet to be developed but they are few in number and have a minimal impact on the overall picture. Once tapped, the oil and gas flowed to the surface under its own pressure delivering great wealth to its owners, who did not have to hire miners to work with a pick and shovel in gruelling conditions underground. In a sense, it provided a form of false liquidity, insofar as it yielded an extreme net energy gain far removed from muscle power. Its polarity also gave it a certain *boom or bust* character. The successful, or lucky, enterprises grew to form some of the world's largest companies, underlining the critical role of oil in fuelling the world's exploding economy and population.

In North America came the Standard Oil empire of the Rockefellers. It was primarily a marketing and refining company which gradually built up a controlling stake in the industry by acquisition until its over-weaning power was broken up by the Government in 1911 under anti-trust legislation. Its daughters, including Exxon, Mobil and Chevron, to name the largest, grew to have a major role in the industry throughout the world in their own right. They competed both with Gulf Oil and Texaco, which came to power following the great discoveries in Texas around 1900, and with two prominent European companies: Shell (an Anglo-Dutch undertaking) and British Petroleum (BP). Together, these companies, which were dubbed *The Seven Sisters,* controlled world supply, but there were of course many other smaller companies, especially in North America. Some of the families controlling the major oil companies took a prominent position in banking, including the Federal Reserve Bank of New York.

Although the foundations of the industry were largely laid in the United States, there were other early successes, notably in Romania, Azerbaijan, Peru, Trinidad, Mexico, Venezuela, Burma and Indonesia, and of course the Middle East itself. The plume of oil that blew into the sky from a well at Masjid-i-Sulaiman in the Zagros foothills of Iran on 26th May 1908 opened what was to prove the world's most prolific province.

Hegemony

The energy released by oil gave a great boost to the Industrial Revolution, leading to the rapid growth of industry, capitalism, trade and empire as rival powers fought for hegemony. The progress was however accompanied by radical social changes as peasants became industrial workers, many to work in gruesome slums, while wealth was accumulated by a financial elite. There were great advances in medical science with far-reaching consequences, helping the population to expand sixfold in parallel with oil supply. Infant mortality was reduced, tropical diseases were mitigated and longevity for at least the more successful societies increased radically. There were also remarkable technological achievements, including for example the aeroplane which became a means of mass transport taking holiday-makers to distant beaches as well as providing the bomber for military attacks on cities far from the line of battle. The space age followed, again stimulated by military objectives, making it possible even to visit the Moon. Television spread around the world capturing audiences to watch often less than edifying programmes of entertainment and news, with far-reaching social implications.

Banks were lending more than they had on deposit, confident that *Tomorrow's Economic Expansion,* driven by cheap oil-based energy, was collateral for *Today's Debt.* Of critical importance was the role of the pound sterling as a world trading currency, being the principal asset of the British Empire that was reaching its splendid peak, spanning the world. The use of this currency throughout the world delivered a massive hidden tribute to the banks of London's financial centre. They were mainly of foreign origins, and later extended their dominion to New York where they influenced the Government to allow them to form the Federal Reserve Bank in 1913. It has played a dominant and very profitable role, printing dollar bills for the American economy to this day, and also having a large indirect global impact. The currency was at first nominally backed by gold, but that measure of security was progressively eroded and finally abandoned.

It was mainly a competitive free-trade environment in which different countries vied for prominence. Britain led the pack in the early years of the last century, having had a pioneering role in the Industrial Revolution. Germany was overtaking Britain in industrial terms but lacked the hidden financial benefits of Empire. These competitive economic factors set the scene for the First World War, opening in 1914, although there were other catalysts, including a certain rivalry with Russia, which was eyeing the Middle East and Balkans. Whereas earlier wars had been practical means of settling disputes in short fixed battles, advances in weapons technology and the provision of railways to constantly supply the front with men and munitions transformed the war into an appalling conflict of attrition. The stalemate was broken in October 1917, when the United States was persuaded to join France, Russia and Britain, although many of its people, being of Irish and German origins, had little sympathy for the British Empire.

We may note in passing that its entry coincided with both the announcement of the Balfour Declaration for a Jewish homeland in Palestine, which had been negotiated by Zionists a year earlier. The sinking of the *Lusitania,* off Ireland, with the loss of American lives, mobilised public opinion for war.

The peace treaties that followed the War placed blame squarely on Germany, which was subjected to heavy reparations, following the example that it had itself set following its victory over France in the Franco-Prussian War, some 50 years before. The treaties also provided for the demolition of the Ottoman Empire of Turkey, which had been an ally of Germany, leading to the creation of a number of new somewhat artificial States in the Middle East, whose oil rights were allocated to the victors.

The war released new social pressures in Europe as the down-trodden workers sought better recognition for their work through various moves towards socialism, some drawing inspiration from the Soviet Union. The most extreme case was in Germany, which saw the growth of *national socialism,* built out of the misery and suffering imposed by the reparations, as well as a sense of injustice that the principle of *Peace without Victory,* enunciated by President Wilson as a basis for ending the war, when the German army was still on foreign soil, had not been respected. The new movement sought to rebuild German grandeur and counter the perceived rival Communist threat from the east, winning support from industrialists and bankers.

America enjoyed an economic boom, having been spared the direct ravages of war, but over-reached itself in 1929 when the Stock Market crashed, ushering in the *First Great Depression.* One-third of its workforce had to join the bread line, prompting almost socialist responses from the Government in the form of the so-called *New Deal.*

Germany, which also suffered greatly in the Depression, was revitalised during the ensuing decade under its new government, when it sought to rebuild the nation by the integration of various German communities in eastern Europe, and to gain for itself more living space, termed *lebensraum.* It also adopted the principles of eugenics, whereby superior humans could be bred in the same way as can race-horses, which were attracting scientific attention at the time. It led to an outbreak of extreme anti-Semitism, causing great suffering and loss of life.

At first, it came to an understanding with the Soviet Union over the definition of spheres of influence. The governments of France and Britain initially responded with appeasement, perhaps seeing merit in confining the Communist threat, which to some extent inspired the socialist movements in their own countries, but, in due course, evidently felt that their interests were at risk and decided to resume hostilities in 1939. Italy and Japan, which had been allies of Britain and France in the First World War, changed sides, having adopted similar political structures to that of Germany.

The opening years of the Second World War moved in Germany's favour, seeing the fall of France and German

Armies at the gates of Moscow and Cairo, while Japan took much of the Far East, including the oilfields of Sumatra. But in 1942, the United States again joined the conflict, following a Japanese attack upon its naval base at Pearl Harbour on Hawaii. This incident mobilised public opinion for war that was promptly declared. The tide of war turned, leading to the eventual defeat of Germany and Japan in 1945, the latter being consummated by the dropping of atomic bombs on Hiroshima and Nagasaki, although the country was suing for peace.

Although Britain emerged victorious in military terms, it was defeated in other respects, being forced to give up its once splendid empire and surrender the pound sterling to the US dollar which became the world's new trading currency.

Victory did not exactly bring peace and tranquillity because before long the Soviet Union and its former allies were glowering at each other across a divided world in what became known as the Cold War. The United States cemented its position with a hugely profitable military-industrial establishment, and expanded its control of the world economy under the mantle of defending it from Communism. Direct military engagements were confined to conflicts in Vietnam and Korea.

One of the mechanisms for control was to speculate against weak currencies before arriving with offers of dollar loans, provided that the country concerned would liberalise trade. As a result, product and profit, based on cheap labour, were exported with the active support of the local elites, while the country itself remained burdened by foreign debt, denominated in dollars. Ecuador, for example, found itself having to dedicate its substantial oil revenues in their entirety to service foreign debt, leaving the *campesino* as badly, if not worse, off than before. It even adopted the dollar as its national currency in place of the *sucre*.

This chapter ended with fall of the Soviet Government in 1991, which left the United States as a somewhat lonely super-power, having no perceived enemies to justify its hegemony. Meanwhile, Europe had come together in an attempt to form another economic empire under the aegis of the European Union, which seeks to expand eastward, even contemplating the admission of Turkey and the Ukraine.

Middle East Tension

The foregoing outlines some important aspects of the general situation, but it is necessary to consider in greater detail the Middle East in view of its critical role in controlling the supply of oil, upon which the world's modern economy has come to depend. As already described, the carve-up of the Ottoman Empire at the end of the First World War defined the new countries, which together with the ancient State of Iran (Persia), came to face the tensions deriving from the creation of Israel in 1948 and the consequential suppression of the indigenous Palestinians. Superimposed on this were the eternal conflicts between the *Shi'ah* and *Sunni* branches of Islam. The former recognises the direct descendents of the Prophet Mohammad, who was born in Mecca in AD 570, as their spiritual inspiration, while the latter are more pragmatic, placing emphasis on the words and writings of the Prophet rather than his line of descent. While the doctrinal distinctions may not be very significant, the divisions certainly have far-reaching political implications.

The role of Zionism deserves comment as it is clearly significant in the modern world although for some reason it is a sensitive subject. Briefly, it seems that a kingdom was established in the valley of the Jordan River some 3,000 years ago. Whether or not its people were survivors of a flood is not known, but in any event they developed a monotheistic religion. That soon carried political implications as rival factions claimed closer divine links, carrying greater political power. One such faction enlisted support from the Romans, who absorbed the territory into their Empire in 47 BC. The next luminary to appear was Jesus of Nazareth, claimed by some to be no less than the son-of-god. This offended the Judaic Establishment, which in due course persuaded the Roman Governor, to have him crucified as a subversive. But his disciples continued to win acclaim providing the foundation of the Christian Church, which spread throughout the western world, having an enormous influence, although it itself fragmented into factions which were often in conflict with each other.

The Romans evidently found it a difficult place to administer and eventually lost patience in AD 135, when they sacked Jerusalem, killing its inhabitants or driving them into exile. The survivors settled in the cities of the countries of their adoption, where, lacking land rights, they concentrated on trade, which soon led them to finance, banking and usury. They declined integration, frowning on inter-marriage, and continued to practice their particular religion. Even so, the enclaves survived well enough without any particular tension: for example, Baghdad developed for a time as a centre of Jewish culture despite lying in the heart of the Muslim world.

However, the Industrial Revolution of the nineteenth century brought new power to the financiers, which may have been one of the factors contributing to anti-Semitic persecutions, notably in France, Russia and eastern Europe. The persecution triggered a reaction leading to the birth of Zionism, as the oppressed developed the notion of securing a homeland, and attracted the sympathetic ear of influential bankers.

Britain, however, remained in control of Palestine, which was classified as a Protectorate, and limited the rate of immigration to levels that could be absorbed into the existing Palestinian population, which had lived there for more than

2,000 years. However, various Zionist terrorist movements developed during the Second World War aiming to oust the British. One such group, known as Irgun Zwai Leumi and led by a Polish immigrant, grew in strength, and in 1947 blew up the King David Hotel in Jerusalem, killing several British officers who were stationed there. Next year, a war-weary Britain handed over responsibility for the place to the United Nations, paving the way for the unilateral declaration of the State of Israel, a year later. Sympathy for the creation of the new State was encouraged as a form of atonement for previous anti-Semitism. Immigration was stepped up, and the indigenous Palestinians found themselves progressively dispossessed and driven into refugee camps in neighbouring countries.

The situation became the focus for much ensuing conflict and tension. Israel succeeded in securing massive financial and military aid, especially from America, which may have served to stimulate anti-Israeli reactions among the neighbouring States, especially Egypt. These passions may have been based on more than mere sympathy for the dispossessed, possibly seeing Israel as a western foothold in the region undermining aspirations for a new Middle East Islamic hegemony, as espoused by certain Arab leaders. These pressures culminated in several wars with Israel. None of this would have particularly mattered to the outside world, save for the Middle East's control of world oil supply.

The First Oil Shock of 1973 arose out of this situation, when certain Arab countries decided to embargo the export of oil to the United States to counter its support for Israel. While no more than a political gesture, lasting only a few months, it did demonstrate the key role of oil in the world economy. Prices rose fivefold, prompting a serious economic recession lasting for years.

But generally speaking, the world came to terms with the Arab–Israeli conflict as it simmered over the last decades of the twentieth century, before the tensions were reignited as described below.

The Geopolitics of Oil

The impact of Israel has been only one of the elements affecting the political evolution of the Middle East in recent years. The control of the region's oil was perhaps the greater underlying element.

One of the first steps in this progression was the rise to power of Reza Shah Pahlavi in Iran in 1925, who aimed to modernise the country. He entertained certain sympathies for Germany in the Second World War, and was forced to abdicate under British pressure in 1941 in favour of his son, Mohammad. The latter continued to rule in an autocratic manner, although pressures for a more liberal regime developed in the post-war years.

The new political situation eventually threw up an ageing aristocrat by the name of Mohammad Mossadegh, whose party came to power in 1951. It felt that the time for change had come, and successfully passed a bill in Parliament for the nationalisation of BP's exclusive Iranian concession. It was thought unjust that the company should deliver more to its shareholders than it paid to the country in whose land the oil lay.

This was not well received by Washington, which succeeded in influencing the Iranian army to depose him. Clearly, its control of a substantial percentage of world oil supply was a critical factor explaining why the nature of the government of this remote country should be of any particular interest. But the political pressure for change grew and erupted in the so-called Islamic Revolution of 1978–1979, which led to the fall of the Shah. This gave the Second Oil Shock as prices soared, prompted by panic buying as traders feared a resumption of shortage. The American Embassy was briefly occupied and hostages were taken in an effort to secure the return of the Shah for trial. Although negotiations subsequently resolved the issue, Iran came to be regarded as a hostile power by the United States.

Meanwhile, Iraq under its President, Saddam Hussein, was in conflict with Iran over the demarcation of the frontier between the two countries. The problems lay in relation to both the oil-rich region of Khuzestan, a *Sunni* enclave, whose people had historical links with Iraq, and the critical Shatt-al-Arab waterway, which could give Iraq better access to the Persian Gulf, by-passing Kuwait. The frontiers had been arbitrarily drawn by Britain at the end of the First World War as already described, so there may have been some historical justification for Iraq's position. The United States began to see Saddam Hussein as a useful ally in its opposition to Iran. Hostilities between the two countries opened in 1980, and dragged on for six long years with appalling loss of life on both sides. Children were in some cases even used to clear minefields. The United States supplied its new ally with arms, finance and intelligence. It is noteworthy that Saddam Hussein belonged to the *Sunni* sect of Islam, whereas most Iranians and the majority in Iraq itself belong to the *Shi'ah* sect. These religious divisions have had an important influence on Middle East politics.

But the war coincided with a period of low oil prices, partly reflecting the development of the North Sea oilfields. As a result, the Organisation of Petroleum Exporting Countries (OPEC) had difficulty in observing their agreed quotas to limit production to support price. In 1985, Kuwait, announced a 50% increase in its reported reserves on which quota was based, although nothing particular had changed in the oilfields. It also began to pump oil from its end of the South Rumaila oilfield that straddled the ill-defined boundary with Iraq, possibly taking some of Iraq's rightful share. These actions gave Iraq legitimate reasons for complaint. It was, at the time, on friendly terms with the United States, whose Ambassador gave tacit support for a military solution, by issuing a statement to the effect that *border disputes between Arab countries were of no concern to the United States*. Saddam Hussein, evidently misreading the mandate for a limited action, mounted a full-scale invasion of Kuwait on 2nd August 1990.

This proved to be too much for Washington, which may have come to fear that Iraq might gain excessive power over oil supply. When diplomatic efforts failed, the United States mobilised an army under General Schwarzkopf to oust Iraq's troops from Kuwait. It successfully did so during the early months of 1991 in what was known as the First Gulf War, but not before the retreating Iraqi army had fired the oilfields. Some 2 billion barrels of oil went up in smoke. The army was however ordered to withdraw at the gates of Baghdad in recognition that further intervention would stir up a hornets' nest.

Saddam Hussein was no longer the good friend he had once been, and the United Nations was persuaded to impose sanctions limiting Iraq's oil exports. While causing much suffering to the Iraqi people, they had the effect of lifting world oil prices which was well received both in the oilfields of Texas and by the other OPEC. The sanctions were however relaxed from time to time *for humanitarian reasons* when prices rose uncomfortably high.

Resource Nationalism

Christ famously proclaimed *Give unto Caesar that which is Caesar's*, but the concept of national ownership of mineral resources had not arisen. The landlords of Spain's Latin America Empire were effectively tenants of the Spanish King, being required to pay a rental, known as a royalty. The currency was divisible into units of eight, being known as *pieces of eight*, one of which was for the King. This was the origin of the $12\frac{1}{2}\%$ Royalty that characterises many oil concessions to this day.

The United States, whose southern States had been part of the Spanish Empire, inherited this practice, giving the landowner title to such minerals as might underlie his land, and continues to do so, save for offshore concessions, where naturally there is no landowner. But in most countries, the Government came to claim title to mineral rights, for which it would in turn grant concessions to national and foreign companies for fixed terms and stated conditions, normally including a drilling commitment.

Even this came to offend certain countries, which were reluctant to grant outright concessions, preferring instead the so-called *Production Sharing Agreement*, whereby the State retained ownership, while the company was entitled to recoup investment and profit from sales. It was paid in oil divorcing the oil operations from the normal tax system.

The concept of national ownership was not new, having been widely practiced in the socialist epoch of Europe following the Second World War, when Britain, for example, nationalised its rail system. So far as oil is concerned, the first major nationalisations had occurred in Russia in 1928 and in Mexico ten years later. Iran followed in 1951, as already described, when it nationalised BP's long-standing exclusive concession, although later relenting under US pressure to the creation of the so-called Consortium, whereby a group of mainly American companies were able to take over 60% of what had been BP's exclusive position. But the Consortium's rights too were later extinguished.

To step back, we should recall that the discovery of the East Texas Fields in the 1930s resulted in a glut of oil and a calamitous fall in prices, which adversely affected the other producing provinces. The Government decided to intervene to cut production to support price, entrusting the task to the Texas Railroad Commission. Since most oil was then moved by rail, it was in a position to enforce the quotas, whereby wells could be produced for only a given number of days in the month.

This precedent inspired Perez Alonso, the oil minister of Venezuela, who concluded that his country, like Iran, was not getting a fair share of the proceeds of oil production. He opened negotiations with the major Middle East producers that led to the formation in 1959 of the OPEC in order to support world prices by limiting production to agreed quotas, such being based on reported reserves and population levels.

But it was not enough, and before long, the other major producers moved towards outright nationalisation: Iraq in 1972; Kuwait in 1975; Venezuela in 1976 and Saudi Arabia in 1979. These moves were primarily intended to do no more than increase the oil revenue to the producing countries, although appealing to a new sense of nationalism, exploited by politicians.

A new and very different chapter of resource nationalism may now be opening, led perhaps by Russia, as governments begin to recognise the inevitable depletion of national resources, and move to preserve what is left for their own use. Such moves, however logical, run in the face of the current doctrine of globalism, whereby the resources of any country are deemed to belong to the highest bidder. The logic of Britain allowing foreign companies to export their substantial share of the country's dwindling North Sea oil reserves, despite its own growing dire needs, is certainly open to debate in terms of national interest.

A New Century Dawns

The twenty-first century dawned against this background. It was in a sort of inter-regnum, or perhaps the lull before the storm. The Cold War had ended, and the world economy was booming, with India and China in particular seeing rapid growth, accompanied by soaring energy demand. It also coincided with the culmination of another technological revolution based on the computer. The Internet opened new doors to information although which lead to the truth remains unsure.

There were no particular tensions anywhere, but this may not have been an advantage in political terms. *If you don't have an enemy, make one* is apocryphal advice that politicians ignore at their peril. The governments of Europe and the United States were in a sense left without a mission, which they feared, perhaps intuitively, could degenerate into weakness and loss of purpose, carrying adverse financial consequences in relation especially to collateral for debt.

Think-tanks in Washington had already began to evaluate the situation, formulating in 1997 a report entitled *Project for the New American Century,* which identified America's critical dependence on Middle East oil, its own supply having been in decline since 1970 thanks to natural depletion. The continuing Israeli–Arab conflict was seen as a related issue in this connection, several of the authors of the report having dual citizenship.

A turning point came on 11th September 2001, when two airliners crashed into prominent buildings in New York, and the Pentagon was also struck. Many curious features of the incident have since attracted comment by scientists and others. In any event, the incident won popular support for the ensuing war in the same way as had the sinking of the *Lusitania* and the attack on Pearl Harbour mobilised popular feeling for the earlier wars.

Barely had the dust settled before the United States declared a *War on Terror*, and moved to attack Afghanistan, which lies on a proposed pipeline route from the Caspian. It subsequently installed a new Government under Hamid Karzai, a former consultant to Union Oil of California, although resistance fighters fight on, as they did in occupied France during the Second World War. The pipeline was not in fact built, as Caspian oil and gas resources turned out to be lower than at first hoped.

The United States then turned its attention to Iraq in the heart of the Middle East oil province, and, on 17th March 2003, President Bush gave its Government an ultimatum to stand down or face attack. Three days later, an Anglo-American force successfully invaded the country, the military plans evidently having been laid long before. The pretext for the invasion were claims that the country had remaining stocks of chemical and biological *weapons of mass destruction,* which in the event proved to be unsubstantiated.

Although the military operations themselves were highly efficient, they led to desperate conditions close to civil war, as the *Shi'ah* and *Sunni* factions fought each other for power, while the Kurds in the north pressed new claims for independence, which in turn has encouraged a Khurdish minority in Turkey to rise in parallel. Saddam Hussein himself was executed in December 2006 on the grounds that he had sanctioned the execution of 148 people in the town of Dujail, who were rebelling against the government. This number was a good deal less than the million or so innocent citizens who have lost their lives as a direct or indirect consequence of the invasion. President Bush later justified it with the words: *our energy supply was at risk*. The United States has also continued to build up its long-standing threats against Iran, stationing a naval force in the Persian Gulf, claiming that the country poses a nuclear threat.

Meanwhile, the production of North Sea oil and gas headed into steep decline, meaning that Europe had become heavily dependent on imports from Russia, especially of gas. This carries considerable political implications that are not lost to the Russian Government, which may come to prefer to conserve its resources for its own use rather than subsidise its industrial competitors with cheap energy. There is also the issue of the new power of the transit countries, including the Ukraine, through which the pipelines from Siberia pass. In this connection, it may be noted that Turkey, which the European Union seeks to incorporate, also attains the status of a transit country, both for oil from the Caspian that is piped to a terminal at Ceyhan on its Mediterranean coast, and for proposed new pipelines that would deliver Middle East gas to Europe. NATO was established as a defensive pact whereby it would go to the defence of any of its members subject to attack. Later, the rules were modified to a more proactive stance: first, permitting intervention if any member was threatened; and later if its *vital interests* were perceived to be at risk.

These oil-related pressures were not confined to the Eastern Hemisphere, for the Government of Venezuela under Hugo Chavez has recognised the political significance of this country's large oil endowment. It has found allies in Ecuador and Bolivia, as well as sympathy elsewhere in the region, encouraging moves to a new resource nationalism. In a sense, it could be said that he moves in the footsteps of Simon Bolivar, the Liberator of Latin America in the nineteenth century.

Oil prices have soared in recent years. At first, the development was seen as a temporary phenomenon due to OPEC or other supply constraints, but now it is increasingly being recognised as a more fundamental development imposed ultimately by natural depletion. The history of oil prices through 2010, duly adjusted for inflation, is shown in the figure. Prices surged to $147 in mid-2008, which prompted an economic and financial collapse that cut demand such that prices fell back before edging up to about $125 in early 2012. It is to be noted however that the prices are stated in 2010 dollars to give a comparative picture.

There have been growing tensions in many countries, including Burma (Myanmar), Tibet, Pakistan, Sudan, Kenya

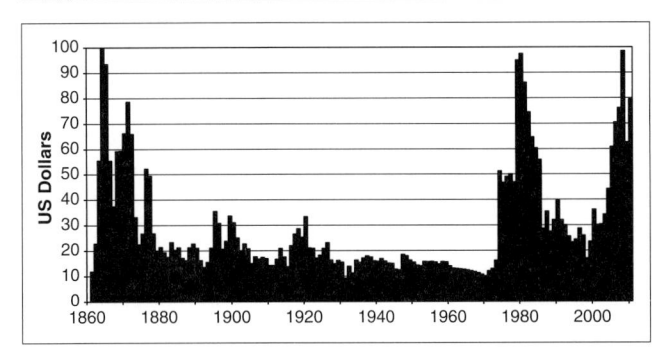

Fig. 77.1 Oil price

and Nigeria, which have experienced assassinations and insurgencies, as the governments of the day come under increasing domestic pressures. These pressures gathered weight in 2011 with popular risings in Tunisia, Egypt, Libya, Bahrain, Syria and the Yemen, collectively known as the *Arab Spring*. The most extreme case was oil-rich Libya, where the rebels were supported by military intervention from primarily Britain and France under a United Nations Resolution. This action, which led to the death of the country's long-standing leader, was justified on humanitarian grounds. But the country's rich oil endowment, providing an important supply for Europe, and the fact that it was reverting to the gold standard may help explain the external interest in the country's future.

Exactly the causes for the uprisings of the *Arab Spring* are unclear but they may reflect new economic difficulties, arising in part from soaring oil and food prices, for which a disaffected populace blame their governments.

Over the past years, the expansion of US domestic debt has matched the cost of its oil imports, meaning that the country effectively obtained its supply for free, benefiting also from the fact that oil trade has been denominated in dollars, helping support the currency. Those days seem to have come to an end.

It is too soon to identify the precise causes for the current financial situation, but it may be that the world's financiers begin to question the assumptions of the *First Half of the Age of Oil*, when cheap and abundant oil-based energy fuelled economic expansion, allowing the banks to lend more than they had on deposit.

The *Second Half of the Age of Oil*, which now dawns, will be marked by the decline of oil supply, meaning that past debt based upon it loses collateral. Whereas the physical decline of oil will be gradual at no more than 2–3% a year, the perception that expansion gives way to contraction can come in a flash, leading to strong reactions on sensitive short-term markets. The following graph illustrates the rise and fall of all categories of oil and gas, measured in billions of barrels of oil equivalent a year, as imposed by natural depletion. The resource estimates are no more than approximate due to the unreliable nature of the statistics, and no doubt there will be short-term departures due to economic and geopolitical circumstances, but the underlying pattern is now beyond dispute. It is unlikely if the recent development of so-called Shale Oil and Gas by artificial fracturing will affect the profile significantly, as the wells are costly and have a short life tapping unsatisfactory reservoirs. Furthermore, the accident at the Fukushima nuclear plant in Japan is relevant as it had an impact on the development for alternative nuclear energy.

Looking Ahead

The foregoing somewhat disjointed review of history sets the scene for trying to imagine what may follow as a consequence of the depletion of oil and gas.

To say it again, there is now no serious doubt left that we come to the end of the *First Half of the Age of Oil*. It was characterised by the rapid expansion of just about everything, including food supply which allowed the population to grow sixfold in parallel. The *Second Half* will prove to be a new world, very different from the one we have known. A debate rages as to the precise date of peak production but misses the point when what matters is the vision of the long decline on the other side of it.

It would indeed be a brave man to claim to forecast what course it will take, but some of the essential elements can be identified. If we are indeed on the brink of the *Second Great Depression*, as seems likely, it will surely dwarf the first, which was little more than the bursting of a speculative bubble. This one results from a much more fundamental factor, namely the depletion of the critical energy supply on which modern society depends. While some commentators shrug it off in the belief that we can turn to coal, nuclear power or renewable energy and continue our present life style, the deep significance of Peak Oil is being increasingly recognised, even by the International Energy Agency, the watchdog of the OECD Governments, and the International Monetary Fund.

We could perhaps make an analogy with the failure of a dam, imagining that someone reports having heard an ominous crack before a small physical break is observed. The management denies that it is anything serious, promising to send a team to investigate. But then a heavy rainstorm causes a small landslip which releases a surge, proving too much for the weakened dam, which fails, releasing a torrent that causes huge damage and loss of life downstream. Today, we are at an early stage of such a progression, with a growing trickle of commentary drawing attention to weaknesses in the modern edifice and the grave implications of change.

The European Union is already facing serious tensions, with some of its member countries in extreme financial difficulties. The United States is facing an economic down-

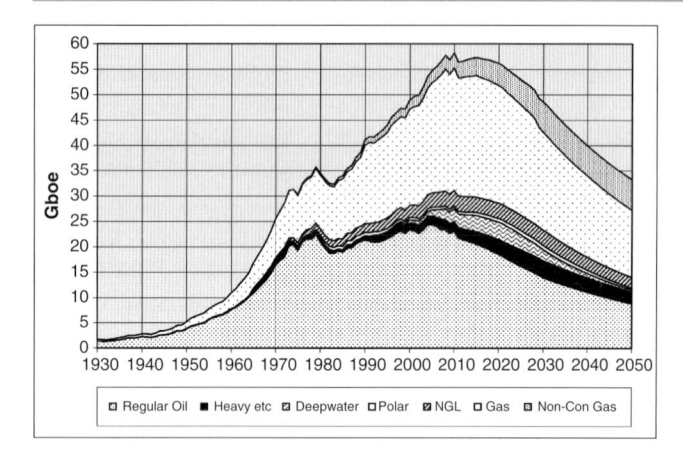

Fig. 77.2 Oil & gas production profiles 2010 base case

turn, and in a sense has farther to fall in view of the height of its past success. Much of its manufacturing base has already moved overseas to exploit cheap labour, and it finds itself heavily in debt to foreign sovereign funds making its currency vulnerable. Its hugely costly foreign military operations may help shore up the imagery of its past hegemony, underpinning its financial foundations. But it looks now as if it seeks to withdraw from its engagements in Iraq and Afghanistan, which have indeed been extremely costly. In assessing political responses, it is never easy to distinguish deliberately planned actions from those that arise almost intuitively from the juxtaposition of circumstances.

A failure in a country at the apex of the world economy could lead to growing cracks everywhere else, many carrying extreme geopolitical responses as people feel defrauded when their hopes of prosperity, deemed to be almost a matter of intrinsic right, are dashed. Food costs have risen radically through much of the world and economic recession has lifted unemployment in many places. There are many such signs, which increasingly attract comment. While the dam in our analogy has not yet ruptured, its structural weakness is now widely perceived.

Whatever the short-term developments, we can say with assurance that *Petroleum Man* will be virtually extinct by the end of this century. Since he has played such a prominent part in the modern world, it poses the question of whether *Homo sapiens* will survive without him. If we look back at the geological record, we see that simple species tended to survive for long periods of time, whereas more adapted forms thrived in a particular niche, only to die out when it closed for whatever reason. In other words, it is necessary to ask the thorny question of whether *Homo sapiens* will be as wise and knowledgeable as his name implies, and find a way to avoid extinction by stepping back to simplicity within the sustainable limits of the planet. It is by all means a serious question.

But this is certainly not necessarily a doomsday message, for one can readily imagine the dawning of a new more benign age as people finally come to live in better equilibrium with themselves, their neighbours, and above all the environment in which Nature has ordained them to live. The challenge is in facing the transition that threatens to be a time of great social, economic, national and geopolitical tension.

A New Reality

The first step on this path is to face reality, and properly evaluate what the oil and gas supply situation truly is. Measuring the size of an oilfield early in its life poses no particular scientific challenge: the problems have lain in the reporting, as this book has tried to emphasise. It is possible that universities may begin to assemble the data and analyse it objectively, following the lead of Uppsala University in Sweden, a country that has already announced a policy of weaning itself of oil dependency by 2020. Even international institutions, charged with monitoring energy supply, may abandon political obfuscation and come to provide an umbrella for new national policies. It is noteworthy in this connection that the International Energy Agency already in 1998 had forecast a shortfall in oil supply by 2010, save for the entry of what it called *Unidentified Unconventional*, being evidently a euphemism for shortage. But this led to political difficulties and the subsequent model returned to *business-as usual*. It now urges that we should *leave oil before it leaves us*.

So far, most of the major oil companies have been in heavy denial, fearing an adverse reaction by the stock market, but there are signs of change. Already, several have tacitly recognised an imminent peak in production without quite bringing themselves to say it outright. The next step may come when an imaginative Chief Executive sees a commercial advantage in realism. He could for example stand up and say that his company offers a well-managed and highly profitable contraction, proposing to sell off minor marketing operations and secondary refineries, admitting that supply constraints impose downstream over-capacity. Some are already moving tacitly in this direction, also shedding staff through early retirement or in other ways, and by out-sourcing services.

Experience to-date suggests that Governments and international institutions are reluctant to face the reality because it is bad politics to bring forth a problem without offering a solution. They have addressed Climate Change concerns, which they find win them haloes, being a voluntary gesture to improve the environment that attracts popular support. It is evidently easier than facing the raw necessity of declining energy supply. More positive responses are however arising at the local level, as exemplified for example by the Transition Town Movement, led by Totnes in Britain, which is winning international acclaim.[1] It simply proposes that individual towns should move towards a more sustainable future, relying on local produce and services. Of especial importance has been a move towards local currencies that return trade to the reality of barter, denying the banking community its take from usury and financial machination, which have created so much apparent money out of thin air, based, as it now transpires, on false premises.

There are already moves to greater regional independence. One can imagine for example that the rocky county of Cornwall in the west of England rediscovers its identity, even its ancient language, and moves on a sustainable path. It might introduce a local currency and raise local taxes to support local services, while requiring residence permits to restrict immigration such that a natural birth rate would soon deliver a reduced population of a size that the county could support. In the United States, there may be moves for individual States to secede from the Union as is their constitutional right, although this outcome may appear extreme from today's vantage point. However, some form of positive isolationism may make a return as the American people deploy their well deserved attributes of enterprise to find a sustainable future on their own rich lands. The cost of its military establishment must far exceed any rational requirement for territorial defence.

But ironically, perhaps the trigger for change might be provided if the United States does finally attack Iran. It would likely lead to turmoil and chaos in the Middle East, with perhaps widespread conflict between the Shi'ah and Sunni factions. Oil production would surely fall in such circumstances, driving oil price to extreme heights. But it would have two unintended benefits: first, it would leave more oil in the ground for the future when it will be desperately needed; and, more important, it would give Governments around the world every possible opportunity to impose draconian new measures to cut consumption, and bring in renewable energies from tide, wave, wind, sun, biomass and geothermal sources, even encouraging nuclear energy.

An alternative less dramatic step would be for the leading countries of the world to sign up voluntarily to a *Depletion Protocol*, whereby they would agree to cut imports at a rate matching world depletion, currently running at only 2–3% a year. It is noteworthy that the Portuguese Parliament passed a resolution to this effect in 2011, setting a valuable precedent. One could imagine that the European Union for example might decide to issue import permits that would be granted only on condition that the vendor declares the oilfield from whence the oil came and grants the right of technical audit to reveal how much is left. This would make it possible to determine an accurate Depletion Rate.

The adoption of such a policy would reduce world prices by bringing demand into balance with supply, and thereby prevent profiteering from shortage while allowing the poor countries to afford their minimal needs. Each country could manage its obligations as best suited to its particular social environment: some might sell to the highest bidder under open market principles, whereas others might have a rationing system, identifying the particular needs of different elements in their communities. The idea of tradable energy quotas becoming a form of currency has already been mooted. The Kyoto Treaty has set a useful political precedent even if unlikely to have much impact on the climate itself. Whereas climate concerns call for universal responses, an oil depletion protocol would soon benefit the countries that adopted its principles, giving them an advantage over countries that continue to live in the past.

A stock market crash could also trigger a range of new strategies and perceptions undermining unrealistic aspirations to wealth and affluence as people seek sustainable futures.

Whatever the mechanism providing the political cloak for action, Governments will have to face the new situation that unfolds. The ideal would be to have modest shock of some sort at an early date to force the change, because the longer the reaction to the forces of Nature is delayed, the more severe and difficult become the responses.

The impact of such a shock would reverberate around the world, no doubt being accompanied by much tension, revolt and revolution, attended by regrettable loss of life. Urban dwellers may become especially vulnerable, but it would force changes of attitude and behaviour, spelling what has been termed *the end of suburbia*. The bicycle could stage a come-back, and the window box could be used to grow tomatoes. Gradually, the city-dwellers could rediscover their roots in the country, perhaps locating a distant relative, and begin to return to a sustainable pattern of life. Parking lots in cities could be converted to grow vegetables. Many would not make it, but hope and positive aspirations would begin to develop in the minds of the survivors. New forms of enjoyment could be identified: no longer would people find pleasure in the superficial challenges of the shopping-mall but would wake to a more benign smile of having harvested food that they had actually grown and nurtured. The loudspeakers and television sets could be switched off allowing people to again meet, sing, dance and talk in each other's houses or the local pub, as they did in Ireland not so long ago before affluence struck the country. Village life could stage a come-back. There could be a return to more basic values, which do still remain, however submerged.

These sound rather idealistic visions, and reality will not deliver them universally. Some societies will probably fail absolutely; some will cling to outmoded ideas and practices; some will fight and die, but in the wider spectrum of things there can be survivors, who better adapt, as has always been the case.

---

[1] See Hopkins R., 2008, *The Transition Handbook*, Green Books ISBN 978-1-900322-18-8

At the same time, it is of course also possible that Nature's patience is reaching its limit. She might react by releasing a pandemic to decimate the population, whose natural resistance has been eroded by the excessive use of antibiotics, or she might unleash severe climate change, with sea-level rising to flood most major cities. The climate has changed many times in the geological past, and the extent to which the present changes are man-made is less than sure, but the outcome is the same whatever the cause.

The hope is that there is an early relatively mild event, sufficient to trigger the new direction. The worst outcome would be a general malaise as people try to prolong outdated life-styles, practices and perceptions. Certainly, it is sure that the longer the transition is delayed, the more severe the outcome.

Time is short. The grandfather, or great-grandfather, of an old man living today experienced how life was before cheap energy from coal, oil and gas changed the world. Europe could barely support its population, finding it necessary to import farm nutrients in the form of *guano* (bird-excrement) from South America in sailing ships, while massive emigration to the New World helped remove the pressure. An extreme example is Ireland where the population halved through death and emigration when the potato crop failed due to a fungal disease in the 1850s.

This is a relatively short span of time in historical terms and will be matched by equally radical changes in the near future. The grandson of a young man today will have to face the rigours of a low energy world when the stocks of coal, oil, gas and prime uranium will be nearly exhausted. Food prices are already soaring, partly reflecting increased oil costs, leading to riots and political disturbances in many countries. Logic suggests that the Planet will be hard pressed to support many more than it did before these new sources of energy distorted the natural order of things.

This book, in a humble way, has tried to draw attention to these issues. It does not pretend to have offered an accurate analysis, recognising the appallingly unreliable nature of the data, but that in itself may be useful in posing a challenge for others to come forth with better detailed information and evidence with which to correct the mistakes. It has tried to look ahead, but recognises that forecasting the future accurately has rarely been successfully accomplished. Nevertheless, as events unfold in whatever manner they do, it may be useful to compare them with the scenarios outlined herein. In particular, emphasis needs to be given to the local and national situations, the sum of which delivers the world panoply.

References

(Number [#] refers to author's archive; [B#] is a book reference)

A

AAPG Explorer, 1993, *Future energy needs not being addressed*; AAPG Explorer, June 1993 [#156]

AAPG, 2001, Discussions and reply re *"Energy resources – cornucopia or empty barrel"*; AAPG Bulletin, Vo.85, No.6, June 2001 [#1333]

AAPG, 2001, *USGS Assessment study methods endorsed*; AAPG Explorer, Nov. 2001 [#1711]

Abernathy, V.D., 2001, *Carrying capacity: the tradition and policy implications of limits*; Science & Environmental Politics, http://www.esep.de/articles/esep/2001/article1.pdf

Abraham, K.S., 1996, *Sifting the stew of Iraqi sales, US politics and future supplies*; World Oil, Mar. 1996 [#427]

Abraham, K.S., 1997, *Venezuela bets on heavy crude for long term*; Oil & Gas Journal, Jan. 1997 [#542]

Abraham, K.S., 1998, *Debate grows over accuracy of industry statistics;* World Oil, May 1998 [#983]

Abraham, K.S., 1998, *Weak demand, future reserve concerns color global picture*; World Oil, Dec. 1998 [#931]

Abraham, S., 2001, *We can solve our energy problems*; The Wall Street Journal, May 18, 2001 [#1591]

Aburish, S.K., 1994, *The rise, corruption and coming fall of the House of Saud*; 226p, Bloomsbury, London [#B15] (ISBN 0-7475-1468-2)

Adam, D., 2004, *Oil chief: My fears for planet – Shell Boss's "confession" shocks industry;* The Guardian UK, June 17, 2004 [#2368]

Adams, T., 1991, *Middle East reserves*; Oilfield Review [#27]

Adelman, M.A. & Lynch, M.C., 1997, *Fixed view of resource limits creates undue pessimism*; Oil & Gas Journal, Apr. 7, 1997 [#615-2]

Adelman, M.A. & Lynch, M.C., 1997, *More reserves growth*; Oil & Gas Journal, June 9, 1997 [629]

Adelman, M.A., 199?, *Modelling world oil supply*; The Energy Journal, 14/1 [#327]

Adelman, M.A., 2001, *Oil-Use predictions wrong since 1974*; The Wall Street Journal, Mar. 6, 2001 [#1772]

Ahlbrandt, T.S., Charpentier, R.R., Klett, T.R., Schmoker, J.W., Schenk, C.J. & Ulmishek, G.F., 2000, *Future oil and gas resources of the world*; Geotimes, June 2000 [#1503]

Ahlbrandt, T., 2000, *USGS world petroleum assessment 2000*; USGS [#1577]

Ahmed, N.M., 2002, *The war on freedom*; Media Messenger, pp 398

Alahakkone, R.R., 1990, *Conventional supply of oil will slow down*; Oil & Gas Journal, Feb. 5, 1990 [#81]

Alberta Energy & Utilities Board., 1998, *Non-conventional oil production for Alberta* [#708]

Aleklett, K. & Campbell, C.J., 2003, *The peak and decline of world oil and gas production;* Minerals & Energy Vo.18 No. 1

Aleklett K & C.J.Campbell, 2006, *The influence of markets and technology on regional oil and gas reserves*; in *The Gulf Oil & Gas Sector*; Emirates Centre for Strategic Studies 221-248

Aleklett K., 2010, *The Peak of the Oil Age – Analyzing the world oil production reference scenario in World Energy Outlook, 2008*. Energy Policy

Aleklett K., (and others), 2010, *The Peak of the Oil Age*; Energy Policy 38/3.1398-1414

Aleklett K, 2012, *Peaking at Peak*. ISBN 978-1-4614-3423-8

Alezard N., J.H.Laherrère & A. Perrodon, 1992, *Réserves et resources de pétrole et de gaz des pays Mediterranées*; Revue de l'Energie 441, Sept. 1992 [#140]

Al-Fathi S.A., 1994, *Opec oil supply outlook to the year 2010*; Wld Petrol. Congr., Stavanger [#236]

Alhajji, A.F., 2001, *Middle East: Investment levels rise higher*; World Oil, August, 2001 [#1488]

Al-Husseini S., 2004, *Why higher oil prices are inevitable this year, rest of decade*; Oil & Gas Journal, Aug 2

Ali.M, 1999, *Oil Crisis*; Al Arab, 19 November [1020]

Al-Jarri A.S. & R.A. Startzman, 1997, *Worldwide supply and demand of petroleum liquids*; SPE 38782 [#699]

Alternative Energy Inst., 2001, *Turning the Corner: energy solutions for the 21st Century*, ISBN-0-9673118-2-9

Amiel, B., 2002, *Bush's victory is the voice of an angry America;* The Daily Telegraph, Nov.11, 2002 [#2018]

Amuzwgar J., 1999, *Managing the oil wealth*; Tauris 266pp (ISBN 1-86044 292-6)

Anderson Forest, S., 2000, *Industrial Management: Energy – "There Is Not Enough Gas Around"*; Business Week, September 18, 2000 [#1388]

Anderson Forest, S., 2000, *Unnatural Demand for Natural Gas*; Business Week, April 3, 2000 [#1644]

Anderson R.B., 1999, *Gas tanks on full; supplied heading for empty*; The Oregonian April 10 [#1046]

Anderson, B., 2001, *The creation of a Palestinian state is essential for the war against terrorism*; The Spectator, October 20, 2001 [#1671]

ARAMCO, 1980, *Aramco and its world: Arabia and the Middle East*; (Ed. Nawwab I.I., P.C.Speers, & P.F.Hoye) Aramco. ISBN 0-9601164-2-7.

Arlington Institute, The, 2003, *A strategy: moving America away from oil*; The Arlington Institute, Aug., 2003 [#2307]

Arnold R., G.A.Macready & T.W.Barrington, 1960, *The first big oil hunt: Venezuela 1911-1916*; Vantage Press, New York 353p [#B3]

Arthur C., 2003, *Fossil Fuels Exhausted-Study Predicts Depletion of Oil and Gas;* Triangle Free Press, Dec.30, 2003 [#2310]

Arthur, C., 2003, *Oil and gas running out much faster than expected, says study;* The Independent, Oct.2, 2003 [#1962]

Attanasi, E.D., Mast, R.F., Root, D.H., USGS, 1999, *Oil, gas field growth projections: wishful thinking or reality?*; Oil & Gas Journal, April 5, 1999 [#1596]

Attarian J., 2000, *The Age of Impiety;* Culture Wars, Feb. 2002 [#2040]

Attarian J., 2001, *The Unsustainability of Economism: Economism is not feasible economically;* The Social Contract, Summer 2001[#2028]

Attarian J., 2002, *Economism vs Earth;* The Social Contract, Fall 2002 [#2038]

Attarian J., 2002, *Malthus Revisited;* The World & I p.257-271 [#2039]

Attarian J., 2002, *The Coming End of Cheap Oil;* The Social Contract Vol.XII, No.4, 2002 [#2026]

Azar C. & Lindgren K. & Andersson B.A., 2003, *Global energy scenarios meeting stringent CO2 constraints-cost effective fuel choices in the transportation sector;* Energy Policy, Elsevier, 2003 [#2311]

B

Baer R., 2003, *The Fall of the House of Saud;* The Atlantic Monthly; May 2003 [#2146]

Bahree B. & Herrick T., 2003, *Officials say OPEC can't keep a lid on rising crude-oil prices;* Wall Street Journal, Feb. 28, 2003 [#2112]

Bahree B., 1999, *Global demand for oil is set to grow in 2000 at faster pace*; Wall St. Journ Aug 11 [#1004]

Bahree, B. & Fialka, J., 2002, *Oil industry ponders Iraq risks;* Wall Street Journal Aug. 30 2002 [#1913]

Bahree, B., 2004, *Soaring global demand for oil strains production capacity*; The Wall Street Journal, Mar. 22, 2004 [#2296]

Bahree, B., 2004, *World oil supply faces stress in months ahead*; The Wall Street Journal, July 14, 2004 [#2396]

Bainerman, J., 2004, *Is the world running out of oil?;* The Middle East, Issue No. 344, Apr., 2004 [#2294]

Bakhtiari A.M.S., 1999, *The price of crude oil;* OPEC Review 13/1 March

Bakhtiari A.M.S., 2000, 1999, *IEA, OPEC oil supply forecasts challenged: Oil & Gas Journ, April 30*

Bakhtiari, A.M.S., 1999, *The price of crude oil*; OPEC Review, Vol. 23, No. 1, March 1999 [#1296]

Bakhtiari, A.M.S., 2001, *2002 to see birth of New World Energy Order*; Oil & Gas Journ, Jan 7, 2002 [#1763]

Bakhtiari, A.M.S., Shahbudaghlou, F., 2001, *IEA, OPEC oil supply forecasts challenged*; Oil & Gas Journal, Apr. 30, 2001 [#1295]

Bakhtiari, A.M.S., 2003, *Middle East oil production to peak within next decade;* Oil & Gas Journal, July 7, 2003 [#2220]

Bakhtiari, A.M.S., 2003, *Middle East production to peak with next decade*; Oil & Gas Journ. July 14

Bakhtiari, A.M.S., 2003, *North Sea oil reserves; half full or half empty?;* Oil & Gas Journal, Aug.25, 2003 [#1943]

Bakhtiari,A.M. S., 2004, *World oil production capacity model suggests output peak by 2006-07*; Oil & Gas Journal, Apr. 26, 2004 [#2315]

Baldauf S., 1998 *World's oil supply may soon run low*: Christian Science Monitor 23 Sept [#840]

Ball J. 2004, *As prices soar, doomsayers provoke debate on oil's future;* Wall St. Journ. 21 Sept.

Banerjee, N., 2001, *Fears, Again, of Oil Supplies at Risk*; The New York Times, October 14, 2001 [#1723]

Banerjee, N., 2001, *The High, Hidden Cost of Saudi Arabian Oil*; The New York Times, Oct 21, 2001 [#1725]

Banks H., 1998, *Cheap oil: enjoy it while it lasts*; Forbes June 15 [#748]

Bardi U. and G. Pancani, *Storia petrolifera del bel paese*, La Balze (ISBN 88 7539 126 2)

Bardi U., 2003, *La Fine del Petroleo*, Riuniti, (ISBN 88-359-5425-8)

Bardi U., 2011, *The Limits to Growth Revisited*. (ISBN 978-1-4419-9415-8)

Barker, R., 2004, *Keeping hot air out of energy reserves*; Business Week, May 10, 2004 [#2316]

Barkeshli F., 1996, *Oil prospects in the Middle East and the future of the oil market*; Oxford Energy Forum, 26 August 1996 10-11 [#483-1, #493]

Barrlett D.L. & Steele J.B., 2003, *Why America is running out of Gas;* Time Magazine Jul.13, 2003 [#2204]

Barron's, 1994, *What Next for oil prices*; Barron's Feb 28 1994 [#209]

Barry R.A., 1993, *The management of international oil operations*. PennWell Books. (ISBN 0-87814-400-5)

Bartlett A.A., 1978, *Forgotten fundamentals of the energy crisis*; Am. J. Phys. 46/9 Sept [#681]

Bartlett A.A, 1998, *Reflections in 1998 on the twentieth anniversary of the publication of the paper: "Forgotten fundamentals of the energy crisis"*, Negative Population Growth [#1240]

Bartlett A.A., 1990, *A world full of oil*; The Physics Teacher Nov. [#696]

Bartlett A.A., 1994, *Reflections on sustainability, population growth and the environment*; Population & Environment 16/1 [#682]

Bartlett A.A., 1997, *Is there a population problem?* Wild Earth Fall 1997 [#714]

Bartlett A.A., 1998, *Reflections on sustainability, population growth and the environment – revisited*; Renewable resources Journal Winter 1997/8 [# 697]

Bauquis P.R., 2003, *A Reappraisal of energy supply-demand in 2050 shows big role for fossil fuels, nuclear but not for nonnuclear renewables;* Oil & Gas Journal Feb.17, 2003 [#2120]

Bauquis, P.-R., 2001, *A Reappraisal of Energy Supply and Demand in 2050*; Oil & Gas Science and Technology, Vol.56, No.4, pp 389-402, 2001 [#1754 & 1843]

BBC News, 2004, *Profile: Hugo Chavez;* BBC News, April 14, 2002 [#2023]

BBC, 2000, *The last oil shock:* The Money Programme Nov.8, 2000

Beaumont P. and J. Hooper, 1998, *Energy apocalypse looms as the world runs out of oil*; Observer Newspaper 26.7.98 [#792]

Beaumont P. and J. Hooper, 1998, *Running on empty;* Bremerton Sun, Oct.23 [#985]

Becker R., 2002, *The battle for Iraqi oil*; Nexus, Dec 2002

Beeby-Thompson A., 1961, *Oil pioneer*; Sidgwick & Jackson, London

Begin J-F, 2003, *La fin du Petrole*; l'actualite 1April [#2067]

Bell S., 1994, *Distribution of resources knowledge helps locate undiscovered reserves*; Petrol. Eng/Int. [#327]

Bentley R.W et al., 2000, *Perspectives on the future of oil*, Energy Exploration & Exploitation 18/2-3

Bentley R.W., 1997, *Oil shock imminent if heavies are slow or expensive to produce*; Energy World **250** 20-22 [# 617-3]

Bentley R.W., 1997, *The future of oil*; Seminar briefing, Reading University [#604 & 1857& 1858]

Bentley R.W., 2003, *What does the Economist know that oil companies don't?;* letters to the editor@economist.com, June 30, 2003 [#2199]

Bentley, R.W., Booth, R.H., Burton, J.D., Coleman, M.L., Sellwood, B.W., & Whitfield, G.R., 2000, *World Oil Supply: Near & Medium Term-The Approach Peak In Conventional Oil Production*; The University of Reading, August 22, 2000 [#1413]

Bentley, R.W., Booth, R.H., Burton, J.D., Coleman, M.L., Sellwood, B.W., & Whitfield, G.R., 2000, *Perspectives on the Future Oil*; Energy Exploration & Exploitation, Vol.18, Nos.2&3, 2000 [#1414]

Bentley, R.W. & Whitfield, G.R., 2001, *Advice on Nuclear Energy Issues Relevant to NERC Science*; The University of Reading, Version 2, August 28, 2001 [#1357]

Bentley, R.W., 2000, *Letter to G R Davis re: Topics related to the Oil Depletion and reporting,* Royal Society of Chemistry; April 19, 2000 [#1639]

Bentley, R.W., 2001, *Letter to The Economist re: "Sunset for the oil business?"* (#1756); The Economist, p 22, November 24, 2001 [#1757]

Bentley, R.W., 2001, *Submission to the Cabinet Office Energy Review*; Draft, The Oil Depletion Analysis Centre, September 6, 2001 [#1653]

Bentley, R.W., 2002, *Global oil & gas depletion: an overview*; Energy Policy 30, p 189-205, 2002 [#1789 & 1838]

Bentley, R.W., 2002 *Oil forecasts, past and present;* Energy Exploration & Exploitation vol. 20, no. 6, pp 481-492, Multi-Science Publishing, 2002

Bentley R.W and G.A. Boyle. 2007, *Global oil production: forecasts and methodologies.* Environment and Planning B: Planning and Design, vol. 34; (on-line version available via the publisher.)

Bentley, R.W., S.A. Mannan, and S.J. Wheeler, 2007, *Assessing the date of the global oil peak: The need to use 2P reserves;* Energy Policy, vol. 35, pp 6364-6382, Elsevier, 2007

Berthelsen, John, 2003, *Asia starts to gasp for energy;* Asia Times Online Ltd., 2003 [#1941]

BGR, 1995, *Reserven, Ressourcen und Verfügbarkeit von Energierohstoffen* 1995; BGR Hannover 498p [#B19]

Bilkadi Z., 1996, *Babylon to Baku*; Stanhope-Seta. ISBN 0952-881608 230p.

Binney, G., 2001, *The Petro-Population Parallel*; The Ecologist, Vol 31, No.6, July/August, 2001 [#1677]

Bjørlykke K., 1995; *From black shale to black gold*; Science Spectra 2, 1995 44-49

Blakey E.S., 1985, *Oil on their shoes*; Amer. Assoc. Petrol. Geol. 192p [#B5]

Blakey E.S.. 1991, *To the waters and the wild*; Amer. Assoc. Petrol. Geol. [#B6]

Blanchard R.D., 2005. *The Future of Global Oil Production.* ISBN 978-0786423576

Bohnet M., 1996, *Promotion of conventional and renewable sources of energy in developing countries*; in Kürsten M. (Ed) *World Energy – Charging Scene*: ISBN 3-510-65170-7

Bookout J.F., 1989, *Two centuries of fossil fuel energy*; Episodes 12/4 [#331]

Borovik S., 2003, *Energy Resources;* Nov. 27, 2003 [#2286]

Bourdaire J.M, 1993, *Le pertrole dans l'economie mondiale*; Energie Universale, Sept. 1993 [#180]

Bourdaire J.M, 1998, *World energy prospects to 2020*; Brit. Inst. Energy Economics, 2 July [#770]

Bourdaire J.M, 2000, *Petrole – Les fondaments de l'economie petroliere*, Encyclopoedia Universalis, France [#1234]

Bourdaire J.M. et al. 1985, *Reserves assessment under uncertainty – a new approach*; Oil & Gas Journ. June 10. 135-140.

Bourdaire, J.M., 2000, *Energy and Economics: "Moving Backwards To The Future"*; July 2000 [#1361]
Bowen J.M., 1991, *25 years of UK North Sea exploration*; in Abbotts J United Kingdom Oil & Gas fields; Geol. Soc Mem14 [#913]
Bradley R.L., 1999, *The growing abundance of fossil fuels*; The Freeman [#1130]
Brauer B, 2001, *A Waning Honeymoon in Kazakhstan;* New York Times [#1195]
Breton T.R. & J.C.Blaney, 1991, *Production rise, consumption fall may turn Soviet oil exports higher*; Oil & Gas Journ. Nov 18 1991 110-114 [#7]
British Petroleum Co, 1959, *Fifty years in pictures*; publ. BP, London 159 p.
British Petroleum Co, 1979, *Oil crisis – again*; BP paper [#590]
British Petroleum Co., *BP Statistical Review of World Energy*; Published annually by BP, London.
Brown D., 1998, *How much oil is really there?*; AAPG Explorer April [#796]
Brown L.R., 2001, *Eco-economy*; W.E.Norton & Co., 333pp (ISBN 0-393-3214193-2)
Brown L.R., 2001, *The state of the world*; W.E.Norton & Co., 268pp (ISBN 0-303-30963-0)
Brown L.R., 2003, *Deflating The Bubble Economy Before It Bursts;* Earth Policy Institute Sept.4, 2003 [#1979]
Brown L.R., 2004, *Outgrowing the Earth*; W.H.Norton (ISBN 0-393-06070-5)
Brown L.R., 2006, *Plan B 2.0* Norton (ISBN 0-393-32831-7)
Brown L.R., 2012, *Full Planet, Empty Plates* (ISBN 978-0-393-34415-8)
Brown, D., 2000, *So, Are We Running Out of Oil?*; AAPG Explorer, p 46, 47, April, 2000 [#1540]
Brown, R.A., 2003, *Critical Paths to the Post-Petroleum Age;* Argonne National Laboratory, 2003 [#1940]
Browne E.J.P., 1991 *The way ahead – hydrocarbons for the 1990s*: Amer. Assoc. Petrol. Geol. London Conference. preprint [#24]
Browne E.J.P., 1991, *Upstream oil in the 1990s: the prospects for a new world order*; Oxford Energy Seminar, Sept. 1991. Publ. The British Petroleum Company, London [#42]
Bruges J., 2007, *The Big Earth Book – Ideas and Solutions for a Planet in Crisis;*(ISBN 978-1-901970-87-6)
Bryant, A., 2000, *The New War Over Oil;* Newsweek, October 2, 2000 [#1404]
Buderi R., 1992, *Oil's downhill skid may be ending*; Business Week Feb. 1992 [#8]
Bundesministerium für Wirtschaft, 1995, *Sources of energy in 1995: reserves, resources and availability*; Report 383 [#483-1]
Burns, S., 2001, *World running out of cheap oil that fuels economies*; Houston Chronicle, July 16, 2001 [#1322]
Burrows, Shauna C., 2002, *Earth Emergency – a call to action;* Positive News, August, 2002 [#1887]
Business Week, 1993, *The scramble for oil's last frontier*; Business Week Jan 11 1993 [#150]
Business Week, 1997, *The new economics of oil* Nov 3 [#939]
Business Week, 1999, *Sheik Yamani on the coming oil price crisis;* April 14 [#1011]
Business Week, 2000, *Is Big Oil Getting Too Big?*; Business Week, October 30, 2000 [#1415]
Byman D., 1996, *Let Iraq collapse!*; The National Interest **45** 48-60.

C

Cambridge Energy Research Associates, 1991, *The capacity race: the long term future of world oil supply*; James Capel Report 1991, [#46]
Campbell C.J., and E.Ormaasen, 1987, *The discovery of oil and gas in Norway: an historical synopsis*; in Spencer A.M. (Ed) Geology of the Norwegian oil and gas fields, Norwegian Petrol. Soc.493p
Campbell C.J., 1988, *Oil: a case of short-sighted vision*; Energy Day, Dec 17 1998 [#885]
Campbell C.J., 1989, *Oil price leap in the early nineties*; Noroil 17/12 35-38
Campbell C.J., 1991, *The Golden Century of Oil 1950-2050*: *the depletion of a resource*; Kluwer Academic Publishers, Dordrecht, Netherlands; 345p.
Campbell C.J., 1992, *The depletion of oil*; Marine & Petrol. Geol., v.9 Dec. 1992, 666-671
Campbell C.J., 1993, *The Depletion of the world's oil*; Petrole et technique. No.383. 5-12 Paris
Campbell C.J., 1994, *An Oil Depletion Model: a resource constrained yardstick for production forecasting*; Rept., Petroconsultants S.A., Geneva.
Campbell C.J., 1994, *Scrambling for oil;* Time Magazine July 11 [#265]
Campbell C.J., 1994, *Scraping the barrel*; The Economist Aug. 6th 1994 [#259]
Campbell C.J., 1994, *The imminent end of cheap oil-based energy*; SunWorld 18/4 17-19
Campbell C.J., & J.H.Laherrère, 1995, *Gauging the North Sea*; Platt's Petroleum Insight April 10 1995.
Campbell C.J., & J.H.Laherrère, 1995, *The world's supply of oil 1930-2050*; Report Petroconsultants S.A., Geneva Mem 2/1-6
Campbell C.J., 1995, *Cassandra or prophet*; Petroleum Economist Oct. 1995 [#362]
Campbell C.J., 1995, *Proving the unprovable*; Petroleum Economist, May 1995
Campbell C.J., 1995, *The next oil price shock: the world's remaining oil and its depletion*; Energy Expl. & Exploit., 13/1 19-46
Campbell C.J., 1996, *Oil Shock*; Energy World, June 1996
Campbell C.J., 1996, *The resource constraints to oil production: the spectre of a pending chronic supply shortfall*; in Kürsten M. (Ed) *World energy – a changing scene*; Proc.7th Int. Symposium BGR, Hanover, E.Schweizerbart'sche Verlagsbuchhandlung, Stuttgart 227p.

Campbell C.J., 1996, *The status of world oil depletion at the end of 1995;* Energy Exploration and Exploitation; March 1996

Campbell C.J., 1997, *As depletion increases, energy demand rises;* Petroleum Economist Sept 113-116

Campbell C.J., 1997, *Better understanding urged for rapidly depleting reserves*: Oil & Gas Journ. April 7th 1997. [#615-2]

Campbell C.J., 1997, *How to calculate depletion of hydrocarbon reserves;* Petroleum Economist Sept. 111-112

Campbell C.J., 1997, *The Coming Oil Crisis;* Multi-Science Publishing Co. & Petroconsultants 210p.

Campbell C.J., 1998, *L'avenir de l'homme de l'age des hydrocarbures;* Geopolitique 63, Oct

Campbell C.J., 1998, *Reserves controversy;* Oil & Gas Journal April 20

Campbell C.J., 1998, *Running out of gas: this time the wolf is coming;* The National Interest, spring [#661]

Campbell C.J., 1998, *The enigma of oil prices in times of pending oil shortage;* World Oil Prices: oil supply/demand dynamics to the year 2020; Centre for Global Energy Studies, 208-225

Campbell C.J., 1998, *The future of oil;* Energy Exploration & Exploitation 16/2-3

Campbell C.J., and J.H.Laherrère, 1998, *The end of cheap oil;* Scientific American March 80-86

Campbell C.J., 1999, *A new mission for geologists as oil production peaks;* Swiss Association of Petroleum Geologists and Engineers, 4/1 19-34

Campbell C.J., 1999, *Oil madness;* Geopolitics of Energy; January

Campbell C.J., 1999, *The dating of reserve revisions;* Petroleum Review. July 44-45

Campbell C.J., 1999, *The extinction of Hydrocarbon Man?* Times Higher Educat. Supplement, Nov 12.

Campbell C.J., 1999, *The imminent peak of oil production;* Presentation at House of Commons, oilcrisis.com [#1101]

Campbell C.J., 2000 *Depletion and denial: the final years of oil supplies* USA-Today Nov

Campbell, C.J., 2000, *Myth of spare capacity setting the stage for another oil shock;* Oil & Gas Journal, p 20, 21, March 20, 2000 [#1544]

Campbell C.J., 2000, *The imminent peak in world oil production;* Middle East Economic Databook 101-

Campbell C.J., 2000, *Deep water heroism keeps oil imports at bay;* Tomorrow's Oil 2/8

Campbell C.J., 2000, *Depletion: the Democratic and Popular Republic of Algeria;* Tomorrow's Oil 2/7

Campbell C.J., 2000, *Italy* Tomorrow's Oil 2/3 April

Campbell C.J., 2000, *The Myths of Oil – A Letter to the Editor;* Petroleum Review, October 2000 [#1408]

Campbell C.J., 2000, *Myth of spare capacity setting the stage for another oil shock;* Oil & Gas Journ. Mar 20 [#1144]

Campbell C.J., 2000, *Oil depletion in the US Lower 48*: Tomorrow's Oil 2/6

Campbell C.J., 2000, *Opec's spare capacity and strategy;* Tomorrow's Oil 2/9.

Campbell C.J., 2000, *The myths of oil;* Petroleum Review, Oct.

Campbell C.J., 2000, *What do the USGS numbers really mean?* Tomorrow's Oil 2/7

Campbell C.J., 2000, *World's oil endowment and its depletion* Tomorrow's Oil 2/2 February

Campbell C.J., 2001 *Oil, Gas and Make-believe;* Energy Exploration & Exploitation 19/2-3 117-133

Campbell C.J., 2001 *The imminent oil crisis – the missing element in the environment debate;* Tomorrow's Oil 3/2 Feb

Campbell C.J., 2001 *The Oil Peak Oil: turning point;* Solar Today, 15/4 July-Aug

Campbell C.J., 2002, *Petroleum and people;* Population and Environment 24/3 November

Campbell C.J., 2002, *The imminent oil crisis;* in Grob. G, ed.. Blueprint for the clean, sustainable energy age; Verlag eco-performance, Geneva

Campbell C.J., F.Lisenborghs, J.Schindler and W.Zittel, 2002, *Oelwechsel;* Deustcher Taschenbuch (ISBN 3-423-24321-X)

Campbell, C.J., 2002, *Bell signals death knell for oil: Times* Higher Educational Supplement [#1865]

Campbell C.J., 2003 *Industry asked to watch for regular oil production peaks, depletion signals;* Oil & Gas Journ. July 14 38-45

Campbell C.J., 2003, *The ageing of oil;* Petroleum Review, November p 44-45

Campbell C.J., 2003, *The Essence of Oil & Gas Depletion;* Multi-Science, 342p (ISBN 0-906522-19-6)

Campbell C.J., 2003, *Iraq and oil – Scenarios for the future;* Energy World 203 March [#2117]

Campbell C.J., 2004, *Le declin des reserves est depuis longtemps une evidence aux yeux des specialiostes;* Le Temps 29 April 2004

Campbell C.J., 2004, *The Urgent Need for an Oil Depletion Protocol;* in Centro Pio Manzu, Trans. Conf. The Economics of the Noble Path, Rimini, Oct. 2003

Campbell C.J., 2005, *Oil Crisis;* Multi-Science Publishing, (ISBN 0906522-39-0)

Campbell C.J., 2005 in M. Dassu M & L. Annunciata (Eds) Aspenia, *Oil, down to a dribble*.

Campbell C.J., 2006, *El final de la primera parte de la Era del Petroleo;* Vanguardia Enero-Marzo

Campbell C.J., 2006, *La prima meta dell'era del petroleo e' finita;* Prometeo 24/93 30-37

Campbell C.J., 2006, *The Dawn of the Second Half of the Age of Oil;* Soundings #34 Autumn 2006

Campbell C.J., 2007, *Oil is running out and Ireland needs to follow Sweden's example and come up with a plan to wean itself off it;* Irish Examiner 20 June 2007

Campbell C.J., 2007, *Living with less oil and gas;* Sustainability v.2 p.28-30

Campbell C.J. and G.Strouts 2007, *Living through the Energy Crisis* (ISBN 0-9547855-1-7)

Campbell C.J., 2008, *A long term view of the Oil Age;* Ariskos 5/6, Inst Galego de Estudos de Seguranca Internacional e da Paz.

Campbell C.J., 2008, *The Oil Depletion Protocol: A Response to Peak Oil*; Globalizations 5/1
Campbell C.J. & S.Heapes, 2009, *An Atlas of Oil & Gas Depletion*. ISBN 978-1-906600-42-6
Campbell C.J., 2010, *The Second Half of the Oil Age dawns*. Swiss Derivatives Review. 43
Campbell C.J., 2011, (Editor) *Peak Oil Personalities*. ISBN 978-1-908378-06-4
Campbell C.J., 2012, *The Anomalous Age of Easy Energy* in Inderwildi O. & King D, (Eds) Energy, Transport & the Environment; Springer ISBN 978-1-4471-2717-8
Campbell C.J., 2012, *Recognition of Peak Oil*: WIRE's Energy Environ 2012 1:114–117
Carmalt S.W. and B.St.John, 1986, *Giant oil and gas fields;* in Future petroleum provinces of the world; Amer. Assoc. Petrol. Geol. Mem.40 [#43]
Cattaneo C., 2003, *Natural gas in dangerous decline, says analyst;* National Post, June 11, 2003 [#2179] [#2213]
Catton W.R., 1982, *Overshoot*, (ISBN 0-252-00988-6)
Cavallo A., 2004, *Spare Capacity(2003) and Peak Production in World Oil;* Natural Resources Research, Oct. 2003 [#2314]
Cavallo A., 2004, *The illusion of plenty;* Bulletin of the Atomic Scientists, Jan./Feb., 2004 [#2313] [#2322]
Cavallo A.J., 2002, *Predicting the peak in World oil production:* Natural Resources Research 11/3, Sept, 2002 [#1998] [#2009] [#2063] [#2151]
Cavallo, A.J., 1995, *High Capacity Factor Wind Energy Systems;* Journal of Solar Energy Engineering Vol.117, May, 1995 [#2065]
Cavallo, Alfred, 2003, *Spare Capacity and Peak Production in World Oil,* Natural Resources Research, Oct., 2003[#1932]
Centre for Global Energy Studies, 1990, *The Gulf crisis* 228pp
Centre for Global Energy Studies, 1992, *The costs of future North Sea oil production*; CGES Rept. v3[#205]
Centre for Global Energy Studies, 2001, *Non-OPEC Production: The Going Is Getting Tougher*; Global Oil Report, Vol.12, Issue 6, November/December, 2001 [#1755]
Centre for Global Energy Studies, 2003, *Global Oil Report-Market Watch-Oil Reserves gained and lost in 2002;* Feb.27, 2003 [#2298]
Charpentier R.R., 2002, *Locating the summit of the oil peak*; Science 295 [#1841]
Charrier B., 2000, *Energy – what future?* Clean Energy 2000 [#1109]
Cheney R., 1999, *Speech to Institute of Petroleum*, London; http://web.archive.org/web/20010810115257; http://www.petrroleum.co.uk/speeches.htm
Chomat P, 2004, *Oil Addiction – The World in Peril;* Universal (ISBN 1-58112-494-5)
Clarke D., 2007, *The battle for barrels: Peak Oil myths and world oil futures*. ISBN 10-1-84668-012-3
Clark P, P. Coene and D. Logan, 1981, *A comparison of ten U.S. oil and gas supply models*; Federal Reserves Institute, Washington [#103]
Clark W.R., 2005, *Petrodollar Warfare*, New Society (ISBN 0 86571-514-9)
Cleveland C.J. & R.K.Kaufmann, *Forecasting ultimate oil recovery and its rate of production: incorporating economic forces into the models of M. King Hubbert*; The Energy Journal 12/2 [#26]
Cleveland C.J., 1992, *Yield per effort for additions to crude oil reserves in the Lower 48 United States 1946-1989;* Amer. Assoc. Petrol. Geol. 76/7 948-958
Clover, C., 2001, *The Balance Of Power*; The Daily Telegraph, February 10, 2001 [#1624]
Cochet Y, 2003, *Petrole apocalypse*; Fayard 275p (9-782213 622040)
Cocks, D., 2001, *There will be life – but not as we know it*; The Australia Financial Review, Friday, July 27, 2001 [#1703]
Coghlan, A., 2003, *'Too little' oil for global warming;* Newscintist.com, Oct.1, 2003 [#1969]
Cohn, L. & Crock, S., 2001, *What To Do About Oil*; Business Week, October 29, 2001 [#1693]
Coleman J.L., 1995, *The American whale oil industry: a look back to the future of the American petroleum industry;* Non-renewable Resources 4/3 273-288 [#603]
Conant M.A., 1999, *The Universe of Oil* – ISBN 1-896091-41-5
Cook R.C., 2004. *Oil, jihad and destiny*; Opportunity (ISBN 1-93847-62-9)
Cook W.J., 1993, *Why Opec doesn't matter anymore*; US News & World Report 13/12/93 [#299]
Cook, L.J., 2001, *Safe- For Now*; Forbes, November 26, 2001 [#1739]
Cooper, C. & Herrick, T., 2001, *Oil Giants Struggle to Spend Profits amid Shortage of Exploration Sites*; The Wall Street Journal Online, July 30, 2001 [#1313]
Cooper, C. & Herrick, T., 2001, *Pumping Money – Major Oil Companies Struggle to Spend Huge Hoards of Cash*; The Wall Street Journal, July 30, 2001 [#1678]
Cope G., 1998, *Have all the elephants been found?* Pet. Review 52/614 24-26
Cope G., 1998, *Will improved oil recovery avert an oil crisis*; Pet. Review June 52/616 22-23
Cornelius C.D., 1987, *Classification of natural bitumen: a physical and chemical approach*; in Meyer R.F. (Ed) *Exploration for heavy crude oil and bitumen*; AAPG Studies in geology #25
Corzine R., 1993, *Warning over Saudi output*; Financial Times, 4/11/93 [#282]
Coy, P., 2001, *The Energy Forecast*; Business Week, August 27, 2001 [#1656]
Crandell J.D. et al, 1994, *Depends on oil prices*; World Oil, Feb 1994[#200]
Crawford J.H., 2002. *Carefree cities*. ISBN 90-7527-042-0
Creswell J., 1996, *Global demand fuels need for more drilling*; Press & Journal Aberdeen [#404]
Creswell J., 1999, *How big oil is slashing*; Press & Journal Aberdeen [#920]

Cummins C. & Schroeder M., 2004, *Shell's Watts Draws Fire;* Wall St. Journ., Jan.14, 2004 [#2318]
Cummins C. & Warren S. & Schroeder M., 2004, *Shell Cuts Reserve Estimate 20% As SEC Scrutinizes Oil Industry;* Wall St. Journ., Jan.11, 2004 [#2316]

D

Daly, M.C., Bell, M.S. & Smith, P.J., 1996, *The remaining resource of the UK North Sea and its future development*; in Glennie K & A Hurst NW Europe Hydrocarbon Industry; Geol. Soc. [#912]
Darley, J., 2004, *High Noon for Natural Gas* – ISBN 1-931498-53-9
Davidson, J.D. & Rees-Mogg, W., 1988, *Blood in the streets – investment profits in a world gone mad;* Sidgwick & Jackson, London 385p
Davidson, J.D. & Rees-Mogg, W., 1994, *The great reckoning;* Pan Books 602pp
Davidson, K., 1998, *Some experts say oil demand will soon exceed supply again;* Sta Barbara News, Sept. 2, 1998 [#839]
Debraine, L., 2004, *Le "peak oil" paraît désormais tout proche;* Le Temps, Apr. 29, 2004 [#2337]
Deffeyes, K.S., 2001, *Hubbert's peak – the impending world oil shortage;* Princeton University Press 208pp (ISBN 0-691-09086-6)
Deffeyes, K.S., 2005, *Beyond Oil – The View from Hubbert's Peak;* Hill & Wang, N. York 202p.
Deffeyes, K.S., 2010, *When Oil Peaked.* ISBN 978-0-8090-9471-4
Deitzman, W.D. *et al.*, 1983, *The petroleum resources of the Middle East;* U.S. Dept. of Energy, Energy Information Administration Report. DOE/EIA-0395 May 19 83 169p.
Dekanik I., and others, 2007, *A Century of Oil (ISBN 9531 82066-X)*
Dell, P.R., 1994, *Global energy market: future supply potentials;* Energy exploration and exploitation, 12/1 59-72 [#377]
Demaison, G. & Huizinga, B.J., 1991, *Genetic classification of petroleum systems;* Amer. Assoc. Petrol. Geol., 75, 1626-43
Department of Energy, 1995, *US crude oil, natural gas and natural gas liquids reserves;* 1994 Annual Report [#429]
DeSorcy, G.J. *et al.*, 1993, *Definitions and guidelines for classification of oil and gas reserves;* Journal Canadian Petrol. Technology, 32/5, 0-21 [#338]
Dittmar M, 2009, *The Future of Nuclear Energy: Facts and Fiction.* www.theoildrum.com Aug 5-Nov 10. Also a 2011 update
Djurasek, S., 1998, *The coming oil crisis;* NAFTA 49/6 167-168 [#884]
Doherty, J., 2004, *Half empty?;* BARRONS, Mar. 15, 2004 [#2285]
Douthwaite, R. (Ed), 2003, *Before the wells run dry;* Green Books (ISBN 1 84351 037 5)
Dowthwaite, R., 1999, *The ecology of money;* Green books 78pp (ISBN 1-870098-81-1)
Drew, L.J., 1997; *Undiscovered Petroleum and Mineral Resources;* Plenum Press [#2329]
Drucker P.F., *Post-capitalist Society;* Harper Collins (ISBN 0-88730-664-6)
Duncan, R.C., 1996, *The Olduvai Theory: sliding towards the post-industrial stone age;* Inst. Energy & Man, Seattle [#491]
Duncan, R.C & Youngquist, W., 1998, *Is oil running out?;* Science 282 [#882]
Duncan, R. C. & Youngquist, W., 1999, *Encircling the Peak of World Oil Production;* Natural Resources Research, Vo. 8, No. 3, p 219-232, 1999 [#1462]
Duncan, R.C., 1999, *Oil production per capita;* Oil & Gas Journal 92/20, May 17, 1999 [#1009]
Duncan, R.C., 1999, *World energy production: kinds? amounts? per capita? who;* Inst Energy & Man [#969]
Duncan, R.C., 2003, *Three world oil forecasts predict peak oil production;* Oil & Gas Journal, May 26, 2003 [#2251]

E

Easterbrook, G., 1998, *The oil crisis really;* Los Angeles Times, June 7, 1998 [#764]
The Economist, 1993, *A shocking speculation about the price of oil;* The Economist, 87-88, Sept. 15, 1993 [#189]
The Economist, 1995, *The future of energy;* Oct. 7, 1995 [#379]
The Economist, 1996, *From major to minor;* May 18, 1996 [#437]
The Economist, 1996, *Pipe dreams in central Asia;* May 4, 1996 [#472]
The Economist, 1996, *The oil buccaneer;* Nov., 1996 [#521]
The Economist, 2000, *In praise of Big Oil;* The Economist, Oct. 21, 2000 [#1425]
The Economist, 2001, *Oil Depletion – Sunset for the oil business?;* The Economist, p 97/98, Nov. 3, 2001 [#1756 & 1837]
The Economist, 2003, *The End of the Oil Age;* Oct. 25, 2003 [#2303]
The Economist, 2003, *There's oil in them tar sands;* June 26, 2003 [#2187]
Edwards, J.D., 1997, *Crude oil and alternate energy production forecasts for the twenty-first century: the end of the hydrocarbon era;* Amer. Assoc. Petrol. Geol., **81**/8, 1292-1305 [#631]
Edwards, R., 1998, *High level risk;* New Scientist [#768]

Edwards, R.H., 1999, *The imminent peak in world oil production;* Gulf Business Economic Databook [#1148]

Edwards, W.R., 1997, *Energy supply;* Oil & Gas Journal, July 9 [#629]

Ehrenfeld, T., 2002, *Iraq: It's the oil, stupid – what if this is the beginning of an oil war?;* Newsweek, Sept. 30 [#1992]

Ellis-Jones, P., 1988, *Oil: a practical guide to the economics of world petroleum;* Woodhead-Faulkner, 347pp (ISBN 0-85941-398-5)

Emerson, T., 2002, *The thirst for oil;* Newsweek, Apr. 8 [#1831]

Energy Exploration & Exploitation, 1998, *The world's non-conventional oil and gas resources – review* [#793]

European Commission, 1995, *For a European Union Energy Policy;* Green Paper [#605-1]

European Commission, 2000, *The European Union's oil supply;* Report, Oct. 4

Exxon Mobil, 1999, *Tomorrow's energy needs;* Wall Street Journal [#1196]

F

Farago, L., 1973, *The Game of Foxes;* Bantam Books, New York 878p

Farrington, S., 2004, *The skeptic: politicians take note: OPEC can't cool oil;* Dow Jones Newswires, May 21, 2004 [#2352]

Ferguson A.R.B, 2000, *The empirical pricing of oil;* OPT 10 Apr [#1160]

Ferguson A.R.B, 2000, *The right price of oil;* OPT 27 Feb [#1091]

Ferguson A.R.B, 2000 *The social and ecological consequences of globalisation;* Optimum Population Trust [#1207]

Ferguson A.R.B, 2001, *Planning for the demise of cheap energy;* Optimum Population Trust [#1209]

Ferguson, A.R.B., 1999, *Ice Age, Glacial and Interglacial;* Optimum Population Trust (U.K.), 9 November, 1999 [#1471]

Ferguson, A.R.B., 1999, *Judging the Oil Experts;* Optimum Population Trust (U.K.), 1 December, 1999 [#1468]

Ferguson N., 2009, *The Ascent of Money: A Financial History of the World,* ISBN 978-0-141-03548-2

Ferrier R.W., 1982, *The history of the British Petroleum Company: Volume 1, the developing years 1901-1932;* Cambridge University Press 801p [#B7] (ISBN 0 521 426447 4)

Financial Times Energy, 2001, *Expecting the improbable: Middle East Capacity and Global Oil Demand;* Energy Economist Briefing, March, 2001 [#1610]

Flavin C., 2000, *Energy for a new century;* Worldwatch Mar/Apr [#1129]

Fleay B., & J.H.Laherrère, 1997, *Sustainable energy policy for Australia; submission to the Department of Primary Industry and Energy Green Paper 1996;* Paper 1/97 Institute for Science and Technology Policy, Murdoch University, W.Australia [#606-1]

Fleay B., 1995, *The decline in the age of oil;* Pluto Press 152p. (ISBN 1 86403 021 6)

Fleming D., 1999, *Need to prepare for rising oil prices*: Times Aug.22 Letter [#996]

Fleming D., 1999, *The spectre of OPEC;* Sunday Telegraph March 21 [#953]

Fleming D., 2000, *After oil;* Prospect Nov

Fleming D., 1999, *The Next Oil Shock?;* Prospect, p 12-15, April, 1999 [#1549]

Fleming D., 2003, *The Wages of Denial;* The Ecologist April 2003 [#2144]

Fleming D., 2005, *Energy and the common purpose;* ISBN 0-9550849-1-1

Fleming D., 2011, *Lean logic: a dictionary for the future and how to survive it.* ISBN 978-0-955-0849-6-6

Foley G and A van Buren, 1978, *Nuclear or not,* Heinemann (ISBN 0 435 54770 4)

Fowler, R.M., Burgess, C.J., Otto, S.C., Harris, J.P. & Bastow, M.A., 2000, *World Conventional Hydrocarbon Resources: How Much Remains To Be Discovered, Where Is It?;* Robertson Research International Limited, 2000, [#1384]

Fromkin D., 1989, *A peace to end all peace;* Avon Books 635p. ISBN 0-380-71300-4.

G

Garb F.A., 1985, *Oil and gas reserve classification, estimation and evaluation;* Journ. Petrol. Technol. March 373-390

Gardner T., 2000, *US ups estimate of non-US recoverable oil by 20 pc;* Reuters [#1132]

Gauthier D.L. et al. 1995, *1995 national assessment of United States oil and gas resources – results, methodology and supporting data*: U.S. Geological Survey CD-ROM

Geological Museum. *Britain's offshore oil and gas.* ISBN 0-565-01029-8

Georgescu-Roegen, N., 1995, *La decroissance;* Sang de Terre, Paris 254pp

Geoscience Horizons, 2003, *The Rise and Fall of the Hubbert Curve: Its Origins and Current Perceptions;* Geoscience Horizons, 2003 [#1978]

Geoscientist, 1996, *World oil supplies,* 6/1 [#440]

Gerth J., 2003, *Oil experts see long-term risks to Iraq reserves;* New York Times, Nov.30, 2003 [#1981]

Ghadhban T.A. et al., 1995, *Iraq oil industry: present conditions and future prospects*; Report Iraq Oil Ministry [#344]

Giles, J., 2004, *Every last drop*; Nature, Vo.429, June 17, 2004 [#2403]

Giraud A., 1993, *1973-1993 – L'eclairage du passe, l'approche du futur*; Profils IFP 94/1 [#216]

Gold T., 1988, *Origin of petroleum; two opposing theories and a test in Sweden*; Geojournal Library **9** 85-92.

Goodman M. & N.C.Chriss, 1979, *Mexico oil estimates inflated, experts say*; Los Angeles Times |May 18 1979 [#14]

Goodstein D., 2004, *Out of Gas*; ISBN-0-393-05857-3

Goodstein D., 2003, *Energy, Technology and Climate: Running out of gas;* New Dimensions In Bioethics, 2003 [#2267]

Grace J.D., R.H.Caldwell and D.I.Heather, 1993, *Comparative reserve definitions: U.S.A., Europe and the former Soviet Union*; Journ. Petrol. Technol. Sept.1993 866-872 [#174]

Grant L., 2005, *The collapsing bubble;* Seven Locks (ISBN 1-931643-58-)

Gray D., 1989, *North Sea outlook and its sensitivity to price*; Petrol. Revue. Jan 1989 [#35]

Greene, D.L. & Hopson, J.L., 2003, *Running out of and into oil: analyzing global oil depletion and transition through 2050*; U.S. Department of Energy [#2308]

Gretener P., 2010, *The Vanishing of a Species?* ISBN 978-1-897093-82-5

Griffin D.R., 2004, *The New Pearl Harbor;* (ISBN 1-84437-036-4)

Griffin D.R & P.D.Scott, 2007, *9/11 and the American Empire* (978-1-56656-659-9)

Grob G., 1999, *New total approach to energy statistics and forecasting*; CMDC-WSEC [#1043][1819]

Gulbenkian N., 1965, *Portrait in oil*; Simon & Schuster, New York [#B9]

Gunther F., 2000, *Vulnerability in agriculture: energy use, structure and energy futures*: INES Conference, Stockholm

H

Haggett, S., 2003, *Oilpatch: Gigantic natural gas finds now rare;* Calgary Herald, Aug.2, 2003 [#1934]

Haggett, S., 2003, *Oh, How the patch has changed: Technology, finances revamp landscape;* Calgary Herald, Aug.2, 2003 [#1933]

Halbouty M.T., 1970, *Geology of giant petroleum fields*; Amer. Assoc. Petrol. Geol. Mem.14

Haldorsen. H.H., 1996, *Choosing between rocks, hard places and a lot more: the economic interface*; in Doré A.G and R.Sinding Larsen (Eds), *Quantification and prediction of hydrocarbon resources*; NPF Sp. pub. 6., Elsevier ISBN 0-444-82496-0.

Hall C. & Tharakan P. & Hallock J. & Cleveland C. & Jefferson M.; *Hydrocarbons and the evolution of human culture;* Nature, Vol.426, Nov. 20, 2003 [#2275]

Hammer A., 1988, *Hammer, witness to history*; Coronet Books 752pp

Hardman R.F.P., 1998, *The future of Britain's oil and gas industry*; Inst. Mining Eng. March [#711]

Hardman R.F.P., 2001, *New Petroleum Provinces of the 21st Century?*; 20 July, 2001 [#1341]

Hardman R.F.P., 2004, *Lessons from oil and gas exploration in and around Britain*; in: Gluyas, J.G. & Hichens, H.M. (eds), 2003, United Kingdom oil and gas fields, Commemorative Millennium Volume, Geological Society, London, Memoir, 20, 5-16 [#2290]

Harigal G.G., 1998, *Energy in a changing world*; Pugwash Meeting 243 [#946]

Harper, F.G., 1999, *Ultimate Hydrocarbon Resources in the 21st Century*; BP Amoco, AAPG Birmingham 1999 [#1399]

Harvey H., 1993, *Innovative policies to promote renewable energy*; Advances in Solar Energy, Amer. Solar Energy Soc[#220]

Hatfield C.B., 1995, *Will an oil shortage return soon?* Geotimes Nov. 1995 [#406]

Hatfield C.B., 1997, *A permanent decline in oil production*; Nature 388 [#632]

Hatfield C.B., 1997, *Oil back on the global agenda*; Nature 367 121 [#589]

Hatley A.G., 1995, *The oil finders*; Centex 267pp (ISBN 0-9649416-0-0).

Haun J.D. (Ed), 1975, *Methods of estimating the volume of undiscovered oil and gas resources*; Amer. Assoc. Petrol. Geol. Studies in Geology No.1.

Hawken P., 1993, *The Ecology of Commerce – a declaration of sustainability*; Publ. Harper Business, New York 250p.

Hawkes N., 1999, *Huge reserve of gas will fuel 21st Century*, The Times [#897]

Hecht, Jeff, 2002, *You can squeeze oil out of a stone;* New Scientist, August 2002 [#1894]

Heinberg, R., 2003, *The Party's Over*; New Society Publishers 274p (ISBN 0-86571-482-7)

Heinberg, R., 2006, *The Oil Depletion Protocol*; New Society (ISBN 10 0-86571 563-7)

Heinberg R., 2007, *Peak Everything*. ISBN 978-1-905570-13-3

Heinberg R. & D. Lerch (Eds), 2010, *The Post-Carbon Reader*. ISBN 978-0-9709500-6-2

Henderson, S., 2002, *The Saudi Way;* Wall Street Journal, Aug. 12, 2002 [#1885]

Herrera, R., 2004, *Herrera: Impressions from Berlin*; Petroleum News, May 30, 2004 [#2351]

Hersh, S.M., 2001, *King's Ransom – How vulnerable are the Saudi royals?*; The New Yorker, www.newyorker.com, October 24, 2001 [#1682]

Hicks B. & C.Nelder, 2008, *Profit from the Peak*; John Wiley (ISBN 978-0-470-12736-0)

Higgins G.E., 1996, *A history of Trinidad Oil*; Trinidad Express Newspapers, 498pp (ISBN 976 8160 07 1)
Hiller K. 1997, *Future world oil supplies – possibilities and constraints*; Erdol Erdgas Kohle **113**/9 349-352
Hiller K., 1998, *Depletion midpoint and the consequences for oil supplies*; WPC 15. [#780]
Hiller K., 1999, *Verfugbarkeit von Erdol;* Erdol, Erdgas, Kohle Jahrgand *115/2 [#918]*
Hines, C., 2004, *Oil drought could be our saviour*; The Guardian, July 19, 2004 [#2397]
Hiro D., 2007, *Blood of the Earth: The Battle for the World's vanishing Oil Reserves.* ISBN 10-56025-544-7
Hirsch, R.L., 2005, *Supply and demand shaping the peak of world oil production*; World Oil Oct.
Hobbs G.W., 1995, Oil, gas, coal, uranium, tar sand resource trends on rise; Oil & Gas Journal Sept 4 1995 [#374]
Hobson G.D., 1991, *Field size distribution – an exercise in doodling?* Journ. Petrol Geol 14/1 [#312]
Holder K., 1998, *Worldwide oil shortage will soon affect us*; Evansville Courier 31 May [#751]
Holder K., 1999, *Oil depletion problem being ignored*; Evansville Courier & Press, December 16 [1029]
Holley D., 1997, *China's thirst for oil fuels competition*; Los Angeles Times July 29 [#635]
Hollis R., 1996, *Stability in the Middle East – three scenarios*; Pet. Review May p 205 [# 442]
Holmes B. & Jones N., 2003, *Brace yourself for the end of cheap oil;* New Scientist, Aug.2, 2003 [#2238]
Homer-Dixon T., 2001, *The Ingenuity Gap*; ISBN 0-676-97296-9
Homer-Dixon T., 2006, *The Upside of Down;* Island Press (ISBN 1-59726-064-9)
Hopkins R., 2008, *The Transition Handbook* (ISBN 978-1-900322-18-8)
Hotelling H., 1931, *The Economics of exhaustible Resources*; Journ. Politic. Economy 1931 [#190]
Hovey, H.H., 2001, *DJ Matt Simmons: Energy Crisis is "Extremely Serious"*; Dow Jones News, April 6 2001 [#1287]
Howard K., 2000, *Running on Empty*; Autocar 23. Feb.
Howe J., 2003, *The End of Fossil Energy –* ISBN 0-9743404-0-5
Hoyos C., 2003, *Energy companies see a big future for gas. But will the West's increasing dependance imperil it's fuel security?;* Financial Times Aug. 15, 2003 [#2242]
Hubbert, K., 1956, *Nuclear energy and the fossil fuels; Drilling and Production Practice 1956*, American Petroleum Institute, 1957 [#2328]
Hubbert M.K., 1962, *Energy resources, a report to the Committee on Natural Resources*; Nat. Acad. Sci. Publ. 1000D
Hubbert M.K. 1969, *Energy resources*; in Cloud P. (Ed) *Resources and Man*; W.H.Freeman
Hubbert M.K., 1949, *Energy from fossil fuels*; Science **109**, 103-109
Hubbert M.K., 1956, *Nuclear energy and the fossil fuels*; Amer. Petrol. Inst. Drilling & Production Practice. Proc. Spring Meeting, San Antonio, Texas. 7-25. [#187]
Hubbert M.K., 1971, *Energy resources of the Earth*; in *Energy and Power*, W.H.Freeman
Hubbert M.K., 1976, *Exponential growth as a transient phenomenon in human history;* in Strom Ed., Scientific Viewpoints, American Inst. of Physics 1976 [#952]
Hubbert M.K., 1980, *Oil and gas supply modeling*; in Gass S.I., ed. proceedings of symposium, U.S. Dept. of Commerce June 18-20, 1980 [#492]
Hubbert M.K., 1981, *The world's evolving energy system*; Amer. J. of Physics 49/11 1007-1029
Hubbert M.K., 1982, *Technique of prediction as applied to the production of oil & gas*; in NBS Special Publication 631. U.S. Dept. Commerce/National Bureau of Standards, 16-141.
Hubbert M.K., 1998 *Exponential growth as a transient phenomenon in human history*; Focus [#902][#1909]
Huddleston B.P., 1998, *The availability of reserve estimates depending on the purpose*; Pet. Eng. Int. Sept [#842]
Humphrys J., 2002, *We're planning a war, but don't mention the oil;* The Sunday Times Dec.8, 2002 [#2034] [#2068]
Huxley, A., *Falling Off Huberts Peak;* NZ Listener, March 8, 2003 [#2307]

I

Inman M, 2010, *Mining the truth on coal supplies;* Nat. Geographic News. Sept.8 2010
Imbert P., J.L.Pittion and A.K.Yeates, 1996, *Heavier hydrocarbons, cooler environment found in deepwater*; Offshore April [#446]
Independent Petroleum Association of America, 1993, *The promise of Oil and Gas in America*; final report of IPAA Potential Resources Task Force [#225]
Industrie et Environnement, 2003, *A Challenge Yet To Be Taken Up; The End Of Oil,* Industrie et Environnement, No. 275, June 12, 2003 [#1931]
Institute of Petroleum, 1993, *Valuable Saudi upstream data published*; Pet. Review Dec 1993 [#283]
Institute of Petroleum, 1995, *The UK continental shelf in 2010: is this the shape of the future?* Report [#481-1]
International Energy Agency, *World Energy Outlook* (published annually)
International Energy Agency, 1998, *World Energy Prospects to 2020*; Report to G8 Energy Ministers, March 31 (www.iea.org/g8/world/oilsup.htm)
Ion D.C., 1980, *The availability of world energy resources;* Graham & Trotman 345pp (ISBN 0 86010 193 2)
Ismail I.A.H, 1994, *Future growth in OPEC oil production capacity and the impact of environmental measures*; Energy Exploration and Exploitation 12/1 17-58 [#378]

Ismail I.A.H, 1994, *Untapped reserves, world demand spur production expansion*; Oil & Gas Journ. May 2, 1994 95-102 [#224]
Ivanhoe L.F., 1976, *Evaluating prospective basins*; in three parts – Oil & Gas Journ. Dec 13. 1976 [#108]
Ivanhoe L.F, 1980, *World's prospective petroleum areas*; Oil & Gas Journ. April 28 1980 [#91]
Ivanhoe L.F, 1984, *Oil discovery indices and projected discoveries*; Oil & Gas Journ. 11/19/84
Ivanhoe L.F, 1985, *Potential of world's significant oil provinces*; Oil & Gas Journ. 18/11/85 [#144]
Ivanhoe L.F, 1986, *Oil discovery index rates and projected discoveries of the free world*; in Oil & Gas Assessment. Amer. Assoc. Petrol. Geol. Studies in Geology #21, 159-178
Ivanhoe L.F, 1987, *Permanent oil shock*; AAPG 71/5 [#309]
Ivanhoe L.F., 1988, *Future crude oil supply and prices*; Oil & Gas Journ. July 25 111-112 [#97]
Ivanhoe, L.F., 1986, *Limitations of Geological Consensus Estimates of Undiscovered Petroleum Resources*; AAPG Studies in Geology #21, September, 1986 [#1799]
Ivanhoe L.F., & G.G.Leckie, 1991, *Data on field size useful to supply planners*; Oil & Gas Journ, April 29 1991 [#30]
Ivanhoe L.F., 1995, *Future world oil supplies:there is a finite limit*; World Oil Oct.1995 [#381]
Ivanhoe L.F., 1997, *Get ready for another oil shock;* Futurist Jan-Feb 1997
Ivanhoe L.F., 1997, *Updated Hubbert curves analyze world oil supply*; World Oil Nov. [#508]
Ivanhoe L.F, 2001, *Hubbert Center Newsletter 2001/1, Petroleum positions of Saudi Arabia, Iran, Iraq, Kuwait, UAE Middle East Region* [#1232]
Ivanovich, D., 1999, *World may learn to wean itself from oil*; Houston Chronicle, Sunday, October 24, 1999 [#1546]

J

Jaffe A.M. and R.A. Manning, 1999, *The Shocks of a World of Cheap Oil*; Foreign Affairs 79/1[1031]
James, M., 1953, *The Texaco Story – the first fifty years 1902-1952*: Publ. The Texas Company. 115p
Jefferson M., 1994, *World energy prospects to 2010*; Petrole et Techn. 389 [# 262]
Jenkins D.A.L., 1987, *An undetected major province is unlikely;* Petrol. Revue Dec 1987 p 16 [#336]
Jennings J.S., 1996, *The millennium and beyond*; Energy World **240** June.
Jochen V.A, & J.P.Spivey, 1997 *Using the bootstrap method to obtain probabilistic reserves estimates from production data*; Petroleum Engineer, Sept. [#651]
Johnson K., 2003, *Is a great Iraqi oil field fading away?;* Wall St. Journ., June 23, 2003 [#2172]
Johnston C., 2008, *After the Crash: An Essay-Novel of the Post Hydrocarbon Age*. ID 2033772 www.iulu.com
Johnston D., 1994, *International petroleum fiscal systems and production sharing contracts*; PennWell 325pp (ISBN 0-87814 426-9)
Johnston D., 1998, *Oil and tax*; Encycl. Human Ecology, draft [#867]

K

Kassler, P., 1994, *Two global energy scenarios for the next thirty years and beyond*; World Petrol. Congr. Stavanger. [#233]
Kaufmann R.& C.J.Cleveland, 1991, *Policies to increase US oil production likely to fail, damage the economy, and damage the environment*; Ann. Rev. Energy Environ.1991 [#210]
Kaufmann R., 1991,*Oil production in the Lower 48 States*: Res. & Energy 13 [#96]
Kaufmann R., 1998, *How much oil remains and why should you care?*; [#934]
Keegan W., 1985, *Britain without oil;* Penguin 128pp
Kemp A.G & L.Stephen, 1996, *UKCS future beyond 2000 depends on oil price and reserve trends*; World Oil Oct [#540].
Kennedy P., 1994, *Preparing for the 21st century*; Fontana 428pp (ISBN 0-00 686298 5)
Kenney J.F., 1996, *Impeding shortages of petroleum re-evaluated*; Energy World **250** June 1992.
Kerr J., 1998, *The next oil crisis looms large – and perhaps close*; Science 281 [#790]
Khalimov E.M., 1993, *Classification of oil reserves and resources in the former Soviet Union*; Amer. Assoc. Petrol. Geol. 77/9 1636 (abstract)
Kinzer S., 2003. *All the Shah's Men*; John Wiley (ISBN 0-471-67878-3)
Kjaergaard T.,.1994, *The Danish revolution 1500-1800*; Cambridge 314pp
Klare M.T., 2002, *Oiling the wheels of war*; The Nation, Nov. 2002 [#2094]
Klare M.T., 2002, *Resource wars: the new landscape of global conflict*; Owl Books. pp289
Klare M.T., 2004, *Blood and Oil;* Hamish Hamilton 265p (ISBN 0-8050-7313-2)
Klemme H.D. & Ulmishek G.F., 1991, *Effective petroleum source rocks of the world: stratigraphic, distribution and controlling depositions factors*; Amer. Assoc. Petrol. Geol 75/12, 1908-185
Klemme H.D., 1983, *Field size distribution related to basin characteristics*; Oil & Gas Journ.Dec. 25. 1983 169-176
Kleveman L.C., 2003, *The New Great Game;* Atlantic (ISBN 1-84354-121-1)

Knickerbocker B., 1998, *West's balancing act: economy, nature*; Christian Science Monitor 10/8/98 [817]
Knoepfel H., 1986, *Energy 2000*; Gordon & Breach 181pp
Knott D., 1991, *Calm surface, deep currents: is industry storing up problems?*; Offshore, Dec 1991, 25-27 [#6]
Knott D., 1996, *Reserves debate*; Oil & Gas Journ. Jan 29. 40 [#526]
Kopytoff, V., 2004, *Peering into oil's future. Experts try to predict when the world will start running low on the natural resource that keeps all the engines running*; San Francisco Chronicle [#2271]
Krauss C., 1997, *Mexican data suggest 30% overstatement of oil reserves*; New York Times 18th March [#577]
Krayushkin V.A., et al., 1994, *Recent application of the modern theory of abiogenic hydrocarbon origins: drilling and development of oil and gas fields in the Dneiper-Donets Basin*; 7th Int. Symposium on the observation of the continental crust through drilling., Sante Fe, New Mexico, proc. 1994 [#455]
Kunstler J.H., 2005, *The Long Emergency*; Atlantic Books 307p

L

Labibidi, M.M. al, 2000, *Depletion of petroleum resources*; Pres. Clean Energy 2000 Conf. [#1064]
Laherrère J.H., Campbell, C.J., Duncan, R.C., & McCabe, P.J., 2001, Discussions & Reply acc. to *"Energy Resourses – cornucopia or empty barrels?"* in AAPG Bulletin, V. 85, No. 6 (June 2001), p 1083-1097, 2001[#1306]
Laherrère J.H. & D.Sornette, 1998, *Stretched exponential distributions in nature and economy*; The European Physical Journal B2 529-539 [#1829]
Laherrère J.H., 1990, *Hydrocarbon classification rules proposed*; Oil & Gas Journ. Aug.13. p.62
Laherrère J.H., 1990, *Les Reserves d'hydrocarbures*; BIP 6629 [#167]
Laherrère J.H., 1992 *Reserves mondiales restantes et a decouvrir*: ATFP Conf. Paris 18.4.91 Revue de Presse TEP No3 20.1.92 [#168]
Laherrère J.H., 1993, *Le petrole, une ressource sure, des reserves incertaines*; Petrol et Technique, 383, October 1993 [#208]
Laherrère J.H., 1994, *Nouvelle approche des reserves ultimate – application aux reserves de gaz des Etas-Unis*; Petrole et Technique, Paris 392. 29-33
Laherrère J.H., 1994, *Published figures and political reserves*; World Oil, Jan 1994 p. 33. [#207]
Laherrère J.H., 1994, *Reverves mondiales de petrole: quel chiffre croire?*; Bull. Inform. Petrol. 7727, 7728, 7729
Laherrère J.H., 1994, *Study charts US reserves yet to be discovered*; American Oil & Gas Reporter 37/9 99-104.
Laherrère J.H., 1995, *World oil reserves: which number to believe?*; OPEC bull. Feb draft. [#256] Final [#346]
Laherrère J.H., 1996, *Distributions de type 'fractal parabolique' dans la nature*; C.R. Acad. Sci. Paris 322 IIa 535-541 [#449]
Laherrère J.H., 1996, *Upstream potential of the Middle East in a World context*; Proc. Oil& Gas project finance in the Middle East IBC Dubai Conf. May 1996 [#565]
Laherrère J.H., 1997, *Production decline and peak reveal true reserve figures*; World Oil Dec. 77 [#658]
Laherrère J.H., 1998, *Development ratio evolves as true measure of exploitation*; World Oil, [#673]
Laherrère J.H., 1999, *Erratic reserve reporting*; Petroleum Review Feb [#914]
Laherrère J.H., 1999, *Reserve growth: technological progress, or bad reporting and bad arithmetic* Geopolitics of Energy [#1065]
Laherrère J.H., 1999, *World oil supply – what goes up must come down – but when will it peak?*; Oil & Gas Journ. Feb 1 57-64 [#968]
Laherrère J.H., 2000, *Oil reserves and potential of the FSU*, Tomorrow's Oil 2/7
Laherrère J.H., 2002, *Forecasting future oil production*; Presentation for the BGR, January, 2002 [#1764]
Laherrère J.H., 2002, *Is FSU oil growth sustainable?*; Petroleum Review April, 2002 [#1810]
Laherrere J.H., 2003, *Forecast of oil and gas supply to 2050;* Petrotech 2003 New Delhi [#2059]
Laherrère J.H., A. Perrodon and G. Demaison, 1993, *Undiscovered petroleum potential: a new approach based on distribution of ultimate resources*; Rept. Petroconsultants S.A., Geneva
Laherrère J.H., A. Perrodon, and C.J.Campbell, 1996, *The world's gas potential*, Report, Petroconsultants
Laherrère J.H., 2007, *Etat des reserves de gaz des pays exportateurs vers l'Europe;* Club de Nice (www.iehei.org/Club_de_Nice/2007)
Lanier, D., 1998, *Heavy Oil – A Major Energy Source for the 21st Century*; UNITAR Centre for Heavy Crude and Tar Sands, No. 1998.039, 1998 [#1697]
Lattice Group, 2002, *Initial response to the governments consultation on UK energy policy;* Lattice Group, June 2002 [#2049]
Lavelle M., 2003, *Living Without Oil;* US News & World Report, Feb.17, 2003 [#2101]
Leach, G., 2001, *The coming decline of oil*; Tiempo, Issue 42, December, 2001 [#1766]
Leblond B., 2003, *ASPO, ODAC see conventional oil production peaking by 2010;* Oil & Gas Journal, June 23, 2003 [#2191]

Leckie G.G., 1993, *Hydrocarbon reserves and discoveries 1952 to 1991*; Energy Exploration & Exploitation, 11/1, 1993 [#214]

Leggett J., 1999, *The carbon war*; Penguin 338pp

Leggett J., 2005, *Half Gone*; 312p Portobello (ISBN 1 84627 004 9)

Lencioni L. et al., 1996, *Breathing new life into an abandoned field*; Hart's Petroleum Engineer International, Dec. [#552]

Leonard R.C., 1996, *Caspian Sea regional hydrocarbon development: opportunities and challenges*; 4th Kazakstan Int. Oil & Gas projects conf. [# 539]

Levitch R.N., 1998, *Making sustainable energy available to all*; Geol. Soc. Amer. [#938]

Lewis, C., 2001, *Natural Gas Hydrates – E&P Hazard or Large Business Opportunity?*; Hart's E&P, 2001 [#1320]

Lindstedt G., *Olja*; (ISBN 91-7588-484-4)

Liesman S., 2000, *OPEC oil cuts raise threat of shortage*; Wall St Journ., 20th Jan [#1065]

Liesman S., 1999, *Texaco's strategy: produce less oil more profitably*; Wall St Journ., Oct 27 [#1019]

Littell G.S., 1999, *World crude production: bad statistics produce poor conclusions*; World Oil June [#986]

Lomborg, B., 2001, *Running on Empty?*; The Guardian, 16 August, 2001 [#1350]

Longhurst H., 1959, *Adventure in oil: the story of British Petroleum*; Sidgwick and Jackson, London 286p. [B#8]

Longwell H., 2002, *The future of the oil and gas industry: past approaches, new challenges*; World Energy 5/3 2002 [#2041]

Lönker O., 2004, *The Noose Tightens*; New Energy No 4

Lovelock J., 1988. *The ages of Gaia*; Oxford University Press 252pp. (ISBN 0-19-286090-9. Centre for Global Energy Studies, 1990, *The Gulf crisis* 228pp)

Lübben H. von & J.Leiner; 1988, *Öl: Perspektiven im Upstream Bereich*; Erdöl, Erdgas, Kohle 104/5 Mai 1988 [#276]

Lugar R.G. and R.J.Woolsey, 1999, *The new petroleum*; Foreign Affairs Jan-Feb [#899]

Luhmann, J., 2003, *Turning Point of the oil age already in 2010?*; Neue Zuercher Zeitung, Sept. 2003 p89 [#1951]

LWV (League of Women Voters of Santa Cruz County), 2001, *The Oil Crash and You – Running on Empty!*; Santa Cruz Voter, February, 2001 [#1613]

Lynch M.C, 2001, *Oil prices enter a new era;* Oil and Gas Journal [#1214]

Lynch M.C., 1992, *The fog of commerce: the failure of long-term oil market forecasting*; MIT Center for Int. Studies Sept. 92

Lynch M.C., 1998, *Crying Wolf: warnings about oil supply*; MIT Center for International Studies [#728]

Lynch M.C., 1998, *Farce this time*; Geopolitics of Energy; December-January [#925]

Lynch M.C., 1998, *Imminent peak challenged*; Oil & Gas Journal 28.1.98 p.6 [#657]

Lynch M.C., 1999, *The debate over oil supply: science or religion*; Geopolitics of Energy, Aug. [#1133]

Lynch, M.C., 2001, *Closed Coffin: Ending the Debate on "The End of Cheap Oil" – A commentary*; M.C Lynch, Chief Energy Economist, DRI-WEFA, Inc., September, 2001 [#1675]

M

Mabro R., 1996, *The world's oil supply 1930-2050 – a review article*; Journ. of Energy Literature II.1.96 [#469]

Macgregor D.S., 1996, *Factors controlling the destruction or preservation of giant light oilfields*; Petroleum Geoscience 2. 197-217

Mack T., 1994, *History is full of giants that failed to adapt*; Forbes 28 Feb 1994 [#195]

MacKenzie J.J., 1995, *Oil as a finite resource: the impending decline in global oil production*: World Resources Inst [# 394 436]

Mackenzie, W., 1999, *Maturing Gracefully – Overview of the 1999 Probable Developments*; UK Upstream Report, No. 316, September 1999 [#1376]

Maegaard P., 2000, *Mobilizing the market for renewable energy*; Clean Energy 2000n [#1114]

Mansfield P., 1992, *A history of the Middle East*; Penguin Books 373p. [#B15] (ISBN 0-14-016989-X)

Marcel V., 2002, *The Future of Oil in Iraq: Scenarios and Implications;* The Royal Institute of International Affairs, Dec. 2002 [#2075]

Marino J., 1992, *Operations set to grow in Colombia*; BPXpress Jan-Feb 1992 [#3]

Martin A.J., 1985, *Prediction of strategic reserves in prospect for the world oil industry*; Eds.T. Niblock & R. Lawless. Univ. of Durham 16-39

Martinez A.R. et al. 1987, *Study group report: classification and nomenclature system for petroleum and petroleum reserves*; Wld Petrol. Congr. 325-342.

Martoccia D, 1997, *Permanent oil shortage impossible*; Futurist May-June [#935]

Mast R.F. et al., 1989, *Estimates of undiscovered conventional oil and gas resources in the United States – a part of the nation's energy endowment*; U.S. Dept of Interior [#94]

Masters C.D. 1994, *World Petroleum analysis and assessment*; Wld. Petrol. Congr. Stavanger [#226]

Masters C.D. D.H.Root & E.D.Attanasi, 1991, *Resource constraints in petroleum production potential*; Science 253. [#28]

Masters C.D., 1987, *Global oil assessments and the search for non-OPEC oil*; OPEC Review, Summer 1987, 153-169 [#92][#1687]

Masters C.D., 1991, *World resources of crude oil and natural gas*; Review and Forecast Paper, Topic 25, p.1-14. Proc. Wld. Petrol. Congr., Buenos Aires 1991 [#113]

Masters C.D., 1993, *U.S.Geological Survey petroleum resource assessment procedures*; Amer. Assoc. Petrol. Geol. 77/3 452-453 (with other relevant references).

Masuda, T., 2000, *World Oil Supply Outlook to 2010*; For the 5th Annual Asia Oil & Gas Conference, May 28-30, 2000 [#1521]

Mathews J., 1996, *World oil market faces instability and change*; ABQ Journ. 28.2.96 [#423]

Maugeri, L., 2004, *Oil never cry wolf – why the petroleum age is far from over*; Science, May 21, 2004 [#2399]

Maugeri L., 2009, *Squeezing more oil from the ground*; Scientific American, Oct. 2009

McCabe P.J., 1998, *Energy resources – cornucopia or empty barrel*; Amer. Assoc. Petrol. Geol. 82/11 2110-2134

McCarthy, T., 2001, *War Over Arctic Oil*; Time, February 19, 2001 [#1605]

McCluney, W.R., 2004, *Getting to the source*. ISBN 0-9744461-1-4

McClure K., 2002, *An imminent peak in world oil production? The arguments for and against;* 13D Research, July, 2002 [#1908]

McClure K., 2003, *The Case for Falling Production From Existing Oil Fields;* 13D Research, June 19, 2003 [#2217]

McCormack M., 1999, *Twenty-first century energy resources: avoiding crisis in electricity and transportation*; American Chemical Society address, [#967]

McCrone A.W., 2001, *Looking beyond the petroleum age;* San Francisco Chronicle, Mar 4th [#1191]

McCutcheon, H. & Osbon, R. & Mackenzie, W., 2001, *Risks temper Caspian rewards potential*; Oil & Gas Journal, p 22-28, December 24, 2001 [#1761]

McKibben B., 1998, *A special moment in history*; Atlantic Monthly, May [#738]

McCluney W.R., 2004, *Humanity's Environmental Future*; Sun Pine Press (ISBN 0-09744461-0-6)

McCluney W.R., 2004, *Getting to the Source: Readings on Sustainable Values*; (ISBN 9744461-1-40

McMahon, P., 2001, *Power suppliers run dry at worst time*; USA Today, Tuesday, March 27, 2001 [#1626]

McMullen, Program Murray, 1976, *Energy resources and supply;* Wiley [# 692]

McRae H., 1994, *The world in 2020: power, culture and prosperity, a vision of the future*; Harper Collins 302p [#B16]

McRay H, *No end of cheap oil*; Independent 11/11/93 [#191]

McRay H, *The real question is why the price of oil is not even higher*; Independent 07/04/02 [#1821]

McRay H., *Oil looks slippery*; Independent 7/5/93 [#161]

McVeigh, C., 2004, Renewable energy policies in an uncertain world; Sustainable Energy Technologies, June 28-30, 2004 [#2362]

Meacher M., 2004, *Plan now for a world without oil;* Financial Times, Jan.5, 2004 [#2309]

Meadows D.H. and others, 1972. *The limits to growth*; Potomac 205pp (ISBN 0 85644 008 6)

Megill R.E., 1993 *Discoveries lag oil consumption*; AAPG Explorer Aug.1993 [#171]

Meling L.M., 2003, *How and for how long is it possible to secure a sustainable growth of oil supply;* Statoil ASA, Norway 2003 [#2294]

Mellbye P., 1994, *Norway's role, satisfying increasing demand for natural gas in Europe*; World Petrol Congr. Stavanger. [#234]

Middle East Economic Survey, 2003, *Middle East Oil Investment of $500bn needed Over Next 30 Years, says IEA;* Middle East Economic Survey, Nov.3, 2003 [#1970]

Miller, B.M., 1982, *The evolution in the development of petroleum resource appraisal procedures in the U.S. Geological Survey*, U.S. Geological Survey [#2331]

Miller K.L., 2002, *Shell's Game – its scenario spinners have made an art of prediction;* Newsweek Sept. 16/23, 2002 [#1985]

Miller R.G., 1992, *The global oil system: the relationship between oil generation, loss half-life and the world crude oil resource*; Amer. Assoc. Petrol. Geol. 76/4 489-500 [#302]

Milling M.E., 2000, *Oil in the Caspian Sea*; Geotimes [#1173]

Mineral Resources of Canada, 1998, *Canada Energy Resources*; [907]

Minnear M.P., 1998, *Forecasting the permanent decline in global petroleum production*; Thesis, Univ. of Toledo

Mitchell J., 1996, *The new geopolitics of energy*; Royal. Inst. Int. Affairs April 1996 [#475-1]

Mitchell J., 1997, *Renewing the geopolitics if energy*; Energy World 245 Jan. [#555]

Mobbs P., 2005, *Energy beyond Oil*; Matadore (ISBN 1-905237 00 6)

Mobus G., 2011, *The dynamics of an abstract economic system*: Biophysical Economics

Monastersky R., *Geologists anticipate an oil crisis soon*; Science News 154/18 31 Oct [#865]

Monbiot G., 2003, *Bottom of the barrel;* The Guardian Dec.2, 2003 [#1980]

Monty M., 1999, *Use data mining to discover odds of making a billion bbl discovery*; World Oil March [#1063]

Moody-Stuart M, 2000, *Realising the value of scientific knowledge – geosciences in energy industries;* Sir Peter Kent lecture, Geological Society [#1278]

Moorbath S., 2009, *Time and Earth History*, Oxford Magazine, 2nd Week, Trinity Term

References

Morehouse D.F., 1997, *The intricate puzzle of oil and gas "reserve growth"*; EIA, Nat. Gas Monthly [#1103]
Morrison D.R.O., 2000, *Energy in Europe – comparison with other regions*; Clean Energy 2000 [#1112]
Morse, E.L. & Richard, J., 2002, *The Battle for Energy Dominance*; Foreign Affairs Magazine, www.foreignaffairs.org, March/April, 2002 [#1792 & 1839]
Mortished C., 1995, *Shell thinks, then does the unthinkable*; Times 31/3/95 [#334]
Mortished C., 2003, *Oil fear takes gloss off BP deal*; Times Online Feb.12, 2003 [#2076]
Mosley L., 1973, *Power play: oil in the Middle East*; Random House [#B10]
Mowlem M., 2002, *The real goal is the seizure of Saudi oil*; The Guardian 5 Sept [#1901]
Myers Jaffe, A. & Manning, R.A., 2000, *The Shocks of a World of Cheap Oil*; Foreign Affairs, Vol. 79, No.1, p 16-29, Jan/Feb, 2000 [#1434]

N

Nakicenonic, N. & Grubler A. & McDonald A., 1998, *Global Energy Perspectives*; 1998 [#2185]
Naparstek, A., 2004, *Oil game over*; New York Press; Vo.17 No.22, June, 2004 [#2389]
Naparstek, A., 2004, *The terminal decline*; The Sunday Oregonian, July 4, 2004 [#2390]
Nation L., 1995, *Hodel sees looming energy crisis*; AAPG Explorer May 1995 [#358]
National Geographic, 2004, *The end of cheap oil, National Geographic*; June, 2004 [#2371]
National Petroleum Council, 2007, *Hard Truths*; (ISBN 978-0-9799700-0-9)
Nehring R, 1978, *Giant oil fields and world oil resources*; CIA report R-2284-CIA [#16]
Nehring R, 1979, *The outlook for conventional petroleum resources*; Paper P-6413 Rand Corp.21p.
Newman S., 2005. *The final energy crisis.* ISBN 978-0-7453-2717-4
New York Times, 2001, *Despite more drilling, gas production falls short*; The Register – Guard National, Sunday, July 22, 2001 [#1318]
Nicolescu N. & B.Popescu, 1994, *Romania*; in Kulke H (Ed) *Regional petroleum geology of the world*; 21. Borntrager [#517]
Niiler, E., 2000, *Awash in Oil*; Scientific American, September 2000 [#1370]
Nikiforuk A, 2000, *Running on empty, when Canada's natural gas reserves hit the crisis point, who will be left out in the cold?*; Canadian Business Magazine [#1255]
Norwegian Petroleum Directorate, 1997, *The petroleum resources of the Norwegian continental shelf*; report [#620-3]
Norwegian Petroleum Directorate, 1997, *Trends in petroleum resource estimates*; report [#659]
Norwegian Petroleum Directorate, 2000, *Two-thirds left to go*; NPD Diary [#1154]
Nuclear Issues, 2003, *When the oil runs out*; Nuclear Issues Vol.25 No.4 April 2003 [#2168]
Nuclear Issues, 2004, *Peak Oil*; Nuclear Issues, Vo.26 No.6, June 6, 2004 [#2381]

O

Obaid, N.E., 2000, *The oil kingdom at 100*; Washington Inst. For Near East Policy 136pp (ISBN 0-944029-39-6)
Odell P.R., 1994, *World Oil resources, reserves and production*; The Energy Journal v. 15 89-113
Odell P.R., 1996, *Middle East domination or regionalisation*; Erdol, Erdgas, Kohle Heft4 [#435]
Odell P.R., 1996, *Britain's North Sea oil and gas production – a critical review*; Energy Exploration & Exploitation 14/1/
Odell P.R., 1997, *Oil shock – a rejoinder*; Energy World 245 [#554]
Odell P.R., 1997 *Oil reserves: much more than meets the eye*; Petroleum Economist Nov. [#656]
Odell P.R., 1998, *Fossil fuel resources in the 21st Century*; Presentation, Int. Atomic Energy Agency [#869]
Odell P.R., 1999, *Oil and gas reserves: retrospect and prospect*; Geopolitics of Energy Jan [#956]
Odell P.R., 1999, *Predicting the future: what lies ahead for fossil fuels?* Horizon June [#1096]
Odell P, R., 1999, *Fossil fuel resources in the 21st Century*; Financial Times Energy [#1279]
Odum H. and E., Odum, 1981, *Energy basis for man and nature*; McGraw Hill, New York
Offshore, 1996, *Improved recovery grow Norwegian reserves*; Offshore April 1996
Oil & Gas Journal, *World Production Reports*; December each year.
Oil & Gas Journal, 1989, *IPAA, US oil production headed for biggest slide since the 1970s*; Oil & Gas Journ., Nov 6th 1989 [#77]
Oil & Gas Journal, 1994, *Steady rise in oil, gas demand ahead*; Oil & Gas Journ June 6 1994 [#238]
Oil & Gas Journal, 1994, *Worldwide oil flow up, reserves steady in 1994*; Oil & Gas Journ 26/12/94 [#322]
Oil & Gas Journal, 2002, *Drop in Pemex's revised oil reserves figure 'significant'*; Oil & Gas Journal Sept.23, 2002 [#2001]
Oil & Gas Journal, 2003, *Peak-oil, global warming concerns opening new window of opportunity for alternative energy sources*; Oil & Gas Journal, Aug.18, 2003 [#1944]
Oil & Gas Journal, 2003, *Future energy supply*; Editorial, Oil & Gas Journal, Aug.18, 2003 [#1945]
Oil & Gas Journal, 2003, *Running out of oil*; Oil & Gas Journal, Sept. 15, 2003 [#1942]
Oil & Gas Journal, 2003, *Political Oil Supply*; Editorial June 23, 2003 [#2193]
Oil & Gas Journal, 2003, *ASPO sees conventional oil production peaking by 2010*; Jun 30, 2003 [#2195]

Oil & Gas Journal, 2003, *Majors replaced 101% of world oil and gas production in 2002;* Aug.1, 2003 [#2229]

Oil & Gas Journal, 2003, *World's Oil Supply/Peak rate of production;* letters to the editor Oil & Gas Journal, July 28, 2003 [#2254]

O'Mahoney B., 2004, *Oil shortages to hit world economy in 2007;* Irish Examiner, Feb.7, 2004 [#2320]

OPEC Bulletin, 1997, *An early touch of millennium fever: the coming oil crisis?* Oct. p.3

Orphanos A., 1995, *Looking for oil prices to gush;* Fortune 15/5/95 [#359]

Owen J.R., 2006, *Powerless – Our children's future.* ISBN1-4196-3743-6

Owen N.A. et al, 2010, *The status of conventional oil reserves – hype or cause for concern;* Journal of Energy Policy 2010.02.026

P

Parent L., 1989, *Natural gas: life after the bubble;* World Oil Feb. 1989 [#397]

Parker H.W., 2001, *After petroleum is gone, what then?;* World Oil Sept. 2001 [#1830]

Patricelli J.A. & C. L. McMichael, 1995, *An integrated deterministic/probabilistic approach to reserve estimations;* Journ. Petrol. Technol. **47**/1 49-53

Patricelli J.A. & C. L. McMichael-, 1996, *Authors' reply to discussion of an integrated deterministic/probabilistic approach to reserve definition;* Journ. Pet Technol. Dec. 1996 [#396]

Pauwels J-P and F.Possemiers, 1996, *Oil supply and demand in the XXIst century;* Revue de l'energie, 477. [#458]

Pearce F., 2003, *Over a barrel?;* New Scientist, 29 Jan. 2003 [#2071]

Pearce F., 1999, *Dry Future;* New Scientist; 10 July [#1007]

Pearson J.C., 1997, *Estimating oil reserves in Russia,* Petroleum Engineer Int., Sept. [#650]

Perrodon A., 1988, *Hydrocarbons* in Beaumont E.A. and N.H.Foster (Eds) *Geochemistry.* Treatise of petroleum geology, Amer. Assoc. Petrol. Geol. Reprint Series No.8 3-26

Perrodon A., 1991, *Vers les reserves ultimes;* Centres Rech.Explor.-Prod. Elf-Aquitaine 15/2 253-369. [#36]

Perrodon A., 1992, *Petroleum systems, models and applications;* Journ. Petrol. Geol.15/3, 319-326. [#38]

Perrodon A., 1995, *Petroleum systems and global tectonics*: Journ. petrol. Geol. 18/4 471-476 [#3

Perrodon A., 1998, *Production: les premices du declin;* Petrol. Int. 1735 [#870]

Perrodon A., 1999, *Quel pétrole demain,* Technip, Paris 94p

Perrodon A., 1999, *Vers un changement de decor sur la scene petroliere;* Pet. Informations 1738 [#1095]

Perrodon A., J.H.Laherrere and C.J.Campbell, 1998, *The world's non-conventional oil and gas;* Pet. Economist [#683]

Peters K.E., 2000, *Review of the Deep Hot Biosphere by Thomas Gold;* AAPG bull. 84/1 Jan

Peters S, 1999, *The West against the Rest: Geopolitics after the end of the cold war;* Frank Cass, Journal off-print from Geopolitics [#1235]

Peters S., 2002, *Courting future resource conflict: the shortcomings of western response strategies to new energy vulnerabilities;* Political Science Dept., Giessen University [#2033]

Peters, S., 2004, *Coercive western energy security strategies: 'Resource wars, as a new threat to global security;* Geopolitics, Vo. 9 No 1, 2004 [#2291]

Petroconsultants, 1997, *World Petroleum Trends* [#676]

Petroconsultants, 1993, *Strategic petroleum insights;* Report [#192]

Petroconsultants, 1994, *Oil production forecast;* Report [#340]

Petroconsultants, 1996, *World exploration – key statistics 1995/1986,* [#525]

Petroconsultants, 1996, *World petroleum trends* 1996 [#512].

Petroconsultants, 1993, *World Production & Reserve Statistics; oil and gas 1992;* Petroconsultants, London

Petroconsultants, 1997 *Oil and Gas Reserves 1997* [#647]

Petroconsultants, 1998, *Deepwater reserves;* report [#689]

Petroconsultants, 1998, *Oil & gas reserves added 1993-97;* Petroleum Review Oct [#899]

Petrole & Gas, 2003, *The Day When Peak Will Come;* Petrole & Gas No.1765/66, Oct., 2003 [#1966]

Petroleum Economist, 1995, *Interview with CJ Campbell- "Prophet or Cassandra",* Petroleum Economist, September, 1995 [#1436]

Petroleum Economist, 2009, *World Energy Atlas 2009.* ISBN 1.186186-273-3

Petroleum Engineer International, 1997, *SPE/WPC reserve definitions to provide more accurate consistent estimates;* Sept [#652]

Petroleum Review, 1996, *Ample energy reserves for the future;* Petroleum Review Aug. 1996 [#473]

Petroleum Review, 1998, *Insights from the statistics;* Pet. Review Sept. [#820]

Petroleum Review, 2000, *Discovery still lags production despite good 1999 results;* Petroleum Review, September 2000 [#1366]

Petroleum Review, 2002, *Filling the global energy gap in the 21st century;* Petroleum Review Aug. 2002 [#1905]

PetroMin, 2000, *Report: The looming crisis – Are you ready for this?;* PetroMin, November, 2000 [#1417]

Petterson W.C., 1990, *The energy alternative;* Channel 4 Book 186pp.

Pettingill, H.S., 2001, *Giant field discoveries of the 1990s;* The Leading Edge, July, 2001 [#1657]

Petzet A., 1999, *Decline in world crude reserves is first sense '92*; O&GJ December 22 [1026]

Pfeiffer D.A., 2001, *The end of the oil age*; (ISBN 1-4116-0629-9)

Phipps S.C., 1993, *Declining oil giants, significant contributors to U.S. production*; Oil & Gas Journ. Oct.4. 1993 [#181]

Picerno J., 2002, *If we really have the oil;* Bloomberg Wealth Manager Sept [#1899]

Pickens T.B., 1987, *Boone*; Houghton Miffin, Boston [#B1] (ISBN 0-450-42978-4)

Pickler N., 2003, *Lieberman offers plan to reduce US reliance on foreign oil;* Star Tribune May 8, 2003 [#2192]

Pimental D., 1998, *Energy and dollar cost of ethanol production with corn*; M.King Hubbert Center for Petroleum Supply Studies, Newsletter April [#727]

PIW, 1994, *PIW ranks the world's top 50 oil companies*; Petroleum Int. Weekly 12/12/94 [#335]

PIW, 1998, *Crying wolf, warnings about oil supply*; PIW April 6th [#723]

PIW, 1998, *The end of cheap oil*; PIW April 6th [#702]

PIW, 1999, *End of sight for North Sea output growth*; March 22 [#956]

PIW, 1999, *Mexico gains credibility by losing reserves*; March 29 [#975]

Plassart P., 1996, *Norvege: derniers puits de petrole avant le desert;* Le Nouvelle Economiste 1050 [# 462]

Pooley, E., 2000, *Who's Right About Oil?*; Time, October 2, 2000 [#1403]

Pope H, 1997, *Oil and geopolitics in the Caucasus*; Wall St. Journ, April 25 [#598]

Popescu B., 1995, *Romania's petroleum systems and their remaining potential*; Petroleum Geoscience **1** 337-350 [#507]

Popular Science, 2001, *Are We Really Running Out Of Oil?/What's Next: Cars;* Popular Science Special Issue, Summer, 2001 [#1335]

Population Today, 1998, *Population, consumption trends call for new environmental policies*; Population Today 26/4 April [#732]

Porter E., 1995, *Are we running out of oil?* API Discussion Paper 081 [#425 & 479-1]

Poruban, S. & Bakhtiari, A.M.S. & Emerson, S A., 2001, *OPEC's Evolving Role- Analysts discuss OPEC's role*; Oil & Gas Journal, July 9, 2001 [#1327]

Power M., 1992, *Lognormality in observed size distribution of oil and gas pools as a consequence of sampling bias;* Int. Assoc. of Mathematical Geology 24/8 [#806]

Power M., 1992, *The effects of technology and basin specific learning on the discovery rate*; Journ. Canadian Pet. Technology 31/3 [#807]

Powers L.W., 2012, *World Energy Dilemma*, PennWell Publishing

Preusse A., 1966, *Coalbed methane production – an additional utilization of hard coal deposits*; in Kürsten M. (Ed) *World Energy – a changing scene*; E. Schweizerbart'sche Verlagshandlung, Stuttgart ISBN 3-510-65170-7

Priestland D., 2011, *Merchant, Soldier, Sage - A new history of power*. ISBN 978-1-84614-485-1

Pursell D., 1999, *Depletion: the forgotten factor in the supply demand equation, Gulf of Mexico analysis:* Simmons & Co rept. [#982]

Pursell, D.A. & Eades, C., 2001, *Crude Oil And Natural Gas Price Update – Tough Sledding Near Term ... But Optimistic About 2003*, Simmons & Co Internat, Energy Industry Research, March 4, 2001 [#1790]

R

Radler, M., 2001, *World crude, gas reserves expand as production shrinks*; Oil & Gas Journal, p 125, December 24, 2001 [#1762]

Randol W., 1995, *No gushers*; Barron's 6/2/95 [#318]

Rasmusen H.J., 1996, *Bright future for natural gas*; Oil Gas European 2/1/96 [#467]

Rauch, J., 2001, *The New Old Economy: Oil, Computers, and the Reinvention of the Earth*; The Atlantic Monthly, p 35-49, January, 2001 [#1623]

Raymond L.R., 2003, *Challenge, Opportunity and Change*; World Energy Vol.6 No.3 2003 [#2236]

Read, R.D., 2001, *The North Sea: Oil Production Has Peaked! The GOM Model Must Come To The North Sea*; Simmons & Company International, October 18, 2001 [#1670]

Rechsteiner R, 2003, *Grun gewinnt: die letzte Olkrise und danach*; Orell Fussli, 215p (ISBN3-280-05054-5)

Reed S. & Bush, 2003, *The Oil Lord Strikes Again;* Business Week, Oct.27, 2003 [#1936]

Reed S., 2000, *Energy; Business Week*, Jan 10 [#1128]

Reese C., 1998, *We may get another great depression*; Evansville Press 8/20/98 [#818]

Rees-Mogg W., 1992, *Picnics on Vesuvius: steps towards the millennium*; Sidgwisk & Jackson 396p [#B17] (ISBN 0-283-06147-2)

Rees-Mogg W., 1999, *Troubled waters for oil*; The Times 30 Aug. [#991]

Register Guard, 2001, *Production peak coming*; The Register-Guard, November 17, 2001[#1737]

Reich, K., 2001, *Gauging the Global Fuel Tank's Size*; Los Angeles Times, Monday, June 11, 2001 [#1308]

Rempel, H., 2000, *Will the hydrocarbon era finish soon?*; BGR, Presentation at the DGMK/BGR event "Geosciences in Exploration ..."; Hannover, May 23, 2000 [#1608]

Reuters, 2003, *OPEC may discuss trading oil in Euros;* Dec.8, 2003 [#2290]

Rhodes R., and D. Beller, 1999 *The need for Nuclear Power*; Foreign Affairs 79/1[1030]

Ridley M., 1998, *Only hot air fuelled the petrol crisis*; Daily Telegraph 17 Mar [#710]
Rifkin, J., 2002, *The Hydrogen Economy: The creation of the worldwide energy web and the redistribution of power on earth;* Blackwell Publishers 2002 [#1907]
Riley D and M.McLaughlin, 2001, *Turning the corner: energy solutions for the 21st Century*; Alternative Energy Inst. 385pp
Rist C., 1999, *Why we'll never run out of oil*; Discover, June [#994]
Ritson N., 1998, *Maintaining production in the new millennium*; Petroleum Review, July [#714]
Riva J.P., 1991, *Dominant Middle East oil reserves critically important to world supply*; Oil & Gas Journ., Sept 23 1991 [#25]
Riva J.P., 1992, *The domestic oil status and a projection of future production*; US Congressional Research Report 92-826 SPR [#166]
Riva J.P., 1993, *Large oil resource awaits exploitation in former Soviet Union's Muslim republics*; Oil & Gas Journ., Jan 4 1993 [#120]
Riva J.P., 1996, *World production after year 2000: business as usual or apocalypse*; Geopolitics of Energy 18/9 September 2-6. [#503]
Roach J.W., 1997, *Reserves growth*; Oil & Gas Journ June 2[#628]
Roadifer R.E., 1986, *Size distribution of world's largest oil, tar accumulations*; Oil & Gas Journ. Feb.26. 1986 93-98 [#17]
Roberts J., 1995., *Visions and Mirages – The Middle East in a new era.* ISBN 1-85158-429-3
Roberts J., 1996, *IEA studies Middle East*; Petrol. Review, Feb. 1996 [#416]
Robertson J., 1989, *Future Wealth*; Cassell 178p. (ISBN 0-304-31930-9)
Robinson A.B. and Z.W.Robinson, 1997, *Science has spoken: global warming is a myth*; Wall St Journ Dec 4 [#746]
Robinson B., 2002, *Australia's growing oil vulnerability*; [#1859]
Robinson J., 1988, *Yamani – the inside story*; Simon & Schuster 302pp
Robinson, M., 2002, *Venezuela syncrude challenging Mideast oil in U.S.*; Reuters Limited, March, 2002 [#1794]
Rodenburg E., *The decline of oil*; World Resources Inst., handbook [#343]
Roeber J, 1994, *Oil industry structure and evolving markets*; Energy Journal 15 [#351]
Roger J.V., 1994, *Use and implementation of SPE and WPC petroleum reserve definitions*; Wld.Petrol. Congr., Stavanger [#228]
Roland K., 1998, *Perceptions of future, often flawed, shape plans and policies*; Oil & Gas Journ Feb 23 [#701]
Root D., E.Attenasi, and R.M. Turner, 1987, *Statistics of petroleum exploration in the non-communist world outside the United States and Canada*; U.S.G.S. Circ. 981 [#110]
Root D., E.Attenasi, and R.M. Turner-1989, *Data and assumptions for three possible production schedules for non-Opec countries*; Memorandum U.S. Dept. of Interior 26 Sept. 1989 [#100]
Rossant J. & P.Burrows, 1994, *Pain at the pump*; Business weekly July 4 [#246]
Rubelius F., 2007, *Giant Oil Fields – The Highway to Oil;* ISBN 978-91-554-6823-1
Rubin, J. & Buchanan, P., *Why Oil Prices Will Have To Go Higher*; CIBC World Markets Inc. Occasional Report #28, February 2, 2000 [#1401]
Rubin, J., Shenfeld, A., & Buchanan, P., 2000, *The Wall/How High Must Oil Prices Rise?*; CIBC World Markets Inc., Monthly Indicators, October 2000 [#1409]
Rubin, J., 2000, *Running On Empty*; CIBC World Markets, October 2000 [#1411]
Rubin J., 2009, *Why your world is about to get a whole lot smaller*, ISBN 978-1-4000-6850-0.
Rubin, M., 2001, *Indict Saddam*; The Wall Street Journal, Thursday, August 9, 2001 [#1347]
Ruppert M.C., 2004, *Crossing the Rubicon – the Decline of the American Empire at the End of the Age of Oil;* 675p New Society Publishers (ISBN 0-86571-540-8)
Ryan E., 2007, *Giving up the black stuff*; Irish Examiner 24.9.2007

S

S.P.E. & W.P.C, 1999, *Petroleum resource definitions;* J. Petr. Geol [#1100]
Salameh M.G., 2000, *Can the oil price remain high?*; Pet. Review April [#1078]
Salameh M.G., 2002, *Can Caspian oil challenge Middle East supremacy?*; Pet. Review Dec [#1905]
Salameh M.G., 2002, *Filling the global energy gap in the 21st Century*; Pet. Review Aug [#1905]
Salameh M.G., 2004, *How realistic are OPEC's Proved Reserves;* Pet. Review Aug
Salameh M.G, 2004, *Over a barrel,* (ISBN 0-9515968 -1-0)
Sampson A., 1988, *The seven sisters: the great oil companies and the world they created*; Coronet, London. [B#11]
Samuelson R.J, 2001, *The American energy fantasy;* Newsweek [#1228]
Samuelson, R.J., 2001, *The Energy War Within Us*; Newsweek, p 28/29, May 28, 2001 [#1298]
Sarkis, N., 2000, *Petrole, le troisieme choc*; Le Monde Diplomatique Mars 2000 [#1149]
Sauer J.W, 1993, *Crude oil prices: why the experts are baffled*; World Oil Feb. 1993
Scheer H., 2002, *The solar economy*; Earthscan Publications, 347p (ISBN 1 85383 835 7)

Scheer H., 2007, *Energy Autonomy*; (ISBN 978184 407 3559)

Schindler, J. & Zittel, W., 2000, *Der Paradigmawechsel vom Oel zur Sonne*; Natur und Kultur, 1/1, p 48-69, 2000 [#1510]

Schindler, J.& Zittel, W, 2000: *Fossile Energiereserven (nur Erdöl und Erdgas und mögliche Versorgungsengpässe aus Europäischer Perspektive;* (Ottobrunn, Final report by LBST, commissioned by the Deutsche Bundestag)

Schindler, J. & Zittel, W., 2005: *Oil Depletion* in Switching to Renewable Power – A Framework for the 21st Century, ed. by Volkmar Lauber, (London, Earthscan), pp 2-61

Schindler, J. & Zittel, W., 2007: *Alternative World Energy Outlook 2006 in* Advances In Solar Energy – An Annual Review of Research and Development Volume 17", ed. by D. Yogi Goswami, (London, Earthscan) p. 1 – 44

Schoell M. and R.M.K.Carlson, 1999, *Diamonds and oil are not forever*; Nature 6 May [#1049]

Schollnberger, 199?, *Energievorrate unde mineralische rogrstoffe: wie lange noch*; Osterichische Akad. Wissenshft. Bnd 12. [#834]

Schollnberger, 1996, *A balanced scorecard for petroleum exploration*; Oil Gas European 2/1/96 [#466]

Schollnberger, 1996, *Projections of the world's hydrocarbon resources and reserve depletion in the 21 Century;* Houston Geol. Soc. Bull Nov. [#861]

Schrempp, J.E., 2000, *Energy for the Future*; Speech at the opening of the World Engineer's Convention, Hanover, Germany, June 19, 2000 [#1538]

Schroeder W.W., 2002, *Clear Thinking about the Hydrogen Economy;* World Energy, Vol.5 No.3, 2002 [#2135]

Schuler G.H.M., 1991, *A history lesson: oil and munitions are an explosive mix*; Oil & Gas Journ. Nov.18, 1991[#182]

Schuyler J., 1999, *Probabilistic reserves definitions, practices need further refinement;* O&GJ May [#1042]

Schweizer P., 1994 *Victory: the Reagan administration's secret strategy that hastened the collapse of the Soviet Union*; Atlantic Monthly Press, New York 284p (ISBN 0-87113-567-1)

Science et Vie, 1995, *Energie: un fantastique tresor cache au fond de mers*; Science et Vie April 1995 [#350]

Science et Vie, 2001, *Energie*; No 214 Mars

Sciences et Avenir, 1997, *Petrole: la penurie a partir de 2015?* Aug. [#634]

Sciolino E., 1998, *It's a sea! It's a lake! No Its a pool of oil*; New York Times 21 June [#761]

Seago D, 2001, *This energy crisis isn't going away;* The News Tribune [#1224]

See M., 1996, *Oil mining field test to start in East Texas*; Oil & Gas Journ Nov. [#541]

Sell G., 1938, *Statistics of petroleum and allied substances*; The Science of Petroleum v1 1938 [#145]

Seskus T., 2003, *Rocketing Gas Prices May Hinder Oilsands growth;* National Post, Feb.21, 2003 [#2111]

Shah S., 2003, *Crude: The Story of Oil;* Jun.16 2003 [#2197]

Shell International, 2001, *Energy needs, choices and possibilities*; Shell publication

Shepherd R., 2000, *All eyes on the deepwater prize*; Offshore engineer, [#1061]

Shirley K., 1997, *Russia's potential still unrealized*; AAPG Explorer Aug. [#653]

Shirley K., 2000, *Caspian ready to face major tests*; AAPG Explorer Feb [#1077]

Shirley, K., 2000, *Discoveries Are Getting Smaller*; AAPG Explorer, p 6, 10, January, 2000 [#1545]

Sierra J., 1994 *European energy supply security*; Petrole et Tech. 389 [#263]

Simienski A, 2000, *Energy Puzzler on Saudi upstream investment*; Deutsche Bank [#1089]

Simmons M.R., 1994, *It's not like '86*; World Oil Feb. 1994 [#201]

Simmons M.R., 1995, *Despite sloppy prices, fundamentals tighten*; Pet. Eng. Int. Sept 1995 [#392]

Simmons M.R., 1995, *Strong market indicators*; World Oil, Feb 1995 [#330]

Simmons M.R., 1996, *Robust demand strengthens outlook*; World Oil Feb 1996 [#415]

Simmons M.R., 1997, *Are our oil markets too tight?* World Oil Feb. [#558]

Simmons M.R., 1998, *Facts don't support weakening market*; World Oil, Feb. [#671]

Simmons M.R., 1998, *The perils of predicting supply and demand*; Energy conf. N.Orleans [#909]

Simmons M.R., 1999, *1998: a year of infamy*; World Oil, Feb.

Simmons M.R., 2000, *The Earth in balance: has energy capacity maxed out;* Proc. Conf. Bridgewater House, London Nov. 16

Simmons, M.R., 2000, *Energy in the New Economy: The Limits to Growth*; Energy Institute of the Americas, October 2, 2000 [#1402] and [#1418]

Simmons, M.R., 2001, *Digging Out of Our Energy Mess: The Need For An Energy Marshall Plan*; AAPG, June 5, 2001 [#1330]

Simmons, M.R., 2001, *Investing in Energy: An Exercise Not for the Faint-hearted*; Managed Funds Association Forum 2001, New York City, July 11, 2001 [#1336]

Simmons, M.R., 2002, *2001: In like a lion, out like a lamb*; World Oil.com – Online Magazine, February, 2002 [#1779]

Simmons, M.R., 2002, *The Growing Natural Gas Supply Imbalance: The Role that Public Lands Could Play in the Solution;* Subcommittee on Energy & Mineral Resources of the House of Representatives, July 2002, [#1916]

Simmons, M.R., 2003, *Are Oil & Murphys Law about to meet?;* World Oil, Feb. 2003 [#2131]

Simmons, M.R., 2003, *The Dawning of a new oil and gas era: Is a sea change ahead?;* World Energy Vol.6 No.3 2003 [#2237]

Simmons M.R., 2005., *Twilight in the Desert*; Wiley 422p (ISBN 0-474-73876-X0
Simon B., 2003, *Canada Is Losing Ability to Fill US Gas Needs;* New York Times, June 26, 2003 [#2235]
Skrebowski C., 1998, *Iraqi production set to triple* Petroleum Review October [#888]
Skrebowski C., 1998, *Is this the third oil shock?* Petroleum Review October [#887]
Skrebowski C., 1998, *Sisters to wed*; Pet. Review Sept. [#822]
Skrebowski C., 1999, *Fossil fuel resources in the 21st Century;* Pet. Review [#992]
Skrebowski C., 2000, *A silly game with no winners;* Pet. Review June [#1177]
Skrebowski C., 2000, *Discovery still lags production despite good 1999 results* Pet. Review Sept
Skrebowski C., 2000, *How much can OPEC actually produce;* Pet. Review April [#1069]
Skrebowski C., 2000, *The North Sea – a province heading for decline?;* Pet. Review Sept
Skrebowski, C., 2000, *Asking the wrong question about oil reserves*; Petroleum Review, Sept., 2000 [#1364]
Skrebowski, C., 2000, *The perils of forecasting*; Petroleum Review, September 2000 [#1365]
Skrebowski, C., 2004, *Oil field mega – projects 2004;* Petroleum Review Jan. 2004 [#2273]
Skrebowski C., 2010, in *The Oil Crunch – a wake-up call for the UK economy.* Industry Taskforce on Peak Oil and Energy Security. ISBN 978-0-9562121-1-5
Slessor M and J.King, 2002, *Not by money alone*; Jon Carpenter 160pp (ISBN 1-897766-72-6)
Smil V., 2003, *Energy at the Crossroads: Global Perspectives and Uncertainties;* MIT Press, 2003 [#2304]
Smith K.K., 2005, *Powering our Future,* (ISBN 13 978-0-595-33929-7)
Smith M.R., 2001, *Environmentalists can relax. Oil supplies will decline sooner than most geoscientists are prepared to accept*; PESGB, Newsletter Aug.-Sept.
Smith M.R., 2002, *Energy security in Europe*; Petroleum Review Aug. [#1871]
Smith M.R., 2002, *US oil supply vulnerability growing;* Offshore August [#1918]
Socci T, 1999, *Surface temperature changes and biospheric responses in the northern hemisphere during the last 1000 years*; US Global Change Serminar 17 May [#980]
Solomon C., 1993. *The hunt for oil*; Wall Street Journ. Aug.25 1993 [#169]
Sorensen, K., 2001, *If oil runs out*; Information, May 16, 2001 [#1575]
Soros G., 1995, *Soros on Soros: staying ahead of the curve*; John Wiley & Sons, ISBN 0-471-12014-6. 326p.
Soros G., 1998, *The Crisis of global capitalism*; Public Affairs 243p
Speight R, 1998, *Hydrocarbon Man: a threatened subspecies*; Shell International [#930]
Spencer, J.E. & Rauzi, S.L. (Arizona Geological Survey), 2001, *Crude Oil Supply and Demand: Long-Term Trends*; Arizona Geology, Vol.31, No.4, Winter, 2001 [#1753]
Spiegel Der, 2000, *Auswege aus dem Energienotstand*; Der Spiegel, No. 23, 2000 [#1513]
Spiegel Online International, 2010, *Military Study warns of a potentially drastic oil crisis.* Sept 2 2010.
Spring C., 2001, *When the lights go on: understanding energy*; Emerald Resource Solutions 120pp
Srodes J., 1998, *No oil painting*; Spectator 20 Aug. [#789]
Stabler F.R., 1998, *The pump will never run dry;* Futurist Nov. [#943]
Stanley B., 2002, *Oil supply seen set to fall*; Washington Times 28 May [#1856]
Stanton W. 2003, *The Rapid Growth of Human Populations 1750-2000*; ISBN 0-906522-21-8
Starling P., 1997, *Oil market outlook*; Petroleum review Feb. [#544]
Steeg H., 1994, *World energy outlook to the year 2010*; 7th Int. Oil & Gas Seminar, Paris [#251]
Steeg H., 1997, *De nouveaux chocs pétroliers nous menacent ... mais notre insouciance est totale*; Le Temps Stratigique, Geneve [# 642]
Stelzer, Irwin, 2002, *Time to end our reliance on Saudi oil;* The Sunday Times, August, 2002 [#1898]
Stevenson, James, 2003, *Bitumen recovery at risk in Northern Alberta: Conoco;* Calgary Herald, Aug.2, 2003 [#1935]
Stone, R., 2002, *Caspian Ecology Teeters On the Brink*; Science, Vol. 295, January 18, 2002 [#1775]
Stoneley, R., 1993, *Book Review of "The Golden Century of Oil 1950-2050"*; Cretaceous Research, p 250, No. 14, 1993 [#1453]
Stow A.R., 1996, *Consequences of US oil dependence*; Energy 3 Nov [#649]
Strahan D., 2007, *The Last Oil Shock*, John Murray (ISBN 978-0-7195-4623-9)
Suddeutsche Zeitung, 2001, *Kassandra-rufe in Ol*; Suddeutsche Zeitung Online, Wissenschaft [#1244]
Syncrude, 1998, *Mining black gold*; World Oil June 1998 [#758]

T

Takin M., 1988, *Energy cycles: can they be avoided?* Opec Bulletin Oct. 1988 [#305]
Takin M., 1989, *The high cost of misunderstanding Opec*; 14th Congr. Wld Energy Conf. [#306]
Takin M., 1993, *OPEC, Japan and the Middle East*; OPEC Bull. 4/2 (March-April 1993) 17-34. [#158]
Takin M., 1996, *Future oil and gas: can Iran deliver?* World Oil Nov. [#516]
Takin M., 1996, *Many new ventures in the Middle East focus on old oil, gas fields*; Oil & Gas Journ. May 27 [#457]
Takin M., 1994, *How much gas is there in the Middle East*; Pet. Review July 1944 [#307]
Takin, M., 2001, *OPEC-consumer cooperation needed to ensure adequate future oil supply*; Oil & Gas Journal, December 3, 2001 [#1760]

Takin, M., 2003, *Sanctions, Trade Controls & Political Risk 03-Case Study: Iran;* CGES Nov. 5 & 6, 2003 [#2301]

Tanner J, 1990, *Looming shock: Mideast peace could trigger a sharp drop in crude oil prices*: Wall St. Journ. Dec. 19. [#50]

Tanzer, A. & Ghosh, C., 2001, *Insatiable – China and the rest of developing Asia are driving the world oil market*; Forbes, July 23, 2001 [#1321]

Tavernise, S. & Brauer, B., 2001, *Russia Becoming an Oil Ally*; The New York Times, Oct 19, 2001 [#1724]

Tavernise, S., 2001, *Exxon Says Way Is Cleared for Development in Russia*; The New York Times, October 30, 2001 [#1728]

Taylor A., 1999, *Oil Forever*; Fortune November 22 [1021]

Taylor B.G.S., 1997, *Towards 2020: a study to assess the potential oil and gas production from the UK offshore*; Petroleum Review, February [#543]

Tchuruk S., 1994, *Les relations entre les societes et les gouvernements dans l'industrie mondiale du petrole et du gas;* Petrole et Technique 389, July 1994, [#266]

Teitelbaum R.S., 1995, *Your last big play in oil*; Fortune 30 Oct. 1995 [#382]

Third World Traveller, 2000, *Global oil reserves alarmingly overstated; 1999* Censored Foreign Policies News Stories [#1180]

Thomasson, M.R., 2000, *Petroleum Geology: Is There a Future?*; AAPG Explorer, p 3-10, May, 2000 [#1543]

Thurow L., 1996, *The future of capitalism;* Nicholas Brealey, 385pp (ISBN 1-85788-136-2)

Tickell C., 1996, *Climate & history*; Oxford today 8/2 [#428]

Tillerson R.W., 2003, *The challenges ahead;* speech at Institute of Petroleums' IP 2003 [#2102]

Tinker, S.W. & Kim, E.M., 2001, *Research: Energy Policy for the Future*; Geotimes, June 2001 [#1331]

Tissot B and D.H.Welte, 1978, *Petroleum formation and occurrence*; Springer Verlag, New York [#B12]

Toal, 1999, *The big picture;* Oil & Gas Investor 1/99 [#866]

Tolub L and M.A. Erb 2010, *Oil price band for the next decade: Utopia versus Reality;* Swiss Derivatives Review, Issue 43, Summer 2010.

Toman M & J. Darmstadter, 1998, *Is oil running oil?* Science 282 2Oct [#882]

Tomitate T., 1994, *World oil perspectives and outlook for supply-demand in Asia-Pacific region;* World Petrol. Congr. Topic 16. [#232]

Toniatti G., 2003, *Brazil furthers its leadership in deepwater E&P;* World Oil, Nov., 2003 [#1971]

Townes H.L., 1993, *The hydrocarbon era, world population growth and oil use – a continuing geological challenge*; Amer. Assoc. Petrol. Geol. 77/5, 723-730. [#157]

Trainer T., 1997, *The death of the oil economy*; Earth Island Journal Spring [#743]

Traynor, J.J. & Sieminski, A. & Cook, C., 2000, *OPEC – Shortage? What Shortage?*; Deutsche Bank, Global Oil & Gas, November 9, 2000 [#1751]

Tugendhat C and A. Hamilton, 1968, *Oil – the biggest business*; Eyre Methuen. ISBN 0-413-33290-X.

Tull S., 1997, *Habitat of oil and gas in the Former Soviet Union*; Geoscientist 7/1 [#549]

U

Udall, S.L., 1974, *The energy balloon*; 288pp McGraw Hill

Udall, S.L., 1980, *America's Trip in the Energy Swamp*; The Washington Post, January 6, 1980 [#1395]

Udall, S.L., 1998, *The Myths of August*; Rutgers University Press 397pp

Ulmishek G.F.- and C.D.Masters, 1993, *Oil, gas resources estimated in the former Soviet Union*; Oil & Gas Journ, Dec 13. 59-62 [#249]

Ulmishek G.F., R.R.Charpentier, and C.C.Barton, 1993, *The global oil system: the relationship between oil generation, loss, half-life and the world crude oil resource: discussion*; Amer. Assoc. Petrol. Geol. 77/5 896-899.

USGS, 1995, *1995 national assessment of United States oil and gas resources*; USGS circular 1118. [#319]

USGS, 1996, *Ranking of oil basins* Open File [#638]

USGS, 2000, *USGS Petroleum Assessment 2000 – Description and Results*? [#1362]

USGS, 2000, *USGS reassesses potential world petroleum reserves;* Press release [#1135]

USGS, 2002, *2002 petroleum resource assessment of the NPRA, Alaska*; Fact Sheet 045-02 [#1890]

V

Valdmanis R., 2003, *Oil – The worlds largest addiction;* Reuters, July 17, 2003 [#2215]

ven Koevering and N.J.Sell, 1986, *Energy – a conceptual approach;* Prentice Hall 271pp

Vlierboom F.W., B. Collini and J.E. Zumberge, 1986, *The occurrence of petroleum in sedimentary rocks of the meteor impact crater of Lake Siljan, Sweden*; 12th Europ. Assoc. Organic. Geochem. International meeting, Julich Sept 1985; Org. Geochem **10** 153-161

Volsett J., A.Abrahamsen & K. Lindbo, 1994, *An enhanced resource classification – a tool for decisive exploration; preprint*, Wld Petrol. Congr. Stavanger [#227]

W

Walberg E. 2011, *Postmodern Imperialism: Geopolitics and the Great Games* ISBN 978-0-9833539-3-5
Wall Street Journal, 2001, *No Policy, No Win*; The Wall Street Journal, Thursday, August 9, 2001 [#1348]
Wall Street Journal, 2001, *World Demand for Oil To Rise a Million Barrels A Day Despite Slump*; The Wall Street Journal? [#1349]
Wall Street Journal, 2002, *Opec warns of possible shortfall;* The Wall Street Journal, Dec. 20, 2002 [#2087]
Wall Street Journal, 2003, *In Russia, BP Bets on a Partner That Burned It;* Feb. 27, 2003 [#2136]
Wallop M., 1998, *Unless stopped, the global warming movement could cause economic disaster*; World Oil Sept [#8]
Warman H.R 1973, *The future availability of oil*; proc. Conf. World Energy Supplies, by Financial Times, London 11p [#1]
Warman H.R., 1972, *The future of oil*; Geographical Journ. 138/3 287-297
Warren S. & McKay P.A., 2004, *Methods for Citing oil Reserves Prove Unrefined;* Wall St. Journ., Jan.14, 2004 [#2317]
Warren, S., 2004, *Exxon upholds reserve estimates*; The Wall Street Journal, Mar. 11, 2004 [#2280]
Washington Post, 2002, *Russia makes big comeback as oil producer*; 20 March [#1825]
Washington Post, 2003, *Greenspan on the natural gas problem;* June 12, 2003 [#2178]
Washington Post, 2003, *Perilous natural gas shortage;* June 10, 2003 [#2177]
Watt K.E.F., 2003, *A test of six hypotheses concerning the nature of the forces that govern historical processes;* University of California, 2003 [#2231]
Wattenberg R.A., 1994, *Oil production trends in the CIS*; World Oil June
Weibull, *Discoveries abound*; World Oil, August, 2001 [#1486]
Weiner J., 1990, *The next hundred years*; Bantam 312pp
Weiner T., 2003, *Pemex: Mexico's oil giant;* The Oregonian, Jan.22, 2003 [#2083]
Weiss M., 1998, *Today's technology – can it take the heat?* Pet. Eng. Int. Sept [#845]
Wellmer F-W, 1994, *Rerserven und reservenlebensdauer von energierohstoffen*; Energie Dialog, July 1994 [#274]
Wellmer F-W., 1997, *Factors useful for predicting future mineral commodity supply trends*; Geol. Rundsch 86 311-321 [# 735]
Wetuski, J., 2002, *Oil Will Dominate 2030's Energy Picture, But Supply Vulnerability Grows;* Oil and Gas Online, Dec, 2002 [#2047]
Weyant J.P. and D.M.Kline, 1982, *Opec and the oil glut: outlook for oil export revenues in the 1980s and 1990s*; Opec Review Winter 1982 334-365 [#93]
Wheelwright, T., 1991, *Oil and world politics;* Left Book Club 220pp (ISBN 1-875285-05-9)
Whipple D., 1998, *Expert: world faces new oil crisis in next 12 years*; Caspar Star Tribune 6/11/98 [#850]
Whipple D., 2002, *Blue Planet: The end is near;* Science & Technology, Sept.27, 2002 [#1995]
Will G., 1996, *Heavy oil – the jewel in Western Canada's oil play*; Petroleum Review Nov. [#513]
Will G., 1997, *Promise from Canada's east coast*; Petrol. Rev. April [#580]
Williams B., 2003, *Debate grows over US gas supply crisis as harbinger of global gas production peak;* Oil & Gas Journal, July 21, 2003
Williams B., 2003, *Heavy hydrocarbons playing key role in peak-oil debate, future energy supply;* Oil & Gas Journal, July 28, 2003 [#2253]
Williams B., 2003, *Progress in IOR technology, economics deemed critical to staving off worlds oil production peak;* Oil & Gas Journal, Aug.4, 2003 [#2245]
Williams C.J., 2000 *Norway looks beyond oil boom;* Los Angeles Times Feb 26 [#1082]
Williams P., 1998, *Half full or half empty*; Oil & Gas Investor 18/2 Feb [#680]
Williams, S., Jones S. & Smallman A., 2004, *Oil use surges, but output cuts still are likely*; The Wall Street Journal, Mar. 11, 2004 [#2284]
Williamson P.E., 1997, *Potential seen in areas off Northwest, Southeast Australia;* Oil & Gas Journal, Nov. 24 1997 [#2118]
Willingham B.J., 1994, *Energy shortage looms again*; The Nat. Times Mag, Oct/Nov [#261]
Wind Energy Weekly, 1996, *New oil price shock seen looming as early as 2000*; Wind Energy Weekly 15 684 12 Feb. [#895]
Winter M., 2006, Peak Oil Prep – three things you can do to prepare for Peak Oil, Climate Change and Economic Collapse. ISBN 978-0-9656000-4-1
World Energy Assessment, 2000, *Energy and the challenge of sustainability*; United Nations Development Program, 2000 [#1921]
World Energy Council, 2000, *Energy for Tomorrow's World* 175pp
World Energy Outlook, 2002, *Methodology for Projecting Oil Production;* chapter 3 – The energy market outlook, 2002 [#1996]
World Energy Outlook, 2003, *Exploration and development;* chapter 4-oil, 2003 [#1973]
World Oil, 1994, *World not running out of oil;* March 1994 p. 9.
World Oil, 1996, *Global oil output inches forward*; World Oil Feb. 1996 [#414]
World Oil, 2001, *Western Europe: Exploration targets becoming smaller* and *FSU/Eastern Europe: A roaring comeback continues*; World Oil, August, 2001 [#1487]

World Oil, 2002, *Gulf of Mexico Deepwater Map* [#1815]
World Oil, 2002, *Norwegian oil minister predicts another half-century of production;* Aug. 2002 [#2095]
World Resources Institute, 1996, *World Resources – A Guide To The Global Environments – The Urban Environment;* World Resources 1996-1997, Oxford University Press, 1996 [#1334]
Wright T.R., 1999, *Oil prices could rise still more;* World Oil Oct. [#1018]

X

Xiaojie X, 1997, *China reaches crossroads for strategic choices;* World Oil April [#568]

Y

Yamani A.Z, 1997, *Containment is too risky;* Petrol Review May [#594]
Yamani A.Z., 1995, *Oil's global role – the outlook to 2005;* MEES 38/33 [#355]
Yeomans M., 2004, Oil – *Anatomy of an Industry* – ISBN 1-56584-885-3
Yergin D, 1988, *Energy security in the 1990s;* Petroleum Review Nov. 1988 [#88]
Yergin D, 2002, *US energy security lies in diversity of supply;* Oil&Gas Journ, 2 Sept 2002 [#1904]
Yergin D, and J.Stanislaw 1989, *The commanding heights;* Simon & Schuster pp 457
Yergin D., 1991, *The Prize: the epic quest for oil, money and power;* Simon & Schuster, New York, 877p [#B2]. (ISBN 0-671-50248-4)
Youngquist W. & Duncan R.C., 2003, *North American Natural Gas: Data Show Supply Problems;* Natural Resources Research, Vol.12 No.4, Dec. 2003 [#2315]
Youngquist W., 1997, *Geodestinies: the inevitable control of earth resources over nations and individuals;* Nat. Book Co., Portland 500p.
Youngquist W., 1998, *Spending our great inheritance – then what?* Geotimes 43 7 24-27 [#791]
Youngquist W., 1999, *The post-petroleum paradigm and population;* Journ Interdisciplinary Studies 20/4 [#1044]
Youngquist, W., 1998, *Shale Oil – The Elusive Energy;* Hubbert Center Newsletter, No.98/4, Oct, 1998 [#1463] [#2313]
Youngquist, W., 1999, *Comments about "The Coming Oil Crisis";* Journal of Geoscience Education, v. 47, p.295, 1999 [#1432]
Youngquist, W., 2000, *Alternative Energy Resources;* Pre-Conference Papers – Kansas Geological Survey Open-File Report 2000-51, October 2000 [#1421]
Youngquist, W., 2002, *Alternative Energy Sources;* Kansas Geological Survey [#1850]

Z

Zach B.A., 1998, *Canada's petroleum industry* [#898]
Zagar J.J., 1999, *World oil depletion: the crisis on our threshold;* IADC Conf. [#958]
Zeldin T., 1994, *An intimate history of humanity;* Minerva 488pp (ISBN0-7493-9623-7)
Zellner W., 1994, *Steamed about natural gas;* Business Week 10/10/94 [#273]
Zhdannikov, D. & Ayton, R., 2004, *No room for Russia to raise oil exports;* interview, Reuters, May 14, 2004 [#2361]
Ziegler W.H., 1990, *World oil and gas reserves;* unpublished report to EU [#147]
Zischka A., 1933, *La guerre secrète pour le pétrole;* Payot, Paris
Zittel, W., 2000, *Comment on EIA Presentation: Long Term world oil supply;* 2000 [#1386]

Printed in the United States of America